AF333977

ADVANCES IN GEOSCIENCE AND REMOTE SENSING

Edited by
DR. GARY JEDLOVEC

Advances in Geoscience and Remote Sensing,
Edited by Dr. Gary Jedlovec

Republished by InTech

Janeza Trdine 9, 51000 Rijeka, Croatia

Edition 2014
Copyright © 2009 InTech

Notice

Additional hard copies can be obtained from orders@intechopen.com

Advances in Geoscience and Remote Sensing,
Edited by Dr. Gary Jedlovec
p. cm.
ISBN 978-953-307-005-6

Preface

Remote sensing is the acquisition of information of an object or phenomenon, by the use of either recording or real-time sensing device(s), that is not in physical or intimate contact with the object (such as by way of aircraft, spacecraft, satellite, buoy, or ship). In practice, remote sensing is the stand-off collection through the use of a variety of devices for gathering information on a given object or area. Human existence is dependent on our ability to understand, utilize, manage and maintain the environment we live in - Geoscience is the science that seeks to achieve these goals. This book is a collection of contributions from world-class scientists, engineers and educators engaged in the fields of geoscience and remote sensing. The content of the book includes selected topics covering instrumentation (SAR, lidar, and radar), and remote sensing techniques to study atmospheric air quality and clouds, oceanography, in-land lake water quality and sea ice, soil moisture, vegetation mapping, and earthquake damage. In also includes several chapters on web tools for GIS applications, and the use of UAVs for low cost remote sensing studies. This book provides a particularly good way for experts in one aspect of the field to learn about advances made by their colleagues with different research interests and for students to broaden their exposure to the topic area.

Dr. Gary Jedlovec
Earth Science Office, VP61
NASA/MSFC
320 Sparkman Drive
Huntsville, Alabama 35805
USA

Contents

X ____

Enhancing the Unmixing Algorithm through the Spatial Data Modeling for Limnological Studies

Enner Herenio Alcântara [1]
José Luiz Stech [1]
Evlyn Márcia Leão de Moraes Novo [1]
Cláudio Clemente Faria Barbosa [2]
[1] *Remote Sensing Division, Brazilian Institute for Space Research*
{enner,stech,evlyn}@dsr.inpe.br
[2] *Imaging Processing Division, Brazilian Institute for Space Research*
claudio@dpi.inpe.br
São José dos Campos, São Paulo, Brazil

1. Introduction

Conventional water quality monitoring is expensive and time consuming. This is particularly problematic if the water bodies to be examined are large. Conventional techniques also bring about a high probability of undersampling (Hadjimitsis et al. 2006). Conversely, remote sensing is a powerful tool to assess aquatic systems and is particularly useful in remote areas such as the Amazon lakes (Alcântara et al., 2008).

Data collected using this technique can provide a synoptic overview of such large aquatic environments, which could otherwise not be observed at a glance (Dekker et al. 1995). However, remote sensing is not easily applied to aquatic environment monitoring mainly because the mixture of the optically active substances (OAS) in the water. Several approaches have been proposed to cope with this issue such as derivative analysis (Goodin et al. 1993), the continuum removal (Kruse et al. 1993), and spectral mixture analysis (Novo and Shimabukuro, 1994; Oyama et al. 2009).

The two first approaches are more suitable for hyperspectral images, whereas the spectral mixture analysis can be used for both hyperspectral and multispectral images. The Spectral Mixture Model (SMM) has largely been used for spectral mixture analysis, uncoupling the reflectance of each image pixel (Tyler et al. 2006) into the proportion of each water component contributing to the signal. The result of a spectral mixture analysis is a set of fraction images representing the proportion of each water component per image pixel. This technique has been applied to TM/Landsat images to determine the concentration of suspended particles (Mertes et al. 1993), chlorophyll-a concentration (Novo and Shimabukuro, 1994); as well as to MODIS images, to determine the chlorophyll-a concentration in the Amazon floodplain (Novo et al. 2006), to characterize the composition of optically complex waters in the Amazon (Rudorff et al. 2007), and to study turbidity distribution in the Amazon floodplain (Alcântara et al. 2008, 2009a).

Remote sensing data have been extensively used to detect and to quantify water quality variables in lakes and reservoirs (Kloiber et al. 2002). One of the most important variables to monitor water quality is turbidity, because it gives information on underwater light availability (Alcântara et al. 2009b). Although turbidity is caused by organic and inorganic particles, one unresolved issue is to distinguish between them using remote sensing (Wetzel, 2001). The Spectral Unmixing Model (SMM) can, however, be useful to analyze the turbidity caused by inorganic particles and by phytoplankton cell scattering.

This chapter book shows how to improve the well know unmixing algorithm using the spatial data modelling concept and also their applicability in limnological studies.

2. Study Area

The Amazon River basin drains an area of approximately 6×10^6 km², which represents 5% of the Earth surface. The Central Amazon has large floodplains covering around 300,900 km² (Hess et al. 2003), including 110,000 km² of the main stem 'Varzeas' (white water river floodplains; Junk, 1997). At high water, the Amazon River flows into the floodplains, and fills both temporary and permanent lakes which might merge to each other. The 'Lago Curuai' floodplain (Figure 1) covers an area varying from 1340 to 2000 km² from low to the high water. This floodplain is located 850 km from the Atlantic Ocean, near Óbidos (Pará, Brazilian Amazon). Formed by 'white' water lakes (characterized by high concentration of suspended sediments), and 'black' water lakes (with a high concentration of dissolved organic matter, and a low concentration of sediments; Moreira-Turcq et al. 2004). These lakes are connected to each other and to the Amazon River. The floodplain also has 'clear' water lakes filled by both rainwater and by river- water drained from 'Terra Firme' (Barbosa, 2005).

The lakes are connected to each other and to the Amazon River. The Curuai floodplain is controlled by the Amazon River flood pulse (Moreira-Turcq et al. 2004) which creates four states (Barbosa, 2005) in the floodplain-river system (Figure 2): (1) rising water level (January - February), (2) high water level (April – June), (3) receding water level (August – October), and (4) low water level (November – December).

The exchange of water between the floodplain and the Amazon River is shown in Figure 2. When the water level is high, there is very little flow, and the surface water circulation is caused predominantly by wind. In the receding state, the exchange of water between the river and the floodplain is inverted, i.e., the water flows from the floodplain to the river. The water level then drops to the low water state, when the exchange of water between the river and the floodplain is at a minimum. According to Barbosa (2005), during the rising water state, the flow from the river to the floodplain starts at a channel located on its Eastern border, and then migrates to small channels located on its Northwestern side.

The rate of inundation is influenced by the floodplain geomorphology, by the density of floodplain channels, and by the ratio of local drainage basin area to lake area. A majority of 93% of the flooded area in the floodplain is between 2-6 m in depth (using a water level reference of 936 cm). The deepest lake is the 'Lago Grande', and the shallowest are the 'Açaí' and Santa Ninha Lakes (Figure 1-c). An area of about 0.04% of the floodplain is below sea level (with the mean altitude of the floodplain at 9 m above sea level (Barbosa, 2005).

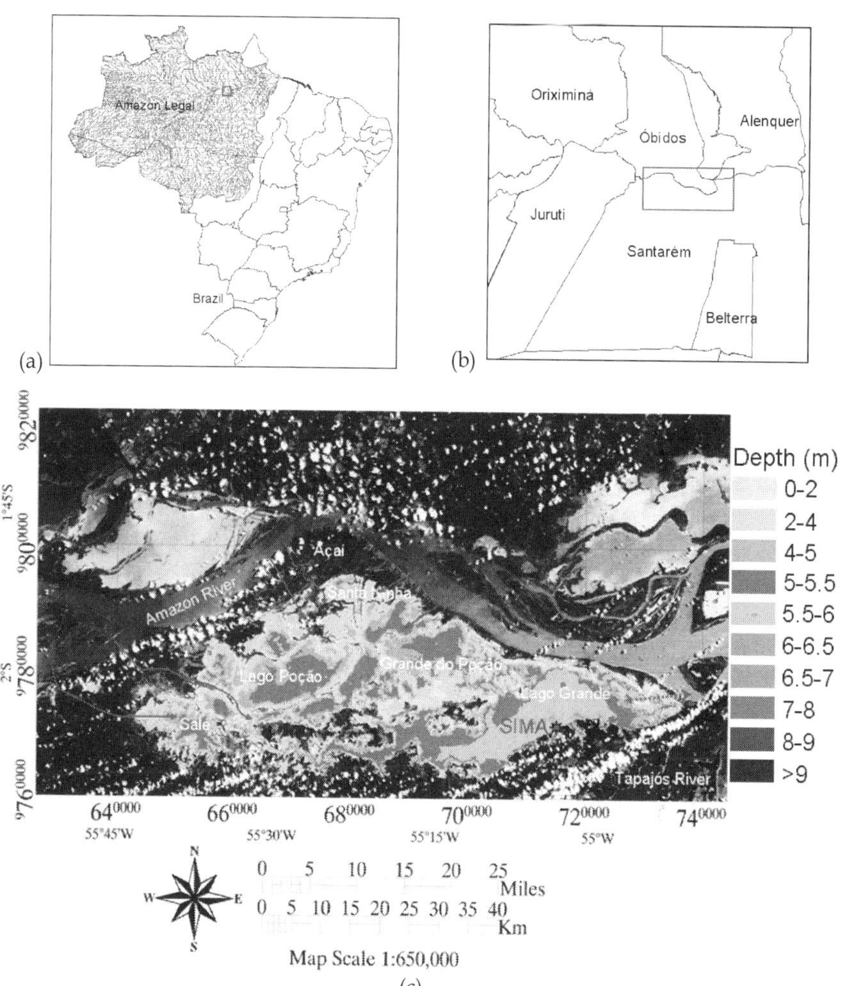

Fig. 1. Legal Amazon location into Brazil (a); (b) Location of Curuai floodplain into Amazon Legal; (c) Landsat-5 Thematic Mapper (normal composition) imagery showing the Curuai Floodplain and their bathymetry. The arrows indicate the main channels of connection Amazon River-floodplain.

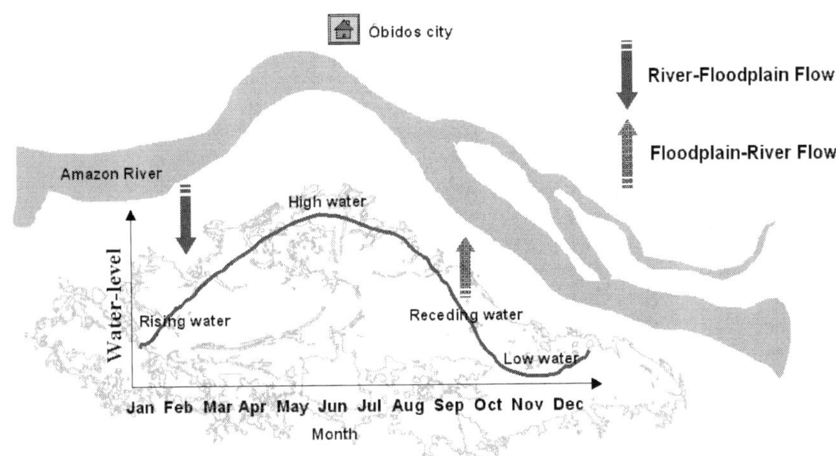

Fig. 2. Theoretical water level dynamic at the Curuai Floodplain (Source: Barbosa, 2005).

3. In-situ Measurements and Remote Sensing Data

The turbidity ground data acquisition was carried out from February 1st to February 14th 2004, during the rising water period. Turbidity measurements were taken at 215 sampling stations using the HORIBA U-10 multi-sensor. This equipment provides turbidity measurements in NTU (Nephelometric Turbidity Unit) with a resolution of 1 NTU. The locations of the sampling stations were determined with the aid of spectral analyses of Landsat/TM images taken at similar water level (Barbosa, 2005). These samples had maximum, minimum and mean values of 569, 101 and 232.29 NTU, respectively.

A Terra/MODIS image, acquired as MOD09 product on February 27th 2004 was used in this study. The spectral bands used in the analyses were band 1 (620-670 nm) and 2 (841-876) with a spatial resolution of 250 m, and bands 3 (459-479 nm) and 4 (545-565 nm) with a spatial resolution of 500 m. The two latter bands were re-sampled to 250 m using MODIS Reprojection Tool software (MRT, 2002).

4. Methodological Approach

The turbidity distribution was assessed using fraction images derived from the Linear Spectral Mixing Model, using four MODIS spectral bands (3 – blue, 4 – green, 1 – red and 2 – near infrared) with a spatial resolution of 250 m. In order to evaluate the turbidity distribution observed in the MODIS fraction images, in-situ measurements acquired during in February 2004 (a few days apart of the MODIS acquisition) were used to apply the Ordinary Least Square (OLS), spatial lag, and spatial error regression models. The kernel estimation algorithm was used to verify the spatial correlation of the in-situ data before performing the regression analyses. A summary of our methodological approach is presented in Figure 3:

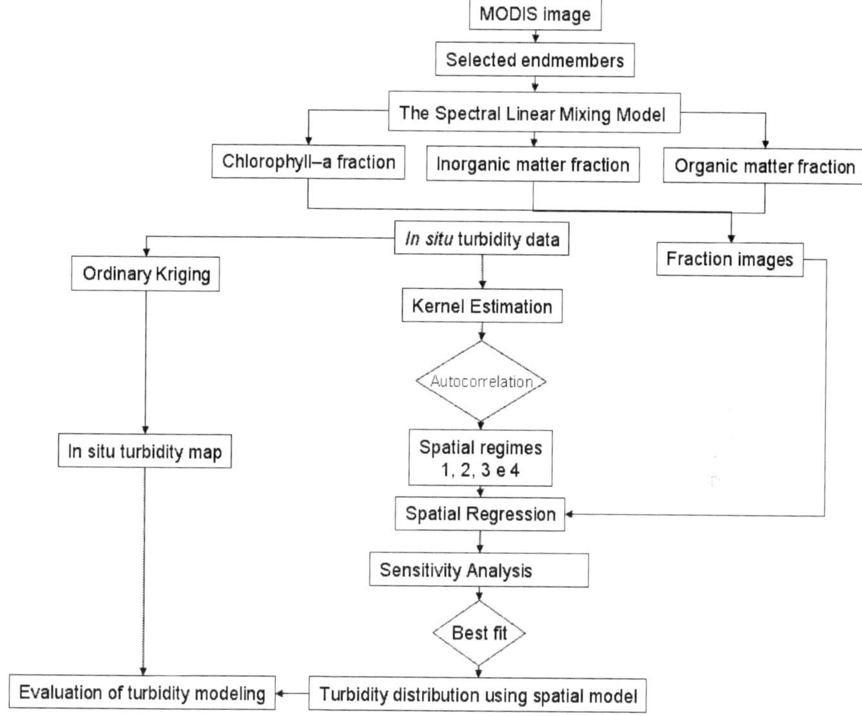

Fig. 3. Flow chart of the methodological approach.

4.1 The SMM

The SMM estimates the proportion of the various surface components present in each image pixel based on their spectral characteristics (Novo and Shimabukuro, 1994). The number of spectral endmembers used in the SMM algorithm must be less than or equal to the number of spectral bands (Tyler et al. 2006). Given these conditions, it is possible to determine the proportion of each component by knowing the spectral responses of pixel components according to equation 1:

$$R_{is} = \sum_{j=1}^{n} f_j r_{ij} + \varepsilon_i \qquad (1)$$

Where: R_{is} is the reflectance at each spectral band i of a pixel with one or more components, r_{ij} is the spectral reflectance of each component j in the spectral band i; f_j is the fraction of the component within the pixel; ε_i is an error term for each spectral band i.

Also, $j = 1, 2, ..., n$ (n = number of components), and $i = 1, 2, ..., m$ (m = number of spectral bands), with the following constraints: $\sum f_j = 1$, and $f_j \geq 0$ for all components.

4.1.1 Endmember selection

As pointed out in (Rudorff et al. 2006), mixtures of dissolved or suspended materials will always occur in natural water bodies. The desired conceptual 'pure' endmembers are hence not accessible for the OAS. Thus, the SMM results will not lead to a complete separation of the fractions. They will rather indicate a relative proportion of each endmember in which the relationship with the actual concentration of a certain OAS will be stronger according to its reflectance spectral dominance.

Some authors have selected the endmembers in a spectral library or laboratory measurements and applied them in the satellite images (Mertes et al. 1993). This approach for selecting the most 'pure pixel' for each component in the water sometimes does not consider the actual characteristics of endmembers found in the local area (Theseira et al. 2003). Thus, some authors collect the endmembers directly in the image, the so called image endmember (Novo and Shimabukuro, 1994).

The endemembers that will be selected in this work are the phytoplankton, inorganic suspended particle and dissolved organic matter laden water. The phytoplankton and the inorganic particle can cause the turbidity in the water and the dissolved organic matter, on the other hand, is a representative manner of non-turbid water. The method of Alcântara et al. (2008) was used to select the endmembers.

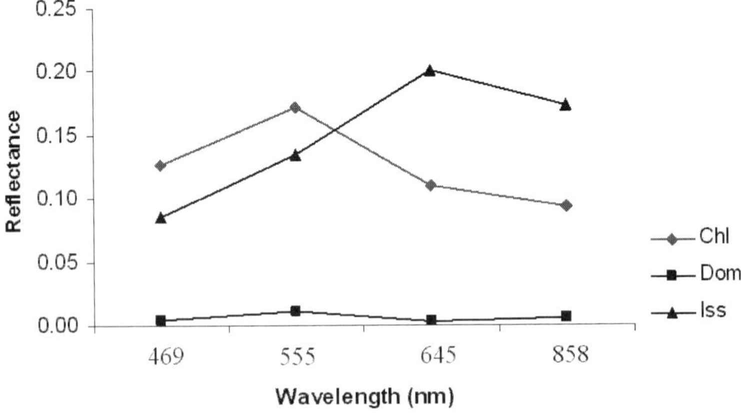

Fig. 4. Image endmembers used as input to run the Linear Spectral Mixing Model on the Terra/MODIS image acquired on 27th February 2004. Chl is the phytoplankton laden water, Dom is the dissolved organic matter laden water, and Iss is the inorganic particle laden water.

Figure 4 shows that the spectral responses of the selected endmembers were quite different. Chl-laden water endmember is characterized by high reflectance in the green band, typical of phytoplankton-laden waters. Dom-laden water endmember presents low reflectance at all wavelengths and the Iss-laden water endmember is characterized by an increasing reflectance towards to visible wavelength, and decreasing slightly at the near infrared.

This model was used to estimate the proportion of each component in the water using the following components: phytoplankton (Chl) laden water, dissolved organic matter (Dom) laden water, and inorganic particle (Iss) laden water. These components (endmembers) were selected based on the dominance of them in the surface water. Then, to estimate the proportion of components into the water we used the following equation:

$$R_{is} = f_{Chl} r_{Chl} + f_{Dom} r_{Dom} + f_{Iss} r_{Iss} + \varepsilon_i \tag{2}$$

Where, R_{is} is the reflectance at the i_{th} band at sampling station s; f_{Chl}, f_{Dom} and f_{Iss} are the fractions phytoplankton laden water, dissolved organic matter-laden water and inorganic matter-laden water, respectively; r_{Chl}, r_{Dom} and r_{Iss} are the reflectance of phytoplankton laden water, dissolved organic matter-laden water and inorganic matter-laden water, respectively and ε_i is the error at the ith spectral bands.

The main idea of this study was to estimate the proportion of each of the three components in the surface water. Ground data of Chl, Dom, and Iss concentration in water was then used to understand the distribution of turbidity in the Curuai floodplain lake. The results obtained with the SMM were compared with in-situ turbidity measurements through the interpolation, using a geostatistical method, called Ordinary Kriging.

4.2 Ordinary Kriging

A turbidity map was generated with the in-situ data, for subsequent comparison with the fraction images generated using the SMM. Thereby, it could be verified to which extent the two data sets matched. To produce such a reference map, Ordinary Kriging was used, interpolating in-situ turbidity measurements, as described in (Isaaks and Srivastava, 1989). Ordinary Kriging is a technique of making optimal, unbiased of regionalized variables at unsampled locations using the structural properties of the semivariogram and the initial set of data values. The calculation of the Kriging weights is based upon the estimation of a semivariogram model, as described by equation 3:

$$\gamma(h) = \frac{1}{2} Var[z(s+h) - z(s)] \tag{3}$$

Where: $\gamma(h)$ is an estimated value of the semivariance for the lag h. The estimation of a semivariogram model relies on two important assumptions: (1) the parameter $\gamma(h)$ exists and is finite for all choices of h and s, and that it do not depend on s. The Ordinary Kriging estimator is presented in:

$$Z(x, y) = \sum_{i=1}^{n} w_i z_i \tag{4}$$

Where n is the number of considered measures, zi are the corresponding attribute values, and wi are the weights.

The experimental semivariogram was fitted with various theoretical models (spherical, exponential, Gaussian, linear and power) using the weighted least square method. The theoretical model that gives minimum RMSE is chosen for further analysis. In this case, the fitted model was based on Gaussian model (Table 1).

Water level	Anisotropy direction	Structure	Nugget	Sill	>range	<range	Model
		1°		114	16436	ε	
Rising	94°	2°	619	7770	17924	16436	Gaussian
		3°		1480	∞	17924	

Table 1. The semivariogram parameters used to interpolate the in-situ turbidity.

The adjustment on Gaussian model suggests the existence of smooth spatial variance pattern in the of study site (Burrough and Mcdonnell, 1998). The reference map was then used to evaluate the result obtained with the SMM. The equation 5 presents the fitted model used to interpolate the turbidity distribution during the rising water phase.

$$\gamma(h) = 619 + 114[Gau(\sqrt{\left(\frac{h_{94°}}{\varepsilon}\right)^2 + \left(\frac{h_{216°}}{16436}\right)^2})] + 7770[Gau(\sqrt{\left(\frac{h_{94°}}{17924}\right)^2 + \left(\frac{h_{216°}}{16436}\right)^2})] +$$
$$1480[Gau(\sqrt{\left(\frac{h_{94°}}{17924}\right)^2 + \left(\frac{h_{216°}}{\infty}\right)^2})] \tag{5}$$

Where $\gamma(h)$ is the semivariance at the lag h, $h_{n°}$ is the semivariance due to angles of anisotropy, ε is the range of the lower anisotropy angle, and 'Gau' is the fitted Gaussian fitted model.

4.3 Spatial regression between the SMM and in situ turbidity

In general, OLS models have been used in research (Tyler et al. 2006). This approach, however, does not consider the spatial autocorrelation of samples within an aquatic system. When spatial autocorrelation is not considered in regression analysis, the significance of parameters can be overestimated, and the existence of large-scale variations might lead to spurious associations (Anselin, 1988). In our study, a method called Kernel Estimation (KE) was applied to in-situ turbidity, thereby testing for the existence of spatial autocorrelation (which is indicative of different spatial regimes).

The KE is a common analysis tool to determine the local density of point events and create a field representation. The Kernel is computed, using a Gaussian function, k, which considers m events (i.e. turbidity samples) $\{u_i,...u_i + m - 1\}$ contained in a bandwidth τ around u and the distance d between the position and the ith sampling (Figure 5), through the functions represented by equation 6:

$$\hat{\lambda}_\tau(u) = \frac{1}{\tau^2} \sum_{i=1}^{n} k\left(\frac{d(u_i,u)}{\tau}\right), d(u_i,u) \le \tau \tag{6}$$

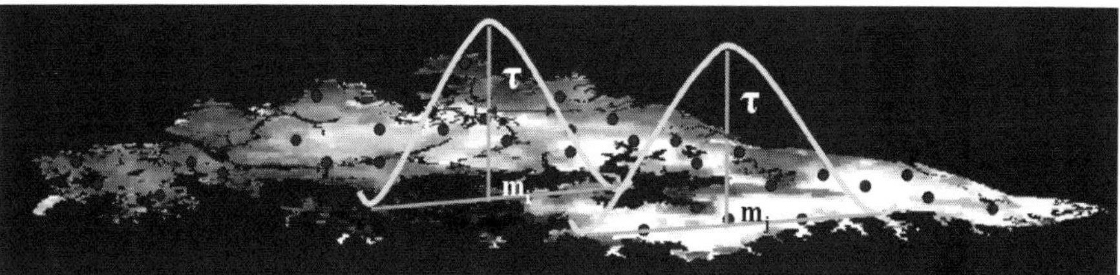

Fig. 5. Kernel estimation of the spatial turbidity pattern.

To improve the kernel estimation we used an adaptive bandwidth version which locally adjusted bandwidth values at different points within the floodplain. Thereby, it ensured that the bandwidth always contained a minimum number of samples, improving estimate precision (Bailey and Gatrell, 1995).

The technique was also applied to the OLS, a model compiled using global in-situ turbidity data to evaluate the spatial regression models. The OLS does not take into account spatial dependence among samples and consequently, different spatial regimes. However, spatial error and lag models were developed for each identified spatial regime, using the KE algorithm, thus including spatial dependence which tends to inflate the variance of OLS regression (Bailey and Gatrell, 1995).

To assess the relationship between the fraction images and in-situ turbidity two approaches were adopted The first, assumes that the variance of the disturbance term is constant; we start with the OLS model:

$$y = X\beta + \varepsilon, \text{ with } \varepsilon \sim N(0,\sigma^2) \tag{7}$$

Where y is an (Nx1) vector of observations on a dependent variable taken at each of N locations, X is an (NxK) matrix of exogenous variables, β is an (Kx1) vector of parameters, and ε is an (Nx1) vector of disturbances.

The second uses two alternative forms of spatial dependence models (Bivand, 1998), the spatial lag model (presented in equation 8) and the spatial error (presented in equation 9).

$$y = \rho W y + X\beta + \varepsilon \tag{8}$$

$$y = X\beta + u, \text{ with } u = \lambda Wu + \varepsilon \tag{9}$$

Where λ is a scalar spatial error parameter, and u is a spatially autocorrelated disturbance vector. These two models can also be related to each other in the Common Factor model.

According to (Bivand, 1998), however the spatial lag and spatial error models can only be combined for estimation if the neighborhood specifications (here: the W matrices of the lag and error components) differ. The spatial lag model is clearly related to a distributed lag interpretation, in that the lagged dependent varible, Wy, can be seen as equivalent to the sum of a power series of lagged independent variables stepping out across the map, with the impact of spillovers declining with successively higher powers of ρ (Bivand, 1998).

Conversely, the spatial error model is a special case of a regression with a non-spherical error term, in which the off-diagonal elements of the covariance matrix express the structure of spatial dependence (Baltaqi, 2003).

Spatial lag and spatial error dependence tests allowed determining which of the two models was more suitable to the data. Subsequently, spatial regression models were created for each spatial regime. Measurements of Log likelihood fit, Akaike information criteria (AIC) and Schwarz criteria (SC) for both models (OLS and spatial models) to verify if the inclusion of spatial parameters improve the regression model.

The Schwarz criterion addresses the problem of selecting one of a number of models of different dimensions (Bailey and Gatrell, 1995). However, some authors (Burrough and Mcdonnell, 1998; Bailey and Gatrell, 1995) suggest the use of AIC to evaluate the best fit. The Akaike information criterion is expressed in equation 10 (Pan, 2001).

$$AIC = -2 * LIK + 2k \tag{10}$$

Where, LIK is the log of the maximized likelihood and k is the number of regression coefficients. A small AIC value suggests a high suitability of the tested model.

4.4 Evaluation of turbidity estimations

Having selected the best model, the next step was to apply it to the Terra/MODIS image to estimate the turbidity distribution. The resulting map was then used as a reference to evaluate the model. Random positions were selected on the image to run correlation analyses so as to assess the potential spatial model and the RSME. At each random geographical position, a $3x3$ pixel window was averaged, obtaining both the measured and modeled value and computing the correlation.

5. Results and Discussion

Figure 6-a shows that the water, of the entire Curuai floodplain lake was rich in inorganic matter (Iss), with a particularly high proportion in the Poção lake (Figure 6-b). The images illustrating the distribution of dissolved organic matter (Dom) revealed that this was particularly apparent in the Salé lake, and in the littoral region (i.e. the region between water and forest). This is mainly due to the fact that some organic matter in decomposition is transported into the floodplain by the water during the rising , phase (Figure 6-c). The phytoplankton (Chl) fraction represented a high proportion of the 'Grande de Curuai' and 'Grande do Poção' lakes (Figure 6-d). As described in (Barbosa, 2005), the water coming from the Amazon River was more enriched with phytoplankton than that in the Curuai floodplain; the proportion of the water rich in Chl was found to increase from the East to the West.

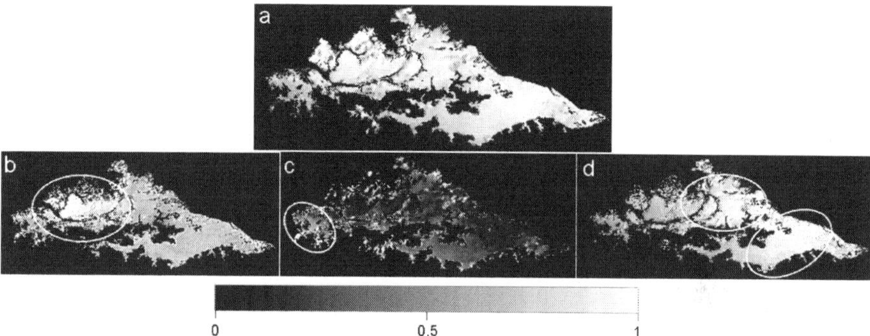

Fig. 6. Terra/MODIS imagery of the Curuai floodplain (a), and proportion images for the tested endmembers: Iss (b), Dom (c) and Chl (d) compiled with the SMM algorithm.

The fraction images (Iss in the red channel, Dom in the blue channel, and Chl in the green channel) unfolded a qualitative picture of the turbidity distribution within the Curuai floodplain lake (Figure 7-a). Figure 7-b shows the SMM error image of the Region 1 (Figure 7-a) which represents the most turbid water within the floodplain. This regions has an area of extremely high turbidity (see the circle in Figure 7-b) which actually is due to cloud cover. Region 2 (Figure 7-a) evidence of the mixture of water masses with dominant proportions of Chl and Iss, and low turbidity (Figure 7-b). Region 3 (Figure 7-a), finally, was dominated by water with a high proportion of Chl, and moderate turbidity. A mixture of water masses containing high concentrations of Chl and Iss lead to a low error of the SMM. Conversely, a high amount of Chl entering the floodplain through the main channel from the Amazon River leads to a moderate error in the unmixing model. Due to the occurrence of a transition zone between the aquatic and the terrestrial environments, higher errors occurred at the edges.

However, these results are qualitative. To obtain suitable quantitative results, an OLS regression model was applied to the SMM results and the in situ turbidity data. Subsequently, we checked for any signs of spatial autocorrelation between samples to prevent the problem of spurious associations, using the result obtained by the kernel estimator algorithm.

The results of the OLS regression model, using all 215 turbidity samples collected, and the fractional abundance of the OAS are poorly correlated ($R^2 = 0.10$, $p < 0.05$). This is presumable due to the fact that the floodplain received water from different sources (rain, black water and white water) when the water level rose. The mixture of water masses caused a high heterogeneity in the spectral response of the surface water, causing high standard deviation in turbidity measurements, what highlights the importance of including the autocorrelation factor in the regression analysis.

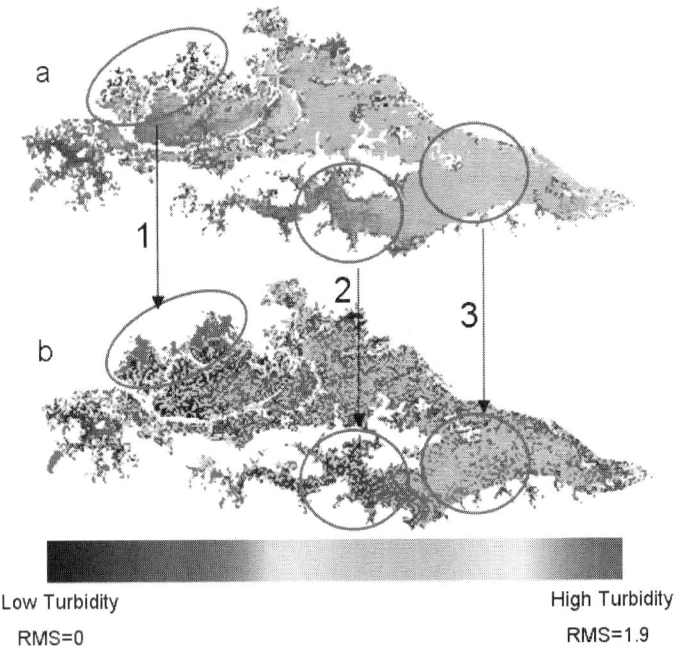

Fig. 7. (a) The composition of the image fractions derived from the SMM (Iss in red channel, Chl in green and Dom in blue), in comparision with the the RMS image (b).

If spatial dependence was verified the OLS method would lose validity (Anselin, 1988). In accordance with Bailey and Gatrell (1995) the use of the non-spatial regression models (OLS), may or may not be suitable because it assumes a stationary water conditions in the period of ground sampling. Also, the OLS model does not consider the presence of spatial autocorrelation among samples distributed within the floodplain. Hence, in order not to overestimate the significance of parameters, autocorrelation must be considered. Likewise, large-scale variations may induce spurious associations.

To check the spatial autocorrelation between turbidity samples, we applied the Kernel estimator, and then separated the samples of fractional abundance of the OAS and in-situ turbidity data in spatial regimes. These cluster the turbidity data from the whole floodplain by their spatial dependence (Bailey and Gatrell, 1995).

The KE revealed that there were four spatial regimes of the Curuai floodplain lake turbidity at the rising water level (Figure 8-a). The kernel also shows four regions of density from high (region 1) to low density (region 4). The region 1 was characterized by the highest spatial dependence in the study area. In this area, the spatial regime is particularly abundant 1(Figure 8-b) and includes the largest number of samples.

Region 2 included a mixture of spatial regimes 1 and 2, and represented the second largest spatial dependence. Region 3 was characterized by regimes 3 and 4, and region 4 included a mixture of all regimes (Figure 8-b). The number of samples grouped in each spatial regime was 64, 54, 51 and 45 for regimes 1, 2, 3, and 4 respectively.

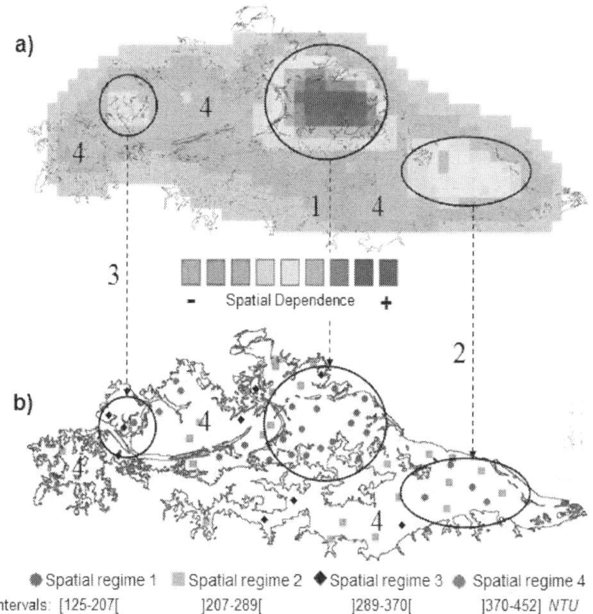

Fig. 8. (a) Spatial dependence of in-situ turbidity samples, as defined by Kernel algorithm, (b) spatial regimes occurring within the Curuai floodplain lakes with corresponding turbidity intervals values (NTU).

The different regimes are brought about by different types of water input when the Amazon River enters into the floodplain. The water arrives in different ways, and at different times. As a result, the water entering the different sections of the floodplain has variable physical and chemical properties.

Table 2 indicates that the application of the spatial regression model increased the R^2 in relation to the OLS model. This is presumably due to the turbidity data being characterized by spatial auto-correlation, as observed in figure 8-a. The OLS model did not take into account the spatial dependence among samples, thereby causing an overestimated significance of the parameters.

	OLS	Spatial regression							
		Spatial Lag Model				Spatial Error Model			
	Global	Re1	Re2	Re3	Re4	Re1	Re2	Re3	Re4
R2	0.10	0.31	0.27	0.61	0.71	0.32	0.30	0.95	0.79
LIK	-295.79	-95.07	-101.88	-35.97	-22.78	-94.93	-101.69	-35.50	-20.70
AIC	599.58	200.15	213.76	81.94	55.57	197.86	211.38	79	49.41
SC	607.54	205.61	219.44	82.93	54.53	202.22	215.92	79.79	48.57

Table 2. Results of the OLS and spatial regression models, with R2 being the coefficient of determination, LIK being the log of the maximized likelihood, AIC being the Akaike information criterion, and SC being the Schwarz criterion.

The approach to estimate the turbidity data in spatial regimes shows that in all regimes the best fit was obtained using the spatial error model. According to Anselin (1988) the fit on the spatial error model suggests that the spatial effect is a noise derived from the interplay of several variables not included in the model.

The spatial regime 1 (Re1) has the poorest correlation compared to the others (R^2 = 0.10, p < 0.001), with LIK, AIC and SC being -295.79, 599.58, and 607.54 respectively. Conversely, the spatial regime 3 (Re3) shows the best results with LIK = -35.50, AIC = 79 and SC = 79.79, and R2 = 0.95, p < 0.001, using the spatial error model (Table 2). The equations of the four spatial regimes (Re1, Re2, Re3, and Re4) are listed below in the equations 11, 12, 13, and 14 respectively:

$$Turbidity = 99.72 + 221.71 Iss + 199.46 Chl + 186.53 Dom \tag{11}$$
$$Turbidity = 311.61 - 254.14 Iss - 23.87 Chl + 22.98 Dom \tag{12}$$
$$Turbidity = 327.19 + 42.62 Iss - 246.85 Chl - 136.06 Dom \tag{13}$$
$$Turbidity = 428.86 - 154.66 Iss - 231.23 Chl - 94.05 Dom \tag{14}$$

Where, Iss is the inorganic-laden water fraction, Chl is the phytoplankton laden water fraction and Dom is the dissolved organic matter-laden water fraction.

Equation 13 indicates the influence of the OAS proportion on this method to quantify turbidity. According to equation 13, the turbidity in this region will be lower when the water has high proportion of phytoplankton than inorganic particles. Conversely, in areas where those types of water are particularly abundant, the modeled turbidity will be smaller. Where the Iss fraction dominates, turbidity will be high. This suggests that the model will perform better when turbidity is determined mainly by suspended inorganic particles.

Before applying the equation 13 to the MODIS image fractions, a sensitivity analysis was carried out using Pearson's correlation analysis for each spatial regime (Figure 9).

This sensitivity analysis revealed statistically significant correlations R^2 = 0.65 (p < 0.05, n = 20) and R^2 = 0.40 (p < 0.05, n = 20) for the spatial regimes 3 [Figure 9 (c)] and 4 [Figure 9 (d)] respectively. In contrast, analyses for the spatial regimes 1 [Figure 9 (a)] and 2 [Figure 9 (b)] were not significant with R^2 = 0.005 (p < 0.05, n = 20) and R^2 = 0.004 (p < 0.05, n = 20) respectively (Figure 9). The best fit was hence found in the spatial regime 3.

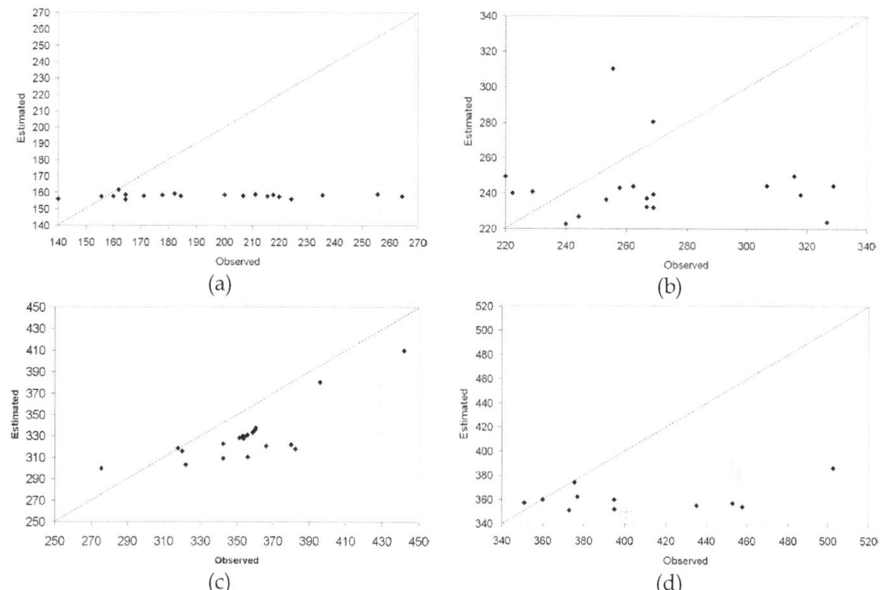

Fig. 9. Correlation between turbidity estimated using the spatial model and observed turbidity: (a) spatial regime 1; (b) spatial regime 2; (c) spatial regime 3; (d) spatial regime 4.

The main reason for this poor statistical performance is probably related to the instantaneous image acquisition and the sampling turbidity design. Both cause a difference in turbidity conditions, which makes a comparison of the image and the in-situ data difficult. In addition, the nonlinearity presented in the relationship between the loads of phytoplankton, dissolved organic matter, and inorganic matter in the water made the unmixing processes difficult.

Figure 10 shows the turbidity distribution resulting from the application of this model to the entire floodplain (Figure 10-a) as well the in-situ turbidity distribution (Figure 10-b). The spatial distribution of the turbidity is similar, especially in regions 1, 2, and 3.

Region 1 is characterized by the highest turbidity compared to the other regions, and by small depths when the water level rises (Barbosa, 2005), and is the main pathway from the Amazon River into the floodplain. The water then flows through many channels, located in the Northwest region of the floodplain, with sufficient energy to keep the inorganic particles in suspension. Winds promote the sediment re-suspension and increment turbidity here.

Region 2 is also characterized by small depths but is protected from wind. As a result, turbidity is related to depth rather than wind perturbation (Carper and Bachmann, 1984; Booth et al. 2000). In Region 3, turbidity varied between 203 and 305 NTU, and the spatial model was able to estimate turbidity values within the range measured in-situ. According to Barbosa (2005), this is one of the regions with the highest depths in the Curuai floodplain lake. It is not subjected to intense wind perturbation due to this characteristic. Instead, suspended particles settle into the bottom of the lake, decreasing the possibility of re-suspension.

Figure 10-c indicates that the spatial model overestimated turbidity in regions 4 and 6, and underestimated it in region 5. The overestimation in 4 can be explained by cloud cover contamination. Region 5 is in a transition zone (natural barrier) that separates the floodplain lake in two larger zones, (i) zone Northwest and (ii) zone Southeast (Barbosa, 2005). This transition zone causes a turbidity underestimation due to mixture of different water masses. The overestimation in region 6 can be related to a large mixture caused by a water inlet owing to a rising water level. The water inlet is particularly pronounced in region 6 (Barbosa, 2005).

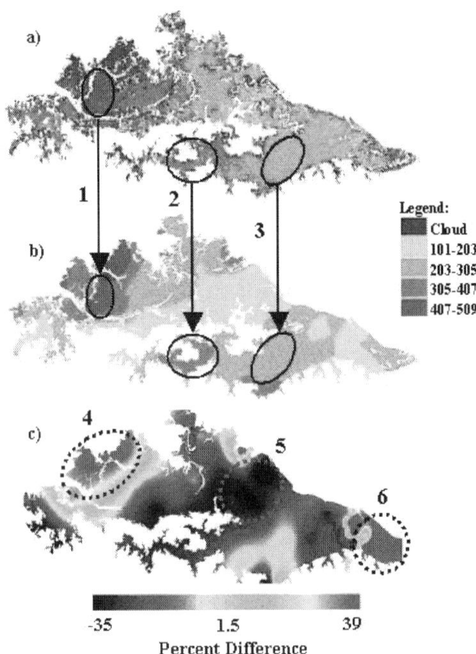

Fig. 10. The turbidity distribution (NTU) in the Curuai floodplain, (a) using the spatial regression model, using the OK algorithm (b) and (c) is in percent difference between (a) and (b).

The samples to evaluate the spatial model were collected the entire lake, except for regions 4 (due to cloud cover), 5 (natural barrier), and 6 (overestimation due to intense water flowing through this channel), as pointed out in Figure 7. The evaluation of the model resulted in a value of $R^2 = 0.90$ (p = 0.05; n = 60) and RSM of 17 NTU for the turbidity model under the spatial regime 3 (Figure 11). Previous studies of turbidity using the SMM in high water level (Alcântara et al., 2008) show that the modeled turbidity had a correlation $R^2 = 0.62$ (p > 0.005; n = 20).

This difference in performance can be explained by the presence of different water types in the floodplain. Their presence makes it difficult to apply one model, adapted to a given regime for the entire region. Another possible error source is the in-situ turbidity-sampling scheme. Due to the size of the lake, it was not possible to get all the samples on a single day. Instead, the data was collected during a 13-day period, in which local turbidity may have been affected by changes in wind intensity, light field, and other environmental factors affecting the lake hydrodynamics. The highest errors were encountered in more turbid waters, as opposed to clear waters, presumably due to a high mixture of the OAS present in the floodplain.

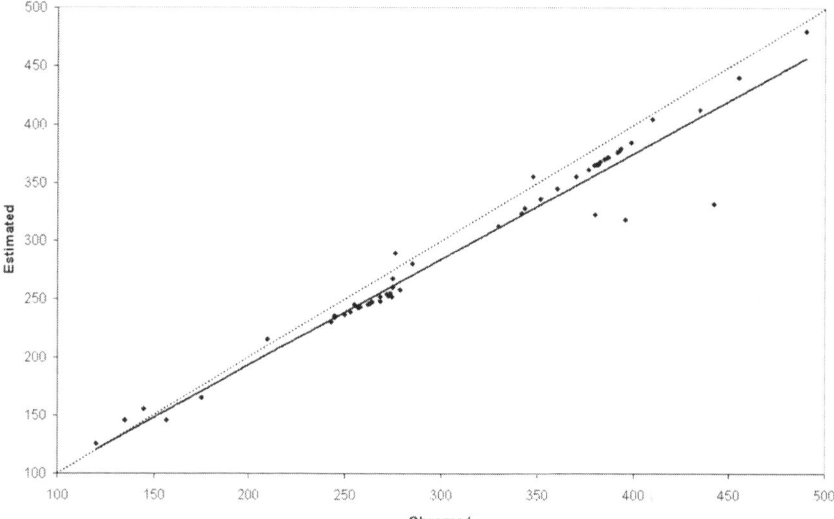

Fig. 11. Evaluation of the turbidity estimated through the spatial model (equation 13) and in-situ turbidity (NTU), interpolated using the OK algorithm.

Since the MODIS image was instantaneously acquired on 27 February 2007, the modeled turbidity does not account for environmental changes, which can affect in-situ conditions. In spite of this drawback, the unmixing model showed a good potential to assess the turbidity in continental aquatic systems. This potential could be improved using hyperspectral remote sensing imagery (Rudorff et al. 2007), since features in spectral responses could then be detected in more detail (Rudorff et al., 2006; Fraser, 1998; Brando and Dekker, 2003).

6. Conclusions

This present work evaluates the suitability of the spectral unmixing model to map the turbidity distribution in the Curuai floodplain. The main conclusions are:
1. The fraction images for the endmembers selected directly from the MODIS image based on dominance of water components allowed assessing the turbidity in the Curuai floodplain lake.
2. Owing to non-linearity in the Amazon floodplain waters, the unmixing model does not work in an optimal way. This is also due to autocorrelation presented in the study area.

Oyama, Y.; Matsushita, B.; Fukushima, T.; Matsushige, K.; Imai, A. (2009). Application of spectral decomposition algorithm for mapping water quality in a turbid lake (Lake Kasumigaura, Japan) from Landsat TM data. ISPRS Journal of Photogrammetry and Remote Sensing 64 (1), 73-85.

Pan, W. (2001). Akaike's information criterion in generalized estimating equations. Biometrics 57 (1), 120-125.

Rudorff, C.M; Novo E.M.L.M.; Galvão L.S; Pereira-Filho W. (2007). Derivative analysis of hyperespectral data measured at field and orbital level to characterize the composition of optically complex waters in the Amazon (in Portuguese). Acta Amaz. 37 (2), 269-280.

Rudorff, C.M.; Novo, E.M.L.M.; Galvão, L.S. (2006). Spectral mixture analysis for water quality assessment over the Amazon floodplain using Hyperion/EO-1 Images. Revista Ambi-Água 1 (2), 65-79.

Theseira, M.A.; Thomas, G.; Taylor, J.C.; Gemmell, F.; Varjo, J. (2003). Sensitivity of mixture modelling to end-member selection. Int. J. Remote Sens. 24 (7), 1559-1575.

Tyler, A.N.; Svab, E.; Preston, T.; Présing, M.; Kovács, W.A. (2006). Remote sensing of the water quality of shallow lakes: A mixture modelling approach to quantifying phytoplankton in water characterized by high-suspended sediment. Int. J. Remote Sens. 27 (7-8), 1521- 1537.

Wetzel, R.G. (2001). Limnology: lake and river ecosystems, 3rd Ed.; Academic Press: San Diego, USA, pp. 1006.

2

Application of remote sensing to the estimation of sea ice thickness distribution

Takenobu Toyota
Hokkaido University
Japan

1. Introduction

Sea ice covers about one tenth of the world ocean surface and plays an important role in the global climate system through its interaction with the atmosphere and ocean. Although its thickness is only several meters at most, it forms an effective thermal insulation sheet due to its high albedo and low thermal conductivity, leading to a significant reduction in the heat flux from the ocean to the atmosphere, especially in winter. Moreover, when sea ice forms, it expels brine into water to produce high density water underneath the ice, leading to the deep vertical convection. This plays a major role in the three dimensional ocean thermohaline circulation of the ocean on a global scale. Another notable feature is its large variability in areal extent on seasonal to annual time scales e.g. in the Southern Hemisphere (Figure 1). Even on a daily time scale, it varies significantly due to the combined effects of wind and ocean swell. In turn this variability also has an impact on both climate and daily weather conditions through the above processes. In addition, the mobility of sea ice leads to the interaction between ice floes and ridging/rafting of the ice (dynamical pile-up), which makes its properties more complex in comparison with other surface types in the cryosphere, such as snow cover or frozen soil.

In these processes, thickness distribution is one of the most important parameters of sea ice. This is firstly because the outgoing heat flux from ocean to atmosphere is highly dependent on the ice thickness distribution, especially in winter when the areal fraction of ice with a thickness of less than 0.4 m essentially controls the net heat input to the atmosphere (Maykut, 1978). This implies that the areal fraction of thin ice controls the rates of thermodynamic ice growth amount and the associated production of the high density water beneath. Therefore, the ice thickness distribution can make a significant difference to both the surface heat budget and deep ocean convection for a given concentration of sea ice. In this context, it is worth mentioning that snow cover on ice also works as a more effective insulator due to its low thermal conductivity. Secondly, and in terms of sea ice dynamics, the thickness distribution determines ice strength, and thus controls ridging/rafting processes and vice versa (Thorndike et al., 1975). These "pile-up" processes are essential to the ice thickening, particularly in the seasonal sea ice zone (SSIZ) where the ice is relatively thin and mobile. From the crystallographic analyses it has been shown that a layered structure due to ice pile-up is prominent in the SSIZ (e.g. Gow et al., 1982; Lange and Eicken,

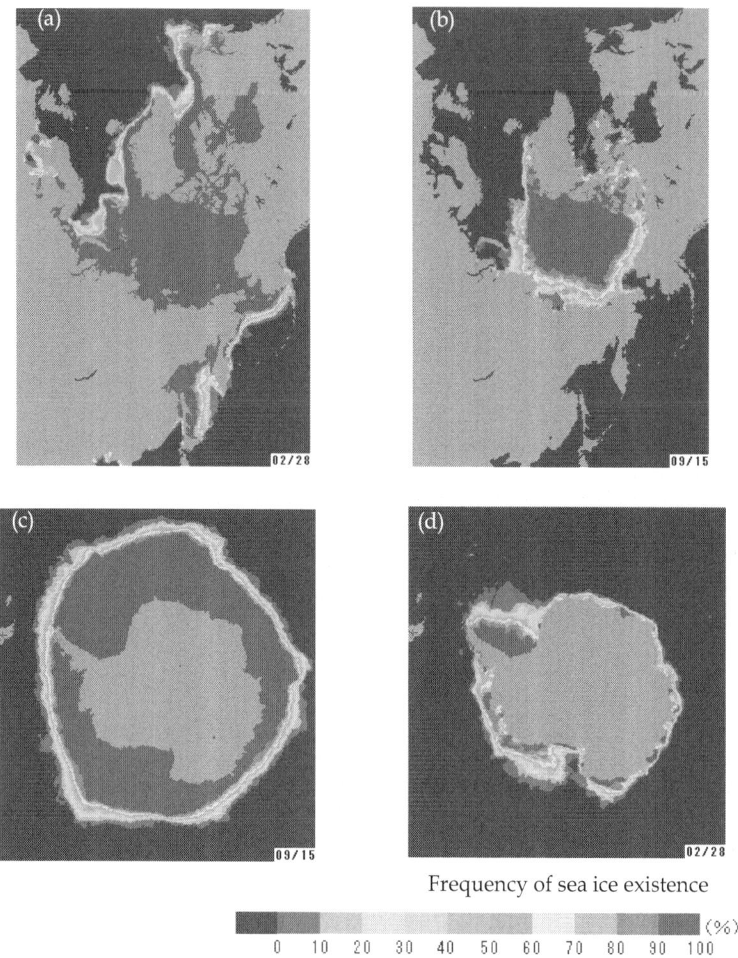

Fig. 1. Mean geographical distribution of sea ice extent (1979-2000) for the northern hemisphere at (a) maximum extent (Feb. 28) and (b) minimum extent (Sep. 15) the southern hemisphere at (c) maximum extent (Sep. 15) and (d) minimum extent (Feb. 28).

(Japan Meteorological Agency, 2005)

1991) and that the mean layer thickness is only about 12 cm both for Antarctic ice (Worby et al., 1996) and that in Arctic peripheral seas e.g. the Sea of Okhotsk (Toyota et al., 2007). Therefore, the observed thickness distribution can provide some important insights into the role of such dynamical thickening processes. Thirdly, the long-term trend and the short-term change in the sea ice thickness distribution is a sensitive indicator of climate change, as observed for Arctic perennial ice (e.g. Nghiem et al., 2007). Therefore, the accurate and consistent estimation of sea ice thickness distribution is one of the most important issues that many researchers in this field have paid attention to for a long time.

Unfortunately, sea ice thickness is also one of the most difficult parameters to measure. It was only two decades ago that the mean hemispheric distribution of ice thickness was first mapped for the Arctic Ocean (Bourke and Garrette, 1987; Figure 2a), and only recently for the Antarctic Seas (Worby et al., 2008; Figure 2b). The most direct and accurate measurement technique is in-situ drilling with an ice auger. In the Arctic, it is said that Nansen first conducted systematic ice thickness measurement with this method during the drift of "Fram" (1893-96) (Wadhams, 2000). In the Antarctic, it was by drilling that the standard measurement protocol was initiated in the late 1970's (Ackley et al., 1979). In addition to being labour-intensive, this technique suffers a problem of representativity. Rothrock (1986) pointed out from a statistical perspective that as many as 560 drilling sites would be required if we need to know the mean thickness at a point with an accuracy of 0.1 m for the Arctic ice. Thus, nowadays this method is used mainly for calibration of other tools. To obtain thickness distribution at a large scale, remote sensing is a requisite tool. Since 1958, when submarine sonar was applied for this purpose in the Arctic Ocean, enormous efforts have been made to estimate it using remote sensing. Unfortunately, submarine sonar coverage of Antarctic ice is not possible.

In this chapter several methods will be introduced to give a brief history of ice thickness measurement, focusing on how remote sensing techniques have been applied and how they have contributed to the improved understanding of sea ice thickness. Finally, the current situation will be summarized and some suggestions will be made about a future work that needs to be done.

2. Measurement techniques

In this section the remote sensing tools that have been used to measure ice thickness to date will be introduced. There are generally two strategies: 1) direct measurement of ice draft or a full depth of ice using an upward-looking sonar or other tools; and 2) relating the observed surface properties with ice thickness using satellite sensors. Here, they will be described in near-chronological order.

2.1 Submarine sonar profiling

This method provides the under-ice profile relative to sea level using an upward beamed sonar on a submarine. By this technique, profiles of ice draft can be obtained along the track of the submarine by recording the distance to the underside of the ice pack and subtracting this from the depth of the sonar, using intervening regions of open water as a reference level. The overall accuracy is about 0.3 m (0.09 m for smooth ice), and the horizontal resolution is 1.3 to 1.5 m with a surface beam diameter of 3.2 m (Wadhams and Horne, 1980). These data are not a complete measurement of ice thickness, as ice draft corresponds to 80 to 95 % of ice thickness, but nonetheless this data set provides the approximate thickness distribution and detailed topography of under-ice with reasonable accuracy, assuming invariability in snow cover thickness and density.

This device was first used during the voyage of the "Nautilus" in early August 1958 which crossed the Arctic Basin, and much of our knowledge about the synoptic ice thickness

(a)   (b)

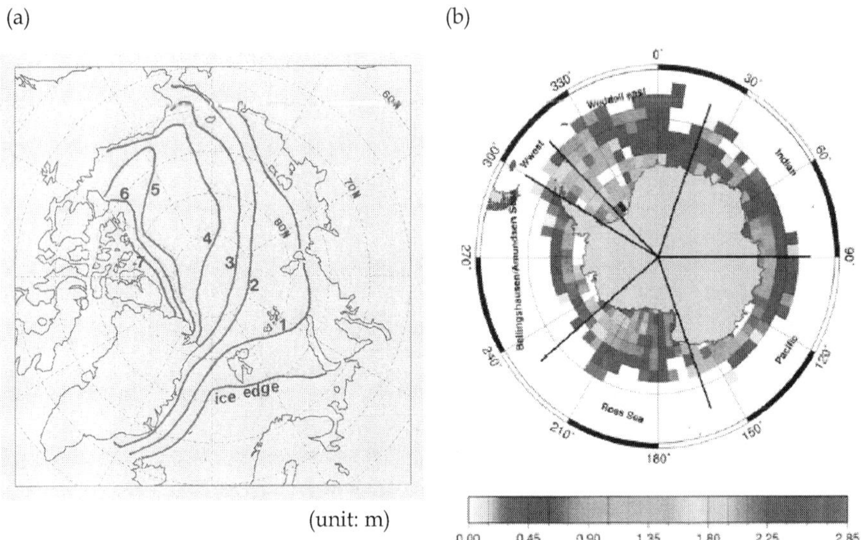

Fig. 2. Climatology of ice thickness distribution for (a) the northern hemisphere (reproduced based on Bourke & Garrett (1987)) (b) the southern hemisphere (cited from Worby et al. (2008)). The average periods are 1960-82 for the Arctic and 1981-2005 for the Antarctic. The data are based on submarine sonar observations for the Arctic and on ship-based repeated visual observations for the Antarctic.

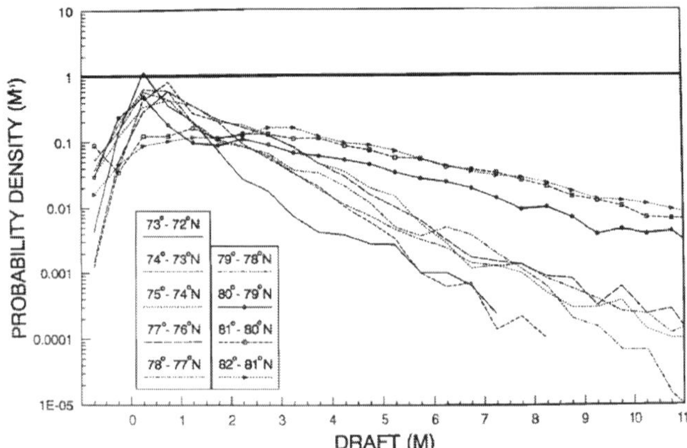

Fig. 3. Probability density of ice thickness in the Greenland Sea. Note that the clear negative exponential tail at the thick end can be seen for all the cases. (cited from Wadhams (1992)).

distribution in the Arctic Ocean is based on the results obtained with this device on subsequent cruises (McLaren, 1989). With the results obtained, the regional statistics of ice thickness distribution and ridge keel profiles were investigated in detail e.g. for the Beaufort Sea (Wadhams and Horne, 1980), in the Fram Strait region (Wadhams, 1983, 1992), and in Davis Strait (Wadhams et al., 1985). One of the most significant findings is that in most cases the thickness distribution follows a negative exponential function, irrespective of regional location (Figure 3). Although the mechanism to explain this feature has not been clarified yet, this feature may provide considerable insight into the ridging process and is therefore a problem worth investigating given that ridging is a key determinant of the thickness distribution. McLaren (1989) examined the general features for the ice over the Arctic Basin from the data sets obtained along the two identical cruises tracks in 1958 and 1970. Bourke and Garrett (1987) then compiled the data obtained during the 17 voyages from 1960 to 1982 and mapped the ice thickness distribution for the first time (Figure 2a). According to their results, the mean ice thickness above 65N is 2.9 m with a standard deviation of 1.8 m. This value was taken to be the standard "baseline" value for the Arctic Ocean before the recent thinning of the Arctic ice that has been reported (e.g. Rothrock et al., 1999).

In a further development of this method, Wadhams (1988) showed that additional information can be gained of the three-dimensional under-ice topography of Arctic ice using a side-scan sonar. Indeed, it was this technique that provided the first evidence for significant thinning of the Arctic ice cover over the past few decades (Wadhams, 1990; Rothrock et al., 1999; Tucker et al., 2001). By comparing the data obtained in 1958-1976 with those in the 1990s, Rothrock et al. (1999) showed that the mean ice draft at the end of the melt season decreased from 3.1 m to 1.8 m in the Arctic Ocean.

Despite all these pioneering results, the drawback to these data is that since this measurement is operated on a military basis, the observation is not necessarily systematic in space and time and time series of thickness data cannot be obtained within given areas (Wadhams, 2000). Therefore, it does not provide sufficient data sets to enable monitoring of change in ice thickness distribution on a seasonal or annual basis. Moreover, it cannot be applied to Antarctic ice because military submarines are prohibited from operating under Southern Ocean sea ice by the Antarctic Treaty.

2.2 Moored ice profiling sonar

This technique measures ice draft using a moored upward-looking sonar system which is tethered to a line attached to the seafloor. The principle is similar to that of submarine sonar. The device transmits a pulse of sound towards the surface, and the ice draft is then obtained by converting the travel time of the pulse to vertical distance allowing for variation in water temperature and pressure and subtracting it from the depth of the device (Figure 4). The footprint at the underside of the sea ice is about 2 m and the overall accuracy is 0.05 to 0.10 m (Melling and Riedel, 1995). The most remarkable merit of this method is that it can provide a time series of ice draft distribution at a fixed site with high temporal resolution (a few seconds to minutes) for several months. Thus it enables the monitoring of daily to seasonal variation in the ice draft distribution. Additionally, it can be deployed on the shallow continental shelf regions, where a submarine cannot operate. This is important because considerable deformation often occurs near the Arctic coastal region, to affect sub-sea ice exploration.

This technique was first applied to the near-shore region of the Beaufort Sea in the late 1970's (Hudson, 1990). From the over-winter measurements, the seasonal change of level ice thickness and the frequency of ridge keels were first revealed. In the same region, Melling and Riedel (1995, 1996) also deployed this system and examined the more detailed characteristics of deformed ice and under-ice topography by combining the ice motion obtained with a Doppler sonar. In Fram Strait, Vinje et al. (1998) examined the annual and seasonal variation of ice draft during the period of 1990 to 1996. By empirically converting ice draft to thickness and combining the ice drift data obtained from buoys, they estimated the annual mean ice volume flux through the Fram Strait to be 2850 km^3 yr^{-1}. Combining this ice thickness data set with satellite-derived ice velocity Kwok and Rothrock (1999) found a relationship between the ice volume flux out of the central Arctic and the large-scale modes of atmospheric circulation (i.e. the North Atlantic Oscillation). In the Sea of Okhotsk, Marko (2003) applied this method to the east of Sakhalin Island and analyzed the ice draft data in conjunction with intense wave energy. Fukamachi et al. (2003, 2006) deployed this system in the coastal region of Hokkaido, Japan, and first revealed the ice thickness distribution including ridged ice in this region from three-year data sets.

In the Antarctic, where submarine sonar data are not available, this method has been an effective tool. However, it should be noted that there are some restrictions regarding deployment: the mooring system has to be operated from the seafloor at a depth deeper than 350-440 m to avoid ice bergs in the sea ice area (Massom et al., 2009), which may damage the instrument (Worby et al., 2001). Strass and Fahrbach (1998) examined the temporal and regional variation of ice draft across the Weddell Sea from six ice profiling sonars deployed between the tip of the Antarctic Peninsula and Kapp Norvegia from 1990 to 1994. Based on these data, Harms et al. (2001) estimated the annual mean ice volume export from the western Weddell Sea to be about 0.05 Sv (~ 1580 km^3 yr^{-1}). In East Antarctica, Worby et al. (2001) analyzed the seasonal variation of ice thickness distribution with this system.

Thus this method has significantly enhanced our knowledge of detailed ice thickness distribution, as a second tool to submarine sonar, for about three decades. It has allowed the estimation of various parameters related to ice thickness and in particular contributed to the understanding the fresh water budget through Fram Strait, which relates to, and has strong implications for climate change. Additionally, this method has enabled detection of ice ridge keels as deep as a few tens of meters (e.g. 29 m in the Beaufort Sea [Melling and Riedel, 1995] and 17 m in the Sea of Okhotsk [Fukamachi et al., 2006]). However, the drawback is that this device is not appropriate for obtaining the spatial distribution of ice thickness on a large scale at a given time, and has difficulties in monitoring ice thickness for more than decadal terms because of logistical constraints i.e. the need to recover and redeploy instruments every few years.

2.3 Video monitoring system

This technique is basically designed as a ship-based measurement. With a downward-looking video camera mounted on the side deck of a ship, ice conditions around the hull are continuously recorded along-track. Ice thickness is then estimated on the video images for ice floes which break up at the bow and turn into a side-up position alongside of the hull (Figure 5). The scale on the video image is determined by lowering a measured stick onto the ice surface while the ship is stationary. The reading error is less than a few centimetres. In

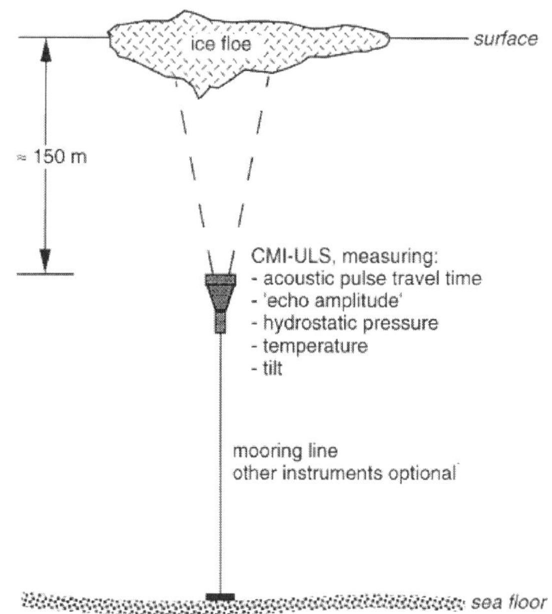

Fig. 4. Schematic representation of an ULS mooring set-up. (cited from Strass (1998))

Fig. 5. Schematic picture of a video monitoring method.

(a) A schematic figure (cited from Shimoda et al. (1997))
(b) A typical picture of ice which was broken and turned into a side-up position. Note the brown color near the bottom, associated with the attachment of ice algae.

this way, spatial ice thickness distribution along the ship track can be determined in a relatively short amount of time. However, it should be noted that deformed ice, which is hard to turn, is beyond the measuring capability of this method, and that it is subject to the ship route which tends to be selected to traverse regions of thinner ice. These facts can cause some bias in the results i.e. a net underestimate in ice thickness distribution estimates. Therefore, this method is not appropriate for the measurement of thick ice distribution. This is in contrast with the measurement by upward-looking sonar, which is relatively free from such artificial bias.

Nevertheless, there are several advantages unique to this method. First, it is useful for obtaining the thickness distribution of level ice, accurate knowledge of which is important to understand the thermodynamic growth and rafting process (stacking of two ice floes without break-up). Secondly, it allows us to measure the overlying snow depth according to the degree of whiteness on the video images. This is particularly effective in the marginal ice zones, where ice is so thin that it is hard to measure snow depth directly. It also allows us to detect brown ice containing ice algae (Figure 5b). Thirdly, this system can be operated at a relatively low cost and does not require complicated calibration. Finally, without any risk of the damage by icebergs, this method can safely be applied in the coastal Antarctic ice area.

The key to this method is the downsizing of the video camera to facilitate its use. This system was first introduced to monitor the sea ice conditions in the Antarctic coastal region during the Japanese Antarctic Research Expeditions (JARE) in late 1980's (Shimoda et al., 1997). In this area, theses observations were continued from 1988 to 2004, and from these results Uto et al. (2006a) found a significant annual variation in the thickness distribution of land-fast ice in the Lützow-Holm Bay near the Syowa Station and discussed the relationship with the break-up events of the land-fast ice there. This system was also deployed in the Sea of Okhotsk in 1991. Monitoring of the ice thickness distribution in the southern Sea of Okhotsk has continued since then from the icebreaker P/V "Soya", in collaboration with the Japan Coast Guard. Based on the data set obtained from 1991 to 2000, Toyota et al. (2004) showed that the thickness distribution follows a near Poisson distribution with an average of 33±10 cm, and proposed a stochastic rafting cycle model, referring to the layered structure. They also checked the accuracy of this method by comparing their data with the ice profiling sonar data obtained in that region (Fukamachi et al., 2003) and confirmed that the thickness distribution of these two methods matches well within the range of a few percent for non-ridged ice. Their group is still continuing these observations over this region, and the annual variations in mean ice thickness and volume are shown in Figure 6. In this figure, an annual variation in ice thickness is apparent with a period of about 7 years, while the trend towards a decrease in ice volume is remarkable, particularly after 2000.

In spite of several drawbacks, a video monitoring system therefore still remains an effective tool for obtaining ice thickness distributions, and in the marginal ice zones in particular. As this method provides direct measurements in a larger sense, it is sometimes used to validate the results of other remote sensing tools. However, the time-consuming task of analysing the images for measurement precludes real time monitoring and this problem should be improved on the computer base in the future.

2.4 Electromagnetic Induction Sounding

This device measures the distance from the instrument to the ice-seawater interface (i.e. ice bottom) using low-frequency (10 to 100 kHz) electromagnetic (EM) induction. The device is composed of transmitter and receiver coils, which are set apart by a few meters. The transmitter generates a primary electromagnetic field which induces eddy currents mainly just below the ice bottom, because the conductivity of sea ice (0-50 mS m^{-1}) is negligible compared to that of seawater (2500 mS m^{-1}). In turn the induced currents generate a secondary electromagnetic field. Then the distance to the ice bottom is calculated from the strength of the secondary electromagnetic field sensed by the receiver coil. Therefore, if this device is used on the ice surface, the data obtained correspond directly to ice thickness. While the accuracy is ±0.1 m for undeformed ice, it degrades to up to 30 % for ridged ice because of the presence of seawater-filled cavities between the ice blocks and the smoothing effect caused by a larger footprint that is approximately equal to the distance to the ice bottom (Haas, 2003). Therefore validation is needed for ridged ice. The primary advantage of this method is that it allows us the measurement of ice thickness from the air, which is much easier than from below the ice. If we mount a laser profilometer on the instrument and measure the height of the instrument above the combined snow and ice surface, ice thickness can be obtained by subtracting it from the EM results and by incorporating knowledge of snow cover depth and density (Figure 7a). In this way, we can use an icebreaker or an aircraft to obtain profiles of ice thickness distribution with this device over a wide scale of 100s of kms. However, it should be noted that this method provides the summation of ice thickness and overlying snow depth. If we want to determine ice thickness itself, snow depth has to be measured by another tool and its density must also be factored in.

Since this technique was first tested for the ground-based observations in the Arctic in the 1970's, numerous measurements were done to obtain ice thickness data in the various regions of the Arctic (Worby et al., 1999). The EM instrument used most frequently is the Geonics EM31 which contains the coils located at each end, 3.66 m apart (Figure 7b), and uses a frequency of 9.8 kHz. The ship-based EM system was first applied to the summer sea ice in the Bellingshausen and Amundsen Seas, Antarctica (Haas, 1998). This study found that the properties of summer ice, in particular seawater-filled gaps close to the ice surface, significantly reduces the accuracy of the EM measurement, although Worby et al. (1999) confirmed that the EM system is more robust when operating under winter and spring ice conditions in the Antarctic. The ship-based EM system was also introduced to the Sea of Okhotsk in 2004, where it was found that the seawater-filled cavities between the ice blocks in this region lower the accuracy of ice thickness estimates significantly, even in winter. Therefore, based on the calibration results from drilling on the ridged ice, Uto et al. (2006b) developed an algorithm that converts conductivity to ice thickness, taking into account the presence of gaps within the ice (Figure 8a). As a result, the measurement error was reduced to about 10 % for ridged ice, and the spatial ice thickness distribution, including ridged ice, was first revealed in this area. The comparison of ice thickness distribution with coincident ship-based video result in the same area is shown in Figure 8b, where reasonable agreement can be seen. However, the ground or ship-based EM measurement has a limitation in spatial coverage and may be biased according to the selection of ice floes or ship-route. For greater representativity, airborne measurement is desirable.

The airborne EM sounding technique was first developed with a helicopter in the Beaufort Sea in the mid 1980's (Kovacs et al., 1987; Kovacs and Holladay, 1989). Multala et al.(1996) subsequently applied the technique to Baltic Sea ice using a fixed-wing aircraft. By validating the data with drill-hole measurements, each study showed the method to be promising. Recently, the technique was developed further and improved using two

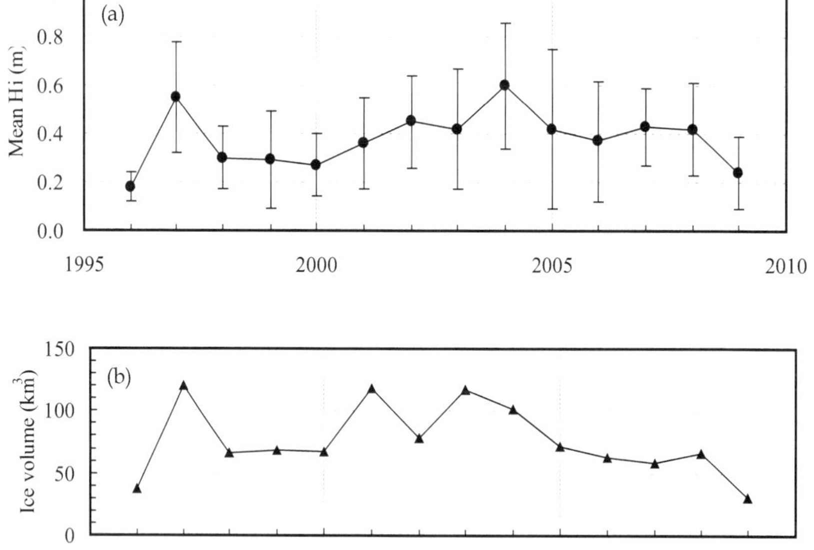

Fig. 6. Annual variations of (a) mean video thickness with a standard deviation (error bars) and (b) ice volume estimated by multiplying mean thickness and sea ice area in the southern Sea of Okhotsk, where sea ice area were offered by Japan Meteorological Agency.

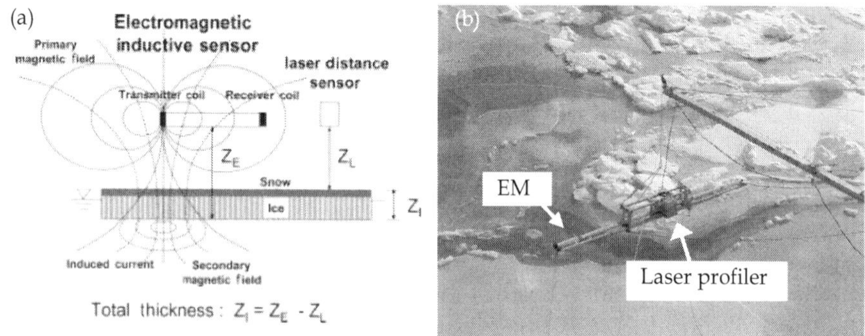

Fig. 7. Schematic picture of EM sounding system.
(a) Illustration of the principle of the measurement (cited from Uto et al. (2006b))
(b) A picture of EM set-up onboard the P/V "Soya" in the Sea of Okhotsk.

frequencies (3.7 and 112 kHz) for the systematic measurement, and since 2001 the monitoring and observation of ice thickness in the Arctic has been operated with a helicopter-borne EM by the German Alfred Wegener Institute for Polar and Marine Research (AWI). The details of the technique are described by Haas et al. (2009). By combining this data set with the ground-based EM measurements acquired since 1991, Haas et al. (2008) showed a rapid thinning of sea ice had occurred in the Arctic Transpolar Drift in the 2000's and suggested a regime shift from multi-year to first-year ice.

Thus, the EM sounding technique has enabled us to remotely measure sea ice thickness from the air, which enhances the measurement considerably in comparison with the previous methods. With a ship- or helicopter-borne system, more detailed and spatial ice thickness distribution was revealed in various regions. Although a ship-based measurement may have some bias, it is effective especially in regions where relatively thin ice is dominant or helicopter operations are prohibited by law. However, it should be noted that it still needs validation, especially for melting ice or regions with significant ridges, and that its total areal coverage is rather limited.

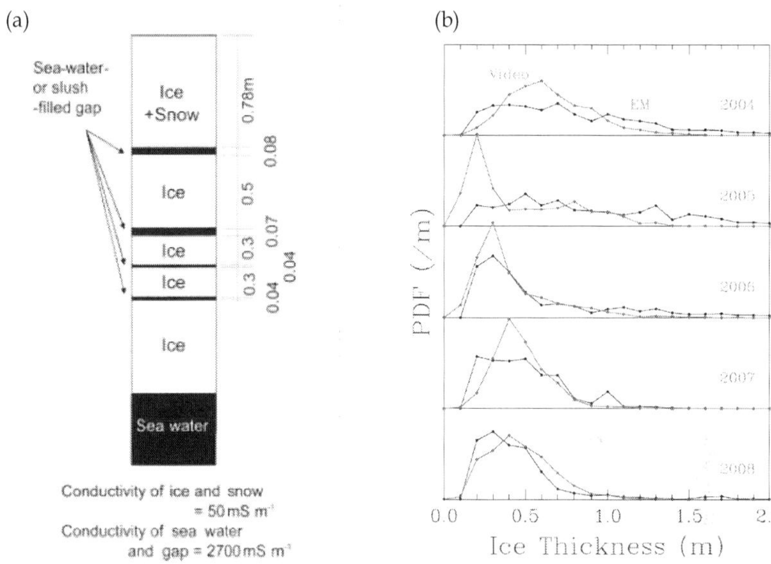

Fig. 8. (a) Model ice structure used for the algorithm converting conductivity to ice thickness developed for the Okhotsk ice. (cited from Uto et al. (2006b)). (b) Comparison of the probability density functions of ice thickness between the results obtained by converting the EM outputs to ice thickness with the algorithm (Figure 8a) and those obtained from video monitoring system in the Sea of Okhotsk during 2004 – 2008.

2.5 Satellite sensors

Previous sections have concentrated on various in-situ and airborne remote sensing tools. Although they have all contributed greatly to our knowledge about ice thickness distribution on a regional scale, ultimately it is the operational monitoring at a global scale via space-borne sensors that is most desirable. In this section, we evaluate the current situation regarding several satellite remote sensing possibilities.

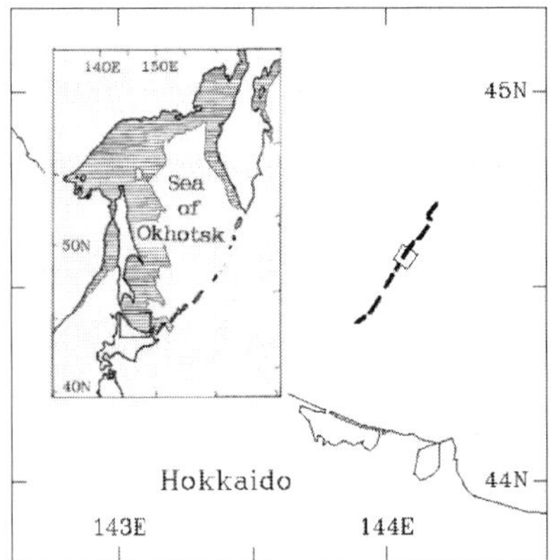

Fig. 12. Location map of the Pi-SAR experiment conducted in mid-February 2005 in the Sea of Okhotsk. Solid lines denote the observation line selected for analysis. (cited from Toyota et al. (2009))

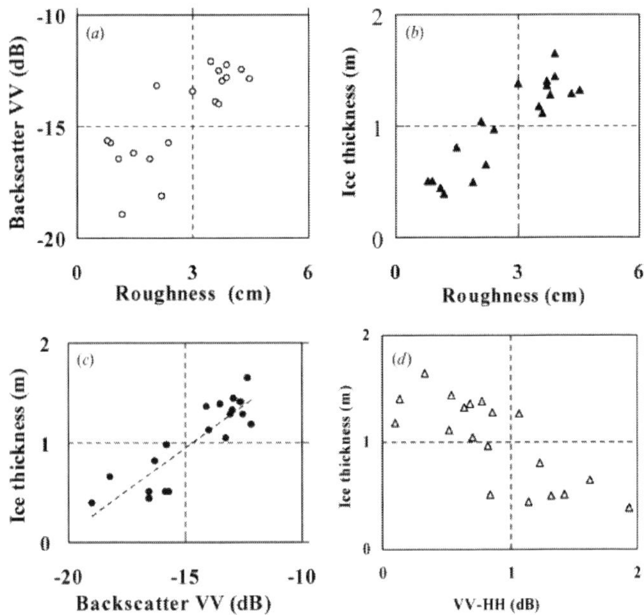

Fig. 13. Scatter plots of the 1 km-averaged values between (a) surface roughness and Backscatter coefficient (VV), (b) surface roughness and ice thickness, (c) backscatter coefficient (VV) and ice thickness, (d) backscattering ratio (VV-HH) and ice thickness.
(cited from Toyota et al. (2009))

In summary, ice thickness retrieval for first-year ice by L-band SAR is promising for a wide range of ice thickness by virtue of the effect on the dielectric constant of surface properties for relatively thin undeformed ice and of surface roughness for thicker ice. On the other hand, multi-year ice has significantly different properties in the Arctic in particular. Due to desalination in summer, the penetration depth becomes deeper for Arctic multi-year sea ice, and then the volume scattering becomes more significant at L-band. In this case, C-band is more appropriate to distinguish between multi-year ice and deformed first-year ice (Rignot and Drinkwater, 1994). Therefore, as pointed out by Rignot and Drinkwater (1994), ultimately the combining of C- and L-band is optimal for the overall classification of ice types. Since the L-band data are now available from the ALOS/PALSAR with polarimetric and ScanSAR modes, combined analysis with C- and L-band is expected in the near future.

2.5.3 Altimeter methods

This sensor measures the freeboard of sea ice (i.e. its height relative to sea level), and freeboard is then converted to ice thickness by assuming that a floating ice floe is in isostatic balance. At present there are two kinds of altimeters available for this type of ice thickness estimation: radar and Laser profiler. Radar altimeters operating at a frequency of 13.8 GHz on the ERS-1 and ERS-2 satellites were launched in 1991 and 1995, respectively. The coverage area for these data extends to 81.5°N and S with a footprint of ~1 km. Since it has been confirmed from the laboratory experiments that reflected echoes come from the interface between sea ice and snow (Beaven et al., 1995), this sensor basically measures the height of snow/ice interface relative to sea level with an accuracy of 10 cm. The relatively large footprint limits the applicable area to closely packed ice area, however the key advantage remains that the data can be obtained irrespective of weather conditions, albeit along a profile. Conversely, the laser altimeter (Geoscience Laser Altimeter System or GLAS) onboard ICESat satellite, launched in January 2003, operates at a wavelength of 1064 nm and covers the area up to 86°N and S. In this case, the sensor measures the elevation of the ice or snow surface within a footprint of 70 m spaced at 170 m and an overall accuracy of 14 cm (Kwok and Cunningham, 2008). Although such a small footprint promises detailed analysis, it should be noted that this sensor is subject to cloud contamination, which may be critical for the polar research, as pointed by Haas (2003). A schematic figure is shown in Figure 14 and the calculation for each sensor is as below:

$$\text{Radar: } h_i = \frac{\rho_w}{\rho_w - \rho_i} h_f + \frac{\rho_s}{\rho_w - \rho_i} h_s \qquad (1)$$

and

$$\text{Laser: } h_i = \frac{\rho_w}{\rho_w - \rho_i} h_{ts} - \frac{\rho_w - \rho_s}{\rho_w - \rho_i} h_s \qquad (2)$$

, where ρ_i, ρ_s, and ρ_w are the densities of sea ice, snow and seawater, respectively, and the notation of h is given in Figure 14. For both altimeters, the key need is to estimate the reference sea level, and to incorporate accurate estimates of the density of seaweater, snow and sea ice, and the snow depth.

This method was first applied in the Arctic Ocean by Laxon et al. (2003), using the ERS radar altimeter. In this case, the reference sea level was analyzed using individual echoes with some corrections for orbits, tides, and atmospheric pressure (Peacock and Laxon, 2004),

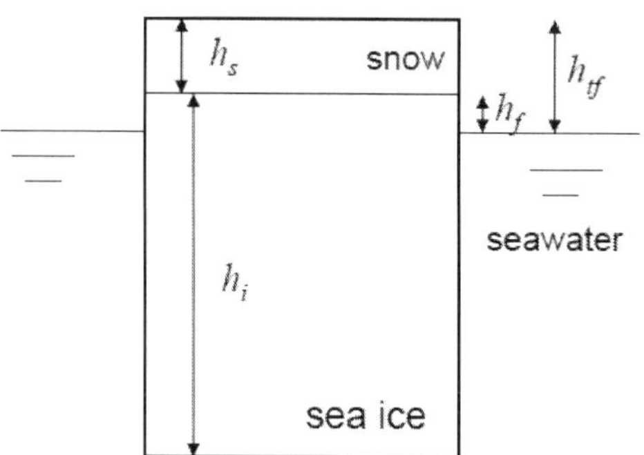

Fig. 14. Schematic picture showing the variables of snow depth (h_s), ice thickness (h_i), ice freeboard (h_f), and total freeboard (h_{tf}).

and the density of sea ice and seawater was assumed to be constant (i.e. 915 and 1024 kg m^{-3}). The density and depth of snow were taken from a monthly climatology. Submarine sonar data were used for validation of the ice thickness product. As a result, the authors succeeded in extracting the geographical distribution of the Arctic ice thickness and showed the availability of this method for ice thicker than 1 m. Furthermore and on the basis of an 8-year data set (1993-2001), they found out that the interannual variation is much larger than predicted by numerical models, and that this large variability could be better explained by the length of melting season than expected changes of atmospheric and oceanic circulation. Similarly the results from the algorithm used by the ICESat GLAS laser altimeter has recently been analyzed by Kwok et al. (2004), Kwok & Cunningham (2008) to the Arctic Ocean, and by Zwally et al. (2008) in the Weddell Sea. Kwok & Cunningham (2008) used the ECMWF outputs for snow data, and tested two cases of sea ice density as a constant value (925 kg m^{-3}) and a function of ice thickness, while Zwally et al. (2008) used AMSR-derived snow depth and constant densities for snow (300 kg m^{-3}) and sea ice (915 kg m^{-3}). The method for estimating the reference sea level was developed individually. As a result, these studies all succeeded in showing the potential usefulness of this method. However, as mentioned by both of the authors, the improved knowledge about snow is crucial because of the sensitivity of laser altimetry to the temporal variation of snow depth and density. In summary, it has been shown that satellite altimeters are useful as a means of estimating ice thickness distributions over a wide scale, and for thick ice in particular. The primary merit is the ability to obtain hemispheric ice thickness distribution on a seasonal time scale, which allows us to monitor trends in ice thickness change and validate the results from numerical models. In turn, this will feed back to an improvement in the way dynamical processes are prescribed in the sea ice models. However, it should also be noted that improved knowledge about densities of snow and sea ice is still needed for this method, since observational data show that they range widely and also evolve over time according to the meteorological conditions. In addition, we need to consider the effect of the seawater-filled cavities between ice blocks on the altimeter estimates. Since such cavities increase the bulk ice density, the further investigation will be needed. An additional concern relates to the footprint size. Since this method is based on the assumption of isostatic balance, research needs to be carried out to determine whether the important underlying assumption of isostatic balance holds at the scale of the footprints.

3. Concluding Remarks

In this chapter, a brief history of ice thickness measurement has been presented, stressing how remote sensing tools have contributed to an improved understanding of the sea ice thickness distribution and its importance in sea ice research. Since Weyprecht first conducted the observation of the ice thickness growth at the Franz-Joseph Island (80°N) during the 1873/74 winter (Zubov, 1945), ice thickness has been one of the most challenging and interesting topics in sea ice research. As seen above, in general the tools have been developed to cover a wider region. Most recently satellite altimeters have allowed us to monitor the ice thickness distribution at a hemispheric scale and on a seasonal time scale, albeit limited to thick ice. Since the response of sea ice associated with recent global warming receives much attention now, it is a notable achievement to be able to discuss the ice thickness trends not only from the products of numerical models but also on the basis of observational results. This will also be helpful for the improvement of physical processes in the models, and validation of their output.

In spite of these advances, it is unlikely that all of the various methods developed to date will be replaced by satellite sensors. The individual tools are appropriate for measurement at their optimum scale. In general, ice thickness varies in a hierarchical fashion, as suggested by Rothrock and Thorndike (1980), and the properties at each scale have an influence on those at a different scale. For example, the ice thickness distribution at a scale of a few to tens of meters is directly involved in the formation of ridges, while it is the thickness properties at a scale of a few kilometres that determine the external forcing for ridging. Thus, a more comprehensive understanding of sea ice behaviours depends upon acquisition of accurate information on the ice thickness distribution at both local and large scales. Considering that ice formation and melting are basically localized processes, the interaction between different scales needs to be further clarified when using the different though complementary tools at different scale; this factor will become more important to the sea ice research in the future.

It should be mentioned that there are other methods widely used for ice thickness measurement. The representative one is visual observation on the basis of ASPeCT protocol, which was designed for the ship-based observations in the Antarctic seas by Australian researchers based on a previous method used by other groups in the 1980's (Worby & Allison, 1999). The observation is conducted every hour from the ship's bridge along-track, including the thickness of sea ice and snow, floe size, and surface properties. Although there may be some problems with accuracy and bias from the ship-track, this is a convenient method and the statistics obtained represent useful general characteristics of the ice cover. A climatology of sea ice thickness obtained in this fashion are presented in Figure 1.2b. The continuation of this type of observing program is to be encouraged.

Finally, I would like to close this chapter by concluding that the history of ice thickness measurement is closely related to the development of sea ice research on both macro- and micro- scale properties, and that many researchers from a variety of backgrounds have been involved in it. Therefore, this is truly an interdisciplinary issue, and further collaboration with many researches is highly desirable in future.

4. Acknowledgments

The author expresses his sincere gratitude to Dr. Robert Massom (Antarctic Climate and Ecosystems Cooperative Research Centre and Australian Antarctic Division) for the critical reading of this manuscript and valuable comments and to Dr. Guy Williams (Hokkaido Univ.) for the proof-reading. They are very helpful for the improvement of the manuscript.

5. References

Ackley, S.F. (1979). Mass-balance aspects of Weddell Sea pack ice, *J. Gloaciol.*, Vol. 24, No.90, pp. 391-405.

Beaven, S.G.; Lockhart, G.L.; Gogineni, S.P.; Hossetnmostafa, A.R.; Jezek, K.; Gow, A.J.; Perovich, D.K.; Fung, A.K. & Tjuatja, S. (1995). Laboratory measurements of radar backscatter from bare and snow-covered saline ice sheets, *Int. J. Remote Sens.*, Vol. 16, No. 5, pp. 851-876.

Bourke, R.H. & Garrett, R.P. (1987). Sea ice thickness distribution in the Arctic Ocean, *Cold Regions Science and Technology*, Vol.13, pp.259-280.

Brown, R. & Armstrong, R.L. (2008). Snow-cover data: measurement, products, and sources, In *Snow and Climate*, Armstrong, R.L. & Brun, E. (ed.), pp. 181-216, Cambridge University Press, ISBN 978-0-521-85454-2, Cambridge, UK.

Cavalieri, D.J.; Gloersen, P. & Campbell, W.J. (1984). Determination of sea ice parameters with the NIMBUS 7 SMMR, *J. Geophys. Res.*, Vol. 89, No.D4, pp. 5355-5369.

Cavalieri, D.J. (1994). A microwave technique for mapping thin sea ice, *J. Geophys. Res.*, Vol.99, No.C6, pp. 12561-12572.

Comiso, J.C. (1986). Characteristics of Arctic winter sea ice from satellite multispectral microwave observation, *J. Geophys. Res.*, Vol.91, No.C1, pp.975-994.

Dierking, W. & Busche, T. (2006). Sea ice monitoring by L-band SAR: An assessment based on literature and comparisons of JERS-1 and ERS-1 imagery, *IEEE Trans. Geosci. Remote Sensing*, Vol. 44, No.2, pp. 957-970.

Fukamachi, Y.; Mizuta, G.; Ohshima, K.I.; Melling, H.; Fissel, D. & Wakatsuchi, M. (2003). Variability of sea-ice draft off Hokkaido in the Sea of Okhotsk revealed by a moored ice-profiling sonar in winter of 1999, *Geophys. Res. Lett.*, Vol.30, No.7, 1376, doi:10.1029/2002GL016197.

Fukamachi, Y.; Mizuta, G.; Ohshima, K.I.; Toyota, T.; Kimura, N. & Wakatsuchi, M. (2006). Sea ice thickness in the southwestern Sea of Okhotsk revealed by a moored ice-profiling sonar, *J. Geophys. Res.*, Vol.111, C09018, doi:10.1029/2005JC003327.

Gow, A.J.; Ackley, S.F.; Weeks, W.F. & Govoni, J.W. (1982). Physical and structural characteristics of Antarctic sea ice, *Ann. Glaciol.*, Vol.3, pp.113-117.

Grenfell, T.C. & Comiso, J.C. (1986). Multifrequency passive microwave observations of first-year sea ice grown in a tank, *IEEE Trans. Geosci. Remote Sensing*, Vol.GE-24, No.6, pp. 826-831.

Haas, C. (1998). Evaluation of ship-based electromagnetic-inductive thickness measurements of summer sea-ice in the Bellingshausen and Amundsen Seas, Antarctica, *Cold Reg. Sci. Technol.*, Vol. 27, pp. 1-16.

Haas, C. (2003). Dynamics versus thermodynamics: The sea ice thickness distribution, In *Sea Ice*, Thomas, D.N. & Dieckmann, G.S. (Ed.), pp. 82-111, Blackwell Publishing, ISBN 0-632-05808-0, Oxford, U.K.

Haas, C.; Pfaffling, A.; Hendricks, S.; Rabenstein, L.; Etienne, J.-L. & Rigor, I. (2008). Reduced ice thickness in Arctic Transpolar Drift favors rapid ice retreat, *Geophys. Res. Lett.*, Vol.35, L17501, doi:10.1029/2008GL034457.

Haas, C.; Lobach, J.; Hendricks, S; Rabenstein, L. & Pfaffling, A. (2009). Helicopter-borne measurements of sea ice thickness, using a small and lightweight, digital EM system, *J. Appl. Geophys.*, Vol.67, pp.234-241.

Hallikainen, M. & Winebrenner, D.P. (1992). The physical basis for sea ice remote sensing, In *Microwave Remote Sensing of Sea Ice*, Carsey (ed.), pp. 29-46, American Geophysical Union, Washington D.C.

Harms, S.; Fahrbach, E. & Strass, V.H. (2001). Sea ice transports in the Weddell Sea, *J. Geophys. Res.*, Vol.106, No.C5, pp. 9057-9074.

Hudson, R. (1990). Annual measurement of sea-ice thickness using an upward-looking sonar, *Nature*, Vol.344, pp.135-137.

Japan Meteorological Agency (2005). *The Results of Sea Ice Observations*, Vol. 23, (2004-2005). (CD-ROM)

Kovacs, A.; Valleau, N.C. & Holladay, J.S. (1987). Airborne electromagnetic soundings of sea ice thickness and sub-ice bathymetry, *Cold Reg. Sci. Technol.*, Vol.14, pp. 289-311.

Kovacs, A. & Holladay, J.S. (1989). Development of an airborne sea ice thickness measurement system and field test results, *CRREL Report*, Vol.89-19, pp. 1-47.

Kovacs, A. (1996). Sea Ice. Part I. Bulk salinity versus ice floe thickness. *USA Cold Regions Research and Engineering Laboratory, CRREL Rep.*, Vol.97-7, pp. 1-16.

Kwok, R. & Rothrock, D.A. (1999). Variability of Fram Strait ice flux and North Atlantic Oscillation, *J. Geophys. Res.*, Vol.104, No.C3, pp.5177-5189.

Kwok, R.; Zwally, H.J. & Yi, D. (2004). ICESat observations of Arctic sea ice: First look, *Geophys. Res. Lett.*, Vol. 31, L16401, doi:10.1029/2004GL020309.

Kwok, R. & Cunningham, G.F. (2008). ICESat over Arctic sea ice: Estimation of snow depth and ice thickness, *J. Geophys. Res.*, Vol. 113, C08010, doi:10.1029/2008JC004753.

Lange, M.A. & Eicken, H. (1991). Textural characteristics of sea ice and the major mechanism of ice growth in the Weddell Sea, *Ann. Glaciol.*, Vol. 15, pp. 210-215.

Laxon, S.; Peacock, N. & Smith, D. (2003). High interannual variability of sea ice in the Arctic region, *Nature*, Vol. 425, pp. 947-950.

Lubin, D. & Massom, R. (2006). *Polar Remote Sensing, Volume 1: Atmosphere and Oceans*, Praxis Publishing Ltd and Springer, ISBN 3-540-43097-0, Chichester, UK and Berlin, Germany.

Marko, J.R. (2003). Observation and analyses of an intense wave-in-ice event in the Sea of Okhotsk, *J. Geophys. Res.*, Vol.108, No.C9, 3296, doi:10.1029/2001JC001214.

Martin, S.; Drucker, R.; Kwok, R. & Holt, B. (2004). Estimation of the thin ice thickness and heat flux for the Chukchi Sea Alaskan coast polynya from Special Sensor Microwave/Imager data, 1990-2001, *J. Geophys. Res.*, Vol. 109, No.C10012, doi: 1010.29/ 2004JC002428.

Martin, S.; Drucker, R; Kwok, R. & Holt, B. (2005). Improvements in the estimates of ice thickness and production in the Chukchi Sea polynyas derived from AMSR-E, *Geophys. Res. Lett.*, Vol. 32, L05505, doi:10.1029/2004GL022013.

Massom, R. (2006). Basic remote-sensing principles relating to the measurement of sea ice and its snow cover, In *Polar Remote Sensing Vol. I: Atmosphere and Oceans*, Lubin, D. & Massom, R. (ed.), pp. 356-380, Praxis Publishing Ltd, ISBN 3-540-43097-0, Chichester, UK and Berlin, Germany, 756 pp.

Massom, R.; Hill, K.; Barbraud, C.; Adams, N.; Ancel, A.; Emmerson, L. & Pook, M.J. (2009). Fast ice distribution in Adelie Land, East Antarctica: Interannual variability and implications for emperor penguins (Aptenodytes forsteri), *Marine Ecol. Progr. Ser.*, Vol. 374, pp. 243-257.

Maykut, G.A. (1978). Energy exchange over young sea ice in the central Arctic, *J. Geophys. Res.*, Vol. 83, No.C7, pp. 3646- 3658.

McLaren, A.S. (1989). The under-ice thickness distribution of the Arctic Basin as recorded in 1958 and 1970, *J. Geophys. Res.*, Vol.94, No.C4, pp. 4971-4983.

Melling, H. & Riedel, D.A. (1995). The underside topography of sea ice over the continental shelf of the Bering Sea in the winter of 1990, *J. Geophys. Res.*, Vol.100, No.C7, pp.13641-13653.

Melling, H. & Riedel, D.A. (1996). Development of seasonal pack ice in the Beaufort Sea during the winter of 1991-1992: a view from below, *J. Geophys. Res.*, Vol.101, No.C5, pp.11975-11991.

Multala, J.; Hautaniemi, H.; Oksama, M.; Lepparanra, M.; Haapala, J.; Herlevi, A.; Riska, K. & Lensu, M. (1996). An airborne electromagnetic system on a fixed wing aircraft for sea ice thickness mapping, *Cold Reg. Sci. Technol.*, Vol.24, pp.355-373.

Nakamura, K.; Wakabayashi, H.; Naoki, K.; Nishio, F.; Moriyama, T. & Uratsuka, S. (2005). Observation of sea ice thickness in the Sea of Okhotsk by using dual-frequency and fully polarimetric airborne SAR (Pi-SAR) data, *IEEE Trans. Geosci. Remote Sensing*, Vol. 43, 2460-2469.

Naoki, K.; Ukita, J.; Nishio, F.; Nakayama, M.; Comiso, J.C. & Gasiewski, Al (2008). Thin sea ice thickness as inferred from passive microwave and in situ observations, *J. Geophys. Res.*, Vol. 113, No.C02S16, doi:10.1029/2007JC004270.

Nghiem, S.V.; Rigor, I.G., Perovich, D.K.; Clemente-Colon, P.; Weatherly, J.W. & Neumann, G. (2007). Rapid reduction of Arctic perennial sea ice, Geophys. *Res. Lett.*, Vol. 34, L19504, doi:10.1029/2007GL031138.

Peacock, N.R. & Laxon, S.W. (2004). Sea surface height determination in the Arctic Ocean from ERS altimetry, *J. Geophys. Res.*, Vol. 109, C07001, doi:10.1029/2001JC001026.

Rignot, E. & Drinkwater, M.R. (1994). Winter sea-ice mapping from multi-parameter synthetic-aperture radar data, *J. Glaciol.*, Vol. 40, No. 134, pp. 31-45.

Rothrock, D.A. & Thorndike, A.S. (1980). Geometric properties of the underside of sea ice, *J. Geophys. Res.*, Vol. 85, No. C7, pp. 3955-3963.

Rothrock, D.A. (1986). Ice thickness distribution – measurement and theory. In: *The Geophysics of Sea Ice* (ed. N. Untersteiner), Plenum, New York, pp. 551-575.

Rothrock, D.A.; Yu, Y. & Maykut, G.A. (1999). Thinning of the Arctic sea-ice cover. *Geophys. Res. Lett.*, Vol.26, No.23, pp. 3469-3472.

Shimoda, H.; Endoh, T.; Muramoto, K.; Ono, N.: Takizawa, T.; Ushio, S.; Kawamura, T. & Ohshima, K.I. (1997). Observations of sea-ice conditions in the Antarctic coastal region using ship-board video cameras, *Antarctic Record*, Vol.41, No.1, pp.355-365. (in Japanese with English summary)

Strass, V.H. (1998). Measuring sea ice draft and coverage with moored upward looking sonars, *Deep-Sea Res. I*, VOl.45, 795-818.

Strass, V.H. & Fahrbach, E. (1998). Temporal and regional variation of sea ice draft and coverage in the Weddell Sea obtained from upward looking sonars, In: *Antarctic Sea Ice, Physical Processes, Interactions and Variability*, Jeffries, M.O. (Ed.), pp.123-139, American Geophysical Union, ISBN 0-87590-902-7, Washington, D.C.

Svendsen, E.; Kloster, K.; Farrelly, B.; Johanessen, O.M.; Johanessen, J.A.; Campbell, W.J.; Gloersen, P.; Cavalieri, D.J. & Matzler, C. (1983). Norwegian remote sensing experiment: evaluation of the Nimbus 7 scanning multichannel microwave radiometer for sea ice research, *J. Geophys. Res.*, Vol.88, No.C5, pp. 2781-2792.

Tamura, T.; Ohshima, K.I.; Markus, T.; Cavalieri, D.J.; Nihashi, S. & Hirasawa, N. (2007). Estimation of thin ice thickness and detection of fast ice from SSM/I data in the Antarctic Ocean, *J. Atmos. Ocean. Tech.*, Vol. 24, pp.1757-1772.

Tamura, T.; Ohshima, K.I. & Nihashi, S. (2008). Mapping of sea ice production for Antarctic coastal polynyas, *Geophys. Res. Lett.*, Vol.35, No.L07606, doi:10.1029/ 2007GL032903.

Tateyama, K.; Enomoto, H.; Toyota, T. & Uto, S. (2002). Sea ice thickness estimated from passive microwave radiometers, *Polar Meteorology and Glaciology*, Vol. 16, pp.15-31.

Thorndike, A.S.; Rothrock, D.A.; Maykut, G.A. & Colony, R. (1975). The thickness distribution of sea ice, *J. Geophys. Res.*, Vol. 80, No. 33, pp. 4501-4513.

Toyota, T.; Kawamura, T.; Ohshima, K.I.; Shimoda, H. & Wakatsuchi, M. (2004). Thickness distribution, texture and stratigraphy, and a simple probabilistic model for dynamical thickening of sea ice in the southern Sea of Okhotsk, *J. Geophys. Res.*, Vol.109, C06001, doi:10.1029/2003JC002090.

Toyota, T.; Takatsuji, S.; Tateyama, K.; Naoki, K. & Ohshima, K.I. (2007). Properties of sea ice and overlying snow in the southern Sea of Okhotsk, *J. Oceanogr.*, Vol.63, No.3, pp. 393-411.

Toyota, T.; Nakamura, K.; Uto, S.; Ohshima, K.I. & Ebuchi, N. (2009). Retrieval of sea ice thickness distribution in the seasonal ice zone from airborne L-band SAR, *Int. J. Remote Sens.*, Vol. 30, No. 12, pp. 3171-3189.

Troy, B.E.; Hollinger, J.P.; Lerner, R.M. & Wisler, M.M. (1981). Measurement of the microwave properties of sea ice at 90 GHz and lower frequencies, *J. Geophys. Res.*, Vol. 86, No.C5, pp. 4283-4289.

Tucker, W.B. ; Weatherly, J.W. ; Eppler, D.T. ; Farmer, L.D. & Bentley, D.L. (2001). Evidence for rapid thinning of sea ice in the weatern Arctic Ocean at the end of the 1980s, *Geophys. Res. Lett.*, Vol.28, No.14, pp. 2851-2854.

Uto, S.; Shimoda, H. & Ushio, S. (2006a). Characteristics of sea-ice thickness and snow-depth distributions of the summer landfast ice in Lützow-Holm Bay, East Antarctica, *Ann. Glaciol.*, Vol.44, pp. 281-287.

Uto, S.; Toyota, T.; Shimoda, H.; Tateyama, K. & Shirasawa, K. (2006b). Ship-borne electromagnetic induction sounding of sea-ice thickness in the southern Sea of Okhotsk, *Ann. Glaciol.*, Vol. 44, pp. 253-260.

Vant, M.R.; Ramseier, R.O. & Makios, V. (1978). The complex-dielectric constant of sea ice at frequencies in the range 0.1-40 GHz, *J. Appl. Phys.*, Vol. 49, No.3, pp.1264-1280.

covers the visible and near-infrared (VNIR), short wave infrared (SWIR) and thermal infrared (TIR), and is effective in studying the Earth's surface land cover, vegetation and mineral resources, etc. We used data from Terra/ASTER original Level 1B VNIR / SWIR/TIR Data (Time of day (UTC): 1:30, June 30, 2003, Path-108/Row-835, and 1:30, July 12, 2004, Path-109/Row-837, the subset coordinate of the UL Geo N45° 08', E141° 36') supplied by the Earth Remote Sensing Data Analysis Center, Tokyo, Japan (©ERSDAC). In the ATCOR software, if a 14-bands ASTER image is loaded the default Layer-Band assignment will be set that input layer 13 (thermal band 13) is set to layer 10 and the output image will be restricted to 10 bands. The reason for this is that from the 5 ASTER thermal bands only band 13 is used in ATCOR. In order to carry out calculation between bands, we re-sampled (layer stacking) this 3 layers with different spatial resolution ASTER VNIR (15 m), SWIR (30 m) and TIR (90 m) data to one layer that has the same spatial resolution (15 m) dataset, and used this dataset input to ATCOR software.

2.2 ATCOR input parameters

With the ASTER data (Path-108/Row-835), ATCOR input parameters include: Solar zenith (degrees): **24.8**; Solar azimuth (degrees): **147.7**; Scene Visibility (km) = **30m**; Model for solar region; fall/spring/rural; various aerosol types: **rural**; Model for thermal region: **fall**. Input satellite data: subset ASTER VNIR-SWIR-TIR, **10-bands** one layer data. In the calibration of ASTER data the Level 1B data is in terms of scaled radiance. The unit conversion coefficients (defined as radiance per 1 DN) are shown in Table 1. Radiance (spectral radiance) is expressed in units of **W/(m2*sr*um)**. The true radiance at sensor can be obtained from the DN values as follows:

$$L = c_0 + c_1 \times DN \tag{1}$$

Where, L is radiance, c_0 (offset) and c_1 (gain) is conversion coefficient; DN is digital number.

Band No.	c_0	c_1
1	-0.1	0.0676
2	-0.1	0.0708
3	-0.1	0.0862
4	-0.1	0.02174
5	-0.1	0.00696
6	-0.1	0.00625
7	-0.1	0.00597
8	-0.1	0.00417
9	-0.1	0.00318

Table 1. The calibration of ASTER original L1B data (Unit: mW/cm2 sr micron)

3. The study area

The study area Sarobetsu Marsh, is the largest registered wetland of 7000 ha located in coastal area of northwestern Hokkaido, Japan (Figure 1) and nominated by the Ramsar Convention on Wetlands in 2005. Test areas are mostly swamp with *Sphagnum* sp., *Moliniopsis japonica* while the western coastal zone area is dominated by isolated small hills covered with broad-leaf trees. The terrain of the study area is very flat with the elevation distributed between 5 m to 15 m. The area was selected because of its natural environment and perseverance by national and municipal governments. On the other hand recently the expansion of a non-moor plant (bamboo grass) has been detected in a swamp. So far, for the monitoring, the classification of the wetland territory into a high moor, low moor and non moor types, to clarify the invasion front of bamboo grass is very important.

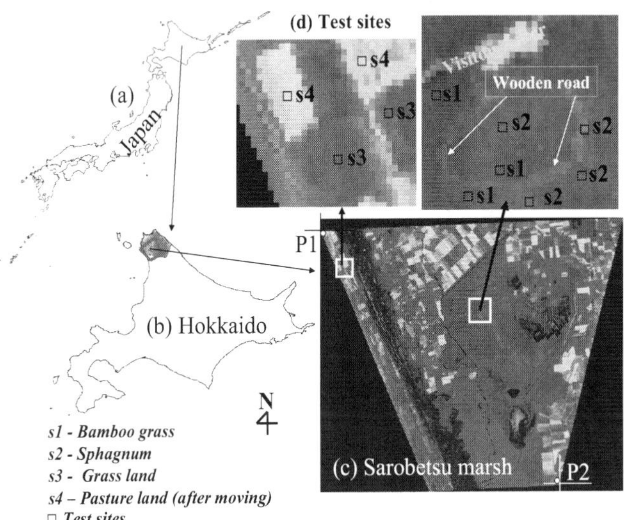

Fig. 1. Location of study area and test sites

4. Result

The atmospheric correction processing evaluated within this chapter is to a major part based on the technology of the ATCOR-3 atmospheric correction procedure [11][12]. In this algorithm, the total signal at the sensor consists of four components: (1)-path radiance; (2)-reflected radiation from the viewed pixel; (3)-scattered radiation from the neighborhood; and (4)-terrain radiation reflected to the pixel. Figure 2. shows the result of ATCOR atmospheric corrected ASTER data in the test area. The satellite scene has to be ortho-rectified to a DEM (digital elevation model) before the ATCOR3 processing starts. The influence of the neighborhood is neglected. A start value for the ground reflectance of the surrounding topography is employed whilst the path radiance component is subtracted from the signal. Shadow cast from surrounding topography is included. However, as our study area is very flat, the topography effect is rare. Haze removal is the important steps prior to the application of imagery. This result shows the ATCOR algorithm is a more effective haze removal and atmospheric correction modeling which combined several improved methods (see Fig. 2(c)).

The global flux on the ground depends on the large-scale (1 km) average reflectance. The global flux in the atmospheric LUT's is calculated for a fixed reflectance=0.15 . This iteration performs the update for the spatially varying average reflectance map of the current scene, if the adjacency range R > 0 [10]. The empirical BRDF correction is areas of low illumination (see Fig. 2(b)).

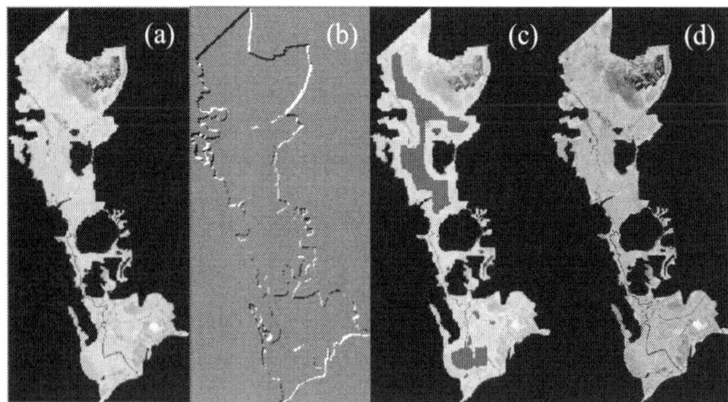

Fig.2. Result of the ATCOR3 correction. (a) original ASTER/L1B data; (b) the illumination azimuth angle ; (c) hazeoverlay; (d) atmospheric corrected ASTER data.

Comparison of the ASTER data before and after ATCOR software atmospheric correction has revealed following results:

(a) The mean values of ASTER band 1 and band 2 decrease after atmospheric correction (Table 2). This means that the visible green and red band has included not only the radiance from a target, but also radiance other than an atmospheric scattering is also included.

(b) Comparison of mean values of NIR and SWIR bands before/after atmospheric correction shows that the radiance values became larger after atmospheric correction. It means that the radiation from the target is absorbed by atmosphere before it reaches the satellite sensor. Atmospheric scattering primarily affects the direction of visible Green and Red band, and atmospheric absorption primarily affects the direction of NIR and SWIR bands.

The most significant interaction that undergoes by the thermal infrared radiation when it passes through the atmosphere is its absorption, primarily due to ozone and water vapor particles in the atmosphere. At the visible shorter wavelengths (i.e. Green or Red band), attenuation occurs by scattering due to clouds and other atmospheric constituents, as well as reflection. The type of scattering in which the energy undergoes is depends upon the size of the particle. Rayleigh scattering occurs when radiation interacts with air molecules smaller than the radiation's wavelength, such as oxygen and nitrogen. The degree of scattering is inversely proportional to the fourth power of the wavelength. When particles are comparable in size to the radiation wavelength, such as aerosols, it results in Mie scattering type [18]. The effect of scattering on the visible wavelengths is significant and must be compensated for when developing empirical relationships through time [19] [20]. Atmospheric scattering primarily affects the direction of short wave radiation. There are four types of atmospheric scattering: Rayleigh, Mie, Raman and non selective. The most significant of these types of scattering is Rayleigh scatter, which effects the short visible wavelengths and results in haze. For ASTER data the scattering is four times as great in Green band of the electromagnetic spectrum as in the NIR band [21],[22].

DN of original ASTER L1B data (before correction)				
Band No.	Min	Max	**Mean**	St dev
1 (Green)	0	255	**50.93**	34.97
2 (red)	0	210	**30.83**	23.15
3 (NIR)	0	179	**69.82**	49.02
4 (SWIR)	0	111	**45.21**	32.59
5 (SWIR)	0	105	**27.82**	20.52
6 (SWIR)	0	144	**29.90**	22.66
7 (SWIR)	0	155	**27.49**	20.41
8 (SWIR)	0	191	**23.83**	18.10
9 (SWIR)	0	133	**21.01**	15.20
DN after ATCOR correction of ASTER data				
1 (Green)	0	229	**26.47**	20.16
2 (Red)	0	255	**30.04**	26.17
3 (NIR)	0	255	**93.53**	66.98
4 (SWIR)	0	224	**90.89**	65.64
5 (SWIR)	0	175	**46.17**	34.13
6 (SWIR)	0	251	**51.75**	39.30
7 (SWIR)	0	241	**42.42**	31.55
8 (SWIR)	0	255	**35.86**	27.33
9 (SWIR)	0	157	**24.34**	17.69

Table 2. Comparison of DN before and after ATCOR atmospheric correction of ASTER data ((Path-108/Row-835, UTC: 1:30, June 30, 2002)

Figure 3(1) and 3(2) shows that the ASD's measurement values and the ATCOR output values have no big difference in the ASTER reflection bands and absorption bands of chlorophyll (i.e. NIR-band and Red-band); the difference has come out in scattering band (i.e. ASTER Green band) and soil reflection bands (i.e. ASTER SWIR bands). However, in the ©EOSDIS AST07 (ASTER surface reflectance products), the values are considerably different in ASTER NIR band. The problem is in low values of NIR after atmospheric correction. In this research, the results of ATCOR software correction were better than those of AST07 products. Figure 3(1) and 3(2) shows the comparison of ATCOR software atmospheric correction result and ASTER surface reflectance products (AST07) data, in non moor and high moor plant samples. In a swamp (high moor plant), the background soil and the open water area will be incorrectly recognized as moving haze of ATCOR software method. After ATCOR correction the ASTER SWIR bands values are becomes larger following the changes of vegetation due to dryness.

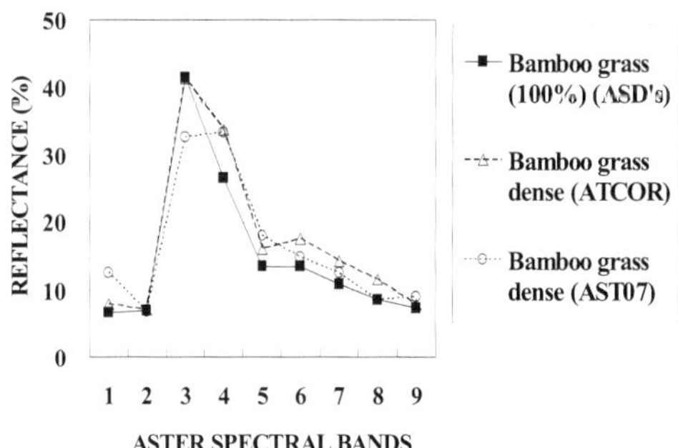

Fig. 3 (1) Comparison of the spectral reflectance of non-moor plant Bamboo grass calculated form ASD's measurement method, ATCOR method and EOSDIS AST07 method.

ATCOR have rectified more correctly scattering with short wavelength visible ((i.e. Band 1 (Green)) and absorption with NIR band (Band 3). The Fig. 4(1) and 4(2) clearly shows that 5% of scattering radiation is contained with the green band and 47% of radiation was absorbed in the NIR band and 17% of radiation was absorbed in the SWIR6 band.

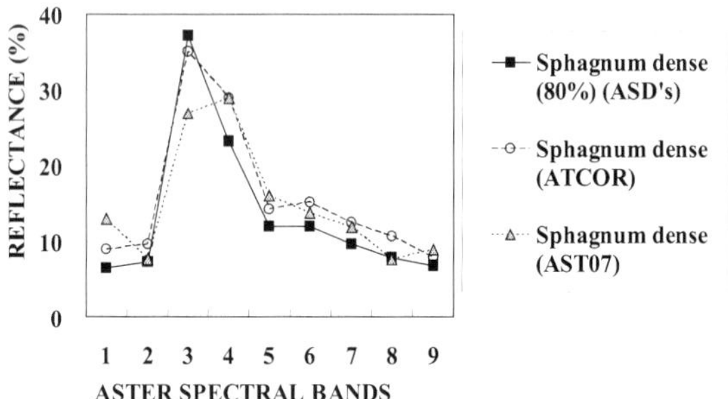

Fig. 3 (2) Comparison of the spectral reflectance of high moor plant *Sphagnum* marsh calculated form ASD's measurement method, ATCOR method and EOSDIS AST07 method.

The input (x: DN of original ASTER L1B data) and output (y: DN of after ATCOR software atmospheric corrected ASTER L1B data) expression of the ASTER data using ATCOR are as follows:

$$\text{Green band (band 1):} \quad y = 0.95x - 28.56 \tag{2}$$

$$\text{NIR band (band 3):} \quad y = 1.473x - 21.12 \tag{3}$$

$$\text{SWIR band (band 6):} \quad y = 1.171x - 1.23 \tag{4}$$

Comparing NDVI from ground ASD's measurement, corrected ASTER data and not corrected original ASTER L1B data (see Figure 4(3)), we found that values from the ground

NDVI and atmospheric corrected NDVI did not greatly differed. However, the value of NDVI of ASTER L1B is smaller than the value of grand NDVI. The formula of the correlation of a NDVI-Corrected value and an original ASTER L1B NDVI value is as follows:

$$NDVI_{corrected} = 1.27 \times NDVI_{L1B} + 0.04 \tag{5}$$

This formula showed that the NDVI value after atmospheric correction became larger than that before atmospheric correction.

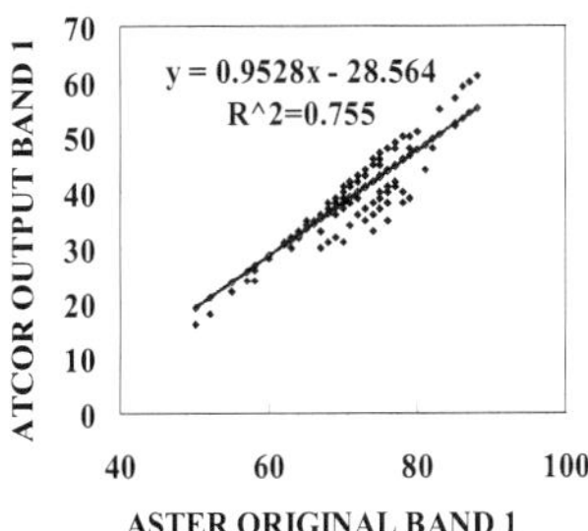

Fig. 4(1) The correlation coefficient of the ASTER original band 1 (Green) and ATCOR output band 1 (Green).

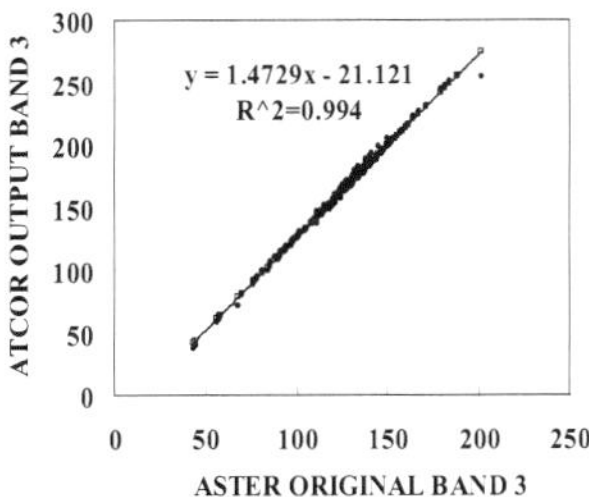

Fig. 4(2) The correlation coefficient of the ASTER original band 3 (NIR) and ATCOR output band 3 (NIR)

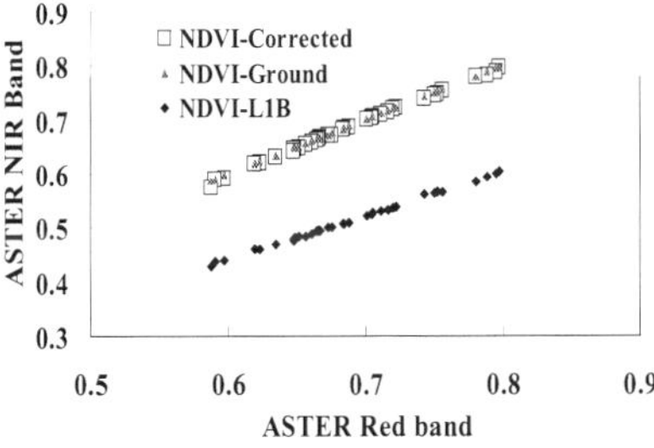

Fig. 4(3) Comparison of NDVI of ground measurement, atmospheric corrected ASTER L1B data and not atmospheric corrected original ASTER L1B data

5. Conclusion

Many techniques have been developed to determine the contribution of atmospheric scattering has on the radiation detected by the satellite sensor. The radiance received from a target against a background surface by the satellite sensor comes from a combination of three sources; first, the intrinsic radiance reflected by the target and then directly transmitted by the atmosphere; secondly, the radiant energy scattered diffusely by the atmosphere which then further interacts with the target background; thirdly, the radiant energy scattered diffusely by the atmosphere. The radiant energy reflected by the target carries the direct energy from the target. The other two sources produce a combined effect.

Atmospheric measurements and modeling involve the theoretical determination of the path radiance contribution of the atmosphere for the particular time of the overpass. To calculate the contribution of the scattering on the reflected radiance requires that many atmospheric variables at the time of the satellite overpass be recorded and input into theoretically derived equations to determine the effect of the atmosphere on each spectral band.

Correction of image data for the effects of atmospheric propagation can be carried out essentially in three ways. It is respectively, 1) based on atmospheric scattering and absorption characteristics of model; 2) based on pre-calibration, on-board calibration against targets of known reflectance method and 3) based on dark-pixel subtraction method. The physic based methods are the most rigorous approach, and also the most difficult to apply. The atmospheric scattering and absorption characteristics area calculated by a computer model requires meteorological, seasonal and geographical variables. In practice, these variables may not all be available with sufficient spatial or temporal resolution, and in particular, estimation of the contribution of atmospheric aerosols is difficult. Other methods mentioned also have a number of requirements that are not easy to satisfy.

In this study, the accuracy of the atmospheric correction with ATCOR software algorithm was based on the usage of ground radiometric measurement data, which were compared also with the radiative transfer code (RTC) that is based on atmospheric corrected ASTER L2B standard products surface reflectance (AST07) data simultaneously.

The NDVI data calculated from ground measurement, atmospheric corrected ASTER L1B data and ASTER surface reflectance product (AST07) data was used to evaluate the accuracy of the ATCOR software atmospheric correction of Terra/ASTER data (Jun 30, 2002). Ground measurements were done using ground radiometric measurement data (ASD's FieldSpec® Pro) at the study area named Sarobetsu Marsh located in coastal area of Hokkaido, Japan.

The study showed that the background soil and leaf area affected the accuracy of ATCOR. It has been found that 5% of scattering radiation is contained with the ASTER Green band and 47% of radiation was absorbed in the ASTER NIR band and 17% of radiation was absorbed in the ASTER SWIR6 band. The ground measurement values and the ATCOR software output values were of no big difference in the ASTER reflection band and absorption bands of chlorophyll (i.e. NIR-band and Red-band). However, the difference was seen in the ASTER scattering bands (i.e. visible Green band) and soil reflection bands (i.e. ASTER SWIR bands). Compared with the data of ASD's measurement, the AST07 (©NASA/EOSDIS ASTER surface reflectance product data (L2B)) values are too low in a NIR band.

For the ASTER original L1B data (Jun 30, 2001), the statistics mean value of Green band and NIR band is 50.9 and 69.8 after atmospheric correction, the statistics mean value of Green band and NIR band is 26.5 and 90.9. (For the ASTER/July 12, 2004 case, the values was 53.8

and 83.5 to 27.5 and 131.4 respectively). The value of NDVI after atmospheric correction is larger than atmospheric correction before, and this rate of change is (NDVI-Corrected) = 1.27 (NDVI-L1B) + 0.04.

Comparison of accuracy of the ATCOR software atmosphere correction of non-moor plant and high moor plant area ASTER imagery showed that the background soil and leaf area affected the accuracy of ATCOR. In the case of a moor plant, the error in ASTER Green band is large.

6. Acknowledgment

This work was supported by Grant-in-Aid for Scientific Research (A) 21370005 and the Global Environmental Research Fund (F-092) by the Ministry of the Environment, Japan.

7. References

[1]   Schott J.R. and Henderson-Sellers, A., "Radiation, the Atmosphere and Satellite Sensors, Contribution in Satellite Sensing of a Cloudy Atmosphere: Observing the Third Planet (Ed: Henderson-Sellers), " 1984, pp45-89.

[2]   Campbell, J. B., "Introdution to Remote Sensing: Second Edition," The Guilford Press, New York. 1996, pp. 444-476.

[3]   W.G.Rees, " Physical Principles of Remote Sensing, Second Edition, " Cambridge University Press. 2001, pp. 76-85.

[4]   Kneizys, F.X., "Atmospheric Transmittance Radiance: computer code " LOWTRANG AFGL-TR-83-0187 (1983).

[5]   Berk, A., L. S. Bernstein, and D. C. Robertson, "MODTRAN. A Moderate Resolution Model for LOWTRAN 7,". GL-TR-89-0122, Phillips Laboratory, Geophysics Directorate, Hanscom Air Force Base, Massachusetts, 1989.

[6]   Berk, A., L.S. Bernstein, G.P. Anderson, P.K. Acharya, D.C. Robertson, J.H., "Chetwynd and S.M. Adler-Golden, MODTRAN Cloud and Multiple Scattering Upgrades with Application to AVIRIS, " Remote Sens. Environ, 1998, pp.65:367-375.

[7]   E. Vermote, D. Tanr, J. Deuz, M. Herman, and J. Morcette, "Second simulation of the satellite signal in the solar spectrum 6S: An overview, " IEEE Trans. Geosci. Remote Sensing, vol. 35, no. 3, 1997, pp. 675 − 686.

[8]   Fraser, R. S, R. A. Ferrare, Y. J. Kaufman, B. L. Markham, S. Mattoo, "Algorithm for atmospheric correction of aircraft and satellite imagery, " Int'l. J. Rem. Sens., 1992, pp.13:541-557.

[9]   Nave, Carl Rod, 2005. Hyperphysics-Quantum Physics, Department of Physics and Astronomy, Georgia State University, CD.

[10]   Gates, David M. (1980) Biophysical Ecology, Springer-Verlag, New York, pp. 611.

[11]   R. Richter, "A fast atmospheric correction algorithm applied to Landsat TM, " Int. J. Remote Sensing 1990, pp.11:159-166.

[12]   R. Richter, "Correction of atmospheric and topographic effects for high spatial resolution satellite imagery", Int. J. Remote Sensing 18:1099-1111 (1997).

[13]   Richter, R. and Schläpfer, D., "Geo-atmospheric processing of airborne imaging spectrometry data. Part 2:Atmospheric/Topographic Correction". International Journal of Remote Sensing, 23(13): 2631-2649 (2002).

[14] Herman, B. M., & Browning, S. R., "A numerical solution to theequation of radiative transfer, " Journal of the Atmospheric Sciences, 22, 1965, pp.559-566.

[15] Slater, P. N., Biggar, S. F., Holm, R. G., Jackson, R. D., Mao, Y., Moran, M. M., Palmer, J. M., & Yuan, B., "Reflectance- and radiance-based methods for the in-flight absolute calibration of multispectral sensors, " Remote Sensing of Environment, 1987, pp.22, 11 – 37.

[16] .Aosier, B. Kaneko, M. Takada, M., Comparison of NDVI of ground measurement, atmospheric corrected ASTER L1B data and ASTER surface reflectance product (AST07) data, Geoscience and Remote Sensing Symposium, 2007 IEEE International; Barcelona,Spain, Publication Date: 23-28 July 2007, pp. 1806-1811.

[17] Buhe Aosier, K. Tsuchiya, M. kaneko, S. J. Saiha, "Comparison of Image Data Acquired with AVHRR, MODIS, ETM+, and ASTER Over Hokkaido Japan," *Advances in Space Research*, Vol. 32, No. 11, 2003, pp.2211-2216.

[18] Drury, S.A., "Image Interpretation in Geology. Allen and Unwin (Publishers) Ltd, London. Forster, B.C. (1980). Urban residential ground cover using Landsat digital data. Photogram, " Eng. Remote. Sen. 46, 547, 1987, pp.58.

[19] Coops, N.C., Culvenor, "Utilizing local variance of simulated high-spatial resolution imagery to predict spatial pattern of forest stands, " Remote Sensing of Environment. 71(3): 2000, pp.248-260.

[20] Jensen, J.R., 1986. 'Introductory Digital Image Processing : A Remote Sensing Perspective', New Jersey : Prentice-Hall.

[21] Crippen, R.E., "The Regression Intersection Method of Adjusting Image Data for Band Ratioing, " *International Journal of Remote Sensing*. Vol. 8. No. 2. 1987, pp. 137 155.

[22] Chavez, P. S., Jr., "An improved   dark-object   subtraction technique for atmospheric scattering correction of multispectral data ," *Remote Sens. Environ.*, 24, 1988, pp. 459-479.

[23] Herman, B. M., & Browning, S. R., "A numerical solution to theequation of radiative transfer, " Journal of the Atmospheric Sciences, 22, 1965, pp.559-566.

[24] Slater, P. N., Biggar, S. F., Holm, R. G., Jackson, R. D., Mao, Y., Moran, M. M., Palmer, J. M., & Yuan, B., "Reflectance- and radiance-based methods for the in-flight absolute calibration of multispectral sensors, " Remote Sensing of Environment, 1987, pp.22, 11 – 37.

* corresponding author Hoshino Buho: e-mail: aosier@rakuno.ac.jp

Environmental hazards in the El-Temsah Lake, Suez Canal district, Egypt

[1]Kaiser, M. F.
Geology Department, Faculty of Science, Suez Canal University
Egypt
[2]Amin, A. S.
Botany Department, Faculty of Science, Suez Canal University
Egypt
[3]Aboulela, H. A.
Marine Science Department, Faculty of Science, Suez Canal University
Egypt

1. Abstract

Chemical and biological analyses were integrated using remote sensing GIS techniques to evaluate the environmental pollution of El-Temsah Lake, in the Suez Canal, in order to provide critical data to enhance development planning and economic projects within the study area. Fifty-six samples were collected from seven sites in the lake from July 2005 to May 2006. Samples were collected in each of four seasons, and included 28 surface sediment samples and 28 water samples. Sediment samples were analyzed for Fe, Pb, Ni, Co, Cu, Mg, K and Na. The results showed an increase in beach sediment pollution from summer to winter. Taxonomic analysis of phytoplankton samples revealed 102 taxa, including 56 Bacillariophyceae, 8 Chlorophyceae, 18 Dinophyceae and 20 Cyanophyceae. Chlorophyll *a* concentrations ranged from 0.3 to 26 µg l^{-1}, with the highest values during the winter and lowest values during the summer. These results suggest that beach sediment pollution is highest in the winter and, at the same time, the water quality conditions in El-Temsah lake favor oxidation conditions which maximize phytoplankton productivity. In contrast, sediment pollution and phytoplankton productivity are lowest during summer, which also corresponds to more alkaline water conditions. The images were rectified and analyzed by ERDAS IMAGINE 8.9. A 1968 topographic map and enhanced 2005 Landsat Thematic Mapper images (30 m resolution) were utilized to determine the coastline positions using ERDAS Imagine 8.9. Image processing techniques were applied using ENVI 4.2 to analyze the ETM+ image data. Image enhancement was applied. Image data was enhanced spectrally to verify surface water pollution detected from chemical and biological analyses and to detect the sources of untreated domestic, industrial and agricultural waste water. In general, the lake has been subjected to successive shrinking due to human activities, primarily through extensive building along the shoreline. The uncontrolled growth of cities is associated with seismic hazards, affecting on buildings and infrastructures, mostly due to insufficient knowledge of earthquakes activity. Seismic epicentres were recorded along the Suez Canal from 1904 to 2006. Widespread moderate to micro earthquakes were identified around the western side of the lake, with scattered events along the eastern side. In general, water pollution in El-Temsah Lake has been mitigated over the last decade due to successive dredging and improved water treatment. Most of untreated water was discharged along the western side of the lake. The eastern part of the lake is less polluted and is, therefore, more suited for fishing, tourism, urban planning and navigation activities, although higher use of eastern portion of the lake could accelerate water and sediment quality deterioration in that region.

2. Introduction

El-Temsah Lake formed in a depression situated in a fault trough (Holmes, 1965 and El Shazley et al., 1974) covered with Nile Delta sediments. The trough originated tectonically as part of the "Clysmic Gulf" that represented the first stage of separation along the Red Sea-Suez Rift during the Late Oligocene-Early Miocene period (El-Ibiary, 1981). The lake covers about 15 km^2 between 30^0 32` and $30^0$ 36` north latitude and 32^0 16` and $32^0$ 21` east longitude, and is located near the middle of the Suez Canal, at a point 80 km south of Port Said. The depth of the lake ranges between 6 and 13 m. Following the construction of High Dam (completed in 1970), lake water quality changed from saline to fresh water due, to a large degree, to the precipitation of gypsum and mud lamina. The lake is the backbone of a tourism industry that attracts a large number of holiday visitors. In addition to the visitors, the tourism and fishing industries employ local citizens and provide a significant portion of the district revenues. Unfortunately, the increasing number of temporary and permanent residents has also created higher volumes of waste, including raw liquid and solid municipal sewage, agricultural runoff and industrial wastewater, all of which end up in El-Temsah Lake. The lake is also a sink for aliphatic and aromatic hydrocarbons originating from shipping activities including accidental and incidental oil pollution, ballast water release and general vessel and facility maintenance. In addition to hydrocarbons, several other potential chemical contaminants originate from various pollutant sources. The concentrations of Pb, Zn, Cu and Cd metals have been shown, for example, to be significantly higher in this area (Abd El Shafy and Abd El Sabour, 1995). Elevated nitrate and nitrite levels are due to the discharge of sewage, fertilizers and pesticides.

Through 1996, water quality along the northern and western boundaries of the lake has experienced substantial deterioration due to the rapid development of tourism projects and continuous waste discharge at Ismailia City (ETPS, 1995). In order to rejuvenate the lake, improve water quality and help re-establish the fishing and tourism industries, local authorities have embarked on a national program, using dredging to remove significant portions of contaminated lake sediment. Moreover, the re-evaluated assessment and mitigation of seismcity risk around any area, especially lake area is important due to the fast growing and developing of great and important constructions. The earliest attempts to review the seismicity and tectonic activity of the overall Suez Canal region were given by Aboulela, 1994. The main objective of the research described here was to assess the environmental status of El-Temsah Lake, focusing on heavy metal analysis of beach sediments and the distribution of phytoplankton. Furthermore, try to make main zoom on the seismicity pattern around concerned area.

The diversity and density of biological communities are impacted by the physicochemical characteristics of the ecosystem, including temperature, pH and dissolved oxygen (DO). In addition, any single species is inevitably impacted by the other species present within the system (Moss, 1998). Phytoplankton species, for example, are strongly influenced by water quality, act as the primary source of autochthonous organic material in many lakes and are the main food for many filter-feeding, primary consumers (Boney, 1989). Monitoring phytoplankton populations, therefore, is often used as a means of evaluating the health of fresh, brackish and saltwater ecosystems. Phytoplankton diversity generally shows substantial temporal variation. In the Suez Canal, 126 diatom species were recorded in the winter and summer from 1969-1971 (Dorgham, 1974); 102 species were identified in 1983 (Dorgham, 1985); 94 species were observed in 1991 (El-Sherief and Ibrahim, 1993). Because of the existing historical data on phytoplankton populations in the Suez Canal, the importance of phytoplankton to overall community structure and the sensitivity of phytoplankton to environmental perturbations, they were used as the biological monitor in this study. Earthquakes epicentres distribution occurred overall the study area and its surrounding were using by recently information recorded data. Geographical information system (GIS) techniques were used to integrate chemical, biological and seismicity data in order to evaluate current conditions and make recommendations for sustainable coastal zone management.

3. Methodology

3.1 Field sampling locations

Six study sites in El-Temsah lake were selected to assess environmental pollution (Fig. 1). Samples were collected during four seasons between June 2005 and April 2006 from the following locations: El-Taween, El- Fayrooz, the Bridge, El-Osra, El-Forsan, Beach clubs. Sediment appearance and composition was dark-gray to blackish, soft, silty sand at the western sites (El-Taween, El- Fayrooz , the Bridge, El-Osra and El-Forsan), but a lighter-colored sand at the eastern El-Temsah sites. During the field visits, it was noticed that the El-Taween area was subject to substantial dumping or leakage of industrial wastes from ships passing through the lake, as well as disposal of domestic wastewater from businesses, particularly clubs, along the shore. The El-Fayrooz and Bridge sites are located near the junction of a western lagoon and El- Temsah lake, and receive large quantities of domestic, agricultural and industrial wastewater. The El-Osra and El-Forsan sites, in the northern part of the lake, receive domestic wastewater. In contrast, the eastern part of the lake (Beach club-1 and Beach club-2) is characterized by much lower pollution exposure. From each of the seven sampling locations, sediment, water and phytoplankton samples were collected from the lake. Sediment was collected at plastic bags and was kept for a week in a fridge before analyses. Subsurface water samples were taken at all sites for phytoplankton identification and counting. Water samples for physico-chemical studies were collected in polyethylene bottles of one liter capacity, and laboratory analysis started within few hours from the time of collection.

3.2 Environmental analyses

3.2.1 Water and sediment analyses

In the field, dissolved oxygen (DO), pH and temperature were determined. Water samples were analyzed for salinity, conductivity, nitrate and phosphorus, according to Standard Methods (APHA 1992). For analysis of trace metal concentrations, the 28 sediment samples were dried at 60^0C for one day, weighed, and then digested with hot concentrated acid under reflux conditions. Metal concentrations in all samples were determined by following the methods described by More and Chapman (1986).

3.2.2 Biological investigations

Phytoplankton samples were preserved in 500 ml glass bottles with lugol's solution (Utermöhl's, 1958). The standing crop was calculated as the number of cells per liter. Qualitative analysis was carried out using the preserved as well as fresh samples. These were examined microscopically for the identification of the present genera and species. The algal taxa were identified according to the following references: Van-Heurck (1986), Prescott (1978), Humm and Wicks (1980) and Tomas (1997). Species richness (alpha diversity) was calculated for each phytoplankton algal group as the average number of species per season. Relative concentration and relative evenness were expressed according to the equations of Whittaker (1972), Pielou (1975) and Magurran (1988).

3.2.3 Seismicity pattern investigation

Seismicity activity occurred around the El-Temsah lake and its surrounding was assessed in this study using information on revealed recorded earthquakes. Seismicity data from 1904 to 2006 were supplied from the National Earthquake Information Center (NEIC), International Seismological Centre (ISC) and the Egyptian National Seismic Network (ENSN).

3.3 Remote Sensing and GIS analyses

Satellite remote sensing is a potentially powerful tool for detecting shoreline change. Enhanced Landsat Thematic Mapper images ETM-7 (30 m resolution) for 2004 (Figure 2) were geometrically corrected and digitized to extract vectors representing shoreline positions in 2004. ERDAS IMAGINE 8.9 (Leica Geosystems, Norcross, GA, USA) was utilized for digitizing and editing data input. The outline of El-Temsah lake in 2004 was compared to the topographic map in 1968. Arc GIS 9.2 (ESRI, Redlands, CA, USA) platforms were used to organize, analyze and manipulate available data and generate new data for heavy metal, seismic and biological analyses.

Image segmentation techniques were used to produce vector maps for the El Temsah coastline to detect the shoreline position and identify changes between 1968 and 2004. In this study, image registration was carried out using ERDAS IMAGINE 8.9. A second-order polynomial was used to provide an adequate transformation for registering a full Thematic Mapper scene to geographical coordinates.

Thematic Mapper band 7 (short-wave infrared) was used to produce a vector map of the shoreline. Image segmentation was used to delineate the land/sea boundary along the coastline. A segmentation approach was also necessary in order to compare the results with previous shoreline maps produced.

4. Results and Discussion

4.1 Metal analyses

Data collected during this study indicate that, in the summer, concentrations of Na, K, Fe and Mg are elevated along the western side of the lake at the Bridge and El-Taween sites (Fig. 2). Low levels were identified along the northeastern side, at the El-Forsan and Beach clubs (Table 1). Concentrations ranged from 9.03 to 28.9 ppm (Na), 2.28 to 10.47 ppm (K), 3.85 to 12.71 ppm (Fe) and 0.42 to 9.69 ppm (Mg). In autumn, concentrations of some metals, including Na, K, Mg, Fe and Co were higher along the western side of the lake, at the Bridge area. Concentrations ranged from 65.74 to 127.63 ppm (Na), 5.27 to 34.19 ppm (K), 6.01 to 32.71 ppm (Mg), 7.97 to 102.44 ppm (Fe) and 0.12 to 0.42 ppm (Co). Mg and Na levels were higher in beach sediments in the winter and spring than during either the summer or autumn, ranging from 841.5 to 2145.12 ppm (Na) and from 29.82 to 676.89 ppm (Mg). Levels of both K (10.4 to 27.12 ppm) and Fe (27.59 to 133.25 ppm) in winter-collected sediment were higher than in the summer. Spring sediment concentrations tended to be very similar to winter levels. The lowest metal concentrations were identified along the northeastern side of the lake; samples collected along the western perimeter had the highest levels in the spring. Concentrations ranged from 8.65 to 18.8 ppm (K), 17.54 to 72.98 ppm (Fe), 29.82 to 340.51 ppm (Mg) and 0 (below analytical detection) to 1.51 ppm (Co). The concentrations of Cu, Pb and Ni were negligible at all study sites.

In summary, the concentrations of metals, including common ions such as Na and Mg, were noticeably elevated along the western corner of the lake at El-Taween, El Fayroos and the Bridge, but declined in the northeastern corner at El-Forsan and the eastern beach clubs. High concentrations of Fe at the El-Taween and Bridge areas are likely due to the disposal of raw sewage from nearby housing, and the impact of industrial waste residues in the Suez Canal. In addition, high concentrations of Mn are most likely associated with agricultural sewage, as well as wastewater from substantial industrial activities associated with ship construction and maintenance.

4.2 Water quality

While average ocean salinity ranges from approximately 32 to 35 ppt, local salinity – particularly near shore - can vary according to freshwater input, evaporation and temperature. Low salinity (31.4 ppt) was recorded when the temperature was low (14.5°C) in the winter. Salinity reached a maximum value (44.9 ppt) during the summer when temperature was highest (31°C), (Table.2). Conductivity varied between 42 μmhos cm^{-1} in the winter and 107 μmhos cm^{-1} in the summer. Freshwater discharged from sewage treatment systems, untreated sources and nonpoint source runoff, enter the lake on the western side, causing a salinity decline (31.4 – 32.2 ppt); the highest salinity (44.9 – 44.8 ppt) was recorded on the eastern side. The pH values recorded in the lake tended to be somewhat alkaline (7.9 – 8.9) in the summer. Dissolved oxygen in the surface water fluctuated between 5.6 mg l^{-1} at El-Osra in the summer, to 10.2 mg l^{-1} at El Fayroos in the winter. Nitrate (NO$_3$-N) concentrations ranged from15.6 μg l^{-1} at El-Taween to 130.4 μg l^{-1} at El-Osra. SiO$_3$ – Si values fluctuated between 2.7 mg.l^{-1} at El Forsan site and 6.8 mg.l^{-1} at the beach clubs. Its highest values occurred in spring and the lowest values in autumn. The importance of silicate lies in its significance for the construction of the cell wall of diatoms. This element is not a limiting factor for diatoms in the lake. Willem (1991) reported was used to construct land use and hazards maps of the study area. The eastern portion of the lake, because of lower pollution levels, is the most appropriate site for fishing, tourism, urban planning and navigation. Fortunately, environmental pollution has shown a decline over the last decade due to successive sediment dredging and improvements in water purification systems.

Figure 1 (A) Location of El-Temsah Lake in the Suez Canal. (B) sample locations within El-Temsah Lake .

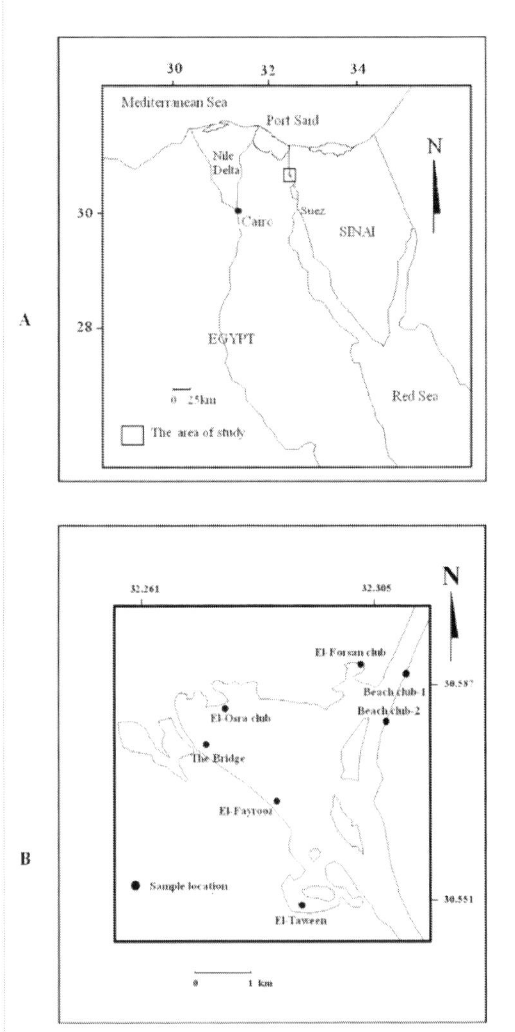

Fig 1. (C) Landsat image for year 2004 of the Suez Canal District.

Fig 2. Enhanced Landsat Thematic Mapper images ETM-7 for year 2004 of the El-Temsah Lake

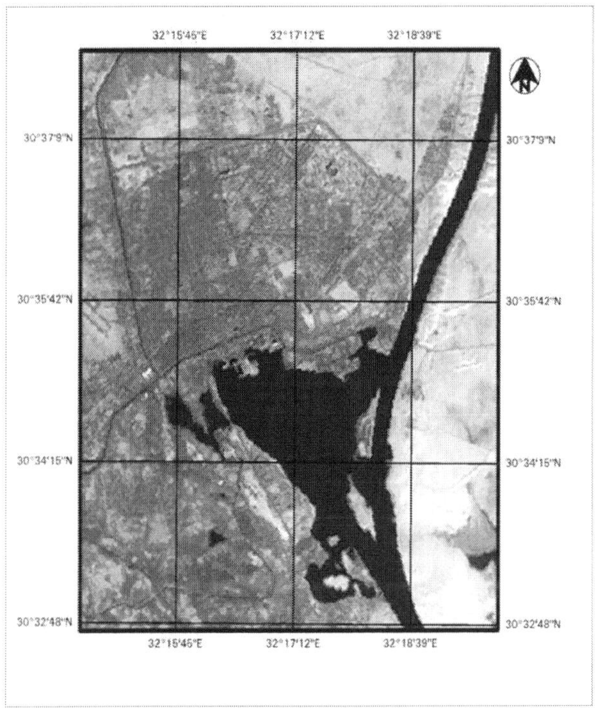

Fig 3. Concentrations of metals in beach sediments during the summer and winter

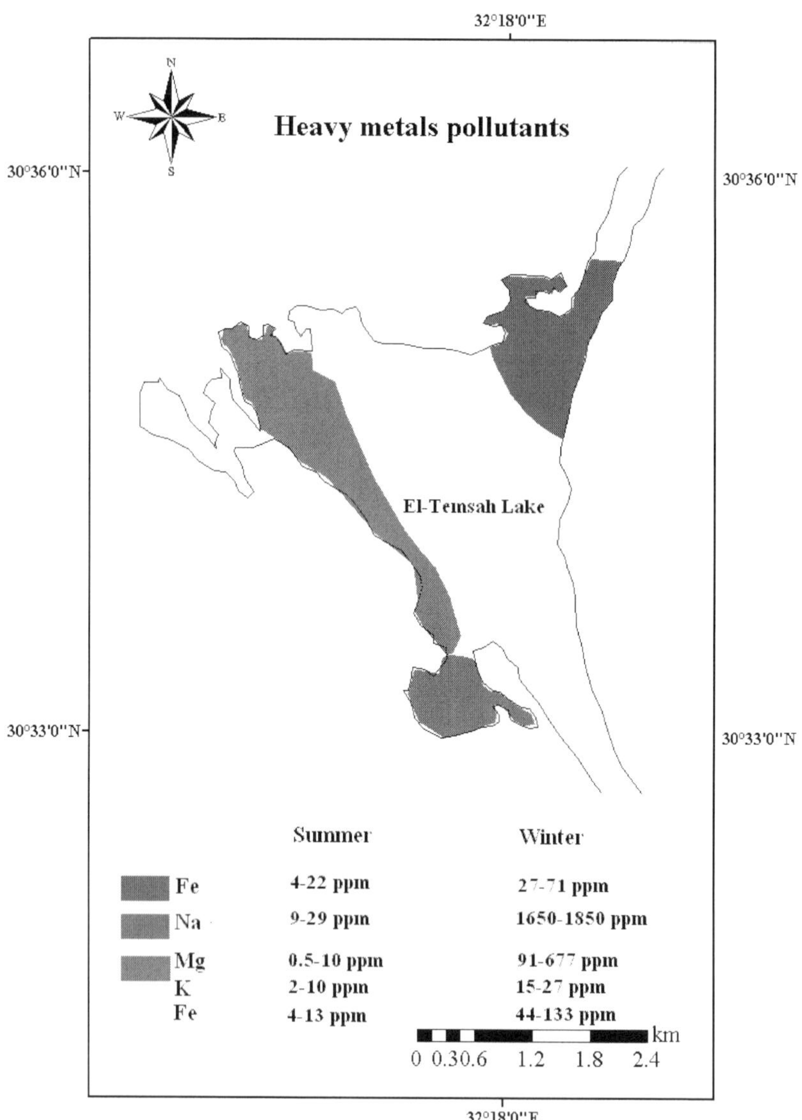

Fig 4. Phytoplankton species richness (Alpha Diversity) in El-Temsah lake during the summer and winter

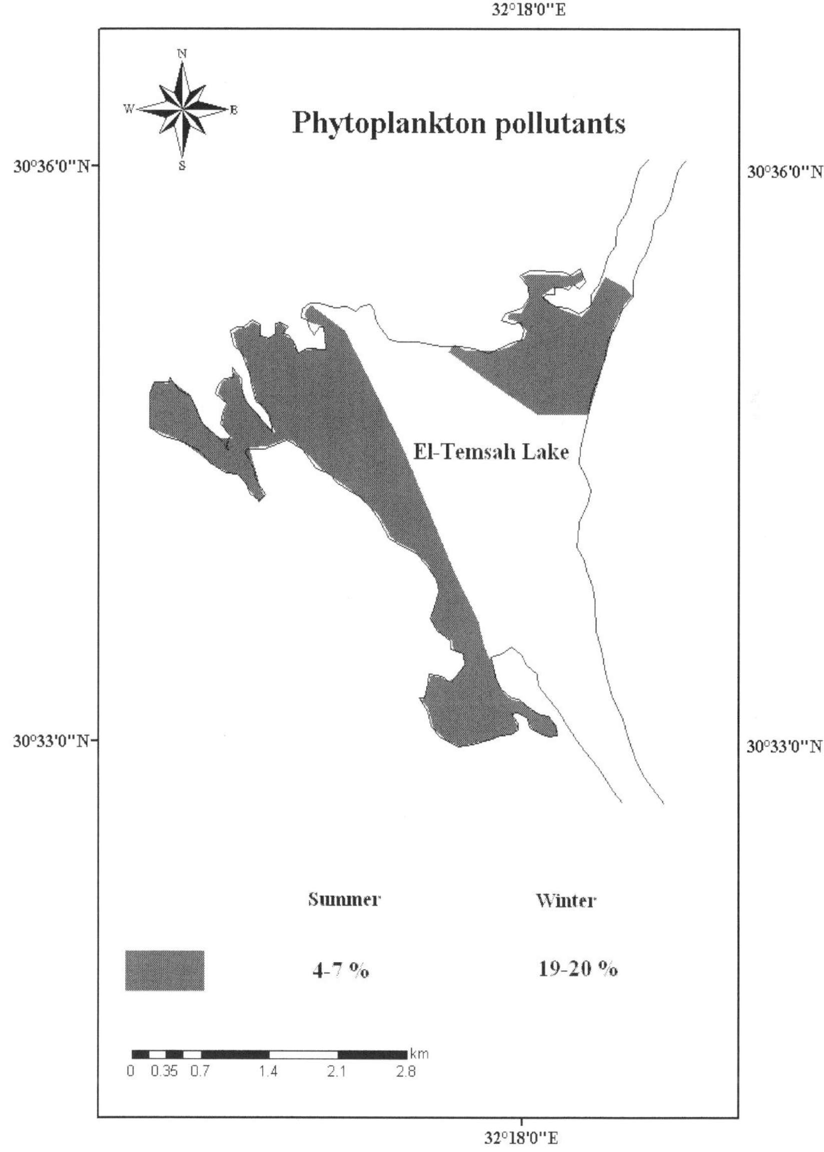

Fig 5. Locations of activities affecting the boundaries of El-Temsah lake

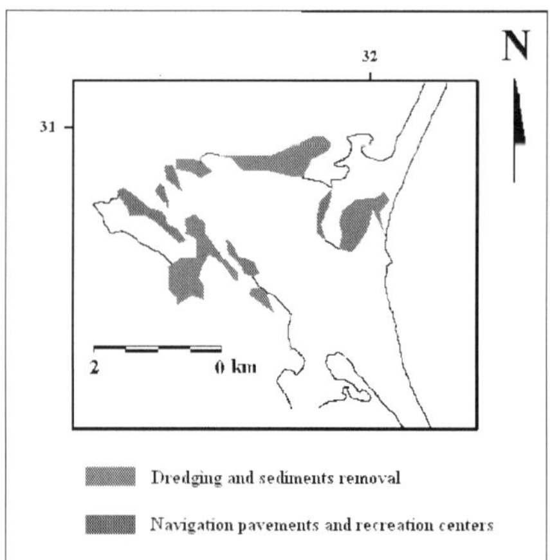

Fig 6. (A): Map showing distribution of recently local and regional recorded earthquakes with magnitude (3.0 ≤ Mb ≥ 4.5) of the study area and its surroundings. (B): Earthquake hazards assessment map of El-Temsah lake.

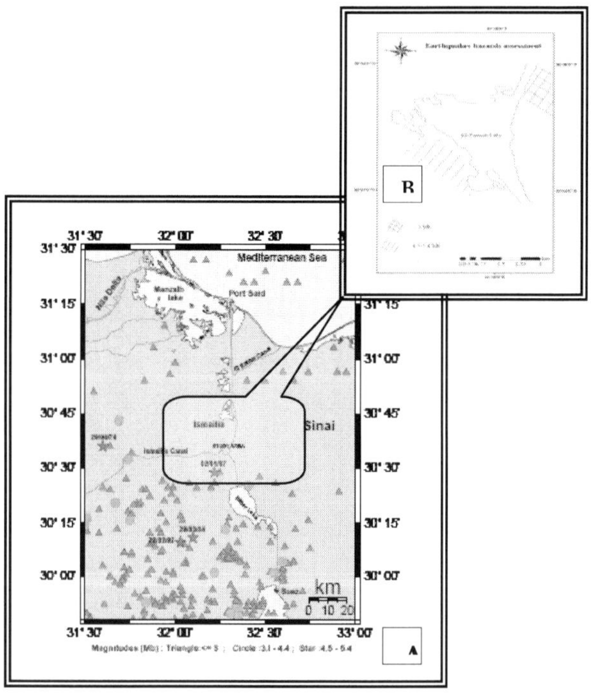

Magnitudes (Mb):   ▲   ≤3   ◯   3.1 -4-4 ;   ★   4.5-5.4

Sample No.	Na (ppm)	K (ppm)	Fe (ppm)	Mg (ppm)	Co (ppm)
	Summer	Summer	Summer	Summer	Summer
El-Taween	28.910	4.82	6.06	4.12	1.51
The Bridge	26.800	6.7	11.5	5.8	1.46
El Fayrooz	27.470	10.47	12.71	9.69	1.29
El-Osra	9.030	2.28	3.85	0.42	0.21
El-Forsan	14.450	3.87	6.71	4.28	0.21
Beach clubs	22.050	7.29	7.48	2.49	0
Sample No.	Autumn	Autumn	Autumn	Autumn	Autumn
El-Taween	65.74	5.27	7.97	6.01	0.25
The Bridge	95.74	20.5	90.89	19.8	0.35
El Fayrooz	113.12	34.19	102.44	23.67	0.42
El-Osra	127.63	24.42	42.81	20.76	0.17
El-Forsan	93.78	10.91	31.05	32.71	0.12
Beach clubs	105.39	20.49	39.4	17.08	0.12
Sample No.	Winter	Winter	Winter	Winter	Winter
El-Taween	1655.88	17.75	44.14	676.89	0.17
The Bridge	1780.9	21.5	98.9	357.9	0.1
El Fayrooz	1844.05	27.12	133.25	153.72	0.17
El-Osra	1731.15	15.41	60.69	91.75	0.13
El-Forsan	2145.12	25.23	71.73	349.29	0.04
Beach clubs	1806.42	10.4	27.59	57.14	0
Sample No.	Spring	Spring	Spring	Spring	Spring
El-Taween	841.5	11.25	25.1	340.51	0.84
The Bridge	870.87	15.89	69.99	245.89	0.5
El Fayrooz	935.5	18.8	72.98	80.21	0.73
El-Osra	870.2	8.85	32.27	46.09	0.17
El-Forsan	1079.5	14.55	39.22	176.79	0.13
Beach clubs	914.6	8.65	17.54	29.82	0

Table 1. Heavy metal analyses result of water samples collected at El-Temsah Lake during summer 2005-spring 2006

Sites / Parameters	El-Taween Min	El-Taween Max	El Fayrooz Min	El Fayrooz Max	The Bridge Min	The Bridge Max	El Osra Min	El Osra Max	El Forsan Min	El Forsan Max	Beach clubs Min	Beach clubs Max
Temperature °C	14	29	14	29	14.5	29.5	14.5	29.5	15	31	15	31.5
pH	7.4	8.5	8	8.8	8.6	8.9	7.9	8.4	7.3	7.9	7.2	7.9
Conductivity $\mu mohs.cm^{-1}$	42.5	52.5	51.5	78	77	85	42	67.1	68	107	79	90.2
Salinity ppt	33.8	41.9	32.2	44.6	33	43.8	31.4	42.7	41.2	44.9	39.9	44.8
Turbidity m	2.6	2.9	2.8	2.9	2.2	2.6	1.8	2.4	0.24	0.45	0.39	0.5
Dissolved Oxygen $mg.l^{-1}$	7.2	8.4	7.3	10.2	6.2	8.9	5.6	9.6	6.3	9.8	7.1	8.8
Nitrate $\mu g.l^{-1}$	15.6	114.1	29.5	130	53.2	112.2	32	130.4	32.2	98.4	28.2	92
Nitrite $\mu g.l^{-1}$	2.7	25.2	1.4	19.2	1.3	22.1	1.1	19.1	1	10	1	13
Ammonium $\mu g.l^{-1}$	118	253	121	188	115.3	159.3	112	157.2	98.3	148	92.2	134.9
Phosphate $\mu g.l^{-1}$	13.2	20.4	12.2	29.6	13.7	28.5	10.5	26.2	12.2	26.8	6.9	25.9
Silicate $mg.l^{-1}$	3.4	6.6	3.8	6.4	3.3	5.8	3.1	5	2.7	6.7	3.4	6.8
Chlorophyll a $\mu g.l^{-1}$	0.3	24	2.8	26	3.2	18.1	4.2	16	3.9	17.9	4.7	19.3
N: P ratio	1.07	5.59	1.8	5.82	2.8	4.75	2	6.74	1.98	4.8	2.03	9.45

Table 2. Physicochemical parameters of water samples collected at El Temsah Lake during summer 2005-Spring 2006

7. References

Abdel-Shafey, H.I. and Abdel-Sabour, A. (1995). Levels of heavy metals in the environment of the river Nile and some Egyptian surface water German Egyptian conference, Ismailia, 6-7, February, pp 36-53

sensor in two ways: radiometrically and geometrically. This means that they can modify the signal's intensity through scattering or absorption processes and its direction by refraction (Sifakis and Deschamps, 1992).

The problem of particulate pollution in the atmosphere has attracted a new interest with the recent scientific evidence of the ill-health effects of small particles. Aerosol optical thickness in the visible (or atmospheric turbidity), which is defined as the linear integral of the extinction coefficient due to small airborne particles, can be considered as an overall air pollution indicator in urban areas (Sifakis, et al., 1998). Air pollution has long been a problem in the industrial nations of the West. It has now become an increasing source of environmental degradation in the developing nations of East Asia. The lack of detailed knowledge of the optical properties of aerosols results in aerosol being one of the largest uncertainties in climate forcing assessments. Monitoring of atmospheric aerosol is a fundamentally difficult problem. The problem of particulate pollution in the atmosphere has attracted a new interest with the recent scientific evidence of the ill-health effects of small particles. Air pollution is one of the most important environmental problems, which concentrates mostly in cities. Aerosols are liquid and solid particles suspended in the air from natural or man-made sources (Kaufman, et al., 1997).

The main objective of the present study is to test the performance of our proposed algorithm for mapping PM10 using Landsat satellite images. In situ measurements were needed for algorithm calibration. We used a DustTrak Aerosol Monitor 8520 to collect the in situ data. We collected the PM10 data simultaneously during the satellite Landsat overpass the study area. An algorithm was developed to determine the PM10 concentration on the earth surface. The efficiency of the proposed algorithm was determined based on the correlation coefficient (R) and root-mean-squares deviation, RMS. Finally, the PM10 map was generated using the proposed algorithm. In addition, the PM10 map was also geometrically corrected and colour-coded for visual interpretation.

(a)

Source: vincentchow, 2009

Source: The Digital Awakening-Haze in Penang

(b)

Source: The Digital Awakening-Haze in Penang Update - 10.48am
Fig. 1. Air pollution was found at (a) Kuching and (b) Penang, Malaysia

2. Remote Sensing

Remote sensing is a technique for collecting information about the earth without taking a physical sample of the earth's surface or touching the surface using sensors placed on a platform at a distance from it. A sensor is used to measure the energy reflected from the earth. This information can be displayed as a digital image or as a photograph. Sensors can be mounted on a satellite orbiting the earth, or on a plane or other airborne structure. Because of this energy requirement, passive solar sensors can only capture data during daylight hours. The major applications of remote sensing include environmental pollution, land cover/use mapping, urban planning, and earth management. We have to understand the basic concept of electromagnetic waves well enough for applying to the remote sensing techniques in our studies. We classify electromagnetic energy by its wavelength. This electromagnetic radiation gives an energy source to illuminate the target except the sensed energy that is being emitted by the target (Figure 2).

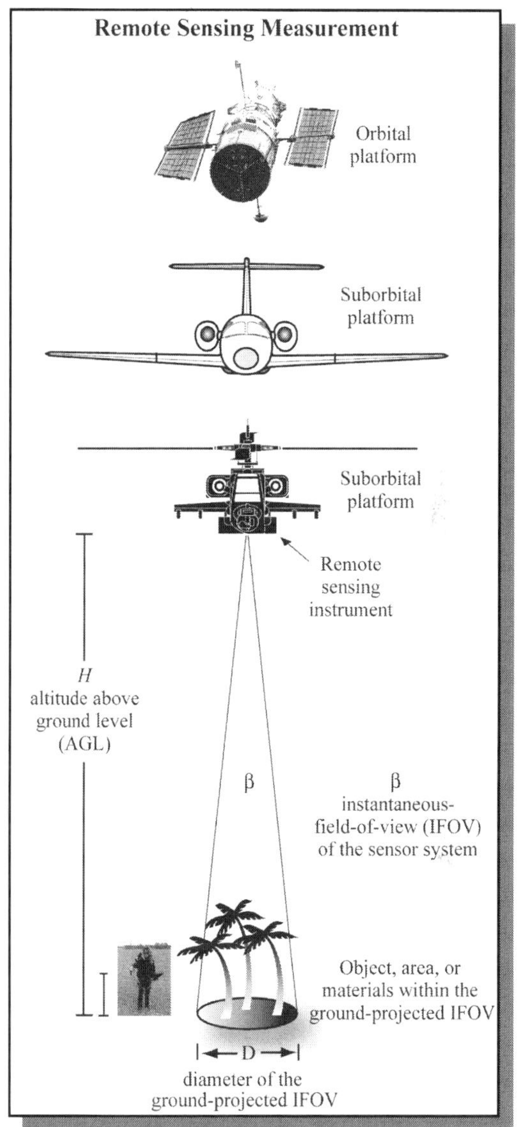

Fig. 2. Remote sensing instrument (Jensen, 2006)

There are two basic types of sensors: passive and active sensors (Figure 3). Passive remote sensors detect reflected energy from the sun back to the sensor; they do not emit energy itself. But active sensors can emit energy or provide its own source of energy and detect the reflected energy back from the target.

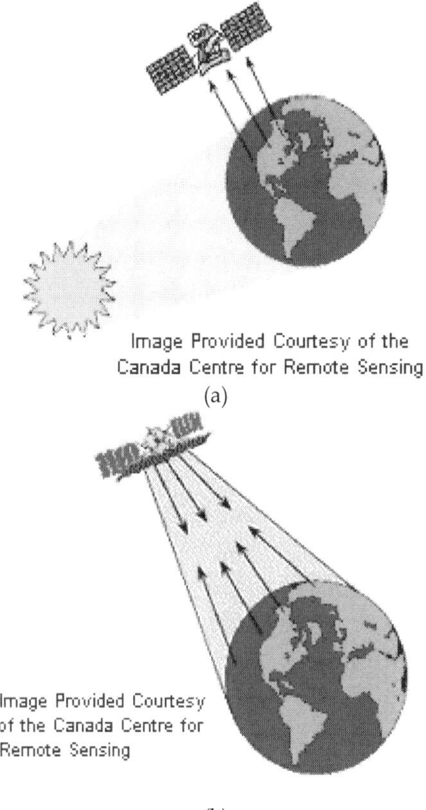

Fig. 3. (a) Passive sensor and (b) Active sensor (Fundamentals of Remote Sensing - http://www.ccrs.nrcan.gc.ca/resource/tutor/fundam/pdf/fundamentals_e.pdf.)

3. Study Area

The study area is the Penang Island, Malaysia, located within latitudes 5° 9′ N to 5° 33′ N and longitudes 100° 09′ E to 100° 30′ E. The map of the study area is shown in Figure 4. Penang Island is located in equatorial region and enjoys a warm equatorial weather the whole year. Therefore, it is impossible to get the 100 % cloud free satellite image over Penang Island. But, the satellite image chosen is less than 10 % of cloud coverage over the study area. Penang Island located on the northwest coast of Peninsular Malaysia.

Penang is one of the 13 states of the Malaysia and the second smallest state in Malaysia after Perlis. The state is geographically divided into two different entities - Penang Island (or "Pulau Pinang" in Malay Language) and a portion of mainland called "Seberang Perai" in Malay Language. Penang Island is an island of 293 square kilometres located in the Straits of Malacca and "Seberang Perai" is a narrow hinterland of 753 square kilometres (Penang-

Wikipedia, 2009). The island and the mainland are linked by the 13.5 km long Penang Bridge and ferry.

Penang Island is predominantly hilly terrain, the highest point being Western Hill (part of Penang Hill) at 830 metres above sea level. The terrain consists of coastal plains, hills and mountains. The coastal plains are narrow, the most extensive of which is in the northeast which forms a triangular promontory where George Town, the state capital, is situated. The topography of "Seberang Perai" is mostly flat. Butterworth, the main town in "Seberang Perai", lies along the "Perai" River estuary and faces George Town at a distance of 3 km (2 miles) across the channel to the east (Penang-Wikipedia, 2009).

The Penang Island climate is tropical, and it is hot and humid throughout the year. with the average mean daily temperature of about 27oC and mean daily maximum and minimum temperature ranging between 31.4oC and 23.5oC respectively. However, the individual extremes are 35.7oC and 23.5oC respectively. The mean daily humidity varies between 60.9% and 96.8%. The average annual rainfall is about 267 cm and can be as high as 624 cm (Fauziah, et al, 2006).

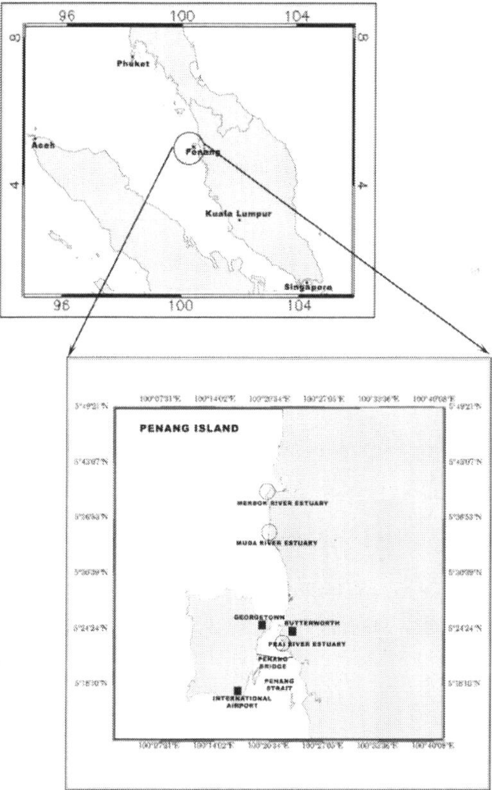

Fig. 4. The study area

4. Satellite Remote Sensing Data

For our research in USM, we use satellite images from passive sensors in our analysed. Images acquired by Landsat TM 5. On March 1, 1984, NASA launched Landsat 5 were used in this study, the agency's last originally mandated Landsat satellite. Landsat 5 was designed and built at the same time as Landsat 4 and carried the same payload: the Multispectral Scanner System (MSS) and the Thematic Mapper (TM) instruments.

Fig. 5. Landsat Satellite TM

5. Algorithm Model

The atmospheric reflectance due to molecule, R_r, is given by (Liu, et al., 1996)

$$R_r = \frac{\tau_r P_r(\Theta)}{4\mu_s \mu_v}$$

$$(1)$$

Where

τ_r = Rayleigh optical thickness

$Pr(\Theta)$ = Rayleigh scattering phase function

μ_v = Cosine of viewing angle

μ_s = Cosine of solar zenith angle

We assume that the atmospheric reflectance due to particle, Ra, is also linear with the τa [King, et al., (1999) and Fukushima, et al., (2000)]. This assumption is valid because Liu, et al., (1996) also found the linear relationship between both aerosol and molecule scattering.

$$R_a = \frac{\tau_a P_a(\Theta)}{4\mu_s \mu_v}$$

(2)

where

τ_a = Aerosol optical thickness

$P_a(\Theta)$ = Aerosol scattering phase function

Atmospheric reflectance is the sum of the particle reflectance and molecule reflectance, R_{atm}, (Vermote, et al., (1997).

$$R_{atm} = R_a + R_r$$

(3)

Where

R_{atm} = atmospheric reflectance
R_a = particle reflectance
R_r = molecule reflectance

$$R_{atm} = \left[\frac{\tau_a P_a(\Theta)}{4\mu_s \mu_v} + \frac{\tau_r P_r(\Theta)}{4\mu_s \mu_v} \right]$$

$$R_{atm} = \frac{1}{4\mu_s \mu_v} \left[\tau_a P_a(\Theta) + \tau_r P_r(\Theta) \right]$$

(4)

The optical depth is given by Camagni and Sandroni, (1983), as in equation (5). From the equation, we rewrite the optical depth for particle and molecule as equation (6)

$$\tau = \sigma \rho s$$

(5)

Where

τ = optical depth
σ = absorption
s = finite path

$$\tau = \tau_a + \tau_r$$

(Camagni and Sandroni, 1983)

$$\tau_r = \sigma_r \rho_r s \tag{6a}$$

$$\tau_p = \sigma_p \rho_p s \tag{6b}$$

Equations (6) are substituted into equation (4). The result was extended to a three bands algorithm as equation (7) Form the equation; we found that PM10 was linearly related to the reflectance for band 1 and band 2. This algorithm was generated based on the linear relationship between τ and reflectance. Retalis et al., (2003), also found that PM10 was linearly related to τ and the correlation coefficient for linear was better that exponential in their study (overall). This means that reflectance was linear with PM10. In order to simplify the data processing, the air quality concentration was used in our analysis instead of using density, ρ, values.

$$R_{atm} = \frac{1}{4\mu_s\mu_v}\left[\sigma_a\rho_a sP_a(\Theta) + \sigma_r\rho_r sP_r(\Theta)\right]$$

$$R_{atm} = \frac{s}{4\mu_s\mu_v}\left[\sigma_a\rho_a P_a(\Theta) + \sigma_r\rho_r P_r(\Theta)\right]$$

$$R_{atm}(\lambda_1) = \frac{s}{4\mu_s\mu_v}\left[\sigma_a(\lambda_1)PP_a(\Theta,\lambda_1) + \sigma_r(\lambda_1)GP_r(\Theta,\lambda_1)\right]$$

$$R_{atm}(\lambda_2) = \frac{s}{4\mu_s\mu_v}\left[\sigma_a(\lambda_2)PP_a(\Theta,\lambda_2) + \sigma_r(\lambda_2)GP_r(\Theta,\lambda_2)\right]$$

$$P = a_0 R_{atm}(\lambda_1) + a_1 R_{atm}(\lambda_2) \tag{7}$$

Where

P	= Particle concentration (PM10)
G	= Molecule concentration
R_{atmi}	= Atmospheric reflectance, i = 1 and 2 are the band number
a_j	= algorithm coefficients, j = 0, 1, 2, … are then empirically determined.

6. Data Analysis and Results

Remote sensing satellite detectors exhibit linear response to incoming radiance, whether from the Earth's surface radiance or internal calibration sources. This response is quantized into 8-bit values that represent brightness values commonly called Digital Numbers (DN). To convert the calibrated digital numbers to at-aperture radiance, rescaling gains and biases are created from the known dynamic range limits of the instrument.

$$\text{Radiance, } L(\lambda) = \text{Bias}(\lambda) + [\text{Gain}(\lambda) \times DN(\lambda)] \tag{8}$$

where
λ = band number.
L is the radiance expressed in $Wm^{-2} sr^{-1} \mu m^{-1}$.

The spectral radiance, as calculated above, can be converted to at sensor reflectance values.

$$\rho^* = \frac{\pi L(\lambda) d^2}{E_0(\lambda)\cos\theta}$$

(9)

Where

ρ^* = Sensor Reflectance values
L (λ) = Apparent At-Sensor Radiance (Wm-2 sr-1µm-1)
d = Earth-Sun distance in astronomical units
  = {1.0-0.016729cos [0.9856(D-4)]} where (D = day of the year)
E0 (λ) = mean solar exoatmospheric irradiance (Wm-2 µm-1)
θ = solar Zenith angle (degrees)

The rescaling gain and offset values for Landsat TM 5 used in this paper are listed in Table 1 (Chander, et al. 2007) and Solar Exoatmospheric spectral irradiances are given in Table 2 (Chander and Markham, 2003).

Landsat TM 5								
Rescaling Gain and Bias								
Processing Date	Mar 1, 1984 - May 4, 2003		May 5, 2003 - Apr 1, 2007		Apr 2, 2007 – Present			
Acquisition Date	Mar 1, 1984 - May 4, 2003		May 5, 2003 - Apr 1, 2007		Mar 1, 1984 - Dec 31, 1991		Jan 1, 1992 - Present	
Band	Gain	Bias	Gain	Bias	Gain	Bias	Gain	Bias
1	0.602431	-1.52	0.762824	-1.52	0.668706	-1.52	0.762824	-1.52
2	1.175100	-2.84	1.442510	-2.84	1.317020	-2.84	1.442510	-2.84
3	0.805765	-1.17	1.039880	-1.17	1.039880	-1.17	1.039880	-1.17
4	0.814549	-1.51	0.872588	-1.51	0.872588	-1.51	0.872588	-1.51
5	0.108078	-0.37	0.119882	-0.37	0.119882	-0.37	0.119882	-0.37
6	0.055158	1.2378	0.055158	1.2378	0.055158	1.2378	0.055158	1.2378
7	0.056980	-0.15	0.065294	-0.15	0.065294	-0.15	0.065294	-0.15

Table 1. Rescaling gains and biases used for the conversion of calibrated digital numbers to spectral radiance for Landsat TM 5

Unit: ESUN = Wm-2µm-1	
Band	Landsat TM 5
1	1957
2	1826
3	1554
4	1036
5	215.0
7	80.67

Table 2. Solar Exoatmospheric spectral irradiances in Wm-2 µm-1 for Landsat TM 5

Landsat TM satellite data set was selected corresponding to the ground truth measurements of the pollution levels. The PCI Geomatica version 10.1 image processing software was used in all the analyses. The Landsat TM 5 satellite images were acquired on 15th February 2001 (Figure 6), 17th January 2002 (Figure 7), 6th March 2002 (Figure 8) and 5th February 2003 (Figure 9).

Raw digital satellite images usually contain geometric distortion and cannot be used directly as a map. Some sources of distortion are variation in the altitude, attitude and velocity of the sensor. Other sources are panoramic distortion, earth curvature, atmospheric refraction and relief displacement. So, to correct the images, we have to do geometric correction. After applying the correction, the digital data can then be used for other processing steps (Anderson, et al. 1976). Image rectification was performed by using a second order polynomial transformation equation. The images were geometrically corrected by using a nearest neighbour resampling technique. Sample locations were then identified on these geocoded images. Regression technique was employed to calibrate the algorithm using the satellite multispectral signals.

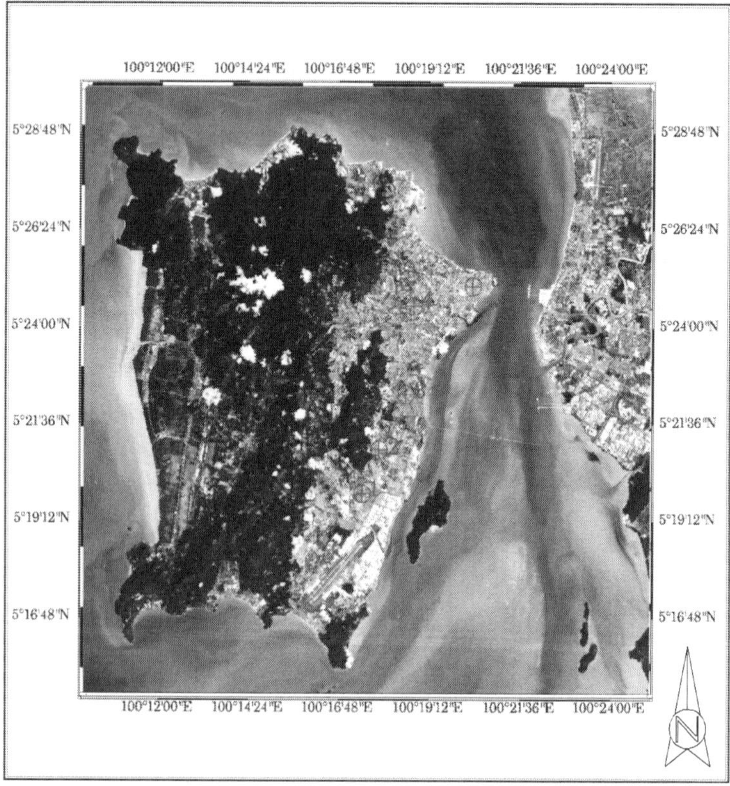

Fig. 6. Raw Landsat TM satellite image of 15th February 2001

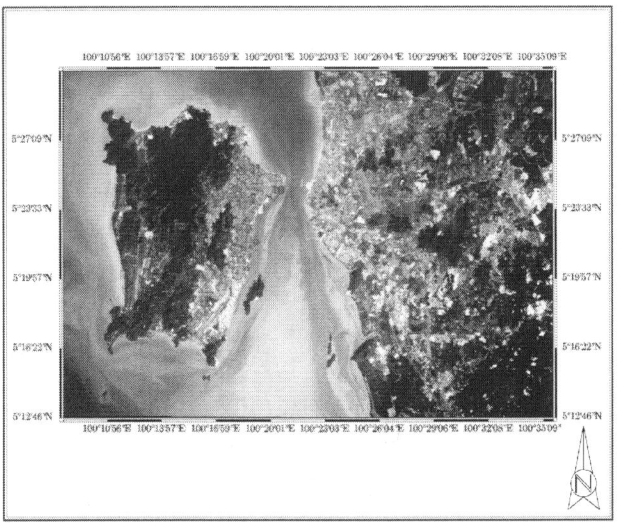

Fig. 7. Raw Landsat TM satellite image of 17th January 2002

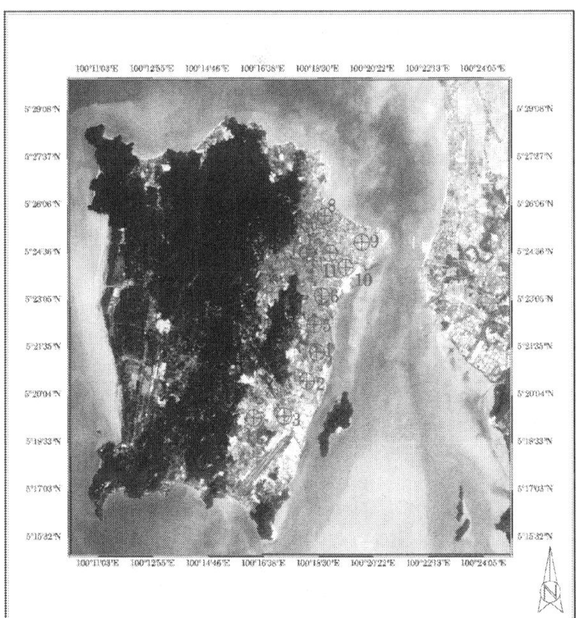

Fig. 8. Raw Landsat TM satellite image of 6th March 2002

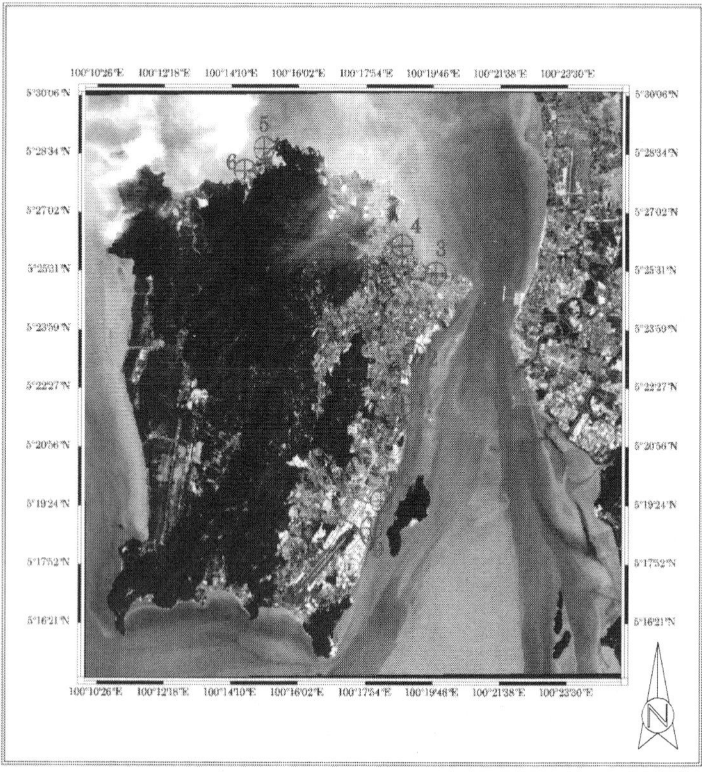

Fig. 9. Raw Landsat TM satellite image of 5th February 2003

It should be noted that the relfectance values at the top of atmospherr is the sum of the surface reflectance and atmospheric relfectance. The signals measured in each of these visible bands represent a combination of surface and atmospheric effects, usually in different proportions depending on the condition of the atmosphere. Therefore, it is required to determine the surface contribution from the total reflectance received at the sensor. In this study, we extracted the surface reflectance from mid-infrared band because the surface reflectance at various bands across the solar spectrum are correlated to each other to some extended. The surface reflectances of dark targets in the blue and red bands were estimated using the measurements in the mid-infrared band (Quaidrari and Vermote, 1999). Over a simple black target, the observed atmospheric reflectance is the sum of reflectance of aerosols and Rayleigh contributions (Equation 10). This simplification, however, is not valid at short wavelengths (less than 0.45 pm) or large sun and view zenith angles (Vermote and Roger, 1996). In this study, a simple form of the equation was used in this study (Equation 11). This equation also used by other research in their study (Popp, 2004).

$$R_s - TR_r = R_{atm} \qquad (10)$$
$$R_s - R_r = R_{atm} \qquad (11)$$

where:

Rs   = reflectance recorded by satellite sensor
Rr   = reflectance from surface references
Ratm   = reflectance from atmospheric components (aerosols and molecules)
T   = transmittance

It should be noted that the relfectance values at the top of atmosphere was the sum of the surface reflectance and atmospheric relfectance. In this study, we used ATCOR2 image correction software in the PCI Geomatica 9.1 image processing software for creating a surface reflectance image. And then the reflectance measured from the satellite [reflectance at the top of atmospheric, $\rho(TOA)$] was subtracted by the amount given by the surface reflectance to obtain the atmospheric reflectance. And then the atmospheric reflectance was related to the PM10 using the regression algorithm analysis (Equation 7). In this study, Landsat TM signals were used as independent variables in our calibration regression analyses. The atmospheric reflectances for each band corresponding to the ground-truth locations were determined. The atmospheric reflectance were determined for each band using different window sizes, such as, 1 by 1, 3 by 3, 5 by 5, 7 by 7, 9 by 9 and 11 by 11. In this study, the atmospheric reflectance values extracted using the window size of 3 by 3 was used due to the higher correlation coefficient (R) with the ground-truth data. The extracted atmospheric reflectance values were regressed with their respective ground -truth data and the proposed algorithm to obtain the regression coefficients. PM10 maps for all the images were then generated using the proposed calibrated algorithm and filtered by using a 3 x 3 pixel smoothing filter to remove random noise.
The data points were then regressed to obtain all the coefficients of equation (7). Then the calibrated algorithm was used to estimate the PM10 concentrated values for each image. The proposed model produced the correlation coefficient of 0.8 and root-mean-square error 16 $\mu g/m^3$. The PM10 maps were generated using the proposed calibrated algorithm. The generated PM10 map was colour-coded for visual interpretation [Landsat TM 5 - 15th February 2001 (Figure 10), 17th January 2002 (Figure 11), 6th March 2002 (Figure 12) and 5th February 2003 (Figure 13)]. Generally, the concentrations above industrial and urban areas were higher compared to other areas.

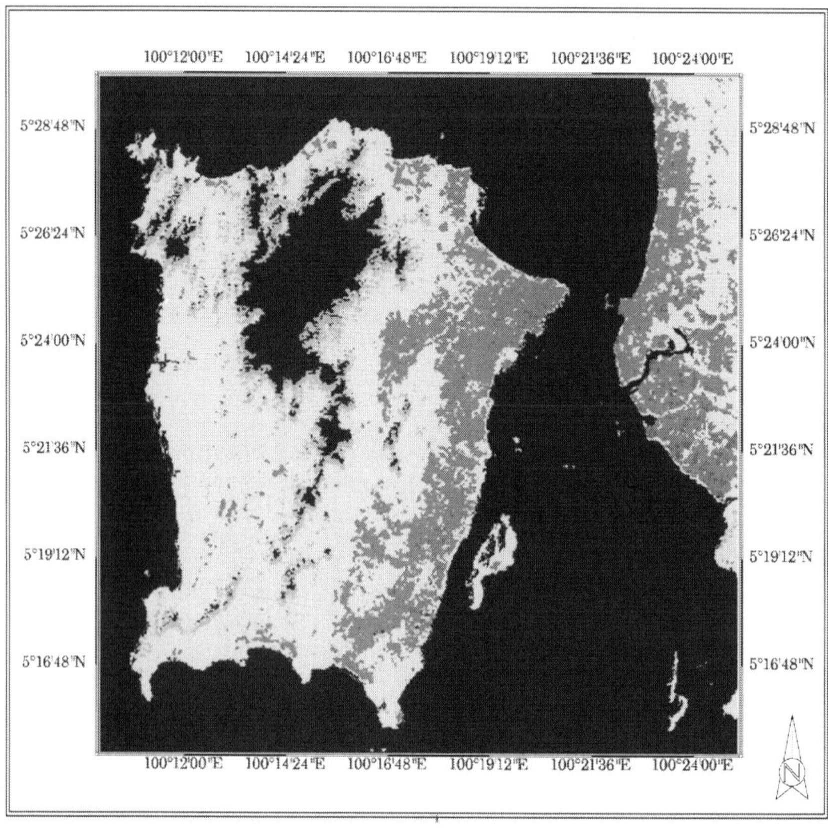

Fig. 10. Map of PM10 around Penang Island, Malaysia-15/2/2001 (Blue < 40 µg/m³, Green = (40-80) µg/m³, Yellow = (80-120) µg/m³, Orange = (120-160) µg/m³, Red = (>160) µg/m³ and Black = Water and cloud area)

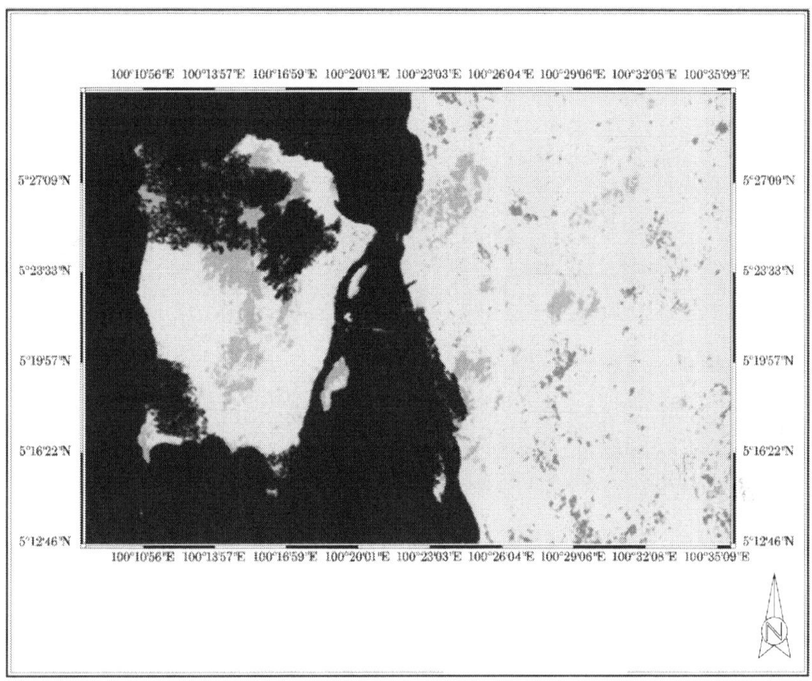

Fig. 11. Map of PM10 around Penang Island, Malaysia-17/1/2002 (Blue < 40 µg/m³, Green = (40-80) µg/m³, Yellow = (80-120) µg/m³, Orange = (120-160) µg/m³, Red = (>160) µg/m³ and Black = Water and cloud area)

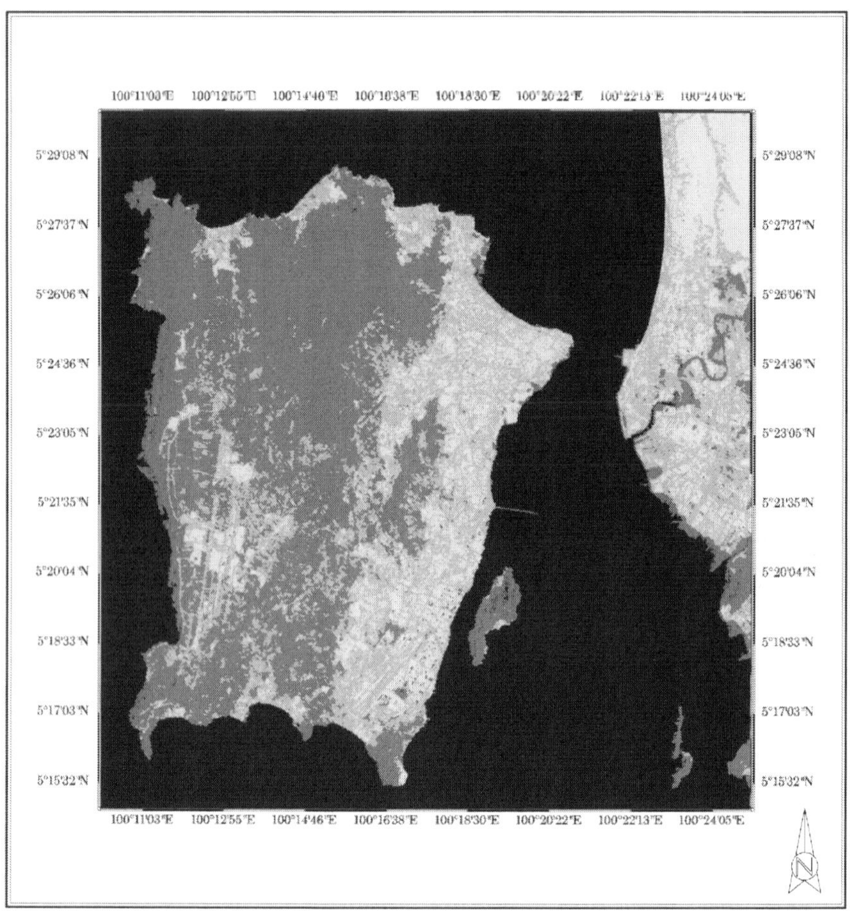

Fig. 12. Map of PM10 around Penang Island, Malaysia-6/3/2002 (Blue < 40 µg/m³, Green = (40-80) µg/m³, Yellow = (80-120) µg/m³, Orange = (120-160) µg/m³, Red = (>160) µg/m³ and Black = Water and cloud area)

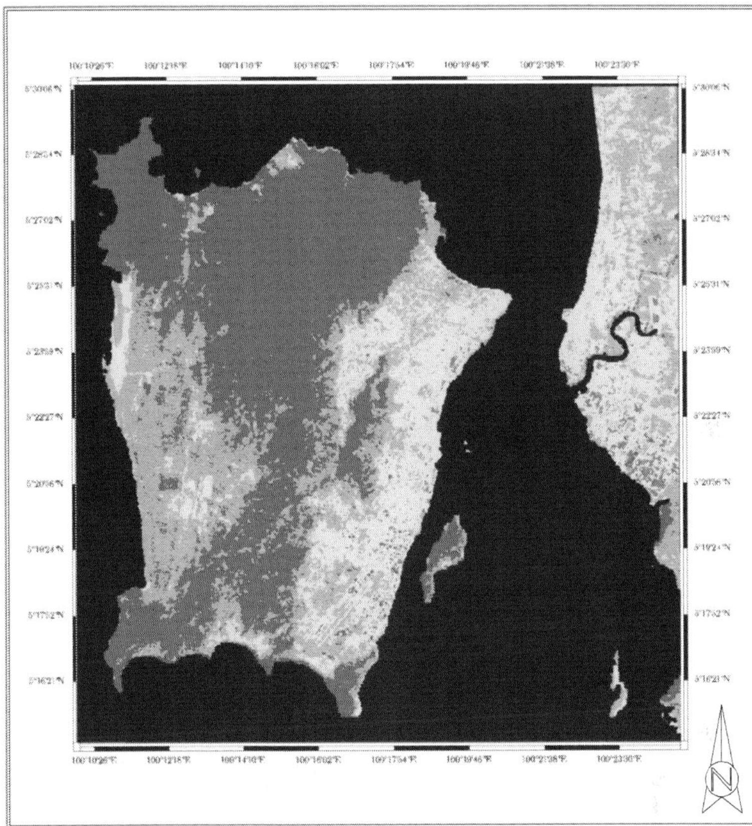

Fig. 13. Map of PM10 around Penang Island, Malaysia-5/2/2003 (Blue < 40 µg/m³, Green = (40-80) µg/m³, Yellow = (80-120) µg/m³, Orange = (120-160) µg/m³, Red = (>160) µg/m³ and Black = Water and cloud area)

7. Conclusion

This study indicates that Landsat TM satellite data can provide very useful information for estimating and mapping air pollution. The proposed algorithm is considered superior based on the values of the correlation coefficient, R=0.8 and root-mean-square error, RMS=16 µg/m³. This technique has been proved to be reliable and cost effective for such environmental study. Further study will be carried out to verify the results.

8. Acknowledgements

This project was supported by the Ministry of Science, Technology and Innovation of Malaysia under Grant 06-01-05-SF0298 " Environmental Mapping Using Digital Camera Imagery Taken From Autopilot Aircraft.", and also supported by the Universiti Sains Malaysia under short term grant " Digital Elevation Models (DEMs) studies for air quality retrieval from remote sensing data". We would like to thank the technical staff who participated in this project. Thanks are also extended to USM for support and encouragement.

9. References

Camagni. P. & Sandroni, S. (1983). Optical Remote sensing of air pollution, Joint Research Centre, Ispra, Italy, Elsevier Science Publishing Company Inc

Dubovik.; O., Holben, B.; Eck, T. F.; Smirnov, A.; Kaufman, Y. J.; King, M. D.; Tanre, D. & Slutsker, I. (2002). Variability of absorption and optical properties of key aerosol types observed in worldwide locations, American Meteorological Society, 590 – 608

Environmental Protection Service Tameside MBC Council Offices, (2008). [Online] available: http://www.tameside.gov.uk/airquality

Fauziah, Ahmad; Ahmad Shukri Yahaya & Mohd Ahmadullah Farooqi. (2006), Characterization and Geotechnical Properties of Penang Residual Soils with Emphasis on Landslides, American Journal of Environmental Sciences 2 (4): 121-128

Fukushima, H.; Toratani, M.; Yamamiya, S. & Mitomi, Y. (2000). Atmospheric correction algorithm for ADEOS/OCTS acean color data: performance comparison based on ship and buoy measurements. Adv. Space Res, Vol. 25, No. 5, 1015-1024

Fundamentals of Remote Sensing, [Online] available:
http://www.ccrs.nrcan.gc.ca/resource/tutor/fundam/pdf/fundamentals_e.pdf

Jensen, J. R. (2006), Remote Sensing of the Environment: An Earth Resource Perspective, 2nd Ed., Upper Saddle River, NJ: Prentice Hall. [Online] available:, http://www.cas.sc.edu/geog/Rsbook/Lectures/Rse/index.html

Kaufman,; Y. J., Tanr, D. C.; Gordon, H. R.; Nakajima, T.; Lenoble, J.; Frouins, R.; Grassl, H.; Herman, B. M..; King, M. D. & Teillet, P. M. (1997). Passive remote sensing of tropospheric aerosol and atmospheric correction for the aerosol effect. Journal of Geophysical Research, 102 (D14):16,815-16,830

Liu, C. H.; Chen, A. J. ^ Liu, G. R. (1996). An image-based retrieval algorithm of aerosol characteristics and surface reflectance for satellite images, International Journal Of Remote Sensing, 17 (17), 3477-3500

King, M. D.; Kaufman, Y. J.; Tanre, D. & Nakajima, T. (1999). Remote sensing of tropospheric aerosold form space: past, present and future, Bulletin of the American Meteorological society, 2229-2259

Penner, J. E.; Zhang, S. Y.; Chin, M.; Chuang, C. C.; Feichter, J.; Feng, Y.; Geogdzhayev, I. V.; Ginoux, P.; Herzog, M.; Higurashi, A.; Koch, D.; Land, C.; Lohmann, U.; Mishchenko, M.; Nakajima, T.; Pitari, G.; Soden, B.; Tegen, I. & Stowe, L. (2002). A Comparison of Model And Satellite-Derived Optical Depth And Reflectivity. [Online} available: http://data.engin.umich.edu/Penner/paper3.pdf

Popp, C.; Schläpfer, D.; Bojinski, S.; Schaepman, M. & Itten, K. I. (2004). Evaluation of Aerosol Mapping Methods using AVIRIS Imagery. R. Green (Editor), 13th Annual JPL Airborne Earth Science Workshop. JPL Publications, March 2004, Pasadena, CA, 10

Quaidrari, H. dan Vermote, E. F. (1999). Operational atmospheric correction of Landsat TM data, Remote Sensing Environment, 70: 4-15.

Retalis, A.; Sifakis, N.; Grosso, N.; Paronis, D. & Sarigiannis, D. (2003). Aerosol optical thickness retrieval from AVHRR images over the Athens urban area, [Online] available:
http://sat2.space.noa.gr/rsensing/documents/IGARSS2003_AVHRR_Retalisetal_web.pdf

UNEP Assessment Report, Part I: The South Asian Haze: Air Pollution, Ozone and Aerosols, [Online] available:
http://www.rrcap.unep.org/issues/air/impactstudy/Part%20I.pdf

Ung, A.; Weber, C.; Perron, G.; Hirsch, J.; Kleinpeter, J.; Wald, L. & Ranchin, T. (2001a). Air Pollution Mapping Over A City – Virtual Stations And Morphological Indicators. Proceedings of 10th International Symposium "Transport and Air Pollution" September 17 - 19, 2001 – Boulder, Colorado

Ung, A.; Wald, L.; Ranchin, T.; Weber, C.; Hirsch, J.; Perron, G. & Kleinpeter, J. (2001b). Satellite data for Air Pollution Mapping Over A City- Virtual Stations, Proceeding of the 21th EARSeL Symposium, Observing Our Environment From Space: New Solutions For A New Millenium, Paris, France, 14 – 16 May 2001, Gerard Begni Editor, A., A., Balkema, Lisse, Abingdon, Exton (PA), Tokyo, pp. 147 – 151, [Online] available:
http://www-cenerg.cma.fr/Public/themes_de_recherche/teledetection/title_tele_air/title_tele_air_pub/satellite_data_for_t

Sifakis, N. & Deschamps, P.Y. (1992). Mapping of air pollution using SPOT satellite data, Photogrammetric Engineering & Remote Sensing, 58(10), 1433 – 1437

Sifakis, N. I.; Soulakellis, N. A. & Paronis, D. K. (1998). Quantitative mapping of air pollution density using Earth observations: a new processing method and application to an urban area, International Journal Remote Sensing, 19(17): 3289 – 3300

The Digital Awakening-Haze in Penang, (2009). [Online] available:
http://www.petertan.com/blog/2005/08/12/haze-in-penang/

The Digital Awakening--Haze in Penang Update - 10.48am, [Online] available:
http://www.petertan.com/blog/2005/08/13/haze-in-penang-update-1048am/

Vincentchow, (2009). Available Online:
http://www.vincentchow.net/289/haze-in-malaysia

Vermote, E. & Roger, J. C. (1996). Advances in the use of NOAA AVHRR data for land application: Radiative transfer modeling for calibration and atmospheric correction, Kluwer Academic Publishers, Dordrecht/Boston/London, 49-72

Vermote, E.; Tanre, D.; Deuze, J. L.; Herman, M. & Morcrette, J. J. (1997). 6S user guide Version 2, Second Simulation of the satellite signal in the solar spectrum (6S), [Online] available:
http://www.geog.tamu.edu/klein/geog661/handouts/6s/6smanv2.0_P1.pdf

6

Some Features of The Volume Component of Radar Backscatter From Thick and Dry Snow Cover

Boris Yurchak
University of Maryland, Baltimore County, GEST
USA

1. Introduction

Radar monitoring of thick snow cover in polar regions with optical thickness (a product of the depth of the snow and the extinction coefficient) of order 1 or more from elevated above-ground and space platforms is of great importance for registration and for understanding glaciology processes caused by climatic change. Applicable to this issue, the volume component of the backscatter coefficient has a notable contribution to the total backscatter (Noveltis, 2005). Although the idea of the radar-cross section (RCS) term and its derivatives, like the backscatter coefficient, aims to separate as far as possible the sensor (radar) and target parameters, this distinction can rarely be fulfilled in the case of spatially extended geophysical targets (SEGT), such as atmospheric clouds and rain, as well as the thick snow cover that is the focus of this paper. Due to this feature, the analysis of backscatter from SEGT strongly depends on the relationship between the technical and physical-geometrical properties of the radar and target, respectively. The main parameters that govern the radar-target configuration for snow sounding are wavelength, antenna characteristics, pulse duration, sounding direction, extinction coefficient and the geometrical depth of the snow. A correct assessment of the volume component of the backscatter coefficient and an understanding of the realm of applicability of any backscattering model is possible only when the size of the scattering volume within a snow slab is known. This parameter depends on the factors mentioned above and should be distinguished for different situations in the practice of radar sounding of snow-covered terrain by Synthetic Aperture Radar (SAR), a scatterometer or an altimeter. This work attempts, at first, to determine the radar-target configurations inherent to volume scattering estimations of the thick snow cover under the different radar applications mentioned above. Next, we analyze the range of applications of the incoherent approach for backscatter magnitude estimation, currently one of the main techniques for snow characteristics assessment. The simplest incoherent approach, based on the so-called "particle" or "discrete" approximation, leads to the dependence of backscatter on the sixth moment of the particle size distribution function (PSDF) and the mean amount of particles in the scattering volume (Siegert & Goldstein, 1951; Battan, 1959; Ulaby et al., 1982). For a medium with losses (such as a thick snow slab) the modification of this approach is referred to as the semi-empirical model (Attema &

Ulaby, 1978; Ulaby et al., 1982; Ulaby et al., 1996). There is some evidence of less backscatter occurring than expected by virtue of the conventional reflectivity factor and the backscatter coefficient for incoherent scatter from homogeneous thick snow slab (e.g., Rott et al., 1993). For dense media (mean distance between particles is less than the wavelength) there are many references in the literature that the backscatter is determined by media inhomogeneities (Naito & Atlas, 1967; Gossard & Srauch, 1983; Fung, 1994). To attempt to evaluate the contributions of these inhomogeneities, the radiative transfer (RT) and the dense media radiative transfer (DMRT) models were developed (e.g., Ulaby et al., 1982; Fung, 1994; Tsang et al., 2007). A review of these models and their modifications provided by Noveltis (2005) stated that in certain aspects, these approaches showed some successes. However, these models have not the close-form solutions that make difficult to use them and analyze results obtained. Kendra et al. (1998), based on experiments with artificial snow of varying depths, concluded that both conventional and dense-medium radiative transfer models fail to adequately explain the observed results.

Finding an appropriate and relatively simple approach to calculate the volume component of backscatter to explain the observable deviations from the classical (incoherent) model is therefore a relevant task. The majority of previous studies, focusing on measurements of a dry snow at temperate latitudes, where the snow depth seldom exceeds 1 m and the corresponding optical thickness is much less than 1, have led to the conclusion that such an approach is not feasible because of the much weaker interaction of the electromagnetic radiation of radar wavebands with ice particles within the snow compared with the backscatter from the soil beneath. The current study focuses on the case of the Greenland ice sheet, however, where the depth of snow significantly exceeds that found in temperate latitudes. This favorable condition, in conjunction with the recent finding that the incoherent approach is only a specific case of a more comprehensive description of electromagnetic wave interactions with spatially-extended individual scatterers (Yurchak, 2009), provides a reason for more detailed investigations of the possibilities of the semi-empirical model for interpreting the observable features of radar backscatter from thick snow cover.

2. Condition of complete burial of the probing pulse into snow medium

To estimate the volume component of backscatter from a snow slab, it is first necessary to understand whether the probing pulse is completely buried within the snow slab or whether the illuminated volume is only part of the probing pulse volume. In weather radar meteorology, applicable to rain and clouds, this problem is known as "partial (or incomplete) beam filling," with a corresponding factor included in the weather radar equation (e.g., Clift, 1985). For sounding thick snow cover, this issue is practically not discussed and is different compared with the sounding of meteorological targets. Complete burial depends on the sounding configuration and snow slab depth. The main criterion for complete burial is the location of the pulse scattering volume with angular size equal to the angular antenna beam width and with spatial length equal to one half of the actual spatial duration of the transmitted pulse into a snow layer. Thus, for complete burial, the angular and radial sizes of the backscatter volume should be matched with the snow slab depth, which is usually known only roughly for a particular geographical region. This condition poses problems for practical applications. Nevertheless, to better understand possible situations where complete burial is feasible, the requirements for the angular antenna beam size and the probing pulse duration will be analyzed separately for different major sounding configurations.

2.1. Condition of complete burial of the angular (transverse) size of the pulse volume

2.1.1. Oblique sounding, flat surface, plane wavefront

The scheme of this configuration is shown in Figure 2.1. Here, and everywhere below, the snow slab is assumed to have limited depth and an unbounded horizontal extent.

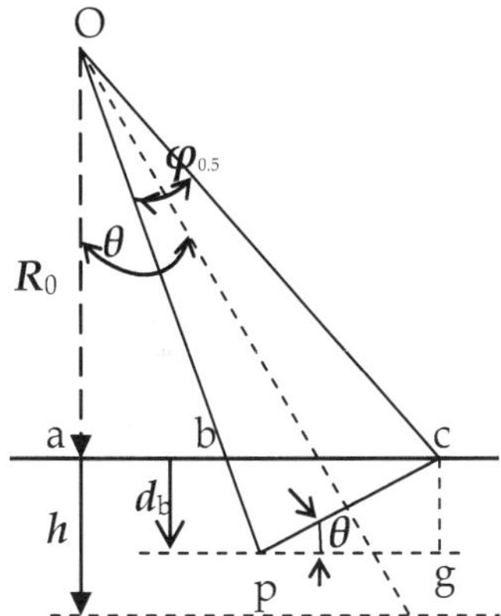

Fig. 2.1. Scheme of sounding for flat surface and plane wavefront

Mathematically, the condition for complete burial in this case is:

$$d_b \le h \tag{2.1}$$

where d_b is the burial depth of the advance point (p) of pulse edge (leading or trailing) when the lagging point (c) reaches the surface and h is the snow depth. As follows from the geometry of Figure 2.1:

$$d_b = cg = pc \times \sin \angle cpg \; ; \; \angle cpg = \theta \; ; \; pc \approx 2 \frac{R_0}{\cos \theta} tg \frac{\varphi_{0.5}}{2} \text{, and hence:}$$

$$d_b = 2 \frac{R_0}{\cos \theta} tg \frac{\varphi_{0.5}}{2} \sin \theta \tag{2.2}$$

For $\varphi_{0.5} \ll 1$ one can state $tg \frac{\varphi_{0.5}}{2} \approx \frac{\varphi_{0.5}}{2}$, and condition (2.1) has the form:

$$\varphi_{0.5} \le \frac{h}{R_0} \frac{1}{tg \theta} \tag{2.3}$$

If $h\sim10$ m and R_0=800 km, then $\varphi_{0.5} << \dfrac{1.25}{tg\theta}10^{-5}$ and for $\theta\sim40^0$, for example, the burial

condition (2.1) is satisfied at roughly $\varphi_{0.5}<<0.001^0$. Obviously, this condition can not be fulfilled for any space-based radar system with a real aperture, but it is possible for the SAR if one assumes that the effective synthetic aperture radar beam illuminating an element of spatial resolution Δx on a flat surface can be equal to an extremely narrow, pencil-like beam with $\varphi_{0.5} \sim \dfrac{\Delta x}{R}$. For example, RadarSAT-1 has $\Delta x \sim 12.5$ m with an orbit height of

approximately 800 km and thus, $\varphi_{0.5}\sim0.9*10^{-3}$ degrees. If one assumes that the typical size of the main lobe of the conventional antenna pattern is of order $\sim1^0$, the complete burial of the transverse size of the pulse volume is possible for $\dfrac{h}{R_0} \geq \varphi_{0.5}tg\theta \sim 2\cdot10^{-2}tg\theta$. This condition

can definitely be fulfilled for airborne and above-surface elevated radars. The estimate obtained above should be considered only a rough approximation, because at nadir sounding (θ=0) and unbounded horizontal extent of snow slab, the wavefront can not be considered planar for the assessment provided. A more precise estimate is given below.

2.1.2. Oblique sounding, flat surface, spherical wavefront

A real wavefront within a snow slab has a spherical shape. The impact of this shape on the estimation of d_b as the look angle θ (Figure 2.1) decreases begins at the moment when the line tangent to the spherical front at point p coincides with the horizontal line pg, Figure 2.2. This situation takes place when θ becomes equal to the angle between pc and the tangent:

$$\theta = \gamma \qquad (2.4)$$

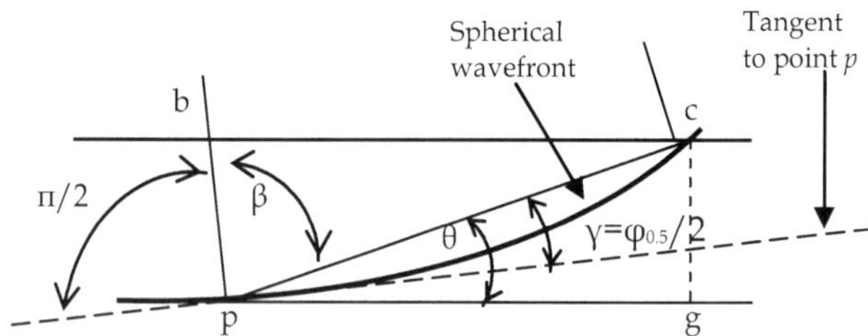

Fig. 2.2. To an assessment of sphericity impact

As follows from geometrical sketch, the auxiliary angle β is equal to: $\beta = \dfrac{\pi - \varphi_{0.5}}{2}$ and, thus,

$\gamma = \dfrac{\pi}{2} - \beta = \dfrac{\varphi_{0.5}}{2}$. Therefore, the impact of sphericity should be taken into account only for

small look angles, when

$$\theta \le \frac{\varphi_{0.5}}{2} \tag{2.5}$$

Thus, the estimate (2.3), provided in the previous subsection, is valid for $\theta > \frac{\varphi_{0.5}}{2}$. An estimate for configurations close to nadir sounding will be carried out in the next subsection.

2.1.3. Nadir sounding, flat surface spherical wavefront

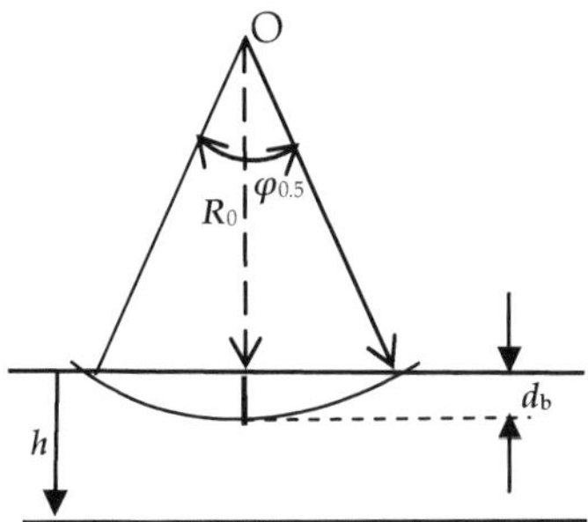

Fig. 2.3. Schematic of sounding in nadir direction

This configuration is shown schematically in Figure 2.3 and the corresponding condition for complete burial is:

$$d_b^{(flt)} \le h \tag{2.6}$$

From the geometrical configuration and $\varphi_{0.5} \ll 1$, it follows that:

$$d_b^{(flt)} = R_0 \left(\frac{1}{\cos \dfrac{\varphi_{0.5}}{2}} - 1 \right) = R_0 \frac{2\sin^2 \dfrac{\varphi_{0.5}}{4}}{\cos \dfrac{\varphi_{0.5}}{2}} \approx \frac{1}{8} R_0 \varphi_{0.5}^2 \tag{2.7}$$

This estimate can be accepted for elevated platforms (for example, airborne radars and those mounted above the surface) but needs to be analyzed further for space platforms due to the sphericity of the surface, as the corresponding area illuminated on the Earth's surface by conventional radar is large.

2.1.4. Nadir sounding, spherical surface, spherical wavefront

In the spherical surface approach, the depth of complete burial in altimeter mode is more than that for a flat surface described above by an increment Δz in the center of the beam. This effect is illustrated in Figure 2.4, and can be written as:

Obviously, that the depth-limited mode is inherent to space based altimeter and scatterometr and the beamwidth-limited mode is realized for SAR. If the "significant part" in the definition of this mode is replaced with a more exact term, the "determined part," the condition (2.15) can be modified and written in the form:

$$\varphi_{0.5} \leq 2\sqrt{2\frac{h}{R_0}\left(1-\frac{\Delta h}{h}\right)} \quad \text{if } h \leq D_p$$

or   (2.15a)

$$\varphi_{0.5} \leq 2\sqrt{2\frac{D_p}{R_0}\left(1-\frac{\Delta h}{D_p}\right)} \quad \text{otherwise}$$

where $\dfrac{\Delta h}{h}$ or $\dfrac{\Delta h}{D_p}$ is the relative part of the scattering volume in which the beamwidth-limited mode is fulfilled. For example, if 90% of the scattering volume ($\dfrac{\Delta h}{h}$ =0.9) is under the beamwidth-limited mode, the condition (2.15a) is $\varphi_{0.5} \leq 2\sqrt{0.2\dfrac{h}{R_0}}$ for $h \leq D_p$.

2.2. Condition of complete burial of radial size of the pulse volume
Suppose that some area of snow cover is illuminated by a radar located in orbit as shown in Figure 2.7. For simplicity, this scenario is depicted with a plane wavefront and flat surface.

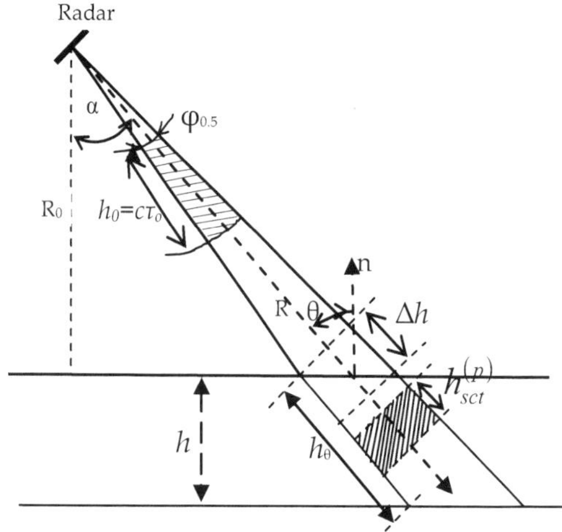

Fig. 2.7. Assessment of the complete burial of radial size of the probing pulse

Let us denote the height of the orbit as R_0, the distance from the radar to the center of the area as R, the anglular width of the main lobe of the antenna pattern as $\varphi_{0.5}$ and, the look angle as a. In addition, we assume that the cross-section of the main lobe is the circular. The illuminated area is assumed to be flat horizontally, and the bottom boundary surface of the snow slab is also flat and parallel to the top surface. Thus, the incidence angle, θ is equal to the look angle, a ($\theta=a$).

The main scaling parameters determining radial propagation are:

1) The radial length of the pulse scattering volume, equal to half of its spatial extent in the snow medium:

$$h_{sct}^{(p)} = \frac{v\tau_0}{2} \tag{2.16}$$

where $v = \dfrac{c}{\sqrt{\varepsilon'}}$ is the wave propagation speed in snow, c is the wave propagation speed in the air, ε' is the real part of the snow permittivity and τ_0 is the duration of the probing pulse;

2) A one-way path in the snow, where the incidence power is decreased by "e" times, usually called the "penetration depth" in the literature

$$D_p = k_e^{-1} \tag{2.17}$$

where k_e is the extinction coefficient, which characterizes the attenuation properties of the medium due to scattering and absorption. One important remark is necessary. Since radar sounding of snow is often performed in off-nadir mode, and the main lobe of the antenna pattern has a finite angular size, it should be

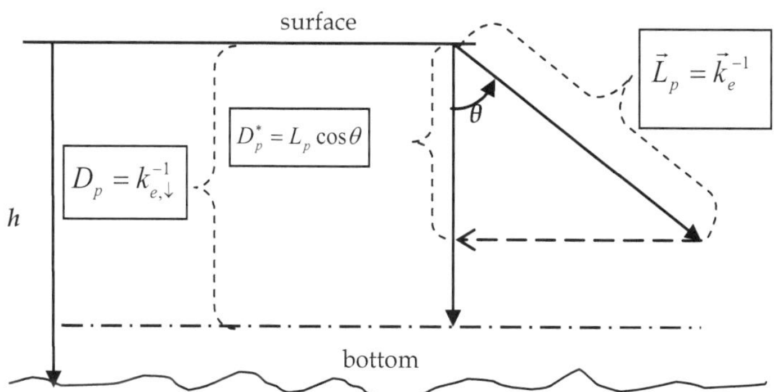

Fig. 2.8. Relationships between the penetration depth (D_p), penetration path (L_p) and propagation depth (D_p^*) in a snow slab

underlined that the extinction coefficient and the "penetration depth" are measured along the direction of the wave (ray) propagation and not only to the vertical. Therefore, calling the term $\propto k_e^{-1}$ as the "…depth" should be considered a little bit confusing. A more appropriate name for this term, from a physical point of view, is the "penetration path" (L_p), keeping in mind that the penetration depth (D_p) is its value in the vertical direction. The depth reached by an electromagnetic wave propagating at an angle θ to the vertical and attenuating by "e" times represents the vertical component of the penetration path, and can be called the propagation depth $\left(D_p^*\right)$, as illustrated in Figure 2.8. Thus, from this point onward in this paper, the following definitions and relationships are used:

$$\text{Penetration path: } \vec{L}_p = \vec{k}_e^{-1} \tag{2.18}$$

$$\text{Propagation depth: } D_p^* = L_p \cos\theta \tag{2.19}$$

$$\text{Penetration depth: } D_p = k_{e,\downarrow}^{-1} \tag{2.20}$$

where $k_{e,\downarrow}$ is the extinction coefficient in the vertical direction. In general, in inhomogeneous medium, the penetration path is a function of distance and direction $L_p = f(R,\theta,\psi)$, where ψ is the azimuthal angle. For homogenous medium $D_p = L_p$. We consider the probing pulse to be short if $h_{sct}^{(p)} \le L_p$; otherwise it is considered long. Obviously, the mode of sounding with current wave (CW) always belongs to the long probing pulse configuration. Also one should distinguish a case with fully scattering snow slabs, when the geometrical snow depth is equal to or less than the propagation depth ($h \le D_p^*$). Otherwise, one has a case of partially (because the wave does not penetrate to the bottom) scattering slow slab. The condition for complete burial, as it follows from Figure 2.7, is

$$h_{sct}^{(p)} \le h_\theta - \Delta h \tag{2.21}$$

where $h_\theta = \dfrac{h}{\cos\theta}$ is the slant snow depth and $\Delta h = 2R_0 tg \dfrac{\varphi_{0.5}}{2} \dfrac{tg\theta}{\cos\theta} \approx R_0 \varphi_{0.5} \dfrac{tg\theta}{\cos\theta}$.

All of the above conditions are summarized in Table 2.1.

#	Condition	Description	Comments
1	$h_{sct}^{(p)} \le L_p$	Short/long probing pulse (P.P.)	Complete burial is satisfied for:
2	$h \le D_p^*$ or $h_\theta \le L_p$	Fully/partially scattering snow slab	a) short P.P. and fully scattering snow slab; b) long P.P. and partially scattering snow slab
3	$h_{sct}^{(p)} \le h_\theta - \Delta h$	Complete burial of the radial size of the probing pulse	

Table 2.1. Conditions for complete burial of the radial size of the probing pulse

Concluding remarks:
If the conditions for complete burial of the radial size of the probing pulse are completed, it can at least be said that one sample of the return signal is formed by the scattering volume

$$V_{sct} = h_{sct}^{(p)} A_{ill} \qquad (2.22)$$

where A_{ill} is the illuminated base of the scattering volume. If, in addition, the conditions for complete burial of the transverse size of the probing pulse (see previous subsection) are also satisfied, that the scattering volume is determined by the entire pulse volume:

$$V_{sct} = V_p = h_{sct}^{(p)} \pi \left(R_0 \frac{\varphi_{0.5}}{2} \right)^2 \qquad (2.23)$$

Otherwise, under $\Delta R \leq \frac{1}{2} R_0 \left(\frac{\varphi_{0,5}}{2} \right)^2$,

$$V_{sct}(\Delta R) \approx h_{sct}^{(p)} 2\pi R_0 \Delta R \qquad (2.24)$$

2.3. An assessment of the scattering volume under incomplete burial condition

If the parameters of the probing pulse and of the snow slab do not satisfy the conditions presented in the table above, the radial size of the scattering volume is determined by the geometry of the snow slab. Several practically important cases for practical application are discussed below.

2.3.1. Flat surface, plane wavefront, long probing pulse and fully scattering snow slab

In this scenario, the illuminated area, A_{ill} on the snow cover (Figure 2.1) is an ellipse, with the minor semi-axis equal to the radar beam cross-section radius $r_{min} = \left(R \frac{\varphi_{0.5}}{2} \right)$ and major semi-axis the same divided by the cosine of the incidence angle: $r_{mjr} = {r_{min}}/{\cos\theta}$. Thus:

$$A_{ill} = \pi \left(\frac{R\varphi_{0.5}}{2} \right)^2 \frac{1}{\cos\theta} \qquad (2.25)$$

If one uses the height of the satellite orbit $R_0 = R\cos\theta$, (2.23) transforms to:

$$A_{ill} = \pi \left(\frac{R_0\varphi_{0.5}}{2} \right)^2 \frac{1}{\cos^3\theta} \qquad (2.26)$$

The pattern of the scattering volume for this case is depicted in Figure 2.9, where the scattering volume is bounded by the surface and bottom planes.

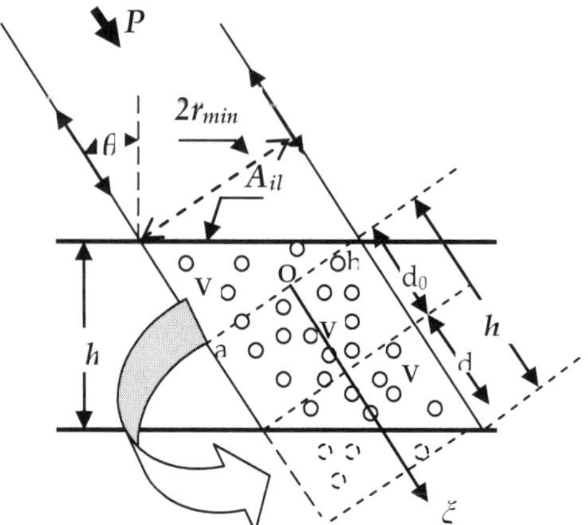

Fig. 2.9. An assessment of the scattering volume for a fully scattering snow slab

In this case, the size of the scattering volume is the sum of the volume of a circular cylinder V_0 with cylindrical element d_0 and two similar volumes of the truncated cylinders V_1 that mutually add up to a completed circular cylinder with cylindrical element d_1 (the bases of all volumes are the same and equal to $A_{ill} \cos\theta$):

$$V_{sct} = V_0 + 2V_1 = A_{ill} \cos\theta(d_0 + d_1) \tag{2.27}$$

Because $d_0 + d_1 = h_\theta$, we have, taking expression (2.25) into account:

$$V_{sct} = A_{ill} h = \pi h_\theta \left(\frac{R\varphi_{0.5}}{2} \right)^2 \tag{2.28}$$

Thus, the equivalent scattering volume is an elliptic cylinder with a base equal to the illuminated area on a surface and with a height element equal to the slab depth (under the assumption that the slab depth is less than the half of the spatial duration of the probe radar pulse). This result was obtained by Matzler (1987).

The duration of the SAR probing pulse duration is equal to several tens of microseconds (e.g., 42 μs for RadarSAT-1 and 37.1 μs for ERS-1). Due to frequency chirp, the compressed probing pulse duration decreases by many times, resulting in a volume radial size equal to only several meters (5-13 m for RadarSAT-1 and 9.7 m for ERS-1, for example). Numerical data are provided based on Alaska Satellite Facility documents ("RadarSAT-1 Standard Beam SAR Images", 1999; and "ERS-1 and ERS-2 SAR Images", 1996).

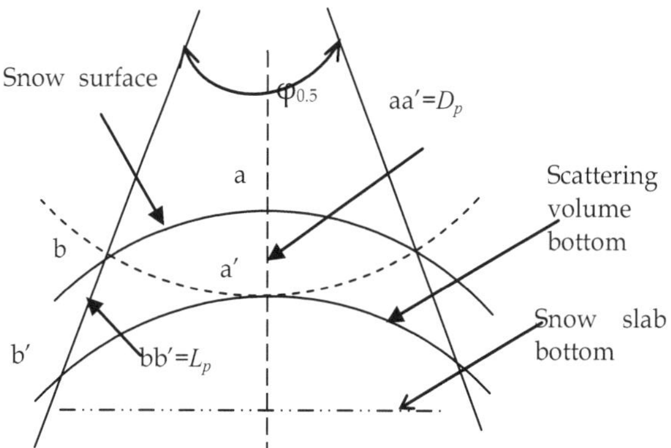

Fig.2.10. Cross-section of the scattering volume for a spherical surface and wavefront. The position of the flattened wavefront at the moment when its central point (a$'$) reaches the penetration depth is depicted with a convex dashed line

Because SAR images represent the backscatter pulse train from the entire path of a wave into a snow slab for any look angle as a point on a surface of some mean brightness (return power), the radial size of the scattering volume can be considered equal to either the slant size of the snow layer or the penetration path (whichever is smaller). That is, although the SAR pulse has a finite compressed spatial duration; it works like a long pulse due to the absence of radial discrimination in the sense of conventional radar terminology.

2.3.2. Long probing pulse, spherical wave, spherical surface, partially scattering slab

For this case, the bottom of the scattering volume is the geometrical placement of points located at a distance L_p from the snow surface along the family of rays within the solid angle of the main antenna lobe, as shown in Figure 2.10. The defining feature of this configuration is that the central and peripheral points of the spherical wavefront pass the length L_p at different times, resulting in different radii of curvature for the top and bottom bounded surfaces.

3. Volume component of the backscatter coefficient

3.1. Semi-empirical model for flat surface, plane wave and long probing pulse

Based on the considerations provided above, one can formulate an estimation of the volume component of the backscatter within the framework of the so-called semi-empirical model (Attema & Ulaby, 1978), as for the case of incomplete burial of the probing pulse with a flat surface, plane wave and a long probing pulse, which better fits the conditions for sounding of a thick snow slab with SAR.

The total radar backscatter from the illuminated area is composed of four components (Fung, 1994):

$$\sigma_t = \sigma_{as} + \sigma_s + \sigma_g + \sigma_{gv} \tag{3.1}$$

where σ_{as} is the radar cross-section (RCS) due to backscatter from the air/snow (top) interface, σ_s is the RCS due to backscatter from the snow volume, σ_g is the RCS due to backscatter from the snow/ground (bottom) interface and σ_{gv} is the RCS due to rescattering between ground and snow volume irregularities. As was summarized by Koskinen (2001), only the volume component is essential for dry snow. Therefore the essential portion of the total backscatter consists of the volume component and the contribution of the ground beneath:

$$\sigma_t = \sigma_s + \sigma_g \tag{3.2}$$

Obviously, the radar becomes sensitive to the properties of the snow only when $\sigma_s \gg \sigma_g$. In accordance with Ulaby et al. (1982), the general form for the backscatter coefficient from a surface is:

$$\sigma_t^0 = \frac{\sigma_t}{A_{ill}} \tag{3.3}$$

This notation presupposes that the illuminated areas on the slab top and on the ground (bottom of the slab) are the same. This assumption applies for small angle divergences of the radar beam, as is valid for strong directed antennae, low refraction on the air-snow interface and a ground surface that is flat and parallel to the surface of the snow slab. Accordingly, the volume component of the backscatter coefficient is:

$$\sigma_s^0 = \frac{\sigma_s}{A_{ill}} \tag{3.4}$$

In the incoherent approach, $\sigma_s = \sum_{i=1}^{N} \sigma_i$; i.e., σ_s represents the total radar cross-section (RCS) of N scatterers contributing backscatter from a volume $V_{sct} = A_{ill} h$ of a snow slab (see subsection 2.3.1). Due to attenuation of the electromagnetic wave upon propagation within a snow mass, the total RCS of the snow slab is equal to:

$$\sigma_s = \sum_{i=1}^{N} \sigma_i \alpha_i \tag{3.5}$$

where the attenuation coefficient, inherent to the i-th particle with a distance of ξ_i from the coordinate origin O, is equal to:

$$\alpha_i = \exp\left(-2k_e \xi_i\right) \tag{3.6}$$

where k_e is the extinction coefficient. The factor α takes into account the two-way distance of forth and back wave propagation. The assessment of the scattering volume described above relates only to the absolute value of the volume. In the case of wave directed propagation and, consequently, directed attenuation, the summation should be performed along the propagation axis, ξ. Because $N >> 1$, the summation in (3.5) can be replaced by integration. As was mentioned above, this integration should be carried out along the direction of wave propagation, i.e., along the ξ axes. Small changes in the propagation direction at the air/snow interface are ignored due to minor differences in the corresponding refraction coefficients. Due to the random spatial distribution of scatterers within the scattering volume, this discussion considers only the mean backscatter characteristics averaged over several illuminated areas, with particles having independent spatial locations and RCSs.

$$\langle \sigma_s \rangle = \left\langle \sum_{i=1}^{N} \sigma_i \alpha_i \right\rangle = \left\langle \int_0^{h_\theta} \sigma(\xi)\alpha(\xi)d\xi \right\rangle = \int_0^{h_\theta} \langle \sigma(\xi) \rangle \alpha(\xi)d\xi \tag{3.7}$$

where $\sigma(\xi)$ is the running RCS (by the unit of a distance along axis ξ, dimension is $\left[\dfrac{L^2}{L} \right]$) of the scattering volume. Assuming that the attenuation properties of snow remain the same along the propagation path $\left(k_e \neq f(\xi) \right)$, one can write:

$$\alpha(\xi) = \exp(-2k_e\xi) \tag{3.8}$$

If the scatterers' RCSs are independent of their locations within the scattering volume (condition of incoherent approach), we get:

$$\langle \sigma_s \rangle = \langle \sigma \rangle \int_0^{h_\theta} \alpha(\xi)d\xi \tag{3.9}$$

where $<\sigma>$ is the mean running RCS of the scattering volume.
Based on (3.4) and taking into account (3.9), the mean backscatter coefficient can be written as:

$$\langle \sigma_s^0 \rangle = \frac{\langle \sigma_s \rangle}{A_{ill}} = \frac{\langle \sigma \rangle}{A_{ill}} \int_0^{h_\theta} \alpha(\xi)d\xi \tag{3.10}$$

Taking into account that h_θ is the radial size of the scattering volume (along axis ξ), $\langle \sigma_s^* \rangle = \langle \sigma \rangle \cdot h_\theta$ is the mean total RCS of snow (ice) particles within the scattering volume (an oblique cylinder) while ignoring the attenuation and $\langle \sigma_v \rangle = \dfrac{\langle \sigma_s^* \rangle}{V}$ is the corresponding mean volume specific backscatter coefficient, we can transform the ratio $\langle \sigma \rangle \big/ A_{ill}$ to the form:

$$\frac{\langle \sigma \rangle}{A_{ill}} = \frac{\langle \sigma \rangle}{A_{ill}} \cdot \frac{h_\theta}{h_\theta} = \frac{\langle \sigma \rangle h_\theta}{A_{ill}h} \cos\theta = \frac{\langle \sigma_s^* \rangle}{V} \cos\theta = \langle \sigma_v \rangle \cos\theta \tag{3.11}$$

Therefore, substituting (3.11) into (3.10), we can state:

$$\left\langle \upsilon_s^0 \right\rangle - \left\langle \sigma_v \right\rangle \cos\theta \int_0^{h_\theta} \alpha(\xi)d\xi \tag{3.12}$$

Taking into account (3.8) and assuming the homogeneity of snow slab $\left(k_e \neq f(\xi,\theta,\psi)\right)$, we can conduct the integration of (3.12) and finally arrive at:

$$\left\langle \sigma_s^0 \right\rangle = \left\langle \sigma_v \right\rangle \frac{1}{2k_e}\left[1 - \exp(-2k_e h_\theta)\right] \cdot \cos\theta \tag{3.13}$$

This expression is the mathematical formulation of the semi-empirical model and has been obtained by Attema & Ulaby (1978) and Ulaby et al. (1982). In this paper, this model will be cited as the "A-U model." For this case, the backscatter coefficient (3.13) depends only on the specific volume backscatter of the snow medium and not on the pulse volume sizes. This remarkable feature is due to (1) "overcomplete" burial of the pulse length into the snow medium, and (2) the backscattering normalization factor A_{ill} (see 3.4) is the basis of the probing pulse.

The obvious imperfection of the model, as applied to sounding of thick snow, is the assumption of a constant extinction coefficient within the snow slab. The problem can be solved by designing an appropriate stratification model for the selected study area and modeling the spatial distribution of the extinction coefficient (e.g., Drinkwater et al., 2001).

The next limitation of the model is due to wave sphericity. Under depth-limited mode conditions (see section 2), the illuminated area changes, and the running RCS in (3.7) can not be assumed to be statistically homogeneous within the scattering volume. The same is true regarding the extinction coefficient in (3.8) as well. Thus, for this case, the A-U model in form (3.13) should be used with care. For the beamwidth-limited mode, the form in (3.13) can be used, taking in mind the "determined part" of the scattering volume (see comments to 2.15a).

The ground component of the backscatter coefficient is localized by the bottom location and can thus be expressed by its backscatter coefficient, taking into account the attenuation:

$$\sigma_g^0 = \sigma_{g,\,max}^0 \exp(-2k_e h_\theta) \tag{3.14}$$

where $\sigma_{g,\,max}^0$ is the backscatter coefficient of the ground bottom surface governed only by the surface properties and its orientation in regards to the incidence of radar illumination..

3.2. Effective snow depth

Given the expression

$$\frac{1}{2}L_p \cos\theta\left[1 - \exp\left(-2\frac{h}{L_p \cos\theta}\right)\right] = H_{eff} \tag{3.15}$$

(3.13) can be presented in the simple form:

$$\left\langle \sigma_s^0 \right\rangle = \left\langle \sigma_v \right\rangle H_{eff} \tag{3.16}$$

where H_{eff} is the effective depth of snow sounding (EDS); i.e., the depth such that the backscatter from which occurs as if without attenuation, with backscatter equal to that which would occur from a slab of larger real depth with attenuation due to absorption and scattering.

Parameter $k_e h_\theta = \dfrac{h_\theta}{L_p} = \tau_\theta$ is the optical thickness of snow (SOT) along the wave propagation direction. It is useful to express the backscatter coefficient through the dimensionless EDS using the geometrical snow depth, h: $H_{eff}^0 = H_{eff} \big/ h$. Given that expression, equation (3.16) can be written in the form:

$$\left\langle \sigma_s^0 \right\rangle = \left\langle \sigma_v \right\rangle h H_{eff}^0 = \left\langle \sigma_s^0 \right\rangle_{max} H_{eff}^0 \tag{3.17}$$

where

$$\left\langle \sigma_s^0 \right\rangle_{max} = \left\langle \sigma_v \right\rangle \cdot h \tag{3.18}$$

is the maximal value of the backscatter coefficient for a given snow depth, h, ignoring both energy losses due to attenuation and scattering and the angular dependence of the scattering volume. The normalized effective depth of snow sounding (nEDS$\equiv H_{eff}^0$) plays the role of a correction factor, and is equal to:

$$H_{eff}^0 = \frac{1}{2\tau_\theta}\left[1 - \exp\left(-2\tau_\theta\right)\right] \tag{3.19}$$

And the main formula of the model (3.2) can be written as:

$$\sigma_s^0 = \left\langle \sigma_s^0 \right\rangle_{max} \cdot H_{eff}^0 + \sigma_g^0 \tag{3.20}$$

It is useful to find the dependence of the correction factor on the SOT magnitude and the incidence angle. Let us consider two extreme cases:

a) "shallow" snow: $\tau_\theta \ll 1$

$$H_{eff}^0 \approx \frac{1}{2\tau_\theta}\left[1 - \left(1 - 2\tau_\theta\right)\right] = 1 \tag{3.21}$$

Given this relation,

$$\left\langle \sigma_t^0 \right\rangle = \left\langle \sigma_v \right\rangle h + \sigma_g^0 \tag{3.22}$$

meaning that the EDS is approximately equal to the geometrical snow depth. Nevertheless, the small magnitude of the volume component compared with the backscatter from the bottom layer ($\langle\sigma_v\rangle h < \sigma_g^0$) makes this dependence difficult to use in practice. For example, a snow layer with $h<\sim1$ m is practically transparent to electromagnetic irradiance of the C-band.

b) "thick" snow: $\tau_\theta \gg 1$

$$H_{eff}^0 \approx \frac{1}{2\tau_\theta} = \frac{L_p \cos\theta}{2h} = \frac{1}{2}\frac{D_p^*}{h} \tag{3.23}$$

Given this relation, and in accordance with (3.18) and (3.20), one gets:

$$\langle\sigma_t^0\rangle = \langle\sigma_s^0\rangle_{max}^* + \sigma_g^0 \tag{3.24}$$

where $\langle\sigma_s^0\rangle_{max}^* = \frac{1}{2}\langle\sigma_v\rangle D_p^*(\theta)$. Equation (3.24) shows that the backscatter depends only on

the penetration path (depth) under a constant incidence angle. In the "thick" snow regime, no additional snow accumulation contributes to the total backscatter due to the saturation effect. On the other hand, in this regime, the backscatter coefficient demonstrates an angular dependence, as the angle of incidence affects the propagation depth (D_p^*). The sensitivity of the backscatter coefficient to the changes in snow thickness takes place in the so-called "intermediate" regime, when $\tau_\theta \sim 1$. Field data gives 20-30 meters of the penetration depth for the C-band (e.g., Hoen & Zebker, 2000), and yield values that can be considered the "working" range for probable snow measurements in this wave band. Values of the nEDS (3.19) can be assessed from the plot of the nEDS as a function of SOT, as shown in Figure 3.1.

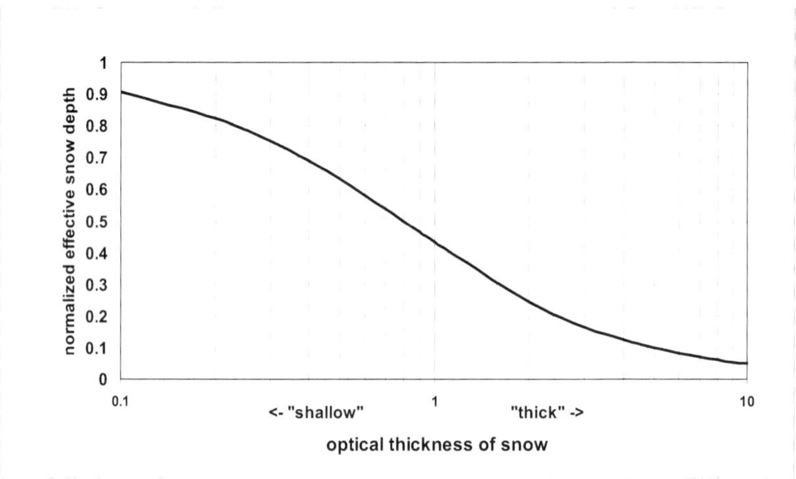

Fig. 3.1. Plot of the normalized effective depth of snow sounding versus the snow optical thickness

To evaluate the angular dependence of the nEDS, we consider the normalized Snow Depth (nSD), which is the depth of snow normalized to the penetration path:

$$nSD = \frac{h}{L_p} \tag{3.25}$$

Comparing nSD with the definition of the SOT yields:

$$nSD = \tau_\theta \cdot \cos\theta \tag{3.26}$$

Substituting (3.26) into (3.19), we have

$$H_{eff}^0 = \frac{1}{2} \frac{\cos\theta}{nSD}\left[1 - \exp\left(-2\frac{nSD}{\cos\theta}\right)\right] \tag{3.27}$$

A plot of (3.27) reduced to zero dB at $\theta=0$ for different nSD is shown in Figure 3.2. As follows from this plot , the angular dependence of the volume component of the backscatter coefficient in a practically appropriate range of angles, 0-50⁰ is rather weak and equal to ~2dB for the thick snow regime. For the intermediate regime, with nSD~1, its range of variation is about 1 dB. Because this range is of the same order as the errors, it is difficult to expect a notable angular dependence in practical measurements inherent to the intermediate regime.

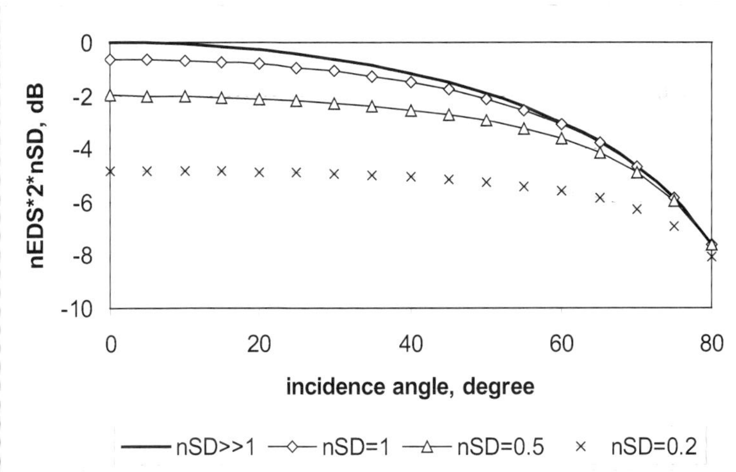

Fig. 3.2. The angular dependence of the normalized effective depth of snow sounding for the "thick" (nSD>>1) and "intermediate" regimes

The counter-phase behavior of the backscatter coefficient and the elevation profiles plotted above can be explained supposing that the snow depth is minimal on the hilltop and increases proportional to the distance from the top of the hill down to its base. This assumption closely matches the field and modeled data provided by Jaedicke et al. (2000).

For quantitative assessment of the semi-empirical model more regular terrain should be chosen. A typical dry snow area is presented on the RadarSAT-1 SAR image of eastern Greenland in Figure 4.2.

As is clearly seen, the magnitude of the backscatter coefficient is about -20 dB in the dry snow zone. This assessment coincides with data in the known literature (e.g., Drinkwater et al., 2001; Partington, 1998; Baumgartner et al., 1999; Forster et al., 1999) which also gives values of -12...-20 dB. We can estimate the corresponding value using the A-U model. We assume that the thick snow regime is valid, and that the snow slab consists primary of particles of roughly the same sizes; i.e., it is a monodisperse medium.

For these conditions (see 3.24), ignoring backscatter from the bottom surface, we get:

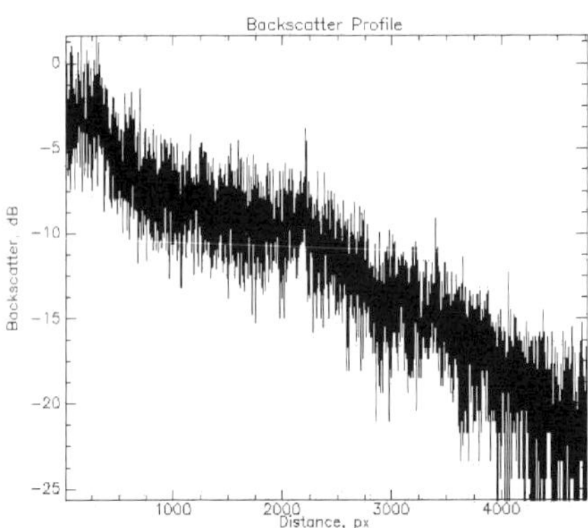
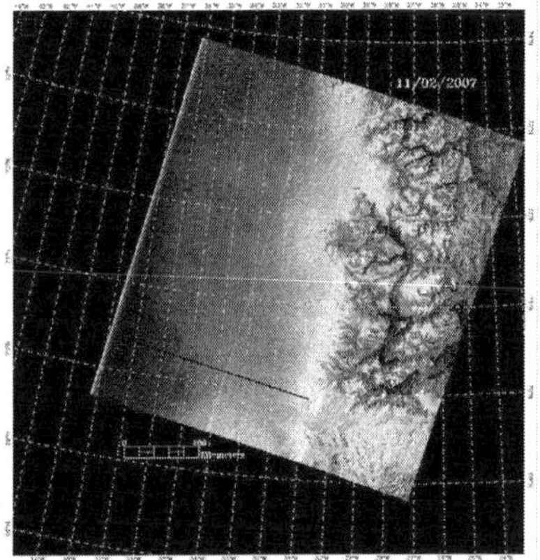

Fig.4.2. Illustration of the backscatter coefficient values (left) range for an arbitrary transect (straight line on RadarSAT-1 SAR signature, right) from the percolation to the dry snow areas within the east margin of the Greenland Ice Sheet (directions are from left to right for the plot of the backscatter profile and contra versa for the SAR signature; date of data acquisition: November 2, 2007; data granule ID: R1_62609_SWB_271)

$$\sigma_s^0 = \sigma_v \frac{L_p \cos\theta}{2} = \frac{1}{2}\sigma_v \frac{\cos\theta}{k_e} \qquad (4.1)$$

Further calculations below aim to discover the specific volume backscatter, σ_v and the extinction coefficient, k_e. The estimates listed below will be conducted with the Rayleigh approach. In accordance with Ulaby et al. (1986), this approach bounds the upper electric size of particle, $\zeta_u = \frac{2\pi r_u}{\lambda}$, with the inequality $\zeta_u |\dot{m}| \le 0.5$, where m is the complex refractive index of the particle matter and r_u is the corresponding upper size of a particle. For ice particles with $|\dot{m}| \approx 1.7776$ observed with C-band radar (λ=5.6 cm), the maximal size satisfying the Rayleigh approach is 2.6 mm. Following known relationships (e.g., Ulaby et al, 1986), the specific volume scattering is:

$$\sigma_v = Z_s \frac{1}{\lambda^4}$$

(4.2)

where

$$Z_s = \frac{2^6}{V} \pi^5 |K|^2 \sum_{i=1}^{N} r_i^6$$

(4.3)

In this formulae

$$|K|^2 = \left| \frac{\dot{m}^2 - 1}{\dot{m}^2 + 2} \right|^2$$

(4.4)

where $\dot{m} = n + j\eta$ is the refractive index of the particle matter and n and η are the real and imaginary parts of the refractive index, respectively. The value of these components can be derived from the value of the complex dielectric permittivity of a material $\dot{\varepsilon} = \varepsilon' + j\varepsilon''$, where ε' and ε'' are the real and imaginary parts of the dielectric permittivity.

$$n = \left[0.5 \cdot \left(\sqrt{(\varepsilon')^2 + (\varepsilon'')^2} + \varepsilon' \right) \right]^{0.5}$$

(4.5)

$$\eta = \left[0.5 \cdot \left(\sqrt{(\varepsilon')^2 + (\varepsilon'')^2} - \varepsilon' \right) \right]^{0.5}$$

(4.6)

For monodisperse media ($r_i = r$) equation (3.3) has the form:

$$Z_s = 2^6 \pi^5 |K|^2 n^0 r^6$$

(4.7)

where n^0 is the particle concentration.
The extinction coefficient can be calculated as a sum of the absorption (k_a) and scattering (k_s) coefficients (Ulaby et al., 1986):

$$k_e = k_a + k_s = n^0 \left[Q_a + Q_s \right]$$

(4.8)

where Q_a and Q_s are the absorption and scattering cross-sections respectively. These parameters, under the Rayleigh approach, can be calculated by the following formulas:

$$Q_a = \frac{\lambda^2}{\pi} \zeta^3 \, \mathrm{Im}\{-K\}$$

(4.9)

and

$$Q_s = 2 \frac{\lambda^2}{3\pi} \zeta^6 |K|^2$$

(4.10)

In accordance with (Ulaby et al., 1982)

$$\text{Im}\{-K\} \approx \frac{3\varepsilon''}{(\varepsilon'+2)^2} \tag{4.11}$$

The real part of the dielectric permittivity for ice, contained in the formulas above, is equal to 3.15 and in practice does not depend on temperature or wavelength. The imaginary part of the dielectric permittivity for ice is not constant with changes in temperature or the illumination frequency. In accordance with Matzler (1987) it can be expressed through the following empirical formula:

$$\varepsilon'' = \frac{A}{F} + BF^C \tag{4.12}$$

where F is frequency in GHz and A, B and C are the empirical coefficients. For a temperature of -15⁰C, which is more appropriate for the dry snow case, the coefficients are equal to: A=3.5*10⁻⁴, B=3.6*10⁻⁵, C=1.2. Now, the expressions obtained for the backscatter coefficient calculation may be combined. Substituting (4.2) in (4.1), we get:

$$\sigma_s^0 = \left(\frac{1}{2}\frac{Z_s}{\lambda^4 k_e}\right)\cos\theta \tag{4.13}$$

Taking into account (4.7) and (4.8)-(4.10), one can get:

$$\sigma_s^0 = \frac{1}{2\lambda^4}\frac{2^6\pi^5|K|^2 n^0 r^6}{n^0\left[\frac{\lambda^2}{\pi}\zeta^3\,\text{Im}\{-K\}+\frac{2\lambda^2}{3\pi}\zeta^6|K|^2\right]}\cos\theta \tag{4.14}$$

Simplifying this expression, we have:

$$\sigma_s^0 = \frac{1}{2}\frac{\cos\theta}{\frac{1}{\zeta^3|K|^2}\text{Im}\{-K\}+\frac{2}{3}} \tag{4.15}$$

An important feature of this expression is its independence with regards to the particle concentration, a reflection of the property inherent to the saturation regime mentioned above in section III. Since the scattering volume is less than the spatial duration of the probing pulse an increasing particle number is equivalent to an increasing particle concentration. Thus, no additional amount of the snow over the snow pack with a thickness greater than the penetration depth can cause a notable increase in the backscatter coefficient. This circumstance is also very useful from a simulation point of view, since there is no microstructure parameter that needs to be assumed. For the numerical calculation listed below we assumed that the snow within the dry snow zone consists of snow grains that are actually grains of ice that look like grains of rice, with a typical size of ~1 mm. Taking into account the fact for ice: $\varepsilon'=3.15$ and $\varepsilon'>>\varepsilon''$ (Tiuri et al., 1984; Matzler, 1987), $|K|^2 = 0.16$.

Given that information, equation (4.15) for an intermediate incidence angle of ~40⁰
($\cos\theta$=0.77) is equal to:

$$\sigma_s^0 = \frac{0.385}{\dfrac{6.25}{\zeta^3}\,\mathrm{Im}\{-K\}+\dfrac{2}{3}} \tag{4.16}$$

Taking into account (4.11) and (4.12), we get:

$$\sigma_s^0 = \frac{0.5775}{\dfrac{1.06}{\zeta^3}\left(\dfrac{A}{F}+BF^C\right)+1} \tag{4.17}$$

The plot of (4.17) is depicted in Figure 4.3 for C-band radar with F=5.3 GHz ($\lambda\approx$ 5.6 cm).

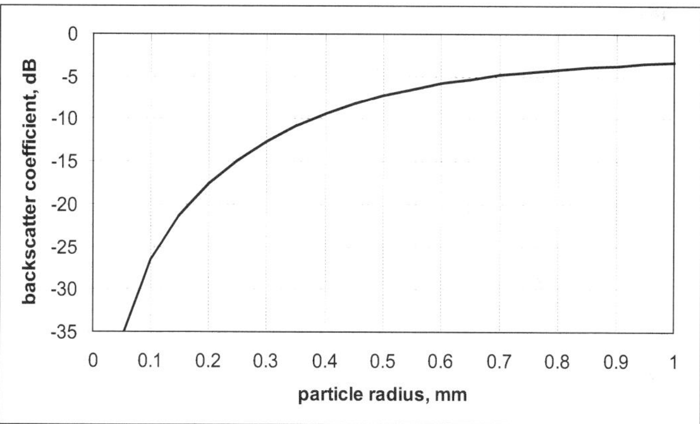

Fig. 4.3. The dependence of the backscatter coefficient on particle size in accordance with
the A-U model for C-band radar and snow depth greater than the penetration depth
("thick" snow regime)

It follows from this plot that the observable backscatter coefficient values of -12…-20 dB can
be caused by particles with radii of less than 0.35-0.17 mm. The current estimates are slightly
lower compared with the same made by Partington (1998), who found that the backscatter
coefficient would range from -20 dB for a mono-distribution of grain sizes with a mean
radius of 0.25 mm to -2 dB for a mean grain radius of 1 mm. Our results showed a 0.17 mm
mean radius for -20 dB and a 1 mm mean radius for -3 dB. The small discrepancy is
probably due to the temperature dependence of the imaginary part of the ice's dielectric
permittivity. The grain sizes values assessed do not coincide with the field data. In
accordance with Table 4.1, the mean particle radius within a ~1 m depth surface layer of dry
snow is 0.5…2.0 mm. As seen in Figure 4.3., these values should produce a backscatter
coefficient of approximately from -8 dB to 0 dB. Thus, there is a discrepancy of ~12 dB
between the observable data and the values forecast by the semi-empirical model. Therefore,

the A-U model overestimates the backscatter coefficient significantly. It should be noted, in addition, that the particle radius can only increases with the snow depth due to the depth-dependent grain radius model (Alley et al., 1982):

$$r^2(h) = r_0^2 + \frac{C \cdot h}{H_a}$$
(4.18)

where r_0 is the mean radius at the surface, C is the crystal growth rate and H_a is the mean annual layer thickness. Given this model, various simulation experiments (e.g., Forster et al., 1999; Drinkwater et al., 2001) and field measurements (e.g., Jezek et al., 1994; Woods, 1994; Lytle & Jezek, 1994), the mean particle radius for the entire scattering volume is even greater than the surface values and the aforementioned discrepancy becomes even more firmly grounded.

#	Mean ice particle radius, mm	Source	Location
1	0.5-2.0	Benson, 1996	Greenland
2	1.0-1.5	Schytt, 1964	Spitsbergen
3	Less than 0.75 mm	Higham and Craven, 1997	Antarctic
4	0.2-0.6	Woods, 1994	Greenland
5	0.1-0.7	Lytle and Jezek, 1994	Greenland

Table 4.1. Literature data on the mean ice particle size near the surface of dry snow cover

Thus, although the A-U model gives a qualitative assessment of the main features of backscatter behavior from a snow slab, the result of quantitative comparisons with field data does not match the theoretical predictions.

5. Enhanced semi-empirical model of the volume component of the backscatter coefficient

The problem highlighted in the previous section can be resolved by considering the statistical properties of the small-scale fluctuations in the particle concentration and its scattering properties within the scattering volume. The appropriate method for doing so that is the so-called "slice" approach, which was primarily suggested in weather radar meteorology (Marshal & Hitchfeld, 1953; Smith, 1964). This approach exploits the known radar feature in accordance with which the particles of a "cloud" located close to the front of the incident radar wave are considered to be approximately at the same distance from the radar and reflect incident electromagnetic wave almost coherently. One can consider that these particles are embedded in a fictitious thin cylindrical volume ("slice"), whose base coincides with the surface of spherical wave front and side-bounded by the main lobe of the antenna pattern. Thus, the scattering volume into a snow slab can be represented as an adjoining series of these slices, as illustrated in Figure 5.1. Each slice is much narrower than the radar wavelength in the wave propagation direction ($\Delta_s \ll \lambda$).

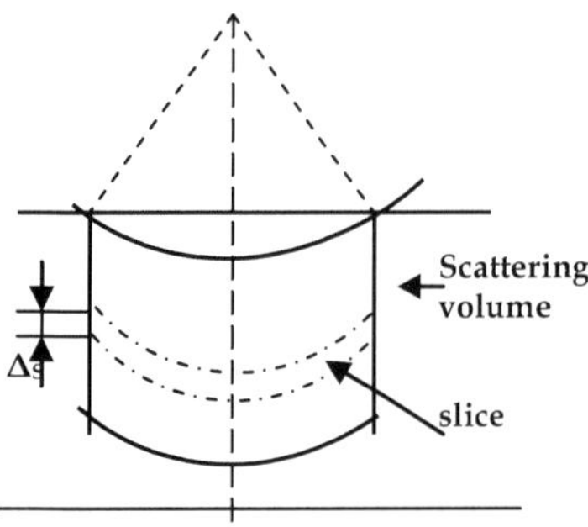

Fig.5.1. Simplified slice approach scheme. The arbitrary position of a single slice is shown on the cross section of the scattering volume

A slice's radial size, Δ_s can actually be considered as a minimal spatial scale of the backscatter property fluctuation, if this scale is much smaller than the wavelength. Using this approach, it was shown by the author (Yurchak, 2009), that the specific volume component of the backscatter from any spatial extended geophysical target (SEGT), included the snow as well, can be presented as the incoherent sum of the radar cross-sections of individual scatterers only if the number of particles in the slices (n) are distributed in accordance with the Poisson law; i.e., the variance of the number of particles ($Var(n)$) is equal to the mean number ($\langle n \rangle$). Otherwise, the "classical" specific volume component should be corrected by a so-called deviation factor $Y(\xi_a, \chi)$:

$$\langle \sigma_v \rangle = \langle \sigma_v \rangle_{class} Y(\xi_a, \chi) \tag{5.1}$$

where

$$Y(\xi_a, \chi) = \frac{\xi_a^2 + \chi}{\xi_a^2 + 1} \tag{5.2}$$

and $\chi = \dfrac{Var(n)}{\langle n \rangle}$ is the Poisson index, $\xi_a = \dfrac{Stdev(a)}{\langle a \rangle}$ is the variation coefficient of the particle radar equivalent length (PREL $a_p \equiv \sqrt{\sigma_p}$, σ_p is the radar cross-section of an individual particle). Formulas (5.1) and (5.2) reflect the fundamental physical principal that the fluctuations of the medium parameters (inhomogeneities) are the cause of the electromagnetic wave scattering (e.g., Atlas, 1964; Fabelinskii, 1968). The parameter ξ_a might be expressed through the measurable parameters of the snow particle size distribution function (PSDF), such as the variation coefficient of particle radius, $\xi_r = \dfrac{Stdev(r)}{\langle r \rangle}$, and the skewness coefficient, Sk:

where $\sigma_v = \dfrac{\sigma_s}{V_p}$ is the specific volume backscatter, and the return signal is proportional to the cosine of the incidence angle and has the envelope of the exponential form distorted by the variability of the statistical, scattering and absorption properties of the snow medium along the wave propagation path. The illuminated area A_{ill} (transverse size of the scattering volume) changes during the burial of the probing pulse into the snow slab due (mainly) to the wave sphericity in the depth-limited mode (see section 2). However, this effect does not impact the accuracy of the specific volume backscatter coefficient presented here, due to its the mutual cancellation in the nominator (via V_{sct}) and the denominator of equation (8.2). As a result of this cancellation, for very short probing pulses, the backscatter envelope depends only on changes in the specific volume backscatter with distance.

8.2. Analytical derivation for commensurate length of the scattering volume and snow depth

In general case, when $h_{sct}^{(p)} \sim \dfrac{h}{\cos\theta}$ and $h_{sct}^{(p)} \sim L_p$, the interactions between a probing pulse with a normalized form $U(R)$ and an extended target such as snow cover is described by the convolution integral (Moore & Williams, 1957):

$$P_r(t) = P_T(t) \otimes P_s(t) \tag{8.3}$$

where $P_r(t)$ is the return signal, $P_T(t)$ is the transmitted pulse profile and $P_s(t)$ is a term that includes the distribution range of the scattering facets, their scattering properties and the effects of the antenna gain. As applied our consideration of volume backscatter, one can write

$$\sigma_s^0(R) = \cos\theta \int_0^\infty dx \cdot U(x-R) \cdot Y(x)\sigma_v(x)\exp\left(-2\frac{x}{L_p(x)}\right) \tag{8.4}$$

If, for example, the probing pulse has a Gaussian form, which is usually an adequate approximation for short pulse systems (e.g., Brown, 1977), the weighting function of the scattering volume is $U(R) = \exp\left(-\beta\dfrac{R^2}{\Delta_q^2}\right)$, where Δ_q is the spatial duration of the scattering volume weight function at level $q = \exp\left(-\dfrac{\beta}{4}\right)$. For level $q=0.5$ (-3 dB), $\beta=-4\ln0.5=2.8$. If, in addition, the microstructural, statistical and attenuation properties of the snow slab do not change along the propagation path within the snow pack, one can write:

$$\sigma_s^0(R) = \cos\theta \cdot Y\sigma_v \int_0^\infty dx \cdot U(x-R) \cdot \exp\left(-2\frac{x}{L_p}\right) \tag{8.5}$$

For short pulses, if their envelope can be represented by the delta-function, i.e., $U(x-R) \sim \delta(x-R)$, equation (8.5) yields the corresponding result obtained earlier (8.3) for the very short pulses configuration. For the long pulses, where $h_{sct}^{(p)} > L_p$ and $\sigma_s^0 \sim \int_0^L dx \exp\left(-2\frac{x}{L_p}\right)$, which yields to the standard form of the A-U model (3.13). For the general case, a closed-form solution of the equation (8.5) can be obtained.

Taking into account the tabulated integral (Abramowitz & Stegun, 1972):

$$\int_0^\infty dx \exp\left(-ax^2 - bx\right) = e^{\frac{b^2}{4a}} \frac{\sqrt{\pi}}{2\sqrt{a}} \Phi\left(\frac{b+2ax}{2\sqrt{a}}\right)\Big|_0^\infty = \frac{1}{2} e^{\frac{b^2}{4a}} \sqrt{\frac{\pi}{a}}\left[1 - \Phi\left(\frac{b}{2\sqrt{a}}\right)\right] \tag{8.6}$$

where $\Phi(x) = \frac{2}{\sqrt{\pi}} \int_0^x \exp\left(-t^2\right) dt$ is the error function, one can get the integral of equation (8.5) for a Gaussian probing pulse:

$$\int_0^\infty dx \cdot U(x-R) \cdot \exp\left(-2\frac{x}{L_p}\right) = \frac{1}{2}\Delta_q \sqrt{\frac{\pi}{\beta}} \exp\left[\frac{1}{\beta}\left(\frac{\Delta_q}{L_p}\right)^2 - 2\frac{R}{D_p}\right] \cdot \left[1 - \Phi\left(\frac{1}{\sqrt{\beta}}\left(\frac{\Delta_q}{L_p} - \frac{\beta}{\Delta_q}R\right)\right)\right] \tag{8.7}$$

If one denotes $\dfrac{\Delta_q}{\sqrt{\beta}} = \Delta$, then (8.5) has the form:

$$\sigma_s^0(R) = \cos\theta \cdot Y\sigma_v \Delta \frac{\sqrt{\pi}}{2} \exp\left[\left(\frac{\Delta}{L_p}\right)^2 - \frac{2}{L_p}R\right] \cdot \left[1 - \Phi\left(\frac{\Delta}{L_p} - \frac{1}{\Delta}R\right)\right] \tag{8.8}$$

Finally, if one denotes: $\sigma_{max}^* = \frac{1}{2}\sigma_v L_p \cos\theta$ (see 3.24), $\left(\frac{\Delta}{L_p}\right) = \gamma$ and $S = \frac{R}{L_p}$, the closed-form solution of equation (8.6) using these dimensionless parameters can be written as:

$$\sigma_s^0(S) = Y\sigma_{max}^* \sqrt{\pi}\gamma \exp\left(\gamma^2 - 2S\right) \cdot \left[1 - \Phi\left(\gamma - \frac{S}{\gamma}\right)\right] \tag{8.9}$$

Distance normalization in (8.9) is chosen through the parameter S for the purpose of comparing the backscatter profile calculated here with the modeled "clear" backscatter profile derived for a very short pulse (8.2) and governed only by attenuation. The equation obtained is close to the volume component of the SV model of Davis & Moore (1993). The distinctions of the current approach are that it uses a slightly different description of the pulse-snow interaction, the end result is written in terms of the backscatter coefficient depth profile, and the state of homogeneity of the snow mass is taken into account through the deviation factor, Y. It is interesting, that equation (8.9) coincides with the backscatter model from the sea surface, obtained by Barrick (1972), exactly by the form with, of course, other meaning of model parameters.

8.3. Comparison of the results obtained with the phenomenological approach

8.3.1. Very short probing pulse

For the limit case when $\gamma \ll 1$, and taking into account that $\Phi(-x) = -\Phi(x)$ and $\lim_{x \to \infty} \Phi(x) = 1$ (Abramowitz & Stegun, 1972), the corresponding backscatter coefficient is:

$$\sigma_s^0(S)_{\gamma \ll 1} = Y\sigma_{max}^* \sqrt{\pi}\gamma \exp[-2S] \cdot \left[1 - \Phi\left(-\frac{1}{\gamma}S\right)\right] = Y\sigma_{max}^* \sqrt{\pi}\gamma \exp[-2S] \cdot \left[1 + \Phi\left(\frac{1}{\gamma}S\right)\right] \approx \quad (8.10)$$

$$2Y\sigma_{max}^* \sqrt{\pi}\gamma \exp(-2S) = \cos\theta\, Y\sigma_v \Delta\sqrt{\pi}\, \exp(-2S)$$

This result confirms relationship (8.2) obtained above based on phenomenological considerations.

8.3.2. Long probing pulse

Taking into account the asymptotic value of the error function $\Phi(x) \underset{x \gg 1}{\approx} 1 - e^{-x^2} \frac{1}{\sqrt{\pi}x}\left(1 - \frac{1}{2x^2} - \ldots\right)$ (Abramowitz & Stegun, 1972), and taking only the first term of the extension, we get the value of (8.9) for $\gamma \gg 1$:

$$\sigma_s^0(S)_{\gamma \gg 1} \approx Y\sigma_{max}^* \sqrt{\pi}\gamma \exp(\gamma^2 - 2S)\exp\left[-\gamma^2 + 2S - \left(\frac{S}{\gamma}\right)^2\right]\frac{1}{\sqrt{\pi}}\left(\gamma - \frac{S}{\gamma}\right)^{-1} \quad (8.11)$$

For long pulses, the available distance variation is much less than the pulse length, i.e., $\left(\dfrac{S}{\gamma}\right)_{\gamma \gg 1} = \dfrac{R}{\Delta} \ll 1$. Hence,

$$\sigma_s^0(S)_{\gamma \gg 1} \approx Y\sigma_{max}^* \exp\left[-\left(\frac{S}{\gamma}\right)^2\right]\left(1 - \frac{S}{\gamma^2}\right)_{\gamma \gg 1}^{-1} \approx Y\sigma_{max}^* = \frac{1}{2}\cdot Y\sigma_v D_p^* \quad (8.12)$$

This expression also confirms the result of the A-U model (enhanced) under thick snow condition (3.24). Thus, an important positive feature of the model (8.9) is that its results coincide with those obtained with phenomenological approaches for extreme values of the scattering volume (much less and much more than the penetration path respectively).

Nevertheless, it should be noted that ignoring the change in the illumination area during the burial of the probing pulse into snow media is, strictly speaking, valid only for a short probing pulse or for the beamwidth-limited mode. For long and intermediate-length probing pulses with wavefront sphericity in the commonly-existing depth-limited configuration, this effect should be taken into account because there are slices (see sections 2 and 5) of different transverse sizes within $h_{sct}^{(p)}$.

8.4. Numerical calculation of the return signal for intermediate length of the probing pulse

A plot of the normalized backscatter coefficient $\sigma^*(S,\gamma) = \dfrac{\sigma_s^0(S,\gamma)}{\sigma_s^0(S_{max},\gamma)}$ (S_{max} is the normalized distance where function (8.7) has the maximum value at a given γ) is shown in Figure 8.1.

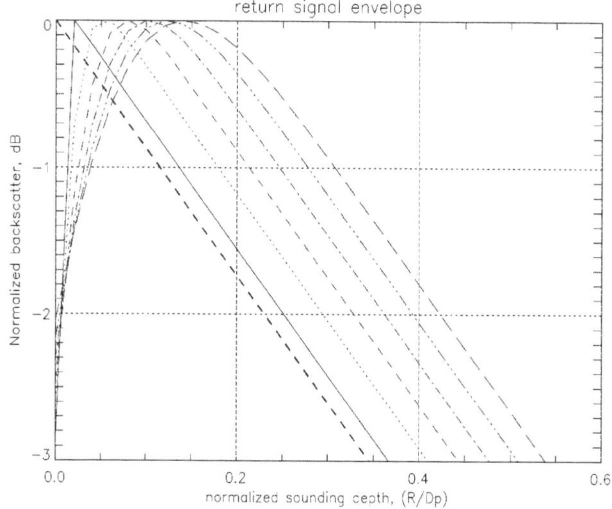

Fig.8.1. Illustration of the reproducibility of the snow backscatter uniform profile taking into account only attenuation (bold dash line) and finite probing pulse length normalized to the penetration depth (parameter $\gamma = \Delta / L_p$).

Legend: solid line, γ=0.01; dot line, γ=0.035; dashed line, γ=0.06; dashed-dot line, γ=0.085; dashed-dot-dot line, γ=0.11; long dashed line, γ=0.135

This figure illustrates the contribution of the pulse length to the return signal envelope pattern. The shorter pulse clearly gives a better approximation of the real backscatter profile. Contemporary altimeters use a probing pulse with a physical duration of 20-50 microseconds and a corresponding effective duration (due to intrapulse frequency chirping) equal to 3.125 nanoseconds (e.g., Rosmorduc et al., 2006). In spatial units, these values are equivalent to $\Delta \approx \dfrac{c\tau_0}{2} \approx 0.5$ m. For $L_p \sim 10\text{-}50$ m, the expected values of the parameter γ range from 0.05 to 0.01. An important feature of this plot is that the value of the maximum of the return signal for any maximum position (S_{max}) and values of parameter γ are the same:

$$\sigma_s^*\left(S_{max},\gamma\right) = \sqrt{\pi}\,\sigma_{max}^* Y \tag{8.13}$$

9. Summary

The semi-empirical model of the volumetric component of the backscatter coefficient, proposed by Attema & Ulaby (1978) (SEM or A-U model) is considered in more detail. The basic advantage of the SEM is its mathematical simplicity and clear physical sense but the SEM, derived using the incoherent approach, does not explain some remote sensing experimental data collected over the dry snow zone of the Greenland ice sheet. We discuss the main configurations of radar sounding of snow, including the relationships between the technical and positional parameters of radar on one hand and the geometrical and electromagnetic properties of snow cover on the other. Based on this consideration, the field of applicability of the SEM is figured out for the further assessments. Some qualitative assessments of the configuration of the scattering volume were carried out using plane and spherical surface approaches. More detailed study of the diffraction of space-bounded spherical electromagnetic wave on a large and strongly absorbing dielectric sphere with a rough surface is needed to evaluate this problem further. Detailed SEM estimates and experimental radar data over the dry snow zone of the Greenland ice sheet are provided for the negative temperatures inherent to the winter seasons. Data analysis shows a significant discrepancy between the expected and actually measured values of the backscatter coefficient. Possible reasons for the observed deviations of the SEM results are discussed based on the enhanced semi-empirical model (ESEM) and analysis of the dependence of the backscatter coefficient on the snow optical thickness and its spectral function. The ESEM is obtained on the basis of the so-called "slice" approach, taking into account partially-coherent backscattering due to finite wave front size and considering the contribution of the statistical characteristics of small-scale fluctuations of concentration and of sizes of the ice particles in a snow pack. The known semi-empirical model is only a special case of the ESEM. Using the ESEM, some observable radar signatures of the Greenland ice sheet, including the dry snow zone, can be interpreted in terms of the statistical characteristics of the snow microstructure. The model obtained retains the mathematical simplicity and physical clarity of the known A-U model and significantly expands its application for the interpretation of experimental data. Available field data, which confirm the consistency of the ESEM with observed values of the backscatter coefficient, are also provided. As a consequence of this analysis, the concept of the normalized snow depth (nSD) is formulated, and its practical application for monitoring snow mass displacement is described. The frequency-dependent core of the backscatter has been analyzed as well. Empirical and theoretical spectral characteristics of the penetration depth are also presented. The result allow for improvements in our understanding of opportunities for a two-frequency method of sounding thick snow covers and promote more accurate interpretations of the known experimental data. The extension of the ESEM for sounding thick snow cover with broadband pulses of short effective length is also discussed, resulting in an exact description of the return signal envelope as a function of the electromagnetic and microstructural characteristics of the snow cover. The validity of the model is supported by the coincidence of the model estimates with assessments obtained for several specific cases using phenomenological approach. In conclusion, this work demonstrates that the enhanced version of the semi-empirical model can be applied successfully in the analysis of radar backscatter from thick snow cover.

10. Acknowledgements

This work was supported by NASA's Cryospheric Sciences Program. The author would like to thank Dr. W. Abdalati for his permanent attention to this work and useful discussions.

11. References

Abramowitz, M., & I.A. Stegun, Eds. (1972). Hanbook of mathematical functions with formulas, graphs, and mathematical tables. New York: Wiley.

Alley, R.B., Bolzan, J.F., & Whillans, I.M. (1982). Polar firn densification and grain growth. Ann Glaciol., vol.3, pp.7-11.

Atlas, D. (1964). Advances in radar meteorology. Advances in Geophysics, vol.10, pp. 317-479.

Attema, E., & Ulaby, F. (1978). Vegetation modeled as water cloud. Radio Science, vol.13, pp. 357-364.

Barrick, D. E. (1972). Remote sensing of sea by radar. In Derr, V. ed. Remote sensing of the troposphere. Washington, DC, U.S. Government Printing Office, pp. 12.1-12.46.

Battan, L.J. (1959). Radar Meteorology, Univ. of Chicago Press, Chicago, Ill.

Baumgarnter F., Jezek K., Forster R.R., Gogineni S.P., & Zabel I.H.H. (1999). Spectral and angular ground-based radar backscatter measurements of Greenland Snow facies. IGARSS'99 Proceedings, 28 June-2 July 1999, Humburg, vol.2, pp. 1053-1055.

Benson, C. S. (1996). Stratigraphic studies in the snow and firn of the Greenland ice sheet. U.S. Army Cold Reg. Res. and Eng. Lab (CRREL), Rep.70, 120 pp. Hanover, NH. (Revised edition of 1962 report).

Brown, G.S. (1977). The average impulse response of a rough surface and its applications. IEEE Transactions on Antennas and Propagation, vol. AP-25, No. 1, January 1977, pp. 67-74.

Clift, G.A. (1985). Use of radar in meteorology. Secretariat of the World Meteorological Organization, Geneva.

Davis, C.H. (1996). Temporal change in the extinction coefficient of snow on the Greenland Ice sheet from an analysis of Seasat and Geosat altimeter data. IEEE Trans. Geosci. Remote Sens., vol.34, No.5, September 1996, pp. 1066-1073.

Davis C.H., & Pozdnyak V.I. (1993). The depth of penetration in Antarctic firn at 10 GHz. IEEE Trans. Geosci. Remote Sens., vol. 31, No.5, September 1977, pp. 1107-1111.

Davis, C.H., & Moore, R.K. (1993). A combined surface- and volume scattering model for ice-sheet radar altimeter. Journal of Glaciology, vol. 39, No. 133, pp. 675-686.

Drinkwater M.D., Long D.G., & Bingham A.W. (2001). Greenland snow accumulation estimates from satellite radar scatterometer data. Journal of Geophysical Research, vol. 106, No. D24, December 2001, pp. 33,935-33,950.

Woods, G.A. (1994). Grain growth behavior of the GISP2 ice core from central Greenland. Technical Report Series 94-002, Earth System Science Center, Penn State University, USA.

Yurchak, B.S. (2009). Radar volume backscatter from spatially extended geophysical targets in a "slice" approach. IEEE Trans. Geosci. Remote Sens., vol. 47, issue 10, (in press).

Zahnen, N., F. Jung-Rothenhausler, H. Oerter, F. Wilhelms & Miller, H. (2002). Correlation between Antarctic dry snow properties and backscattering characteristics in Radarsat SAR imagery. Proceedings of EARSeL-LISSIG_Workshop Observing our Cryosphere from Space, Bern, March 11-13, 2002, pp. 140-148.

Zwally, H.J., & Brenner, A.C. (2001). Ice sheet dynamics and mass balance. In book: Satellite altimetry and earth sciences. L.L. Fu and A. Cazenave, (Eds). International Geophysics series, vol. 69, pp. 351-369.

7

Theoretical Modeling for Polarimetric Scattering and Information Retrieval of SAR Remote Sensing

Ya-Qiu Jin
Fudan University
China

1. Introduction

Synthetic aperture radar (SAR) imagery technology is one of most important advances in space-borne microwave remote sensing during recent decades. Completely polarimetric scattering from complex terrain surfaces can be measured. Fully understanding and retrieving information from polarimetric scattering signatures of natural media and SAR images have become a key issue for the SAR remote sensing and its broad applications.

This paper presents some research progress in Fudan University on theoretical modeling of the terrain surface for polarimetric scattering simulation and Mueller matrix solution, mono-static and bistatic SAR image simulation, new parameters for unsupervised surface classification, DEM inversion, change detection from multi-temporal SAR images, and reconstructions of buildings from multi-aspect SAR images, etc. Some applications are discussed.

2. Model of Vegetation Canopy and Mueller Matrix solution

As a polarized wave $\mathbf{E}_{inc}(\chi,\psi)$ is incident upon the natural media, the scattering field is written as

$$\begin{bmatrix} E_{vs} \\ E_{hs} \end{bmatrix} = \frac{e^{ikr}}{r} \begin{bmatrix} S_{vv} & S_{vh} \\ S_{hv} & S_{hh} \end{bmatrix} \cdot \begin{bmatrix} E_{vi} \\ E_{hi} \end{bmatrix} \equiv \frac{e^{ikr}}{r} \overline{\mathbf{S}} \cdot \mathbf{E}_{inc}(\chi,\psi) \tag{1}$$

where 2×2-D (dimensional) complex scattering amplitude function $\overline{\mathbf{S}}$ can be measured by the polarimetry. The incident polarization is indicated by the elliptic and orientation angles (χ,ψ). Using the Mueller matrix solution of vector radiative transfer equation (Jin, 1994) and Eq. (1), the scattered Stokes vector (four Stokes parameters) can be obtained or measured as (Jin, 2005)

$$\mathbf{I}_s(\theta,\phi) = \overline{\mathbf{M}}(\theta,\phi;\pi-\theta_0,\phi_0) \cdot \mathbf{I}_i(\chi,\psi) \tag{2}$$

The Mueller matrix solution of a layer of scatterers model, as shown in Figure 1, is written as

$$\overline{\mathbf{M}}(\theta,\phi;\pi-\theta_0,\phi_0) = \overline{\mathbf{M}}_0 + \overline{\mathbf{M}}_1 + \overline{\mathbf{M}}_2 + \overline{\mathbf{M}}_3 + \overline{\mathbf{M}}_4 \tag{3}$$

$$\overline{\mathbf{M}}_0 = \exp\left[-d\sec\theta\overline{\mathbf{\kappa}}_e(\theta,\phi)\right]\cdot\overline{\mathbf{R}}(\theta,\phi;\pi-\theta_0,\phi_0)\cdot\exp\left[-d\sec\theta_0\mathbf{\kappa}_e(\pi-\theta_0,\phi_0)\right] \tag{4a}$$

$$\overline{\mathbf{M}}_1 = \sec\theta\int_{-d}^{0}dz'\exp\left[z'\sec\theta\overline{\mathbf{\kappa}}_e(\theta,\phi)\right]\cdot\overline{\mathbf{P}}(\theta,\phi;\pi-\theta_0,\phi_0)\cdot\exp\left[z'\sec\theta_0\overline{\mathbf{\kappa}}_e(\pi-\theta_0,\phi_0)\right] \tag{4b}$$

$$\overline{\mathbf{M}}_2 = \sec\theta\int_{-d}^{0}dz'\exp\left[z'\sec\theta\overline{\mathbf{\kappa}}_e(\theta,\phi)\right]\cdot\int_{0}^{2\pi}d\phi'\int_{0}^{\pi/2}d\theta'\sin\theta'\overline{\mathbf{P}}(\theta,\phi;\theta',\phi')$$
$$\cdot\exp\left[(-d-z')\sec\theta'\overline{\mathbf{\kappa}}_e(\theta',\phi')\right]\cdot\overline{\mathbf{R}}(\theta',\phi';\pi-\theta_0,\phi_0)\cdot\exp\left[-d\sec\theta_0\overline{\mathbf{\kappa}}_e(\pi-\theta_0,\phi_0)\right] \tag{4c}$$

$$\overline{\mathbf{M}}_3 = \sec\theta\exp\left[-d\sec\theta\overline{\mathbf{\kappa}}_e(\theta,\phi)\right]\cdot\int_{0}^{2\pi}d\phi''\int_{0}^{\pi/2}\sin\theta''d\theta''\overline{\mathbf{R}}(\theta,\phi;\pi-\theta'',\phi'')$$
$$\cdot\int_{-d}^{0}dz'\exp\left[(-d-z')\sec\theta''\overline{\mathbf{\kappa}}_e(\pi-\theta'',\phi'')\right]\overline{\mathbf{P}}(\pi-\theta'',\phi'';\pi-\theta_0,\phi_0)\cdot\exp\left[z'\sec\theta_0\overline{\mathbf{\kappa}}_e(\pi-\theta_0,\phi_0)\right] \tag{4d}$$

$$\overline{\mathbf{M}}_4 = \sec\theta\exp\left[-d\sec\theta\overline{\mathbf{\kappa}}_e(\theta,\phi)\right]\cdot\int_{0}^{2\pi}d\phi''\int_{0}^{\pi/2}d\theta''\sin\theta''\overline{\mathbf{R}}(\theta,\phi;\pi-\theta'',\phi'')$$
$$\cdot\exp\left[(-d-z')\sec\theta''\overline{\mathbf{\kappa}}_e(\pi-\theta'',\phi'')\right]\int_{0}^{2\pi}d\phi'\int_{0}^{\pi/2}d\theta'\sin\theta'\overline{\mathbf{P}}(\pi-\theta'',\phi'';\theta',\phi') \tag{4c}$$
$$\cdot\exp\left[(-d-z')\sec\theta'\overline{\mathbf{\kappa}}_e(\theta',\phi')\right]\overline{\mathbf{R}}(\theta',\phi';\pi-\theta_0,\phi_0)\exp\left[-d\sec\theta_0\overline{\mathbf{\kappa}}_e(\pi-\theta_0,\phi_0)\right]$$

where $\overline{\mathbf{\kappa}}_e$, $\overline{\mathbf{P}}$, $\overline{\mathbf{R}}$ are the extinction matrix, phase matrix of non-spherical scatterers, and reflectivity matrix of underlying rough surface, respectively. Five contributions from scatterers and underlying surface are illustrated in Figure 1. $\overline{\mathbf{\kappa}}_e$, $\overline{\mathbf{P}}$ can be expressed by $<S_{pq}S_{st}^*>$ of $\overline{\mathbf{S}}$ in Eq. (1). The co-polarized and cross-polarized backscattering coefficients, σ_c and σ_x, polarization degree m_s for scattered Stokes echo with partial polarization and other functions can be numerically calculated (Jin and Chen, 2002).

Eqs. (4a-c) will be applied to numerical simulation of polarimetric scattering from the terrain surface with physical parameters of scatterers (Jin, 2005), as described in next sections.

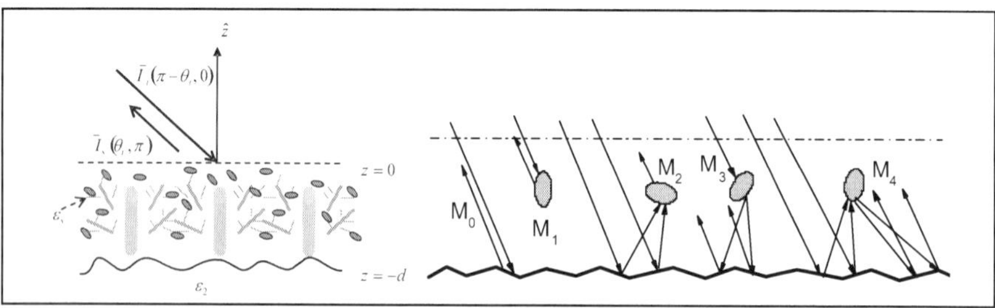

Fig. 1. A model of non-spherical scatterers for vegetation canopy

The Mueller matrix is a 4×4-D real matrix with complex eigenvalues and eigenvectors. To be physically realizable, this matrix must satisfy the Stokes criterion together with several other restrictive conditions. Unfortunately, however, these restrictions do not have any direct physical interpretation in terms of the eigenstructure of the Mueller matrix.

The coherency matrix $\overline{\mathbf{C}}$ is applied to the study of polarimetric scattering of SAR image (Jin and Cloude, 1994). Define the scattering vector as

$$\mathbf{k}_P = \tfrac{1}{2}[S_{vv}+S_{hh}, S_{vv}-S_{hh}, S_{vh}+S_{hv}, i(S_{vh}-S_{hv})]^T = \tfrac{1}{2}[A,B,C,iD]^T \tag{5}$$

where A, B, C, D are sequentially defined in Eq. (5), and the subscript P denotes Pauli vectiorization. The coheremcy matrix is defined as

$$\overline{\mathbf{C}} = <\mathbf{k}_P \mathbf{k}_P^+> = \frac{1}{4}\begin{bmatrix} <AA^*> & <AB^*> & <AC^*> & <AD^*> \\ <BA^*> & <BB^*> & <BC^*> & <BD^*> \\ <CA^*> & <CB^*> & <CC^*> & <CD^*> \\ <DA^*> & <DB^*> & <DC^*> & <DD^*> \end{bmatrix} \tag{6}$$

The eigenvalue and eigenvector of the coherency matrix are defined as

$$\overline{\mathbf{C}} = <\mathbf{k}_P \mathbf{k}_P^+> = \sum_{i=1}^{4} \lambda_i <\mathbf{k}_{Pi}\mathbf{k}_{Pi}^+> \tag{7}$$

All eigenvalues are real and non-negative, and $\lambda_1 \geq \lambda_2 \geq \lambda_3 \geq \lambda_4$.
The coherency matrix is also related with the Mueller matrix as

$$M_{i+1,s+1} = \sum_t \sum_u \frac{1}{2}\mathrm{Trace}(\overline{\sigma}_i \ \overline{\sigma}_t \ \overline{\sigma}_s \ \overline{\sigma}_u)[\overline{\mathbf{C}}]_{tu} \tag{8}$$

where $\overline{\sigma}_i$, $i = 0,1,2,3$ are the Pauli matrices.
Usually, the case of $S_{vh} = S_{hv}$ is simply assumed. The entropy H is defined by the eigenvalues of the coherency matrix as

$$H = -\sum_{i=1}^{3} P_i \log_3 P_i \ , \quad P_i = \lambda_i/(\lambda_1 + \lambda_2 + \lambda_3) \tag{9}$$

The entropy is an important feature since it relates the randomness of scatter media with other physical parameters, such as canopy configuration, biomass, etc.
The coherency matrix is 4×4-D positive semidefinite Hermitian and as such has real non-negative eigenvalues and complex orthogonal eigenvectors. Amplitude and difference of four eigen-values are functionally related with polarimetric scattering process of the terrain canopy. Eigen-analysis of the coherency matrix yields better physical insight into the polarimetric scattering mechanisms than does the Mueller matrix and further, it can be employed to physically identify a Mueller matrix by virtue of the semi-definite nature of the corresponding coherency matrix.
For the first-order solution in small albedo, cross polarization is small and correlations such as the terms $<CA^*>$, $<CB^*>$, $<DA^*>$, $<DB^*>$ etc. in Eq. (6) are always very small, and might be neglected in our following derivations. Actually, for weak assumption of azimuthal symmetric orientation, these correlations have been proved to be zero. Applying this approximation to Eq. (6), the eigenvalues of $\overline{\mathbf{C}}$ have been derived as (Jin and Cloude, 1994)

$$\lambda_{1,2} = \frac{1}{8}[<|A|^2> + <|B|^2> \pm \sqrt{(<|A|^2> - <|B|^2>)^2 + 4<A^*B><AB^*>}] \, , \quad \lambda_3 = \frac{1}{4}<|C|^2> \qquad (10)$$

It is interesting to see that if $<AB^*><A^*B>=<AA^*><BB^*>$ at an ordered case, it yields $\lambda_2 = 0$.

Substituting A,B,C and D of Eq. (6) into Eq. (10), the eigen-values are derived using the power terms of $<|S_{vv}|^2>$, $<|S_{hh}|^2>$, $<|S_{hv}|^2>$ as follows,

$$\lambda_1 = \frac{1}{2}(<|S_{vv}|^2> + <|S_{hh}|^2>) - \lambda_2 \, , \quad \lambda_2 = \frac{1}{2}\frac{<|S_{vv}|^2><|S_{hh}|^2>}{(<|S_{vv}|^2> + <|S_{hh}|^2>)}(1-\Delta) \, , \quad \lambda_3 = <|S_{hv}|^2> \qquad (11)$$

where the configuration parameter

$$\Delta = \frac{<S_{vv}S_{hh}^*><S_{vv}^*S_{hh}>}{<|S_{vv}|^2><|S_{hh}|^2>} \qquad (12)$$

varies from 0 of totally random media to 1 of ordered media, i.e. no-random case.

It can be seen that the first eigen-value λ_1 now indicates the total vv and hh co-polarized power subtracting their coherence defined by the parameter Δ of Eq. (12), which varies from 1 at ordered case to 0 in totally random medium. The second eigen-value λ_2 now takes into account for the vv and hh coherent power. As Δ approaches to one, $\lambda_2 \rightarrow 0$ for ordered media, and as Δ approaches to zero, λ_2 would increase to indicate the loss of the vv and hh power coherency for totally random media. The third eigen-value λ_3 is now due to depolarization caused by the media randomness.

The $(p,q=v,h)$-polarized backscattering coefficient is

$$\sigma_{pq} = 4\pi \cos\theta <|S_{pq}|^2> \qquad (13)$$

Substituting Eq. (13) into Eqs. (11,12), the functions P_i, $i = 1,2,3$, of Eq. (9) are then derived as

$$P_1 = 1 - P_2 - P_3, \quad P_2 = X(1-\Delta)(1-\delta), \quad P_3 = \delta(1-\delta) << 1 \qquad (14)$$

where

$$X \equiv \frac{\sigma_{hh}/\sigma_{vv}}{(1+\sigma_{hh}/\sigma_{vv})^2}, \quad \delta = \frac{2\sigma_{hv}}{\sigma_{vv}+\sigma_{hh}} \qquad (15)$$

Note that all functions P_i, $i = 1,2,3$, can be now calculated by polarized backscattering coefficient σ_{pq}, $p,q = v,h$, not by Eq. (7) from full Mueller matrix solution. Then, the entropy H of Eq. (9) is calculated using P_i of Eq. (14). It can be seen that the function P_1 is related with the total vv and hh co-polarized power $\sigma_{hh} + \sigma_{vv}$; P_2 is due to the difference between σ_{hh} and σ_{vv}; and P_3 is due to depolarization σ_{hv}. They are modulated by the media configuration and randomness via the parameters Δ and δ. As the medium becomes more random or disordered, δ increases and Δ approaches zero. Vice versa, as the medium becomes more ordered, δ decreases and Δ increases. Thus, the relationship between the entropy H and backscattering measurements σ_{pq} ($pq = vv, hh, hv$) is established. Define the co-polarized and cross polarized indexes, respectively, as

$$CPI \equiv 10\log_{10}(\sigma_{hh}/\sigma_{vv}) \ (dB), \quad XPI \equiv \delta \tag{16}$$

The entropy H is directly related with backscattering indexes CPI and XPI via Eqs. (14~16). Figure 2(a) presents an AirSAR image of total power of σ_{vv}, σ_{hh} and σ_{vh} at the L band. The image data from some typical areas such as the lake, an island surface, sparse trees, and thick forest (see the frames in the figure) are chosen, as shown in the discrete points of Figure 2(b), and are compared with theoretical results of H, CPI and XPI. The H is first rigorously calculated by polarimetric data of the Mueller matrix, i.e. the eigenvalues of coherency matrix $\overline{\mathbf{C}}$. Using the AirSAR data σ_{vv}, σ_{hh}, σ_{vh} and the averaged δ in the respective region, the lines are calculated with appropriate Δ to match the center of discrete data.

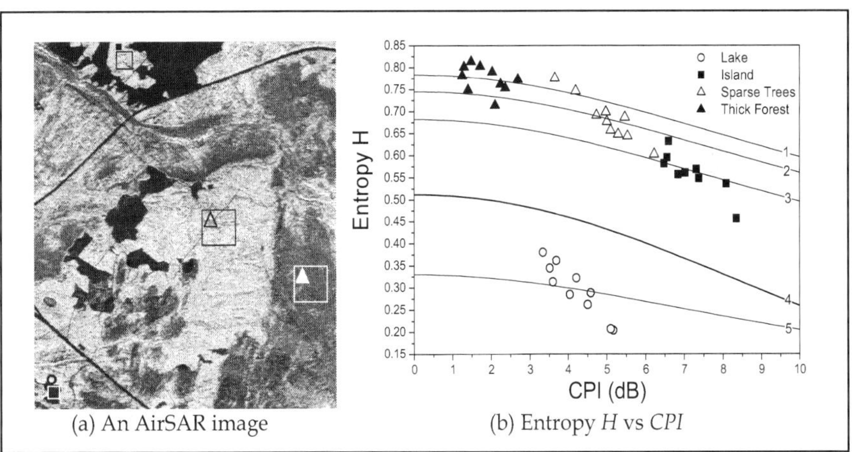

(a) An AirSAR image   (b) Entropy H vs CPI

Fig. 2. An AirSAR image of total power σ_{vv}, σ_{hh}, σ_{vh} at L band and Entropy H vs CPI.
Line 1: $\Delta = 0.257$, $\delta = 0.176$, Line 2: $\Delta = 0.318$, $\delta = 0.156$, Line 3: $\Delta = 0.365$, $\delta = 0.118$
Line 4: $\Delta = 0.$, $\delta = 0.$, Line 5: $\Delta = 0.731$, $\delta = 0.0265$

3. SAR Imaging Simulation (MPA)

Given the sophisticated mechanisms of electromagnetic (EM) wave scattering and SAR signal collecting, it is hard to tackle SAR imagery without the aid of forward simulation of scattering based on physical modeling. Simulation of SAR imaging presents a different mode to study inhomogeneous terrain scene under SAR observation. It may evaluate or predict the SAR observation, interpret the physical scattering behavior, and take account of the parameterized characteristics of terrain surface objects. Some approaches to SAR image simulation in previous literature might be classified into two catalogues: one focuses on the statistical characteristics of SAR images, and the other on physical scattering process. Natural scene is more complicated including randomly, inhomogeneously distributed, penetrable or impenetrable objects, such as vegetation canopies, manmade structures and perturbed surface topography. Particularly, volumetric scattering through penetrable scatterers, e.g. timberland forests, crops, green plants etc., plays an essential or dominant role in SAR imagery. It is meaningful to develop computerized simulation of SAR imaging over comprehensive terrain scene with heterogeneous terrain objects.

we present an novel approach of the Mapping and Projection Algorithm (MPA) to polarimetric SAR imaging simulation for comprehensive scenarios, which takes account of scattering, attenuation, shadowing and multiple scattering of spatially distributed volumetric and surface scatterers (Xu and Jin, 2006). In this approach, scattering contributions from scatterers are cumulatively summed on the slant range plane (the mapping plane) following geometric principles, while their extinction and shadowing effects are cumulatively multiplied on the ground range plane (the projection plane). It finally constructs a general imaging picture of the whole terrain scene via mapping and projection operations. A MPA is then devised to speed up the simulation of the whole process of scattering, extinction, mapping and projection in association with a grid partition of the comprehensive terrain scene. Our SAR simulation scheme incorporates polarimetric scattering, attenuation or shadowing of several typical terrain surfaces, as well as the coherent speckle generation.

3.1 The mapping and projection algorithm (MPA)

Since the slant range is much larger than the synthetic aperture, the incidence angle to the same target is nearly invariant during the interval of radar flying over a length of synthetic aperture. Thus in the imaging simulation, the whole scene is divided into cross-track lines along the azimuth dimension, and then the scattering contribution from the terrain objects lying in each line (included in the incidence plane) are calculated sequentially. It finally constructs a scattering picture or a scattering map of the whole scene.

Following from VRT, we derived a general expression of the scattering power at range r of x-position in azimuth dimension as

$$S(x,r) = \int_{\theta_0}^{\theta_1} \exp\left[-\int_{r_0}^{r} dr' \kappa_e^+(x,r',\theta)\right] P(x,r,\theta) \exp\left[-\int_{r}^{r_0} dr' \kappa_e^-(x,r',\theta)\right] r dx dr d\theta \tag{17}$$

where the phase function P stands for scattering, $\kappa_e^-(x,r,\theta)$, $\kappa_e^+(x,r,\theta)$ for the backward and forward extinction of an differential element dv at the position (x,r,θ) of the imaging space. θ is incident angle. Integrating $S(x,r)$ over the n,i-th pixel i.e. $x \in [x_n,x_{n+1})$, $s \in [s_i,s_{i+1})$, the corresponding discrete scattering map $S_{n,i}$ can be obtained. Considering polarimetric scattering, the computational discrete form of Eq. (17) for both volume and surface scatterer is given as

$$\overline{\mathbf{S}}_{n,i} = \sum_{m=m_n}^{m_{n+1}-1} \sum_{p=p_i}^{p_{i+1}-1} \sum_{q=0}^{Q-1} \prod_{p'=0}^{p} \overline{\mathbf{E}}_o^+(m,p',q)\overline{\mathbf{S}}_o(m,p,q) \prod_{p'=p}^{0} \overline{\mathbf{E}}_o^-(m,p',q) \tag{18}$$

$$\overline{\mathbf{S}}_o(m,p,q) = \begin{cases} v_{ef}\overline{\mathbf{S}}_{\mathrm{vol}}(m,p,q) & \text{for volume scatterer} \\ a_{ef}\overline{\mathbf{S}}_{\mathrm{surf}}(m,p,q) & \text{for surface scatterer} \end{cases} \tag{19}$$

$$\overline{\mathbf{E}}_o^{\pm}(m,p',q) = \begin{cases} \exp\left[-d_{ef}\,\overline{\boldsymbol{\kappa}}_e^{\pm}(m,p',q)\right] & \text{for volume scatterer} \\ 0 & \text{for surface scatterer} \end{cases} \tag{20}$$

where $v_{ef} = \gamma v_0$ is the effective volume of volume scatterer, calculated from particle number γ and volume v_0; $a_{ef} = \Delta x \Delta s$ is the effective area of surface scatterer; d_{ef} is the effective penetration depth. All of them take into account the proportion coefficients of each scatterer in the same discrete unit. The subscript o denotes scattering or extinction of a single scatterer. Eq. (18) presents a general description of SAR imaging which deals with single scattering, attenuation or shadowing of all possible objects in the imaging space.

We further devise a mapping and projection algorithm (MPA) to fast compute Eq. (17) The first step is to partition the scene and terrain objects lying in the horizontal plane xoy into multi-grids. Each grid unit contains its corresponding parts of the terrain objects overhead, e.g. ground, vegetation, building, etc.

As shown in Figure 3, two arrays $\underline{\mathbf{E}}^{\pm}$ and $\underline{\mathbf{S}}$ are deployed for temporary storage of scattering and extinction (the under-bar denotes an array). The mapping plane and projection plane are equally divided into discrete cells corresponding to elements of $\underline{\mathbf{S}}, \underline{\mathbf{E}}^{\pm}$. Attenuation or shadowing effect of each grid unit is counted in the increasing sequence of y axis and cumulatively multiplied into the array $\underline{\mathbf{E}}^{\pm}$.

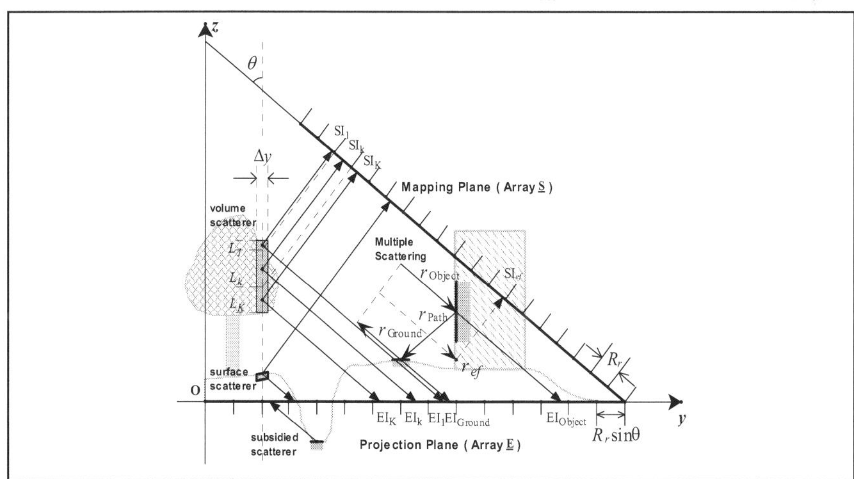

Fig. 3. The MPA algorithm for natural media scene.

The MPA calculating sequence of the terrain objects in one grid unit is: first to calculate the scattering power of the current terrain object and accumulate onto the array $\underline{\mathbf{S}}$, where the attenuation or shadowing from other terrain objects are obtained from array $\underline{\mathbf{E}}^{\pm}$; then, to calculate the attenuation or shadowing caused by the current terrain object itself, and cumulatively multiply onto the array $\underline{\mathbf{E}}^{\pm}$; ultimately when all grid units of the current line are counted, the scattering map is exactly produced in the array $\underline{\mathbf{S}}$.

As shown in Figure 3, a vertical segment (length L) of volume scatterers above the current grid unit is firstly divided into K sub-segments L_k in order to make each sub-segment mapping into one cell SI_k in mapping plane, while its mid-point is projected to cell EI_k in projection plane. After the mapping and projecting, the elements of array $\underline{\mathbf{S}}$ are refreshed as

$$\underline{\mathbf{S}}[\mathrm{SI}_k] := \underline{\mathbf{S}}[\mathrm{SI}_k] + \underline{\mathbf{E}}^{+}[\mathrm{EI}_k] \cdot \overline{\mathbf{S}}_o^k \cdot \underline{\mathbf{E}}^{-}[\mathrm{EI}_k], \quad k = 1,..,K \tag{21}$$

$\overline{\mathbf{S}}_o^k$ is the scattering of the k-th sub-segment. Assuming that volume scatterers always uniformly fill the whole sub-segment, then the effective volume in Eq. (19) is $v_{ef} = \Delta x \Delta y L_k f_s$. Therefore, we have

$$\overline{\mathbf{S}}_o^k = \Delta x \Delta y L_k f_s \overline{\mathbf{S}}_{\mathrm{vol}} = v_{ef} \overline{\mathbf{S}}_{\mathrm{vol}} \tag{22}$$

where f_s is the fractional volume of scatterers. Otherwise, it is necessary to estimate the total number of scatterers γ within the sub-segment, and calculate scattering from Eq. (19). The next step is to refresh the elements of the array $\underline{\mathbf{E}}^{\pm}$ as

$$\begin{cases} \underline{\mathbf{E}}^{-}[\mathrm{EI}_k] := \overline{\mathbf{E}}_{o,k}^{-} \cdot \underline{\mathbf{E}}^{-}[\mathrm{EI}_k] & k = 1,..,\mathrm{K} \\ \underline{\mathbf{E}}^{+}[\mathrm{EI}_k] := \underline{\mathbf{E}}^{+}[\mathrm{EI}_k] \cdot \overline{\mathbf{E}}_{o,k}^{+} & k = 1,..,\mathrm{K} \end{cases} \tag{23}$$

where $\overline{\mathbf{E}}_{o,k}^{\pm} = \exp(-d_{ef}\overline{\boldsymbol{\kappa}}_e)$ is the attenuation of the k-th sub-segment. Due to discrete calculation for attenuation, the array $\underline{\mathbf{E}}^{\pm}$ should store the averaged attenuation. Thus, the ratio of shadowed area of the sub-segment to the whole projection cell should be taken into account, when counting the contribution of the sub-segment to the mean attenuation. The effective depth in $\overline{\mathbf{E}}_{o,k}^{\pm}$ is calculated as

$$d_{ef} = \frac{v_{ef}}{R_x R_r / \tan\theta} = \frac{\Delta x \Delta y L_k}{R_x R_r / \tan\theta} = \tan\theta \frac{\Delta x \Delta y L_k}{R_x R_r} \tag{24}$$

It can be seen from the effective depth d_{ef} that if we redistribute all volume scatterers of the sub-segment and make them cover the whole projection cell under the invariance of density, then the layer depth of redistributed scatterers along the projection line is seen as the effective depth.

In a similar way, larger surface scatterers, like vertical wall surfaces, cliff slopes etc., are segmented according to mapping cells. But, for nearly horizontal surfaces, e.g. ground or roof surface, the facet cut by one grid unit is small enough and is projected into one projection cell, i.e. $K = 1$. Hence, given the normal vector of facet $\hat{n}$, the effective area and its scattering in Eq. (21) are calculated in two ways:

$$\overline{\mathbf{S}}_o^k = a_{ef}\overline{\mathbf{S}}_{\text{surf}} = \begin{cases} \Delta x \Delta y \overline{\mathbf{S}}_{\text{surf}} \cos\phi_2 / \cos\phi_1 & K = 1 \\ \Delta x L_k \overline{\mathbf{S}}_{\text{surf}} \cos\phi_2 / \sin\phi_1 & K > 1 \end{cases} \tag{25}$$

where ϕ_1 is the angle between $\hat{n}$ and z axis; ϕ_2 is the angle between $\hat{n}$ and $\hat{i}$ incidence. L_k is the length of sub-segment. Moreover, the array $\underline{\mathbf{E}}^\pm$ should be set to zero over all projection interval from the first projection cell to the last one, i.e.

$$\underline{\mathbf{E}}^\pm[j] := 0 \quad \text{EI}_1 \le j \le \text{EI}_K \tag{26}$$

The elements of the array $\underline{\mathbf{E}}^\pm$ are initialized to unit matrices at the start point of calculation for each line, while the array $\underline{\mathbf{S}}$ should be saved before cleared to start the next line. Note that each line is strictly computed in the sequence of increasing y in order to precisely count attenuation caused by the terrain objects ahead.

Double and triple scatterings between terrain objects and the underlying ground are regarded as shown in Figure 3. First, the corresponding ground patch of multiple scattering can be located by the ray tracing technique. Then the length of propagation path, i.e. the effective range r_{ef} can be given as

$$r_{ef} = \begin{cases} \left(r_{\text{Object}} + r_{\text{Ground}} + r_{\text{Path}}\right)/2 & \text{for double scattering} \\ r_{\text{Object}} + r_{\text{Path}} & \text{for triple scattering} \end{cases} \tag{27}$$

In the same way, multiple scattering of a segment of scatterers is mapped into the array $\underline{\mathbf{S}}$ by r_{ef}. The same step of dividing into sub-segments is employed if the mapping interval exceeds one cell.

For double scattering, the attenuation or shadowing suffered through the propagation paths of r_{Object}, r_{Ground} are the same with single scattering of the terrain object and ground patch, respectively. It means that the array $\underline{\mathbf{E}}^\pm$ can be used as well. The attenuation during the way from the terrain object to ground patch r_{Path}, is omitted for simplicity. Therefore, the double scattering contribution can be written as

$$\underline{\mathbf{S}}[\text{SI}_{ef}] := \underline{\mathbf{S}}[\text{SI}_{ef}] + \underline{\mathbf{E}}^+[\text{EI}_{\text{Ground}}] \cdot \overline{\mathbf{S}}_{\text{Ground}}^{2+} \cdot \rho \overline{\mathbf{S}}_{\text{Object}}^{2-} \cdot \underline{\mathbf{E}}^-[\text{EI}_{\text{Object}}] + \underline{\mathbf{E}}^+[\text{EI}_{\text{Object}}] \cdot \rho \overline{\mathbf{S}}_{\text{Object}}^{2+} \cdot \overline{\mathbf{S}}_{\text{Ground}}^{2-} \cdot \underline{\mathbf{E}}^-[\text{EI}_{\text{Ground}}] \tag{28}$$

where $\text{EI}_{\text{Ground}}$, $\text{EI}_{\text{Object}}$ are the projection indices of the terrain object and the ground patch respectively. $\overline{S}_{Ground}^{2\pm}$, $\overline{S}_{Object}^{2\pm}$ are the Mueller matrices of the ground patch and the terrain object along the forward and backward directions of double-scattering, respectively. The coefficient $\rho = v_{ef}$ or a_{ef} is calculated similarly to Eqs. (19,22,25).

For triple scattering (object-ground-object), the expression can be directly written as:

$$\underline{S}[\text{SI}_{ef}] := \underline{S}[\text{SI}_{ef}] + \underline{E}^{+}[\text{EI}_{Object}] \cdot \overline{S}_{Object}^{2+} \cdot \overline{S}_{Ground}^{3} \cdot \rho \overline{S}_{Object}^{2-} \cdot \underline{E}^{-}[\text{EI}_{Object}] \tag{29}$$

where $\overline{S}_{Ground}^{3}$ is the Mueller matrix of ground patch along the triple scattering path; $\rho = v_{ef}$ or a_{ef}.

Correct sequence for calculation of each grid unit is first to count single scattering, then multiple scattering, at last its attenuation or shadowing. Notice that the ground patch of multiple scattering could be located at any place around the terrain object. For simplicity, we take the projection cell in the current incidence plane as an approximate substitute. Some other coding techniques are adopted to guarantee the correct computation and make it more efficient.

3.2 Scattering models for terrain objects

For vegetation canopies, VRT model of a layer of random non-spherical particles (Jin, 1994) is employed. Leaves, small twigs or thin stems are modeled as non-spherical dielectric particles under the Rayleigh or Rayleigh-Gans approximation, while branches, trunks or thick stems are modeled as dielectric cylinders. A tree model is composed of crown and trunk. The crown is a cloud with a simple geometrical shape containing randomly oriented non-spherical particles. The trunk is an upright cylinder with the top covered by the crown. Oblate or disk-like particles and elliptic crown are adopted for broad-leaf forest, while prolate or needle-like particles and cone-like crown are for needle-leaf forest. In addition, the crown shapes take a small random perturbation. Similarly, a farm filed is modeled as a layer of randomly oriented non-spherical particles with perturbed layer depth. If the grid units are small enough, the segment of crown or crops cut by one grid unit can be regarded as a segment fully filled with particles. Differences between the trunk and crown are: (a) the trunk is solid within a grid unit i.e. fractional volume is set to 1; (b) double scattering between the trunk and ground is taken into account.

Scattering of the building is seen as the surface scattering from its wall and roof surfaces and multiple interactions with the ground surface. The integral equation method (IEM) is employed to calculate surface scattering. The building is modeled as a parallelepiped with an isoscales triangular cylinder layered upon it. Due to the orientation of wall surface and roof surface, the incident angle must be transferred into the local coordinates of the surface, as well as the polar basis of wave propagation. Geometrical relationships among the wall surfaces, the roof surfaces, as well as the double and triple scatterings between the wall and ground can be found in Franceschetti et al. (2002). Figure 2 illustrates the building model, its image and shadowing in SAR imagery.

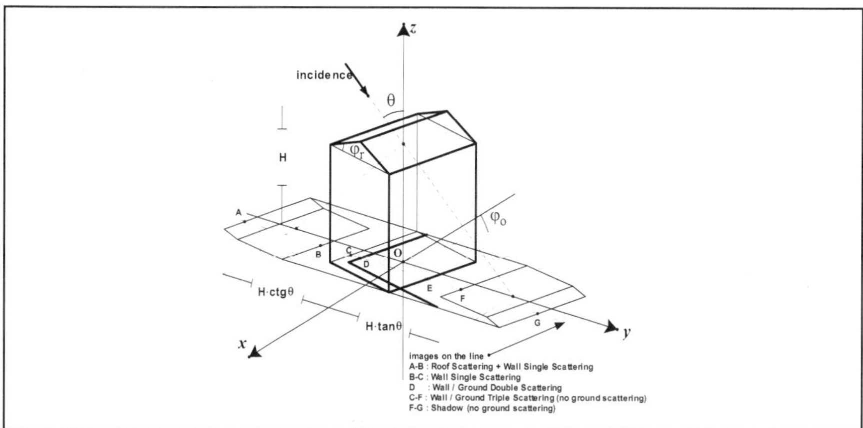

Fig. 4. Image and shadow of a building in SAR imagery

For ground facet, the tangent vectors along x, y axes on the current grid unit are used to construct the facet plane.

The MPA approach is based on the VRT theory for incoherent scattering power, not coherent summation of scattering fields. However, speckle due to coherent interference is one of the most critical characteristics of SAR imagery, especially for studies of SAR image filtering and optimization. Assuming Gaussian probability distribution of the scattering vector $\mathbf{k}_L = \left[S_{hh}, S_{hv}, S_{vh}, S_{vv} \right]^T$, random scattering vector is then generated as follows:

$$\mathbf{k}_L = \mathbf{k}^0 \cdot \overline{\mathbf{C}}^{\frac{1}{2}}, \quad k^0_{\ i} = I_i + jQ_i, \quad i = 1,..,d \tag{30}$$

where I_i, Q_i are independent Gaussian random numbers with the zero mean and unit variance. Given the positive semidefinite property of $\overline{\mathbf{C}}$, its square root can be obtained by Cholesky decomposition.

3.3 Simulation results

The configurations and parameter settings of the radar and platform in simulation are selected follow the AIRSAR sensor of NASA/JPL.

A virtual comprehensive terrain scene is designed, as shown in Figure 5(a), as according to a true DEM of Guangdong province, south China. It contains different types of forests covering on the hill, crops farmland, ordered or random buildings in urban and suburban regions, roads and rivers.

A simulated scattering map (decibel of normalized scattering power, from -50dB to 0dB, pseudo color: R-HH, G-HH, B-HV.) at L band with 12m resolution (pixel spacing is 5m) is displayed in Figure 5(b). The simulated SAR image at L band is shown in Figure 5(c).

Figures 6(a,b) give the simulated scattering map and SAR image at C band, respectively. In this higher band, vegetation scatterings are significantly increased, as well as attenuation is increased and wave penetrable depth is decreased. It causes that trees become thinner while the shadowing becomes darker.

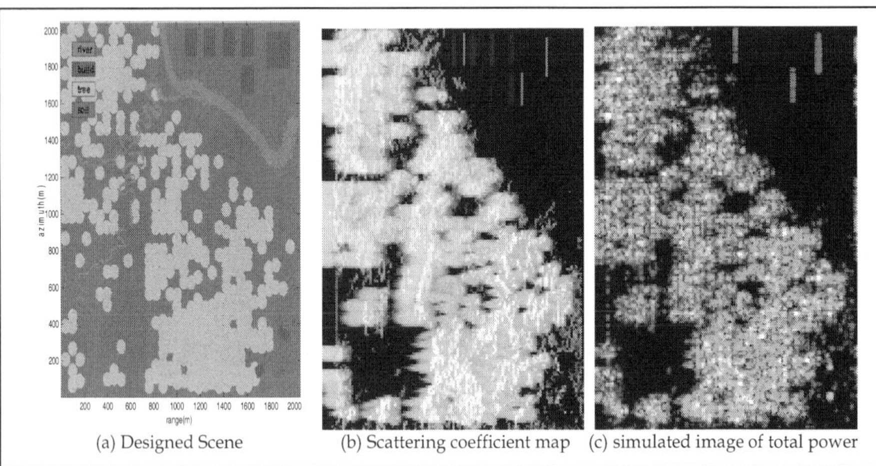

(a) Designed Scene   (b) Scattering coefficient map   (c) simulated image of total power

Fig. 5. Simulated SAR image at L band

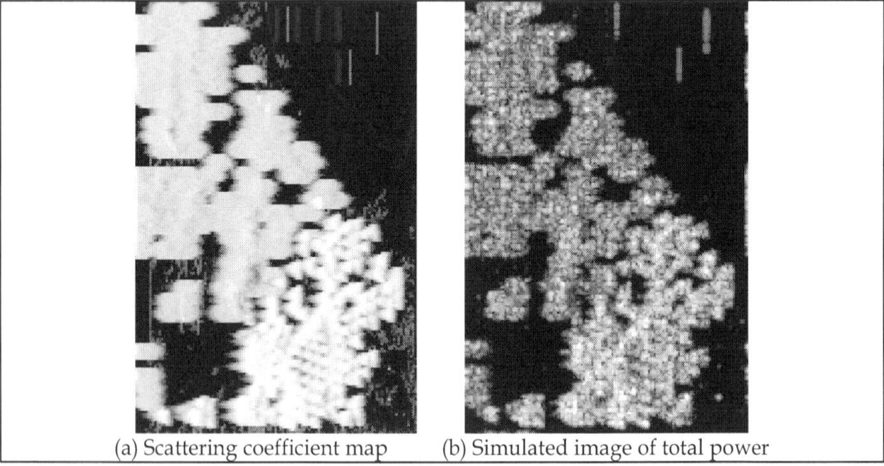

(a) Scattering coefficient map   (b) Simulated image of total power

Fig. 6. Scattering and Simulated SAR image at C bands

At the upper-right corner, the blocks images appear like parallel lines, which reveal the dominance of double scattering in the urban area. However, on the other side below the river, the buildings are oriented nearly 45 degrees to the radar flying direction. As a result, the cross-pol scattering (blue) is stronger, while double scattering is reduced. Timberlands can be classified into two classes: broad leaves in yellow color due to balance between HH and VV scattering, and narrow leaves in green color due to weaker scattering of HH. The narrow leaves trees make weaker attenuation of HH and therefore stronger HH scattering of the shadowed ground, which is the reason why most areas of narrow leaves forest become red. Overlay and shadowing effects caused by mountainous topography are particularly perceptible. Croplands are generally uniform and dense, and appear like patches in the image. Its VV scattering (green) is always stronger than HH. They have distinguishable brightness due to different density, sizes, shapes etc. More detail discussion can be seen in (Xu and Jin, 2006)

3.4 Bistatic SAR imaging

Bistatic SAR (BISAR) with separated transmitter and receiver flying on different platforms has become of great interests. Most of BISAR research is focused on the engineering realization and signal processing algorithm with few on land-based bistatic experiments or theoretical modeling of bistatic scattering. The MPA provides a fast and efficient tool for monostatic imaging simulation. It involves the physical scattering process of multiple terrain objects, such as vegetation canopy, buildings and rough ground surfaces (Xu and Jin, 2008a).

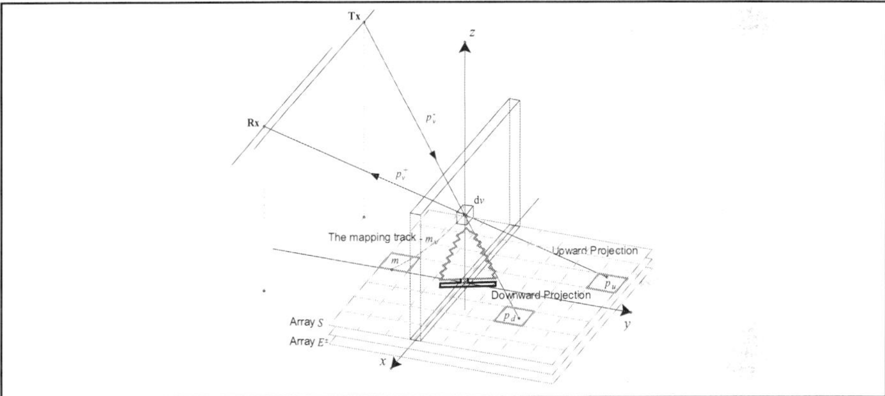

Fig. 7. Bistatic MPA

Similar to mono-static MAP, the bistatic-MAP steps are described as follows,

(1) The arrays $\underline{\mathbf{E}}^+, \underline{\mathbf{E}}^-$ are initialized as unit matrices, $\underline{\mathbf{S}}$ is as zero matrices. Then, visit the grid units, sequentially, along $+y$ direction.

(2) Perform 3D projections of the current grid unit along incidence and scattering directions to the cells p_d, p_u of $\underline{\mathbf{E}}^+, \underline{\mathbf{E}}^-$, respectively.

(3) Determine the mapping position of the current grid unit based on the information of its synthetic aperture and Doppler history, which corresponds to the cell m of the array $\underline{\mathbf{S}}$.

(4) Obtain or calculate the scattering matrix $\overline{\mathbf{S}}_0$ and the upward/downward attenuation matrix $\overline{\mathbf{E}}_0^\pm$ of the current grid unit.

(5) Refresh the elements of $\underline{\mathbf{S}}$ as

$$\underline{\mathbf{S}}[m] := \underline{\mathbf{S}}[m] + \underline{\mathbf{E}}^+[p_u] \cdot \overline{\mathbf{S}}_0 \cdot \underline{\mathbf{E}}^-[p_d] \tag{31}$$

Refresh the elements of $\underline{\mathbf{E}}^{+}, \underline{\mathbf{E}}^{-}$ as

$$\underline{\mathbf{E}}^{-}[p_d] := \overline{\mathbf{E}_0^{-}} \cdot \underline{\mathbf{E}}^{-}[p_d], \quad \underline{\mathbf{E}}^{+}[p_u] := \underline{\mathbf{E}}^{+}[p_u] \cdot \overline{\mathbf{E}_0^{+}} \tag{32}$$

(6) Return to Step (2), and continue to visit the next grid at the same coordinate of y or, if all grids at this coordinate have been visited, step forward to a larger coordinate of y till the whole scene is exhausted.

Figures 7,8 explain the bistatic imaging process of a tree and a building. One of the differences between bistatic and monostatic cases is the shadowing area projected in both incidence and scattering directions. Additionally, due to the split incidence and scattering directions, double scattering terms of object-ground and ground-object are different. One reflects at the ground and diffusely scatters from the object, the other scatters at the object and then reflects from the ground. These two scattering terms have different paths which give rise to separated double-scattering images.

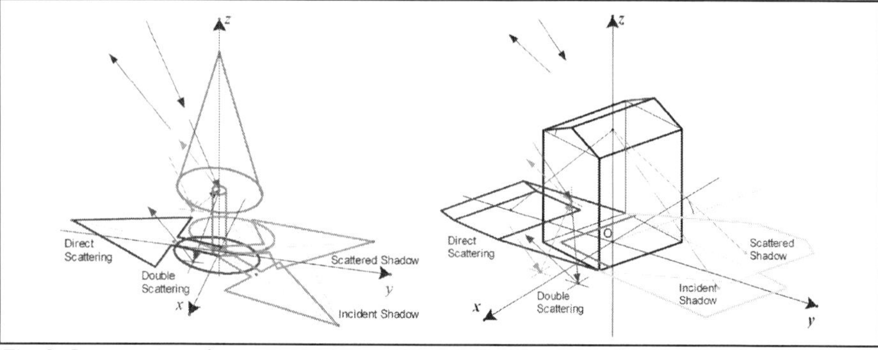

Fig. 8. Composition of a tree and a house in BISAR image

A comprehensive virtual scene as shown in Figure 9(a) is designed for simulation and analysis. Simulated BISAR images of the configurations, AT (across track) and TI (translational invariant) are shown in Figures 9(b,c), respectively.

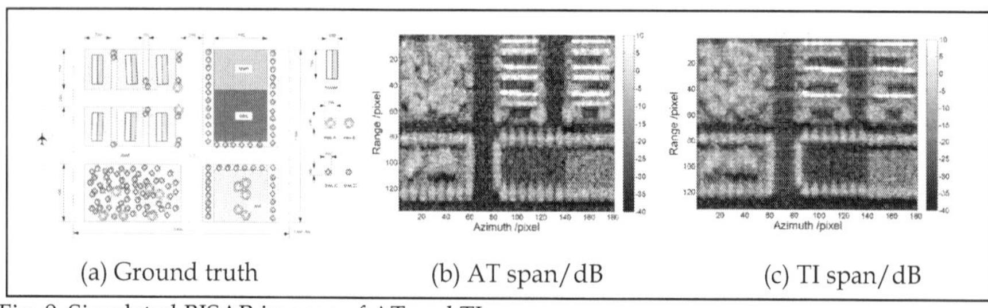

(a) Ground truth   (b) AT span/dB   (c) TI span/dB

Fig. 9. Simulated BISAR images of AT and TI

3.5 Unified bistatic polar bases and polarimetric analysis

It is found in the BISAR simulation that in AT BISAR image, bistatic scattering of terrain objects preserves polarimetric characteristics analogous to monostatic case. while the major disparity is on total scattering power.

However, the polarimetric parameters behavior is different in the image of general TI mode. First, H is generally higher, probably due to the fact that large bistatic angle is more likely to reveal the randomness and complexity of the object, and makes the scattering energy more uniformly distributed on different polarizations. Second, almost all terrain surfaces show $\alpha > 45°$. It is found that scattering energy is concentrated on the 4-th component of Pauli scattering vector $\mathbf{k}_p$ all over the image.

It seems that α might lose its capability to represent the polarimetric characteristics under certain bistatic configuration. Generally speaking, all parameters α,β,γ in monostatic SAR image cannot reflect polarimetric characteristics in the general TI case.

Conventional polarimetric parameters such as α,β,γ may largely depend on angular settings of the sensors rather than instinct properties of the target. It becomes inconvenient to interpret the physical meaning of polarimetric parameters when they are severely involved with the bistatic configuration.

As proposed in many studies, inverse problems of bistatic scattering are usually discussed in a coordinates system determined by the bisectrix. We believe it is important to first transform the conventional bistatic polar bases to a new one defined by the bisectrix. The unified bistatic polar bases are defined as

$$\hat{h}'_i = \frac{\hat{b} \times \hat{k}_i}{\left|\hat{b} \times \hat{k}_i\right|}, \quad \hat{v}'_i = \hat{h}'_i \times \hat{k}_i, \quad \hat{b} = \frac{\hat{k}_s - \hat{k}_i}{\left|\hat{k}_s - \hat{k}_i\right|} \tag{33}$$

where $\hat{b}$ denotes the bisectrix, which is defined as the bisector of the incident and scattered wave vectors in the plane, as shown in Figure 10(a).

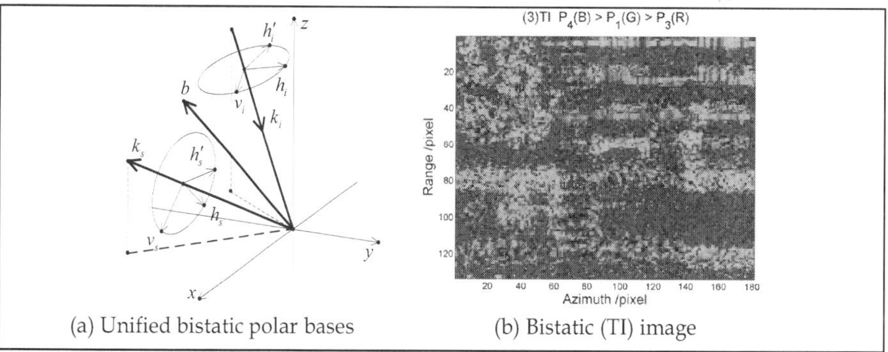

(a) Unified bistatic polar bases   (b) Bistatic (TI) image

Fig. 10. Unified bistatic polar bases and bistatic (TI) pseudo color images coded by NPC after unified bistatic polar bases transform

The relationship between the unified bistatic polar bases and the conventional polar bases can be described by polar basis rotation, i.e. $\mathbf{E}, \overline{\mathbf{S}}$ defined in $(\hat{v}_i, \hat{h}_i, \hat{k}_i)$ are re-defined as

$$a = \tan^{-1}(|S_{vv}| / |S_{hh}|), \quad b = \frac{1}{2}\arg(S_{vv} / S_{hh}), \quad c = \cos^{-1}(\sqrt{2}\,|S_x| / |\mathbf{k}_L|) \tag{38}$$

Rotating the angle ψ of the polarization base along with the sight line, the electrical field vector $\mathbf{E}$ becomes $\mathbf{E}'$ as

$$\mathbf{E}' = [R] \cdot \mathbf{E}, \quad [R] = \begin{bmatrix} \cos\psi & -\sin\psi \\ \sin\psi & \cos\psi \end{bmatrix} \tag{39}$$

In backscattering direction, this rotation makes the scattering vector $\mathbf{k}_P$ to be $\mathbf{k}'_P$, which can be expressed as

$$\mathbf{k}'_P - [U] \cdot \mathbf{k}_P, \quad [U] = \begin{bmatrix} 1 & 0 & 0 \\ 0 & \cos 2\psi & \sin 2\psi \\ 0 & -\sin 2\psi & \cos 2\psi \end{bmatrix} \tag{40}$$

Applying minimization of cross polarization (min-x-pol) to $\mathbf{k}'_P$, i.e. let

$$\frac{\partial |k'_{P,3}|^2}{\partial \psi} = 0, \quad \frac{\partial^2 |k'_{P,3}|^2}{\partial \psi^2} > 0$$

it yields the deorientation angle ψ_m is obtained as

$$\psi_m = \left[\operatorname{sgn}\{\cos(\phi_2 - \phi_3)\} \cdot \frac{\beta}{2} \right]_{\frac{\pi}{2}} \tag{41}$$

It means that such rotation of the angle ψ_m along the sight line makes $\mathbf{k}_P$ to the min-x-pol status as deoriented $\mathbf{k}_P^d$.

New parameters u, v, w are then defined from the deoriented scattering vector of Eqs. (38,40) as:

$$u = \sin c \cos 2a, \quad v = \sin c \sin 2a \cos 2b, \quad w = \cos c . \tag{42}$$

In the case of non-deterministic targets, we obtain the uncorrelated scattering vectors, i.e. eigenvectors, through eigen-analysis of coherency matrix. Here the most significant eigenvector i.e. principal eigenvector is considered as the representative scattering vector of non-deterministic targets. Thus, the deorientation of non-deterministic targets is merely conducted on the principal eigenvector.

4.2 Numerical simulation

Based on the model of Figure 1, polarimetric scattering is calculated, and the scattering vector and Mueller matrix are obtained. A model of a layer of random non-spherical particles above a rough surface for scattering from terrain surfaces, as shown in Figure 1, is applied to numerical simulation of u, v, H at L band in various cases.

Figures 12(a,b) show numerical relationship between the parameters u, v and the particle parameters: Eular angle orientation and particle shape (oblate or prolate spheroids). It can be seen that $|u|$ indicates the non-symmetry of particle's projection along with the sight line, and the dielectric property of the particle, $|\varepsilon_s|$, affects $|u|$, and $\varepsilon_s''/\varepsilon_s'$ affects v.

Figure 12(a) presents the distributions of some typical terrain surfaces on the $|u|-H$ plane, which are obtained from the scattering model of Figure 1. It can be seen that H indicates the complexity of layer structures of terrain surfaces. Different dielectric property of the soil land and oceanic surface also makes u identifiable on the $|u|$ axis.

Figure 12(b) presents the distribution of scattering terms on the plane $|u|-v$, where M_i denotes the i-th order scattering. Concluded from these figures, H indicates the canopy randomness, u is useful for distinguishing different terrain targets and v is helpful to take account of scattering mechanisms of different-orders.

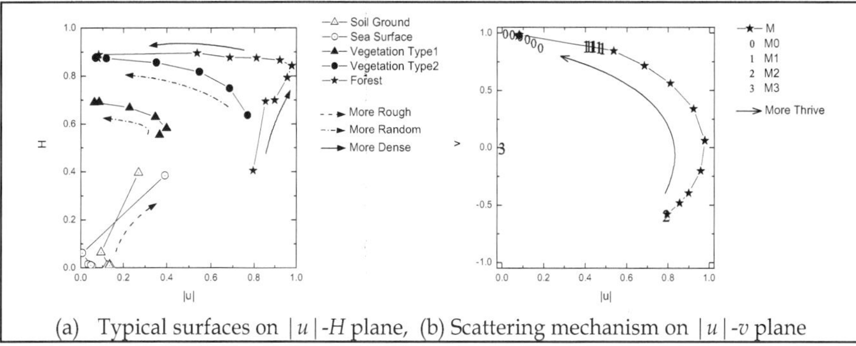

(a)   Typical surfaces on $|u|$-H plane,   (b) Scattering mechanism on $|u|$-v plane

Fig. 12. Distributions of typical surfaces and different scattering mechanism

4.3 Surface classification

The flow-chart of the classification method of PolSAR Data based on deorientation and new parameters u, v, ψ is shown in Figure 13. The classification decision tree of parameters u, v, H is displayed on Figure 14.

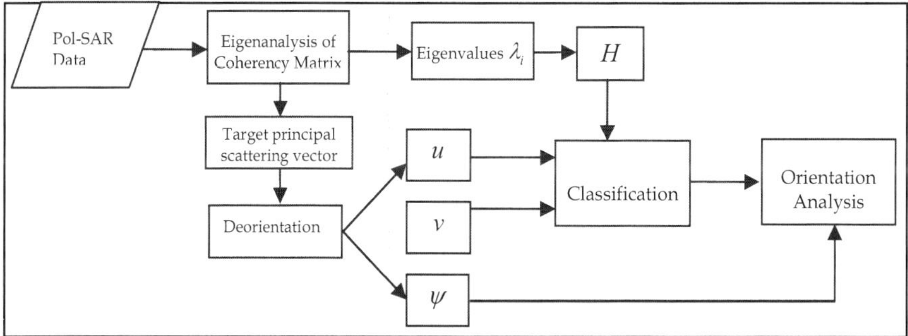

Fig. 13. Flow-chart of the surface classification.

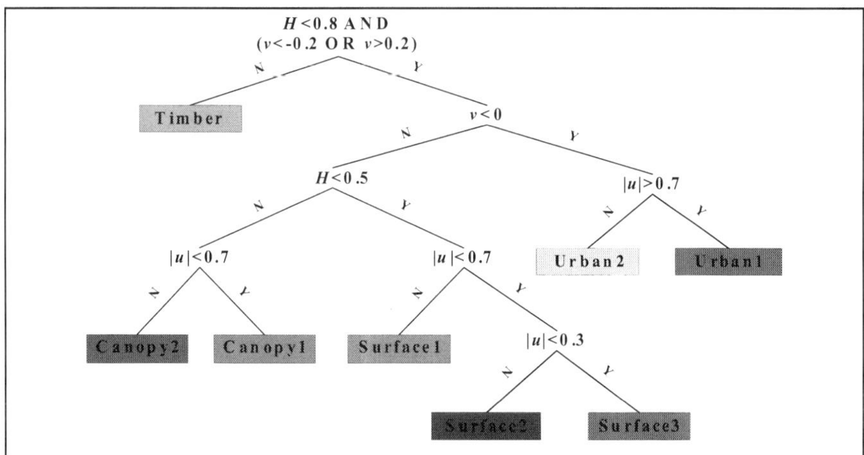

Fig. 14. Classification decision tree of parameters u, v, H .

An AirSAR data at L band of the Boreal area in Canada with rich resources of vegetation is chosen for classification and orientation-analysis, as shown in Figure 15(a). Figure 15(b) is the deorientation classification over this area. Terrain surfaces are divided into 8 classes following the decision-tree: Timber, Urban 2, Urban 1, Canopy 2, Canopy 1, Surface 3, Surface 2 and Surface 1.

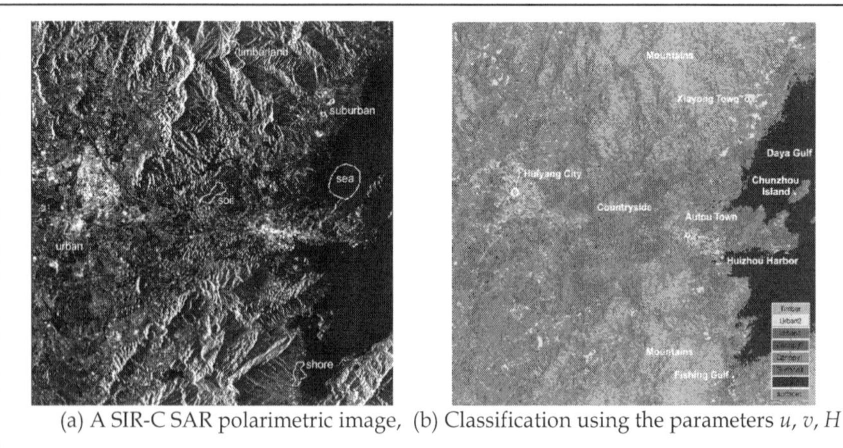

(a) A SIR-C SAR polarimetric image, (b) Classification using the parameters u, v, H

Fig. 15. A SIR-C SAR image and surface classification

Figure 16 shows the data distributions on the planes $|u| - H$ for parameter $v > 0.2$ and $v < -0.2$, and corresponding classification.

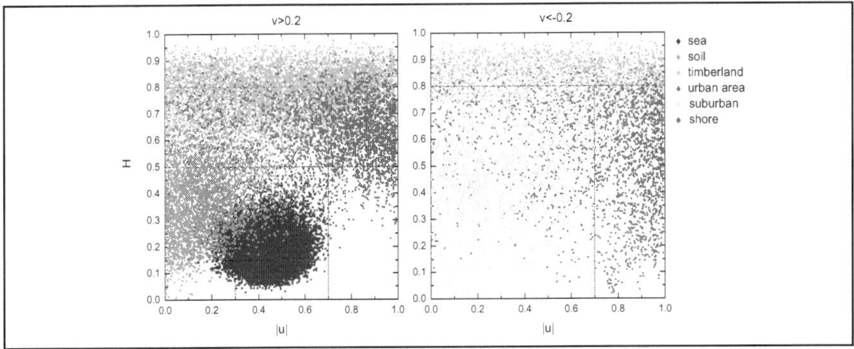

Fig. 16. Classification by u,v,H for 19 classes

The orientation distribution of several classes are selected to be displayed in Figures 17(a,b). It can be seen that

(a) While in the Urban 1 (sparse forest of vertical trunks, instead of artificial constructions), there are uniformly vertical orientations indicating orderly trunks.

(b) Random orientation in the Canopy 1 means that the vegetation canopy in this area might be the disordered bush. Note that the region with the roads inside the forest might show randomness confused by the bush vegetation on the roadsides.

Following the above orientation analysis, the terrain surfaces are further classified into the subclasses and are identified by their types.

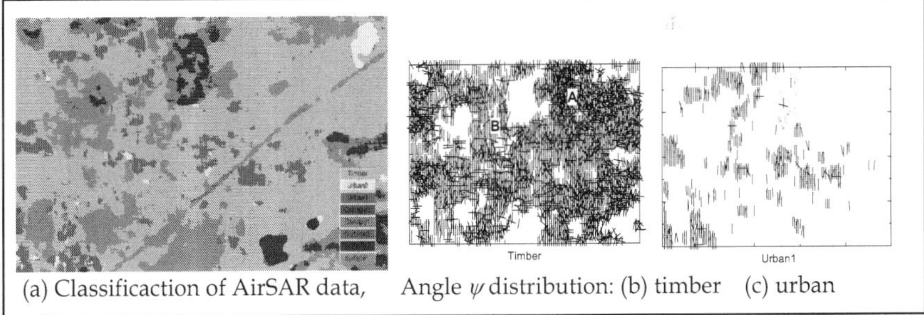

(a) Classificaction of AirSAR data,   Angle ψ distribution: (b) timber   (c) urban

Fig. 17. Classification of AirSAR data, Boreal area.

4.4 Faraday rotation and Surface classification

To obtain the moisture profiles in subcanopy and subsurface, it needs UHF/VHF bands (435MHz, 135MHz) to penetrate through the dense subcanopy (~20kg/m² and more) with little scattering and reach subsurface. However, when the wave passes through the anisotropic ionosphere and action of geomagnetic field, the polarization vectors of electromagnetic wave are rotated, which is called the Faraday rotation (FR). The FR depends on the wavelength, electronic density, geomagnetic field, the angle between the direction of wave propagation and geomagnetic field, and the wave incidence angle.

Assuming that the propagation direction is not changed passing through homogeneous ionosphere, $ds = \sec\theta_i dz$ (where $\hat{z}$ is the normal to the surface), and geomagnetic field keeps constant as one at 400 km altitude, FR is simply written as

$$\Omega_F \approx -2620\rho_e B(400)\lambda^2 \cos\Theta_B \sec\theta_i \quad \text{(radians)} \tag{43}$$

where Θ_B is the angle between the directions of electromagnetic wave propagation and geomagnetic field, ρ_e is the total electron content per unit area (10^{16} electron/m²), and $B(400)$ is the intensity of geomagnetic field (T) at 400 km altitude.

The FR may decrease the difference between co-pol backscattering, enhance cross-polarized echoes, and mix different polarized terms. Thus, the satellite-borne SAR data at low frequency becomes distorted due to FR effect. Since the FR is proportional to the square power of the wavelength, it yields especially serious impact on the SAR observation operating at the frequency lower than L band. The FR angle at P band can reach dozens of degrees.

As a polarized electromagnetic wave passes through the ionosphere, the scattering matrix $\overline{\mathbf{S}}^F$ with FR (indicated by superscript F) is written by the scattering matrix $\overline{\mathbf{S}}$ without FR as follows

$$\begin{bmatrix} S_{hh}^F & S_{hv}^F \\ S_{vh}^F & S_{vv}^F \end{bmatrix} = \begin{bmatrix} \cos\Omega & \sin\Omega \\ -\sin\Omega & \cos\Omega \end{bmatrix} \begin{bmatrix} S_{hh} & S_{hv} \\ S_{hv} & S_{vv} \end{bmatrix} \begin{bmatrix} \cos\Omega & \sin\Omega \\ -\sin\Omega & \cos\Omega \end{bmatrix} \tag{44a}$$

$$\begin{aligned} S_{hh}^F &= S_{hh}\cos^2\Omega - S_{vv}\sin^2\Omega, & S_{vv}^F &= -S_{hh}\sin^2\Omega + S_{vv}\cos^2\Omega \\ S_{hv}^F &= S_{hv} + (S_{hh} + S_{vv})\sin\Omega\cos\Omega, & S_{vh}^F &= S_{hv} - (S_{hh} + S_{vv})\sin\Omega\cos\Omega \end{aligned} \tag{44b}$$

The measured polarimetric data with FR, $\overline{\mathbf{S}}^F$ or $\overline{\mathbf{M}}^F$, are distorted and cannot be directly applied to terrain surface classification. However, fully polarimetric 4×4-D $\overline{\mathbf{M}}$ without FR can be inverted from $\overline{\mathbf{M}}^F$. And it shows that the remained $\pm\pi/2$ ambiguity error does not affect the classification parameters: u, v, H, α, A. Based on intuitive assumption of gradual change of FR degree with geographical position, a method of 2D phase unwrapping method with random benchmark can be employed to eliminate the $\pm\pi/2$ ambiguity error is designed. Some example shows the fully polarimetric SAR without FR at low frequency, such as P band, can be fully inverted (Qi and Jin, 2007).

5. Change Detection of Terrain Surface from Multi-Temporal SAR Images

Multi-temporal observations of SAR remote sensing imagery provide fast and practical technical means for surveying and assessing such vast changes. One direct application is to detect and classify the information on changes in the terrain surfaces. It would be laborious to make an intuitive assessment for a huge amount of multi-temporal image data over a vast area. Such assessment based on qualitative gray-level analysis is not accurate and might lose some important information. How to detect and automatically analyze information on change in the terrain surfaces is a key issue in remote sensing.

In this section, two-thresholds EM and MRF algorithm (2EM-MRF) is developed to detect the change direction of backscattering enhanced, reduced and unchanged regimes from the SAR difference image (Jin and Luo, 2004, Jin and Wang, 2009).

On May 12, 2008, a major earthquake, measuring 8.0 on the Richter scale, jolted southwestern China's Sichuan Province, Wenchuan area. To evaluate the damages and terrain surface changes caused by the earthquake, multi-temporal ALOS PALSAR image data before and after the earthquake are applied to detection and classification of the terrain surface changes. Using the tools of Google Earth for surface mapping, the surface change situation after the earthquake overlapped the DEM topography can be demonstrated in multi-azimuth views as animated cartoon. The detection and classification are also compared with the optical photographs. It is proposed that multi-thresholds EM and MRF analysis may become traceable when multi-polarization, multi-channels, multi-sensors multi-temporal image data become available.

5.1 Two thresholds expectation maximum algorithm

Consider two co-registered images X_1 and X_2 with size $I \times J$ at two different times (t_1, t_2). Their difference image is $X = (X_2 - X_1)$, and denote $X_D = \{X(i,j),\ 1 \le i \le I, 1 \le j \le J\}$ (in some cases, $X = X_2 / X_1$ also can be used).

As usually, one-threshold expectation maximum (EM) algorithm has been employed to classify two opposite changes: the unchanged class ω_n and the changed class ω_c. The probability density function $P(X)$, $X \in X_D$ was modeled as a mixture distribution of two density components associated with ω_n and ω_c, i.e.

$$p(X) = p(X \mid \omega_n)P(\omega_n) + p(X \mid \omega_c)P(\omega_c) \tag{45}$$

Assume that both the conditional probabilities $p(X \mid \omega_n)$ and $p(X \mid \omega_c)$ are modeled by Gaussian distributions. The iterative equations for estimating the statistical parameters and the *a priori* probability for the class ω_n are the following

$$P^{t+1}(\omega_n) = \frac{\displaystyle\sum_{X(i,j)\in X_D} \frac{P^t(\omega_n)p^t(X(i,j)\mid\omega_n)}{p^t(X(i,j))}}{IJ} \tag{46a}$$

$$\text{mean } \mu_n^{t+1} = \frac{\displaystyle\sum_{X(i,j)\in X_D} \frac{P^t(\omega_n)p^t(X(i,j)\mid\omega_n)}{p^t(X(i,j))}X(i,j)}{\displaystyle\sum_{X(i,j)\in X_D} \frac{P^t(\omega_n)p^t(X(i,j)\mid\omega_n)}{p^t(X(i,j))}} \tag{46b}$$

$$\text{variance } (\sigma_n^2)^{t+1} = \frac{\displaystyle\sum_{X(i,j)\in X_D} \frac{P^t(\omega_n)p^t(X(i,j)\mid\omega_n)}{p^t(X(i,j)}[X(i,j)-\mu_n^t]^2}{\displaystyle\sum_{X(i,j)\in X_D} \frac{P^t(\omega_n)p^t(X(i,j)\mid\omega_n)}{p^t(X(i,j))}} \tag{46c}$$

where the superscripts t and $t+1$ denote the current and next iterations.

Analogously, these equations can also be used to estimate $p(\omega_c)$ with μ_c and σ_c^2.

The estimates are computed starting from the initial values by iterating the above equations until convergence. The initial value of the estimated parameters can be obtained by the analysis of the histogram of the difference image. A pixel subset S_n likely to belong to ω_n and a pixel subset S_c likely to belong to ω_c can be obtained by preliminarily selecting two-threshold T_n and T_c on the histogram. This is equivalent to solving the ML boundary T_o for the two classes ω_n and ω_c on the difference image. The optional threshold T_o requires

$$\frac{P(\omega_c)}{P(\omega_n)} = \frac{p(X \mid \omega_n)}{p(X \mid \omega_c)} \tag{47}$$

Under the assumption of Gaussian distribution, Eq. (3) yields

$$(\sigma_n^2 - \sigma_c^2)T_o^2 + 2(\mu_n\sigma_c^2 - \mu_c\sigma_n^2)T_o + \mu_c^2\sigma_n^2 - \mu_n^2\sigma_c^2 + 2\sigma_n^2\sigma_c^2 \ln\left[\frac{\sigma_c P(\omega_n)}{\sigma_n P(\omega_c)}\right] = 0 \tag{48}$$

to obtain optimal T_o.

Figures 18 (a,b) are the ALOS PALSAR images (L band, HH polarization) in February 17 and May 19, 2008 before and after earthquake in Beichuan County, Wenchuan area, respectively. Spatial resolution is 4.7m×4.5m. Figure 18(c) is the difference image of Figures 18(a,b), which seems very difficult to evaluate the terrain surface changes only using man's vision, especially to accurately classify the change classes in large scale.

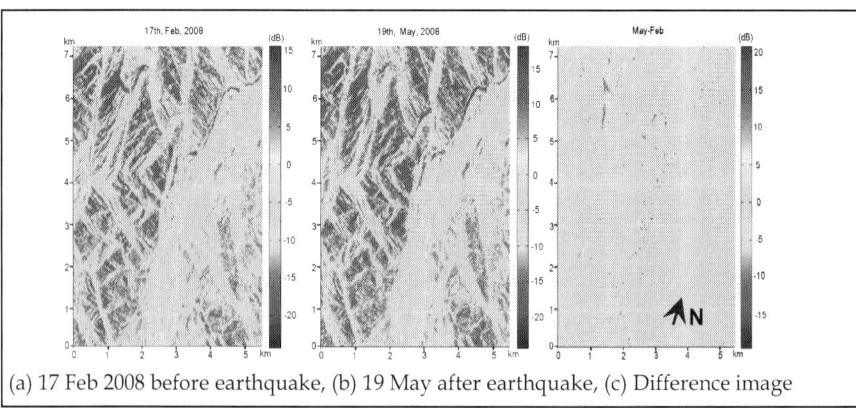

(a) 17 Feb 2008 before earthquake, (b) 19 May after earthquake, (c) Difference image

Fig. 18. ALOS PALSAR image in in Beichuan

The pixels of the difference image (i.e. $X_D = \sigma_D^0 = \sigma_2^0 - \sigma_1^0$) are classified into three classes: ω_{c1} of σ_D^0-enhanced, ω_n of σ_D^0-unchanged and ω_{c2} of σ_D^0-reduced. Thus, the probability density function $P(X)$ is modeled as a mixture density distribution consisting of three components:

$$p(X) = p(X \mid \omega_{c1})P(\omega_{c1}) + p(X \mid \omega_n)P(\omega_n) + p(X \mid \omega_{c2})P(\omega_{c2}) \tag{49}$$

The parameter T_{o1} is first obtained by application of the EM algorithm to the enhanced class ω_{c1} and no-enhanced class ω_{n1} (ω_n and ω_{c2}). Then, we obtain T_{o2} by applying the EM to the reduced class ω_{c2} and no-reduced class ω_{n2} (ω_n and ω_{c1}), where $\omega_{n1} = \omega_n \cup \omega_{c2}$, $\omega_{n2} = \omega_n \cup \omega_{c1}$. Finally, the two results are superposed to get the final classification.

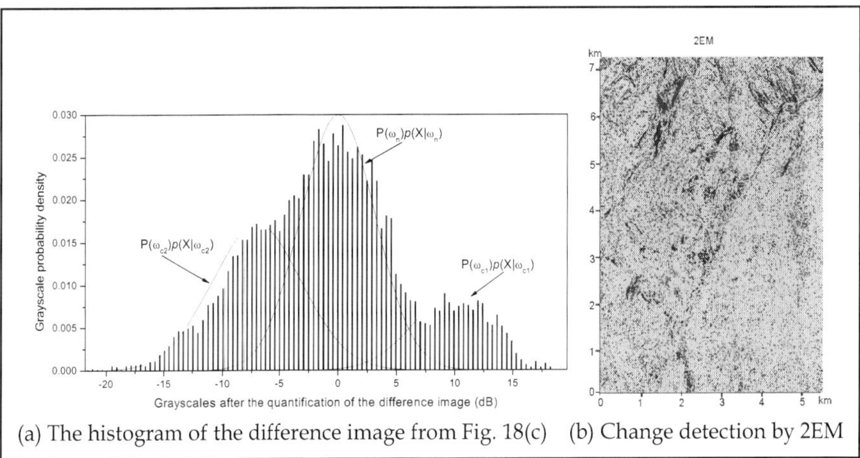

(a) The histogram of the difference image from Fig. 18(c)   (b) Change detection by 2EM

Fig. 19. The histogram of the difference image and change detection using 2EM method

As an example, Figure 19(a) shows the grayscale histogram of a small part chosen from the difference image, Figure 18(c). Three classes of the probability distributions are outlined in this figure. The histogram is normalized in accordance with the probability density distributions.

The initial statistical parameters of the subsets of

$$S_{n1} = \{X(i,j) \mid 0 < X(i,j) < T_{n1}\}, \ S_{c1} = \{X(i,j) \mid X(i,j) > T_{c1} \tag{50a}$$

are first derived, and then derive the parameters of

$$S_{n2} = \{X(i,j) \mid T_{n2} < X(i,j) < 0\}, \ S_{c2} = \{X(i,j) \mid X(i,j) < T_{c2}\} \tag{50b}$$

The initial statistical parameters are related to the classes ω_{n1}, ω_{c1} and the classes ω_{n2}, ω_{c2}, respectively. Then, Eqs. (46a-c) are sequentially used to perform the iterations on the four classes, i.e. the *a priori* probability and other statistical parameters $[\, P(\omega_{n1})\, , \mu_{n1}\, , \sigma_{n1}^2\,]$, $[\, P(\omega_{c1})\, , \mu_{c1}\, , \sigma_{c1}^2\,]$, $[\, P(\omega_{n2})\, , \mu_{n2}\, , \sigma_{n2}^2\,]$ and $[\, P(\omega_{c2})\, , \mu_{c2}\, , \sigma_{c2}^2\,]$. Solving Eq. (48), the thresholds $T_{o1} = 3.1643 \ \mathrm{dB}$ and $T_{o2} = -2.8381 \ \mathrm{dB}$ are obtained.

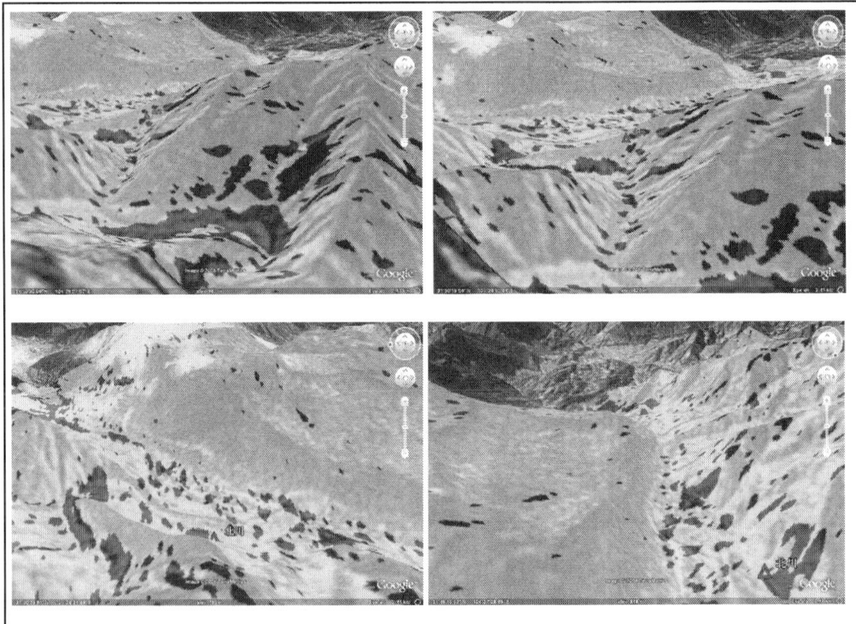

Fig. 21. Multi-azimuth views in animated cartoon showing terrain surface changes on Google Earth mapping

6. DEM Inversion from a Single POL-SAR Image

Co-polarised or cross-polarised backscattering signature is the function of the incidence wave with the ellipticity angle χ and orientation angle ψ. Recently, polarimetric INSAR image data has been utilized to generate digital surface elevation and to invert terrain topography. When the terrain surface is flat, polarimetric scattering signature has the maximum largely at the orientation angle $\psi = 0$. However, it has been shown that as the surface is tilted, the orienattion angle ψ at the maximum of co-polarized (co-pol) or cross-polarized (cross-pol) signature can shift from $\psi = 0$. This shift can be applied to convert the surface slopes. Making use of the assumption of real co-pol and zero cross-pol scattering amplitude functions, the ψ shift is expressed by the real scattering amplitude functions.

Since both the range and azimuth angles are coupled, two- or multi-pass SAR image data are required for solving two unknowns of the surface slopes. This approach has been well demonstrated for inversion of digital elevation mapping (DEM) and terrain topography by using airborne SAR data.

However, scattering signature is an ensemble average of echo power from random scatter media. Measurable Stokes parameters as the polarized scattering intensity should be directly related to the ψ shift. In this section, using the Mueller matrix solution, the ψ shift is newly derived as a function of three Stokes parameters, I_{vs}, I_{hs}, U_s, which are measurable by the polarimetric SAR imagery.

Using the Euler angles transformation between the principal and local coordinates, the orientation angle ψ is related with both the range and azimuth angles, β and γ, of the tilted surface pixel and radar viewing geometry. These results are consistent with Lee et al. (1998) and Schuler (2000), but are more general.

It is proposed that the linear texture of tilted surface alignment is used to specify the azimuth angle γ. The adaptive threshold method and image morphological thinning algorithm are applied to determine the azimuth angle γ from image linear textures. Thus, the range angle β is then solved, and both β and γ are utilized to obtain the azimuth slope and range slope. Then, the full multi-grid algorithm is employed to solve the Poisson equation of DEM and produce the terrain topography from a single pass Polarimetric SAR image (Jin, 2005; Jin and Luo, 2003).

6.1 The ψ shift as a function of the Stokes parameters

Consider a polarized electromagnetic wave incident upon the terrain surface at $(\pi - \theta_i, \varphi_i)$. Incidence polarization is defined by the ellipticity angle χ and orientation angle ψ (Jin 1994). The 2×2-D complex scattering amplitude functions are obtained from polarimetric measurement as

$$\bar{I}_i(\chi,\psi) = [I_{vi}, I_{hi}, U_i, V_i]^T = \left[\frac{1}{2}(1 - \cos 2\chi \cos 2\psi), \frac{1}{2}(1 + \cos 2\chi \cos 2\psi), -\cos 2\chi \sin 2\psi, \sin 2\chi\right]^T \quad (53)$$

$$\sigma_c = 4\pi \cos \theta_i P_n \quad (54)$$

where

$$P_n = 0.5\left[I_{vs}(1 - \cos 2\chi \cos 2\psi) + I_{hs}(1 + \cos 2\chi \cos 2\psi) + U_s \cos 2\chi \sin 2\psi + V_s \sin 2\chi\right] \quad (55)$$

When the terrain surface is flat, co-pol backscattering σ_c versus the incidence polarization (χ,ψ) has the maximum at $\psi = 0$. However, it has been shown that as the surface is tilted, the orientation angle ψ for the σ_c maximum is shifted from $\psi = 0$.

Let $\partial P_n / \partial \psi = 0$ at the maximum σ_c and $\chi = 0$ of symmetric case, it yields

$$0 = -(I_{hs} - I_{vs})\sin 2\psi + U_s \cos 2\psi + 0.5(I_{hs} + I_{vs})' + 0.5U_s' \sin 2\psi + 0.5(I_{hs} - I_{vs})' \cos 2\psi \quad (56)$$

where the superscript of prime denote $\partial / \partial \psi$. It can be shown that

$$0.5(I_{hs} + I_{vs})' \sim (M_{13} + M_{23})\cos 2\psi \sim (\text{Re} < S_{vv}S_{vh}^* > + \text{Re} < S_{hv}S_{hh}^* >)\cos 2\psi \quad (57a)$$

$$0.5U_s' \sim M_{33} \cos 2\psi \sim \text{Re} < S_{vv}S_{hh}^* + S_{vh}S_{hv}^* > \cos 2\psi \quad (57b)$$

$$(I_{hs} - I_{vs}) \sim 0.5(M_{11} + M_{22})\cos 2\psi \sim 0.5(<|S_{vv}|^2 > + <|S_{hh}|^2 >)\cos 2\psi \quad (57c)$$

$$0.5(I_{hs} - I_{vs})' \sim 0.5(M_{13} - M_{23})\cos 2\psi \sim 0.5(\text{Re} < S_{vv}S_{vh}^* > - \text{Re} < S_{hv}S_{hh}^* >)\cos 2\psi \quad (58a)$$

$$U_s \sim 0.5(M_{31} + M_{32}) \sim \text{Re} < S_{vv}S_{hv}^* > + \text{Re} < S_{vh}S_{hh}^* > \quad (58b)$$

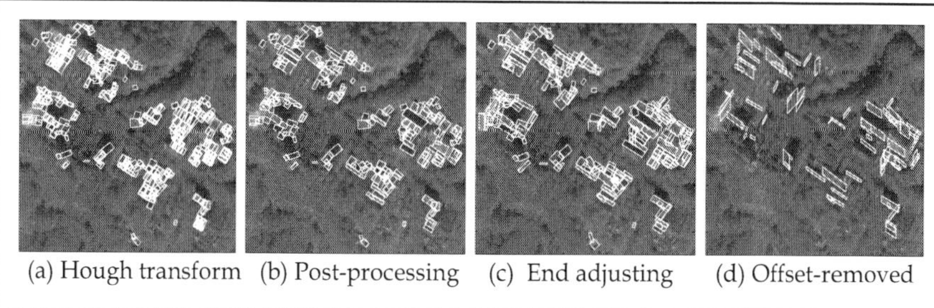

(a) Hough transform   (b) Post-processing   (c) End adjusting   (d) Offset-removed

Fig. 28. Extraction of strip-like building-images

Figures 28(a-d) show the detection from 4 aspects Pi-SAR images. Few building-images are not detected and some false images are wrongly produced. In fact, there is a compromise between over-detection and incomplete-detection. We prefer to preserve more detected building-images, so as to control the non-detected rate, while the false images are expected to be eliminated through the subsequent auto-selection of effective images for reconstruction. However, there always remain some undetectable images, attributable to shadowing of tree canopy, overlapping of nearby buildings or with too complicated wall structures.

The detected parallelogram of a homogenous scattering area could be direct scattering from façade, double scattering of wall-ground, combination of direct and double scatterings, projected shadow of building, or even strong scattering of strip-like objects (e.g. metal fence or metal awning), which is not considered in classification.

Shadowing is identified if scattering power of that area is much lower than the vicinity. Specifically, first set up two parallel equal-length strip windows on its two sides and then calculate the median scattering powers of the three regions. If the middle one is weaker than two sides, it is classified as shadow instead of building image.

The wall-ground double scattering can be differentiated from direct scattering based on polarimetric information. The de-orientation parameter v indicating the scattering mechanism is used to identify double scattering.

To implement target reconstruction from multi-aspect SAR data, calibrating multi-aspect data requires that at least one aspect is calibrated beforehand, and other aspects are then calibrated with respect to this calibrated aspect. A natural object, such as flat bare ground, is usually chosen as a reference target, which is expected to preserve identical scattering for all aspects. Thereafter, the channel imbalance factors are estimated from distribution of the phase difference and amplitude ratio of co-polarized, hh and vv, echoes of the reference target, and are then used to compensate the whole SAR images (Xu and Jin, 2008b).

7.2 Building reconstruction

Complexity of objects-terrain surfaces and image speckles incorporate certain uncertainty for detection of object images. To well describe multi-aspect object image, the parameterized probability is introduced for further proceeding of automatic reconstruction. For convenience, the detected building-images are parameterized. Generally, the edge pixels detected by CFAR are randomly situated around the real, or say, theoretical boundary of the object. It is reasonable to presume that the deviation distances of the edge points follow a normal distribution.

The original edge can be equivalently treated as a set of independent edge points. The line detection approach is considered as an equivalent linear regression, i.e. line fitting from random sample points. According to linear regression from independent normal samples, the slope of the fitted line follows the normal distribution.

All parameters have normal distributions and their variances are determined by the variance of the edge points deviation. After counting the deviation of the edge points in the vicinity of each lateral of all detected building-images and making its histogram, the variance can be determined through a minimum square error (MSE) fitting of the histogram using normal probability density function (PDF). The 4-aspect Pi-SAR images are counted.

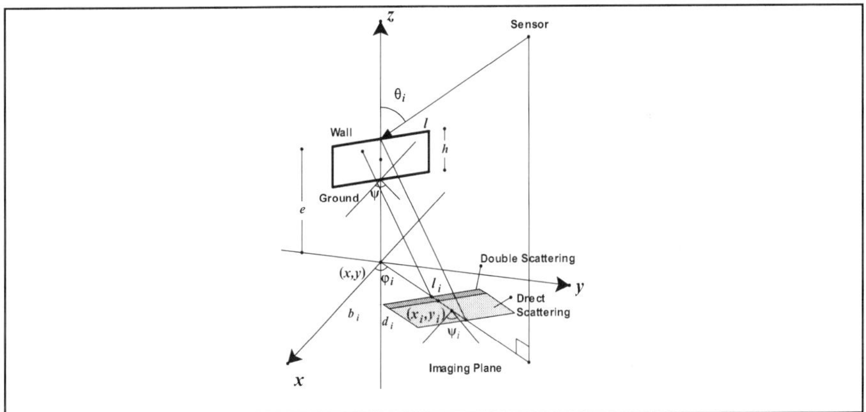

Fig. 29. Geometry of a wall/façade imaging

Given a group of building-images detected from multi-aspect SAR images, the corresponding maximum- likelihood probability can be further used as an assessment of the coherence among this group of multi-aspect images. A large maximum-likelihood probability indicates a strong coherence among the group of multi-aspect building-images, and vice versa.

Multi-aspect co-registration, as a critical pre- processing step, is necessary when dealing with real SAR data.

Given the specification of a region of interest (ROI), e.g. the longitudes and latitudes, the corresponding area in SAR image of each aspect can be coarsely delimited according to its orbit parameters. It can be regarded as a coarse co-registration step. Manual intervention is necessary if the orbit parameters are not accurate enough.

Only parameters of the building-images are needed to be co-registered to the global coordinates, but rather than the original SAR images. It is regarded as a fine co-registration step. In this study, only linear co-registrations are considered. A simple and effective method should be manually choosing ground control points. In general, a featured terrain object or building is first selected as the reference of zero-elevation, and then its locations of different aspects are pinpointed. Coherence among multi-aspect building-images is the basis of automatic reconstruction.

Jin, Y.Q. (1994). Polarimetric scattering from a layer of random clusters of small spheroids. *IEEE Transactions on Antenna and Propagation*, 42(8): 1138-1144, ISSN 0018-926X.

Jin, Y.Q. & Cloude, S. (1994). Numerical eigenanalysis of the coherency matrix for a layer of random nonspherical scatterers. *IEEE Transactions on Geoscience and Remote Sensing*, 32(6): 1179-1185, ISSN 0196-2892.

Jin, Y.Q. (1995). Polarimetric scattering and emission from a layer of random clustters of small spheroids and dense spheres. *Microwave Radiometry and Remote Sensing of the Environment*, ed. by Solimini, D., VSP, 419-427, The Netherlands, ISBN-90-6764-189.

Jin, Y.Q. (1995). Polarimetric scattering from heterogeneous particulate media. *Journal of Applied Physics*, 78(8): 4835-4840, ISSN 1087-3848.

Jin, Y.Q., Zhang, N., Chang, M. (1999). The Mueller and coherency matrices solution for polarimetric scattering simulation for tree canopy in SAR imaging at C band. *Journal of Quantitative Spectroscopy and Radiative Transfer*, 63(2): 699-612,   ISSN 0022-4073.

Jin, Y.Q. & Zhang, N. (2002). Statistics of four Stokes parameters in multi-look polarimetric SAR imagery. *Canadian Journal of Remote Sensing*, 28(4): 610-619, ISSN 0703-8992.

Jin, Y.Q. & Chen, Y. (2002). An improved method of the minimum entropy for refocusing moving target image in the SAR observation. *The Imaging Science Journal*, 50(6): 147-152, ISSN 1368-2199.

Jin, Y.Q. & Chen, F. (2002). Polarimetric scattering indexes and information entropy of the SAR imagery for surface monitoring. *IEEE Transactions on Geoscience and Remote Sensing*, 40(11): 2502-2506, ISSN 0196-2892.

Jin, Y.Q. & Liang, Z. (2003). Iterative solution of multiple scattering and emission from inhomogeneous scatter media. *Journal of Applied Physics*, 93(4): 2251-2256.

Jin, Y.Q. & Chen, F. (2003). Scattering simulation for inhomogeneous layered canopy and random targets beneath canopies by using the Mueller matrix solution of the pulse radiative transfer. *Radio Science*, 38(6): 1107-1116, ISSN 0048-6604.

Jin, Y.Q. & Liang, Z. (2004). Iterative inversion from the multi-order Mueller matrix solution of vector radiative transfer equation for a layer of random spheroids. *Journal of Quantitative Spectroscopy and Radiate Transfer*, 83(4): 303-311, ISSN 0022-4073.

Jin, Y.Q. & Liang, Z. (2004). An approach of the three-dimensional vector radiative transfer equation for inhomogeneous scatter media. *IEEE Transactions on Geoscience and Remote Sensing*, 42(2): 355-360, ISSN 0196-2892.

Jin, Y.Q., Chen, F. & Chang, M. (2004). Retrievals of underlying surface roughness and moisture for stratified vegetation canopy using polarized pulse echoes in the specular direction. *IEEE Transactions on Geoscience and Remote Sensing*, 42(2): 426-433, ISSN 0196-2892.

Jin, Y.Q. & Liang, Z. (2004). Iterative retrieval of canopy biomass and surface moisture by using multi-order Mueller matrix solutions of polarimetric radiative transfer. *Canadian Journal of Remote Sensing*, 30(2): 169-175, ISSN 0703-8992.

Jin, Y.Q. & Luo, L. (2004). Terrain topography inversion using single-pass polarimetric SAR image data. *Science in China (F)*, 47(4): 490-500, ISSN 1009-2757.

Jin, Y.Q., Chen, F. & Luo, L. (2004). Automatic analysis of change detection of multi-temporal ERS-2 SAR images by using two-thresholds EM and MRF algorithms. *The Imaging Science Journal*, 52: 234-241, ISSN 1368-2199.

Jin, Y.Q. & Liang, Z. (2004). Scattering and emission from inhomogeneous vegetation canopy and alien target by using three-dimensional vector radiative transfer (3D-VRT) equation. *Journal of Quantitative Spectroscopy and Radiative Transfer*, 92(3): 261-274, ISSN 0022-4073.

Jin, Y.Q. (2007). Theory and application for retrieval and fusion of spatial and temporal quantitative information from complex natural environment. *Frontiers of Earth Science in China*, 1(3): 284-298, ISSN 1673-7385.

Jin, Y.Q., Xu, F. & Fa, W. (2007). Numerical simulation of polarimetric radar pulse echoes from lunar regolith layer with scatter inhomogeneity and rough interfaces. *Radio Science*, 42, RS3007, doi: 10.1029/ RS2006003523: 1-10, ISSN 0048-6604.

Jin, Y.Q. & Wang, D. (2009). Automatic Detection of the Terrain Surface Changes after Wenchuan Earthquake, May 2008 from ALOS SAR Images Using 2EM-MRF Method, *IEEE Geoscience and Remote Sensing Letters*, 6(2): 344-348, ISSN 0196-2892.

Cao, G.Z. & Jin, Y.Q. (2007). A hybrid algorithm of BP-ANN/GA for data fusion of landsat ETM+ and ERS-2 SAR in urban terrain classification. *International Journal of Remote Sensing*, 28(2): 293-305, ISSN 0143-1161.

Cao, G. & Jin, Y.Q. (2007). Automatic detection of main road network in dense urban area using microwave SAR image. *The Image Science Journal*, 55: 215-222, ISSN 1368-2199.

Chang, M. & Jin, Y.Q. (2002). The temporal Mueller matrix solution for polarimetric scattering from a layer of random non-spherical particles with a pulse incidence. *Journal of Quantitative Spectroscopy and Radiative Transfer*, 74: 339-353, ISSN 0022-4073.

Chang, M. & Jin, Y.Q. (2003). Temporal Mueller matrix solution for polarimetric scattering from inhomogeneous random media of non-spherical scatterers under a pulse incidence. *IEEE Transactions on Antenna and Propagation*, 51(4): 820-832, ISSN 0018-926X.

Dai, E., Jin, Y.Q., Hamasaki, T. & Sato, M. (2008). 3-D stereo reconstruction of buildings using polarimetric images acquired in opposite directions. *IEEE Geoscience and Remote Sensing Letters*, 5(2): 236-240, ISSN 0196-2892.

Fa, W., & Jin, Y.Q. (2009). Image Simulation of SAR Remote Sensing over Inhomo-geneously Undulated Lunar Surface. *Science in China (F)*, 52(4): 559-574, ISSN 1009-2757.

Liang, Z. & Jin Y.Q. (2003). Iterative approach of high-order scattering solution for vector radiative transfer of inhomogeneous random media. *Journal of Quantitative Spectroscopy and Radiative Transfer*, 77: 1-12, ISSN 0022-4073.

Qi, R. & Jin, Y.Q. (2007). Analysis of Faraday rotation effect in satellite-borne SAR observation at P band. *IEEE Transactions on Geoscience and Remote Sensing*, 45(5): 1115-1122, ISSN 0196-2892.

Wei, Z., Xu, F. & Jin, Y.Q. (2008). Phase unwrapping for SAR interferometry based on ant colony optimization algorithm. *International Journal of Remote Sensing*, 29(3): 711-725, ISSN 0143-1161.

Xu, F. & Jin, Y.Q. (2005). Deorientation theory of polarimetric scattering targets and application to terrain surface classification. *IEEE Transactions on Geoscience and Remote Sensing*, 43(10): 2351-2364, ISSN 0196-2892.

Xu, F. & Jin, Y.Q. (2006). Multi-parameters inversion of a layer of vegetation canopy over rough surface from the system response function based on the Mueller matrix solution of pulse echoes. *IEEE Transactions on Geoscience and Remote Sensing*, 44(7): 2003-2015, ISSN 0196-2892.

Xu, F. & Jin, Y.Q. (2006). Imaging simulation of polarimetric synthetic aperture radar remote sensing for comprehensive terrain scene using the mapping and projection algorithm. *IEEE Transactions on Geoscience and Remote Sensing*, 44(11): 3219-3234, ISSN 0196-2892.

Xu, F. & Jin, Y.Q. (2007). Automatic reconstruction of 3-D stereo buildings from multi-aspect high-resolution SAR images, *IEEE Transactions on Geoscience and Remote Sensing*. 45(7): 2336-2353, ISSN 0196-2892.

Xu, F. & Jin, Y.Q. (2008a). Imaging simulation of bistatic SAR and polarimetric analysis. *IEEE Transactions on Geoscience and Remote Sensing*, 46(8): 2233-2248, ISSN 0196-2892.

Xu, F. & Jin, Y.Q. (2008b). Calibration and validation of a set of multi-aspect airborne polarimetric Pi-SAR data. *International Journal of Remote Sensing*, 29(23): 6801-6810, ISSN 0143-1161.

Xu, F. & Jin, Y.Q. (2009). Raw signal processing of stripmap bistatic SAR. *IET-Radar Sonar and Navigation*, 3(3): 233-244, ISSN 1751-8784.

8

Polarimetric responses and scattering mechanisms of tropical forests in the Brazilian Amazon

J. R. dos Santos[1], I. S. Narvaes[1], P. M. L. A. Graça[2] and F. G. Gonçalves[3]
(1) INPE - National Institute for Space Research
Av. dos Astronautas, 1758 CEP.:12.227-010 São José dos Campos – SP., Brazil
Email: {jroberto, igor}@dsr.inpe.br
(2) INPA - National Institute of Amazonian Research
Av. André Araujo, 2936   CEP.: 69060-001   Manaus – AM., Brazil,
Email: pmlag@inpa.gov.br
(3) Oregon State University
Department of Forest Ecosystems and Society, Corvallis, OR 97331   EUA,
Email: fabio.goncalves@oregonstate.edu

Abstract
This chapter discusses the polarimetric responses of PALSAR (L-band) data and scattering mechanisms of some tropical forest typologies based on target decomposition. The fundamentals of polarimetric theory related to both SAR topics under development are summarized. For representation of polarimetric signatures, the cross-section of the forest target (σ) was plotted on a bi-dimensional graphic as a function of the orientation angle, ellipticity angle and the intensity of co-polar components of the radar signal. The analysis of scattering mechanisms was done by the association of entropy and mean alpha angle values of each sample, plotted in a bi-dimensional classification space. This study improves the understanding of the interaction mechanisms between L-band PALSAR signals and structural parameters, supporting the forest inventory in the Brazilian Amazon region.
KEYWORDS: tropical forest, polarimetric signatures, target decomposition, PALSAR, monitoring.

1. Introduction

The ongoing particular and governmental struggle over the Brazilian Amazon's future mirror broader current discourses on the "environment" (Andersen et al., 2002), experts and layperson alike vary widely along continuum perspectives, based on two scenarios: (a) the line of defenders of global ecological services ("conservationists") and (2) the focus of development interests in the countries hosting these forests ("developmentalists"). They both present an important view points and significant arguments. While these discussions occur in the frame of a possible territorial planning, the degradation process with habitat fragmentation is still going on, with an estimative of rate gross deforestation in the Amazon region of around 11,968 km^2 (timeframe of 2007-2008), with an error of approximately 4% (<http://www.obt.inpe.br/prodes/index.html>). Due to the advancement of economic activities in the domain of the Brazilian Amazon tropical rainforest, remote sensing have being used as a fundamental tool to characterize the causes of degradation (conversion of natural vegetation to agriculture and cattle raising, selective logging, charcoal production, etc.), and also to monitor the impact of human activities over these large ecosystems.

Taking into account that in tropical regions there is, all over the year, a high percentage of cloud cover which impedes the inventory and updating of the forest cover by optical data, it is important to use sensor systems which operate at microwave wavelengths. A multi-temporal analysis of TM-Landsat data (timeframe of 12 years) used to estimate the annual gross deforestation rate in the Brazilian Amazon region showed that in average 21.12% of the imaged areas was cloud covered (Santos et al., 2008). Since there is a significant dynamic in environments of the Amazon region due to seasonal variations or from the different conditions resulting from human action, it is important to use radar images to understand the transformation process in the forest landscape.

To briefly introduce a reader on this subject, including concepts and applications of radar data in forestry studies, the following papers are recommended: Ulaby and Elachi (1990), Beaudoin et al. (1994), Henderson and Lewis (1998) and Coops (2002). These authors studied multitemporal SAR (Synthetic Aperture Radar) backscatter for discriminating forest types, discussing those conditions of scattering and attenuation of radar signals at different frequencies interacting with structural vegetation features. Pope et al. (1994) developed indices based on ratios and normalized different of multipolarimetric radar data, which can be related to certain characteristics of vegetation cover, such as e.g. biomass, canopy structure, volume scattering, applied to tropical forest in Central America. Hoekmann and Quiñones (2000), Hawkins et al. (2000) and Santos et al. (2002; 2003) correlated information obtained from radar with forest biomass levels, discussing the saturation of backscatter saturation caused by the high density of tropical forests. Power spectrum analysis was made using autoregressive moving-average (ARMA) models, for evaluation of forest canopy structure from airborne SAR (X–band) data in the Brazilian Amazon (Neeff et al., 2005a). This approach is novel where the radar spectra match those obtained from ground data, establishing the basis for the use of this relationship in spatial structure analysis of ecosystems from remote sensing sources. In more recent studies, and exploring the radar potentialities beyond its traditional polarimetric characteristics, Kugler et al. (2006) and Treuhaft et al. (2006) used interferometric radar data to estimate biophysical parameters in tropical forest areas. Using another approach, Neeff et al. (2005b) explored, through the local maximum filter technique (Wulder et al., 2000) and statistical Markov processes (Cressie, 1993) the interferometric height derived from airborne SAR (X- and P- bands) for the analysis of spatial forest patterns (three dimensional canopy structure and distances of very large trees). The above mentioned studies exemplify the importance of radar information in the analysis of tropical environments.

Within this frame, the objective of this chapter is to explorer the polarimetric mode, by the analysis of the graphic representation of signatures from PALSAR/ALOS data (L-band) in primary forest, secondary succession (initial, intermediate and advanced levels of natural recovery) and also, forests with timber exploitation. Additionally, we also discuss an exploratory analysis of scattering mechanisms of tropical forest typology, in accordance with the alternative procedure of SAR image classification based on target decomposition.

1.1 Theoretical background

In order to understand the theme of this chapter, it is necessary to introduce some concepts about SAR polarimetry and of the classification technique based on target decomposition, as described below.

1.1.1 Concepts of polarized waves

Through different polarizations of radar imaging, the polarimetric signature of a certain target can be obtained, as well as the characterization of the dominant mechanism which controls the interaction of the microwave signal with the structure of the target under consideration.

The principle of polarimetry is related to the state of polarization of the wave, represented by definition as the electric field from the transmitting and receiving antenna. The state of polarization is described as an arbitrary set of axis used to describe horizontal polarized waves in the x axis (H polarization) and at the vertical polarized waves in the y axis (V polarization) along the wave propagation plan (z), which corresponds to the pointing direction of the antenna to the target. In this way, four types of polarization are obtained: HH for transmission and reception at the horizontal plan; VV for transmission and reception at the vertical plan; HV for transmission at the horizontal plan and reception at the vertical plan; and finally, VH for vertical transmission and horizontal reception. Ulaby and Elachi (1990) and Henderson and Lewis (1998) discuss how the electric field of an irradiated electromagnetic wave could be mathematically represented by a polarization ellipse. The polarization ellipse (Figure 1) is described by the three geometric parameters: the orientation angle ($-90° < \psi < +90°$), the ellipticity angle ($+45° > \chi > -45°$) and amplitude (A). The flattening factor from the ellipsis is given by angle χ, and there is a linear polarization when $\chi = 0$; the circular polarization to the left and right when $\chi = - 45°$ and $\chi = + 45°$, respectively. At the condition of $\chi = 0°$ and $\psi = 0°$ there is horizontal polarization and at $\chi = 0°$ and $\psi \pm 90°$ a vertical polarization. The analyst, who will discuss the polarization ellipse based on some literature references, must pay attention at the representation scale from the orientation axis, because graphical plots are presented in two forms: one of them with the polarization power varying between $0°$ and $180°$, and the other varying from $-90°$ to $+90°$, which can cause an interpretation error.

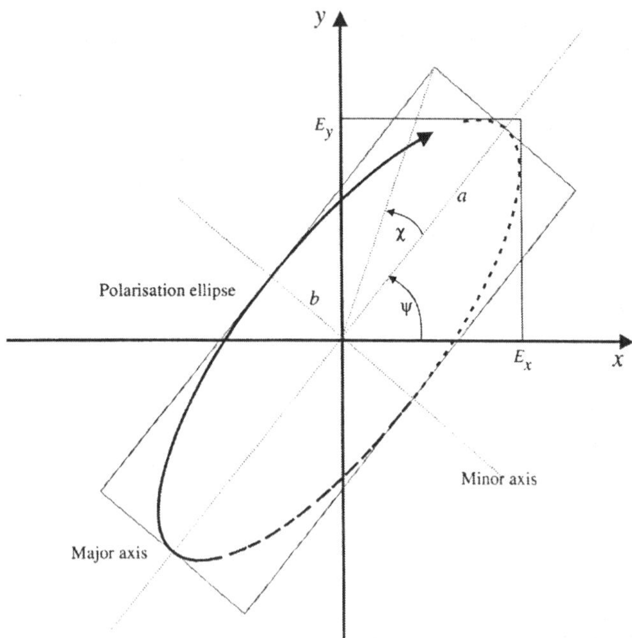

Fig. 1. Polarization pattern in the x-y plane, rotation angle ψ, ellipticity angle χ and auxiliary angle α. The size of the ellipse is governed by the maxima of the horizontal and vertical electrical field components. Source: Woodhouse (2006).

It is important to comment that the description of the wave polarization can be different according to the convention being considered. In the Forward Scattering Alignment (FSA) system, the wave propagation axis changes its direction after interaction with the target (this case represents the waves scattered by forest targets). In the Back Scattering Alignment (BSA) system the x and y axis are aligned in the same direction, and the wave propagation axis (z) is considered parallel, with the advantage of some mathematical simplifications on how the data are described (Woodhouse, 2006). In this way the coordinate system does not influence the nature of polarization, but just the way how the data are described.

There are other ways to represent the polarization state: (1) the graphic representation with the Poincaré sphere (Ulaby and Elachi, 1990), where the circular polarizations to the right and left are represented at the north and south poles respectively, while the intermediate polarization states are located at the regions between these poles. In this representation the meridians correspond to different inclination angles of the largest axis of the ellipse ($\psi = 0$) and the parallels to the ellipticity of the electric field ($\chi = $ arctan ($-B/A$)); (2) the mathematical representation by Jones vector, where the complex data received from each pulse emitted after the wave interacted with a target is demodulated at in-phase (In) and quadrature (Q) and converted to the digital format. Therefore, this mathematic representation is important because there is a direct relation with the polarization ellipse (van der Sanden, 1997; Woodhouse, 2006); (3) the Stokes vector where the polarized waves can be characterized by four real and measurable parameters, with the same physical dimension, but only three are independent and describe the polarization state as:

$$I_0^2 = Q^2 + U^2 + V^2 \tag{1}$$

Where: Q = is the tendency of more vertical (Q > 0) or horizontal polarization (Q < 0);
     U = expresses the tendency to be polarized at +45° (U > 0) and - 45° (U < 0);
     V = polarization referring to the rule of right (V < 0) and left (V > 0) hand rules.

The parameters U and V are sometimes referred as being co-variances in-phase (real) and quadrature (imaginary), respectively between the components of the vertical and horizontal field of the wave. This expression is applied to completely polarized waves (amplitude and phase are invariant in time). Additionally the Stokes vector has a direct relation with the Poincaré sphere, because its ray is the Stokes parameter I_o (Woodhouse, 2006).
It is important to mention that the first imaging radars known as "conventional" recorded only information about the amplitude of the electric field at components x and y, i.e. the maximum magnitude of the electric field from a wave as a measurable indicator of size (van der Sanden, 1997). The amplitude takes into account the texture and brightness of the image, influenced by the surface roughness (macro-, meso- and micro-scales) and by the dielectric constant (moisture content of the target). With the advancement of technology, imaging radars started to record, phase information (the angular position δ of an oscillating movement at a defined instant in time t, as a function of the distance radar to target) in addition to amplitude, being called polarimetric radars. Such phase information provides information about the scattering mechanisms occurring in the imaged scene.
To improve the knowledge of the microwave interaction process with a certain target, it is necessary to understand the types of scatterers involved. The deterministic scatterers, also called coherent or punctual targets, are those where the interaction with electromagnetic waves reflects completely polarized waves, preserving the polarization of the incident wave (ESA, 2009a). Examples of this are urban targets. The non-deterministic targets, also known as incoherent targets or partially polarized or random distributed scatterers, are those that have more than one center of scattering, where the measured signal is the overlapping of a large amount of waves with variable polarization (ESA, 2009a). This is the case of forest targets. The change of polarization state is known as depolarization, caused mainly by multiple scattering due to surface roughness and to volumetric interaction with the target. Besides that, there is an almost specular reflection for smooth undulating surfaces, whose physical concept and representations will be presented below, when we consider the target decomposition approach.

1.1.2 Targets decomposition
In the polarimetric imaging the four complex components derived from the backscattering in each resolution cell (S_{HH}, S_{HV}, S_{VH}, S_{VV}) can be expressed by a scattering matrix [S] which, according to Ulaby and Elachi (1990) is given by:

$$[S] = \begin{pmatrix} S_{VV} & S_{VH} \\ S_{HV} & S_{HH} \end{pmatrix} \tag{2}$$

According to Woodhouse (2006), the scattering matrix [S] is applied to deterministic targets (coherent), and models how the scatterers transform the components of the incident electric field (E^i) in the scattered electric field (E^s).

$$\begin{pmatrix} E_v^s \\ E_h^s \end{pmatrix} = \frac{e^{-ik_0 r}}{R} \begin{pmatrix} S_{VV} & S_{VH} \\ S_{HV} & S_{HH} \end{pmatrix} \begin{pmatrix} E_v^i \\ E_h^i \end{pmatrix} \tag{3}$$

Where R is the distance from the antenna to the target and $k = 2\pi / \lambda$

Non-deterministic targets (incoherent) are represented by second order matrices. When the polarization of the electromagnetic wave is represented by the modified Stokes vector, the relation between the polarization of the incident electric field F^i and the scattered field F^s is given by the Müller matrix [M] (or Stokes matrix) in the FSA system, or also by the Kennaugh matrix [K] in the BSA system (Woodhouse, 2006), according to the equations below:

$$\left\langle F^s \right\rangle = \frac{1}{r^2} [M] \left\langle F^i \right\rangle \tag{4}$$

$$\left\langle \overline{F}^s \right\rangle = \frac{1}{r^2} [K] \left\langle \overline{F}^i \right\rangle \tag{5}$$

Besides these matrices, the scattering of non-deterministic targets can be represented, according to Cloude and Pottier (1996), by power matrices of covariance [C] and coherence [T]. Since the radar image pixels contain many scatterers (especially in the forest targets), with different scattering properties, the covariance and coherence can be used to represent the matrix [S]. The covariance matrix [C] is generated by multiplying the vector-target (Kennaugh) and its joint complex.

$$[C] = \left\langle \vec{K}_c . \vec{K}_c^{*T} \right\rangle \tag{6}$$

where:

$$\vec{k}_c = (S_{hh}, \sqrt{2} S_{hv}, S_{vv})^T ;$$
$$\vec{K}_p = 1/\sqrt{2} [(S_{hh} + S_{vv})(S_{hh} - S_{vv})(S_{hv} + S_{vh}) i (S_{hv} - S_{vh})]^T$$

<...> represents the spatial average.

When one is working with the average value of scattering mechanisms from a pre-defined window in the imaged scene, the characterization of the targets by polarization power is a consequence of a spatial average (normally of pixels), and this operation is called multi-look processing.

On the other hand, the coherence matrix [T], which is related to the physical and geometric properties of the scattering processes, is given by:

$$[T] = \left\langle \vec{K}_p . \vec{K}_p^{*T} \right\rangle \tag{7}$$

Where: $\vec{K}_p$ is the scattering vector written in the Pauli basis (ESA, 2009b)

The matrices [C] and [T] contain the same information about amplitude, phase and correlations, besides its' eigenvalues are real and similar. As a complement, the matrices [M], [K], [C] and [T] are linearly related among them and there are formal methods for transformation among them (Woodhouse, 2006).
The known decompositions to model the nature of coherent targets (represented by the matrix [S]) are those from Pauli, Krogager (SHD), Cameron and Hyunen (Touzi et al., 2004; Zhang et al., 2008), but they are beyond the scope of discussion at this chapter. For non-deterministic targets (represented by second order matrices) there are two decomposition mechanisms implemented at most software packages, which are used for the discrimination of scattering type occurring in forest targets: the decompositions of Freeman-Durden and Cloude-Pottier.
The Freeman-Durden decomposition models the matrix [C] as a contribution of the surface scattering (f_s), the double-bounce (f_d), and volumetric type (f_v) where each type of contribution is represented by the theoretical target, related to the respective type of scattering which occurs at the forest (Freeman and Durden, 1998).
The polarimetric decomposition of Cloude and Pottier (1997) is based on three canonic elements (entropy, anisotropy and alpha angle) defined as a function of the decomposition of eigenvalues and eigenvectors from the matrix [T]:

$$\left\langle |T_3| \right\rangle = \sum_{i=1}^{3} [T_{3i}] = \lambda_1 (e_1 e_1^{*T}) + \lambda_2 (e_2 e_2^{*T}) + \lambda_3 (e_3 e_3^{*T}) \tag{8}$$

Where: λ_i and e_i are eigenvalues and eigenvectors of the matrix [T] respectively.

The entropy (H) is a parameter that indicates the type of target involved in the imaged scene and the number of dominant scattering mechanisms $(H{\sim}0$ = mechanism of unique scattering; $H{\sim}1$ = multiple scattering mechanisms). In a dense tropical forest, the H values are normally elevated, indicating a larger variation of structural arrangements due to the diversity of species composition, the different levels of competition and growth, as well as the inter-specific association among species. In short, the entropy represents the randomness of the scattering.
The alpha angle (a) indicates the type of scattering dominating the scene, where $a = 0°$ (surface scattering), $a = 45°$ (volumetric scattering) and $a = 90°$ (double-bounce scattering). The anisotropy (A) is a complementary parameter that measures the relative importance of λ_2 and λ_3 (Equation 8). When $H > 0.7$, the anisotropy contributes with additional information to characterize the target. Below this value, the anisotropy is noisy and has limited

importance. The intensity of these canonic elements identifies the presence of primary or secondary scatterers, and the dominant type of scattering.

The association of entropy (H) and alpha angle (a) planes is one of the most efficient and usual means of improving the understanding of forest targets. In this classification procedure based on target decomposition, the H - a components are plotted in a plane with finite zones (Figure 2), which represent scattering mechanisms defined by Cloude and Pottier (1997). These attribute spaces help defining the dominant type of scattering which can be multiple, volumetric or superficial. The theory behind this target decomposition method assumes that the formation of an image is a stochastic process, where an image with a high value of entropy ($0.5 < H < 0.9$) presents much more details.

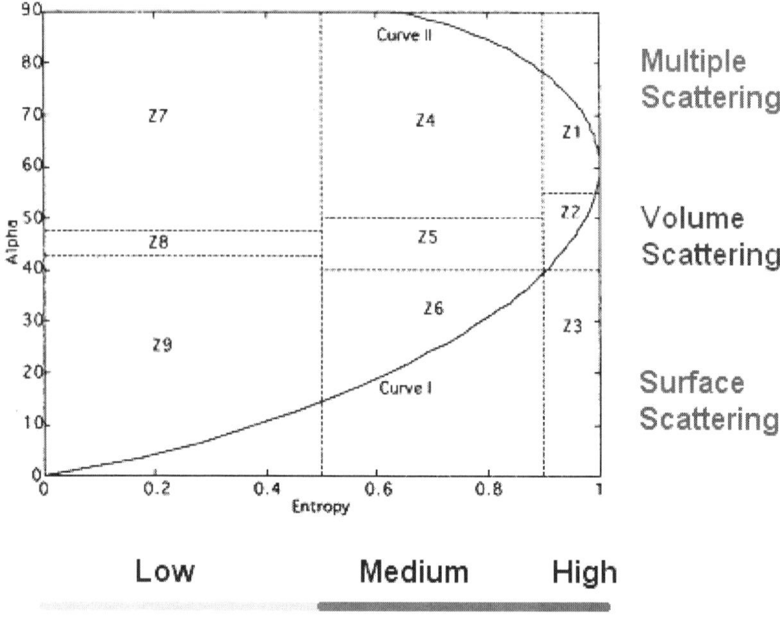

Fig. 2. Schematic diagram of the bi-dimensional classification based on entropy (H) and average orientation alpha angle (a).
Source: Modified from Cloude and Pottier (1997).

In this procedure of classification by target decomposition, the scattering mechanisms are defined in 9 distinct zones (Figure 2), namely: Z1- High entropy multiple scattering; Z2 - High entropy vegetation scattering; Z3 - High entropy surface scatter; Z4 - Medium entropy multiple scattering; Z5 – Medium entropy vegetation scattering; Z6 – Medium entropy surface scatter; Z7 – Low entropy multiple scattering events; Z-8 – Low entropy dipole scattering; Z9 – Low entropy surface scatter. Within the partition of the plan entropy and angle α in several regions which determine and label the intensity of each scattering mechanism, it is important to mention that Z3 is outside de representation areas from curves I and II (Figure 2) and so it is not a non-feasible region (Hajnsek et al., 2003), because the polarimetric SAR systems are not able to discriminate high entropy ($H > 0.9$). The curves I and II represent the borders of values with maximum and minimum observable α angle, as a function of entropy, whose limits are determined by the H-α variation for a coherence matrix with the degenerate minor eigenvalues with amplitude m($0 \leq$ m ≤ 1), according to the Cloude and Pottier (1997).

1.1.3 Polarization synthesis and signatures

An important aspect for the characterization of forest targets is the polarization synthesis, because when the matrix [S] of a target is known, it is possible to synthesize its backscatter (σ) to any combination of transmitted and received polarization (Ulaby and Elachi, 1990; Woodhouse, 2006). This is a good advantage of polarimetric data, synthesized according to equations 9 and 10:

$$P_{rec}(\psi_r, \chi_r, \psi_t, \chi_t) = \frac{K}{r^2} \cdot \left| \vec{F}^r \cdot [K] \vec{F}^t \right| \tag{9}$$

$$\sigma_{rt}(\psi_r, \chi_r, \psi_t, \chi_t) = \lim_{r \to \infty} 4\pi r^2 \left(\frac{P_{rt}^{rec}}{P^{trans}} \right) \tag{10}$$

Where: ψ = orientation angle; χ = ellipticity angle; K = vector of target; r = radius.

The polarization states of the electric fields E_v, and E_h which are associated with the radar backscatter (σ) and the dependence of amplitude on polarization, can be represented graphically as a function of ellipticity (χ) and orientation (ψ) angles of the transmitted wave, defining a three-dimensional surface plot called polarization signature or polarization response (Figure 3). The polarization signatures can be used to characterize targets, considering the more detailed knowledge of the mechanisms responsible for the scattering of the radar signal, and the intrinsic target characteristics (van Zyl et al., 1987). It can also be used for the polarimetric calibration of radar data. It is very important to point out that the polarized signature of a certain target is not unique, since different combinations of scattering mechanisms can give a similar configuration, as will be shown in this chapter. This can be due to the structural composition of the forest target, such as the density of trees, regular spatial distribution of trees, trunk diameter and density of twigs and branches, the moisture content in the leaves and in the soil, the dielectric constant of the target under study, etc. In addition, a target can also present distinct polarimetric behavior shown in its signatures, as a function of variations from radar frequency and/or local incidence angle.

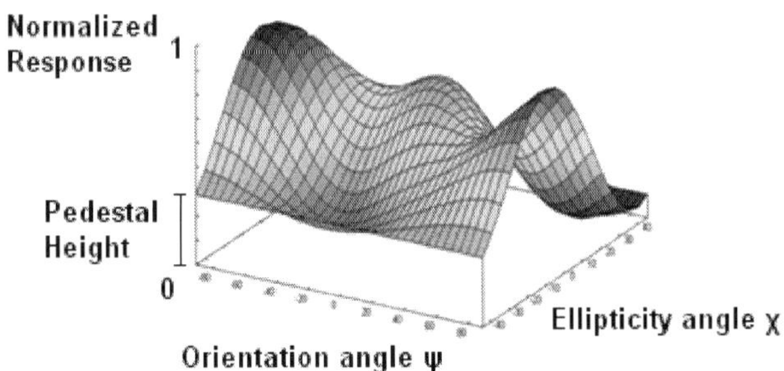

Fig. 3. Polarimetric response of a secondary succession area in the Brazilian Amazon.

It is important to observe in Figure 3 the representation of values of transmitted and received power, where there is a minimum value (pedestal height) of polarization intensity. This minimum value represents the amount of the non-polarized power in the received signal, and thus is related to the polarization level of the scattered wave (CCRS, 2001). If a single target is scattering and the backscattered wave is fully polarized, or if the signature is calculated from a single unaveraged measurement, the pedestal height is zero. But if the signature is calculated from an average of several samples, and there are multiple, dissimilar scatters present or there is noise in the received signal, the pedestal height will be non-zero. Thus the pedestal height is also a measure of the number of different types of scattering mechanism found in the averaged samples. The pedestal of the polarimetric signature may also be a worthy source of information for target characterization (McNairn et al., 2002). In fact, the pedestal corresponds to the ratio of the extreme power received when the antennas are co-polarized or cross-polarized. Therefore, the pedestal provides information equivalent to that obtained with the coefficient of variation by van Zyl et al. (1987). The latter should be more effective because it is not limited to co-polarizations and cross-polarizations (Touzi et al., 2004).

2. ALOS PALSAR data over the Brazilian tropical forest

2.1. Area under study

In order to understand the mechanisms controlling the interaction radar-forest target, a test-site was selected in the Tapajós region (NW Pará State, Brazil), delimited by the geographical coordinates 3° 01′ 60″ - 3° 10′ 39″ S and 54° 52′ 45″ - 54° 59′ 53″ WGr (Figure 4). This region is characterized by a low rolling relief, typical of the lower Amazon plateau and the upper Xingu-Tapajós Plateau. It is dominated by a continuous cover of primary tropical rainforest characterized by the presence on emergent trees and an uniform vegetation cover (Dense Ombrophilous Forests), as well as sections of low to dissected plateaus with few emerging individuals and a high density of palm trees (Open Ombrophilous Forests). The land use is primarily subsistence agriculture, few cash crops, cattle raise and selective logging activities (Figure 5). Areas of secondary succession are also found in different growth stages, resulting from land tracts abandoned after a short period of use, with intensive subsistence agricultural practice or cattle raising.

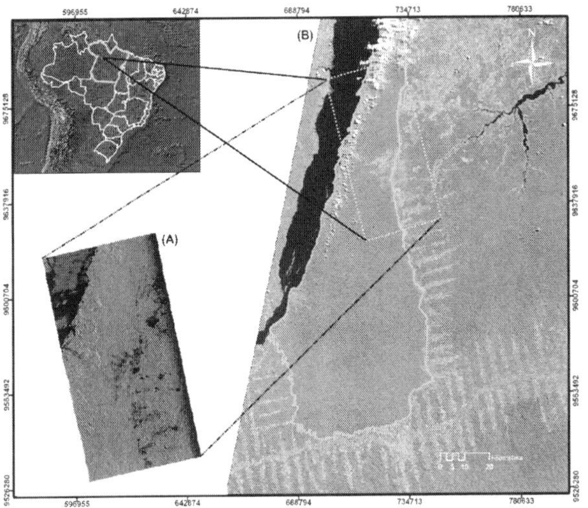

Fig. 4. Study area (dotted rectangle) as shown on LANDSAT/ETM+ (RGB = B3, B4, B5) and ALOS/PALSAR (RGB = HH, HV, VV) images.

Fig. 5. Land use and land cover classes in the Tapajós region.

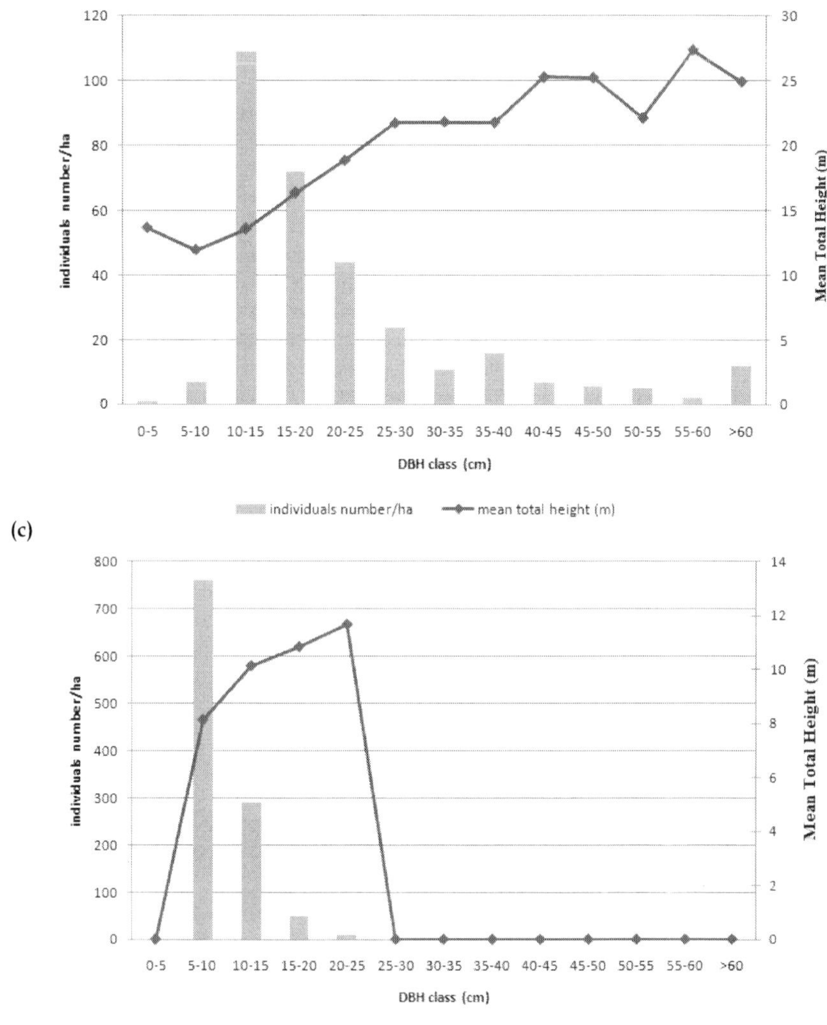

Fig. 6. Structural profile and dendrometric parameters of primary forest (a, c) and initial secondary succession (b, d), respectively.

2.2.3 Results and discussion

In order to familiarize the reader with the physiognomic-structural features of the forest typology of this section from the Amazon region under study, some results of the field survey are briefly presented.

The initial stage "capoeiras" regrowth (ISS) had an average height of 8m and an average DBH of 9cm. Floristic composition was dominated by species such as *Vismia guianensis* (Aubl.) Choisy (Guttiferae), *Cecropia leucoma* Miquel. (Moraceae), *Guatteria poeppigiana* Mart. (Annonaceae), *Aegiphila* sp. (Verbenaceae) and *Cordia bicolor* A. D.C. (Boraginaceae). The intermediate secondary succession (IntSS) had an average height of 11m and an average DBH of 10cm, with a higher abundance of *Cecropia sciadophylla* Mart (Cecropiaceae), *Schizolobium amazonicum* Huber (Fabaceae), *Cassia pentandra* Raddi (Fabaceae), *Vismia japurensis* Reichardt and *Vismia guianensis* (Aubl.) Choisy. (Guttiferae). The advanced secondary succession (ASS) had an average height of 18m and average DBH of 25cm with species *Bagassa guianensis* Aubl. (Moraceae), *Guatteria poeppigiana* Mart. (Annonaceae), *Casearia decandra* Jacq. (Flacourtiaceae), *Cassia pentandra* Raddi (Fabaceae), *Couratari oblongifolia* Ducke & R. Knuth (Lecythidaceae), and *Apeiba albiflora* Ducke (Malvaceae).

The primary forest stands (PF) present 3 to 4 strata with about 360 trees per hectare, whose upper stratum shows individuals with an average height of 21 m and DBH of 27 cm. The botanical diversity of the primary forest was significantly higher compared to areas of secondary succession, being dominated by species such as: *Protium apiculatum* Swart (Burseraceae), *Picrolemma sprucei* Hook (Simaroubaceae), *Rinorea guianensis* (Aubl.) Kuntze (Violaceae), *Cordia bicolor* A.DC. in DC. (Boraginaceae), *Jacaranda copaia* D.Don (Bignoniaceae), *Pouteria* sp. (Sapotaceae) and *Inga* sp. (Fabaceae). The primary forest stands that were affected by timber exploitation (SL) present a decrease of timber volume of around 27.3 m^3/hectare, which corresponds to an average density of exploration of 3.3 trees/ha, in accordance with the forest management plan formerly established by IBAMA/PNUD (1997). At the gaps resulting from the extraction of trees with commercial value and also at the trails where tree trunks were dragged, quickly pioneer species install themselves, such as *Cecropia leucoma* Miquel, *Cecropia sciadopylla* Mart., and *Guatteria poeppigiana* Mart., modifying the environment and providing resources for other species.

The description above suggest that there is a distinct frequency of trees by diameter and height class, as well as the occurrence of characteristic species in each forest types, which model the physical space under study. There is a positive gradient in these two ecological parameters, increasing from initial secondary succession until a mature forest. One observes that at the initial regrowth there is a more uniform canopy formed by 1 or 2 strata of low height, with pioneer species with thinner trunks found repeatedly in the area, from a few botanical families (Figure 6). The older secondary succession stages, the higher is the structural complexity, characterized by a wide variation in tree size, with a stepwise replacement of pioneer species and the formation of multiple canopy layers. A detailed description of the floristic composition and structure of forest typology from the Tapajós region can be found at Gonçalves (2007) Gonçalves and Santos (2008). The considerations above are important to facilitate the understanding of the polarized response.

At Figure 7 the shape features of co-polarization signatures derived from the PALSAR data of the 11 ROIs are presented, which were selected to represent the diversity of the tropical forest in Tapajós region. In all the polarization signature plots presented, the maximum intensity has been normalized to 1.0, according to Ouarzeddine et al. (2007). Before discussing this subject it is important to remind that in theory, in the co-polarization response, taking into account the representation in a plane ($-90° < \psi < +90°$), the highest values of intensity for an ellipticity angle at $0°$ and an orientation angle at $0°$, the polarization is horizontally oriented (Evans et al., 1988; Mc Nairn et al. 2002).

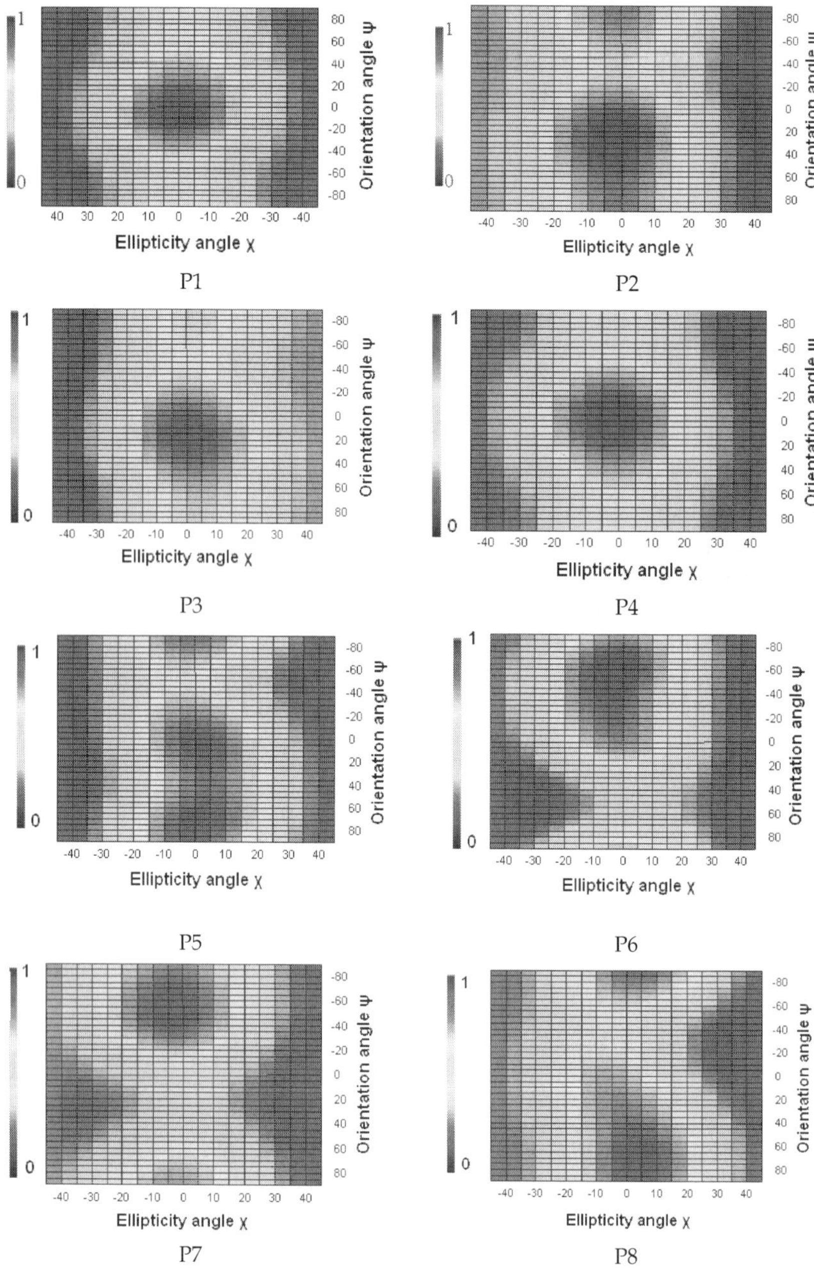
P1
P2
P3
P4
P5
P6
P7
P8
Ellipticity angle χ
Orientation angle ψ

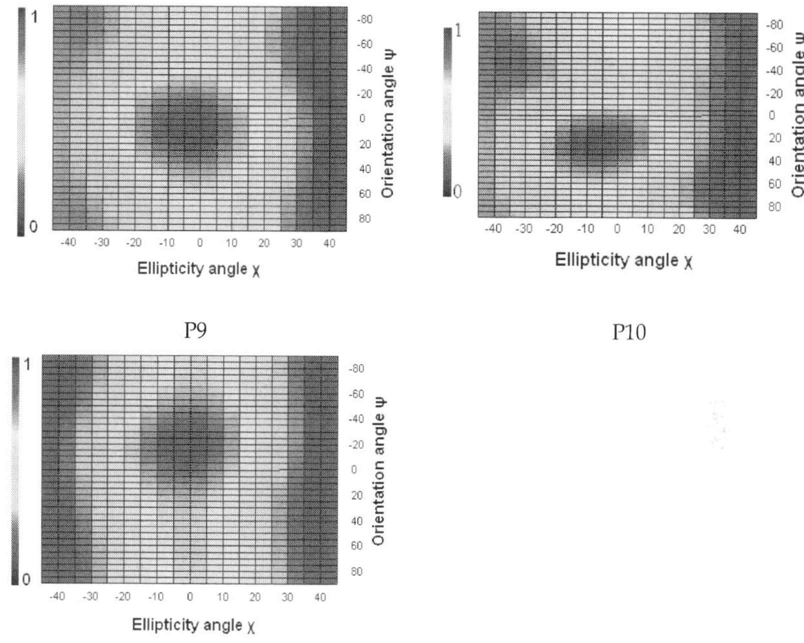

P9   P10

P11

Fig. 7. Polarization responses derived from at PALSAR data acquired in the Tapajós region, Brazil. Targets include primary forest (P1, P2), forest with old selective logging (P3, P4, P5), advanced secondary succession (P6, P7), intermediate secondary succession (P8, P9) and initial secondary succession (P10, P11).

The polarization responses of samples of ASS (P6, P7), IntSS (P8) and ISS (P10, P11) show a strong similarity to the theoretical responses of short and thin conducting cylinders (i.e. radius and lengths much shorter than the wavelength). This suggests that the scattering produced by twigs and small branches has an important contribution for the total backscattering of these regrowth types. The co-polarized response of samples P6 and P7 (both old regrowth) with $\psi = - 70°$ and P8 (intermediate regrowth) with $\psi = \pm 90°$, show stronger returns at VV than HH, consistent with increased double reflections and present diffusers with preferentially vertical orientation. The responses of the other analyzed areas (i.e. P10, P11) indicate the predominance of branches oriented at ±45° in relation to the horizontal plane, possibly due to more homogeneous canopies of these younger recovery sections with low species diversity and a quite uniform growth rate (height intervals of 6 to 9 m corresponds to 90% of the trees) (Figure 6). One must emphasize that, taking into account the distinction of the stands investigated, the polarization responses of sections P7 and P10 are similar to those presented by Ulaby and Elachi (1990), but with maximum co-polarized power at $\psi = - 60°$ and $\chi = 0°$ for P7 sample and maximum co-polarized response at $\psi = 30°$ and $\chi = -5°$ for P10 site and with a thin cylinder oriented vertically.

The polarization signatures of plots P1 and P2 (primary forest), P4 and P5 (forest under exploitation) and also P9 (intermediate secondary succession), are similar those of trihedral corner reflectors, which have a double bounce scattering geometry, when one of the reflector sides is parallel to the line connecting the radar antenna and the corner reflector, as mentioned by Zebker and Norikane (1987). For sections of more complex forest structures, the highest σ values occur in linear polarizations, with a certain independence of the orientation angle (ψ). This can be observed in plots P1, P2, P3 and P4, which represent the mature forest (with or without legal selective logging), whose concentrated section of maximum co-polarized power occurs in a range of $\psi = 0°$ to $\pm20°$ when $\chi = 0°$. As for forest areas affected by timber exploitation, one expects predominance of double-bounce scattering, due of the larger number of gaps resulting from the removal some trees with commercial value and also because this type of mature forest has a well-defined vertical structure in various strata with a much thinner undergrowth. Specifically, for the plot P3 presents most diffusers oriented horizontally ($\psi = 0$) and it has a stronger return at HH than VV, where the maximum for this signature seems slightly shifted from the horizontal position towards the vertical. This effect might be explained by the presence of the topographic slopes in the azimuth direction. In additional, medium or high impact logging practices can increase the structural heterogeneity, causing changes in the polarization response of the forest target investigated.

At the analysis of polarimetric signatures, the height of the pedestals must also be considered because it provides an additional source of information for target characterization, and to provide information similar to that obtained with the variation coefficient (van Zyl et al, 1987; Mc Nairn et al, 2002). Contrary to expected, in the co-polarized response derived from PALSAR, the highest pedestals occur for areas of secondary succession if compared to those of forest typologies, suggesting the existence of a considerable variation on the scattering properties of the adjacent resolution elements. The highest values of 0.313 and 0.333 were found respectively for areas of initial (P10) and intermediate (P8) recovery, respectively. Old-growth forests, with or without timber exploitation (P1, P2, P3 and P4) showed pedestal height ranging from 0.068 to 0.195. Within the characteristics of natural forest stands, the lower branch density can explain the lower values of pedestal height, which results at a lower non-polarized power component in the backscatter, indicating a smaller amount of volume scattering (CCRS, 2001). Note that, in contrast to our findings, previous studies conduced in tropical zones using SAR data have shown that a higher age and diversity of species and consequently, a stronger variability of forest structure, results at an increase of the pedestal size on the polarization responses.

In tropical areas with complex vegetation types, the polarimetric signature of each type is really not unique. One can simply observe that areas of early secondary succession could, for instance, could have its' growth influenced by aspects, such as: previous land use, soil fertility level, bio-ecological level of fragmentation, etc. From measurements performed during field survey we also verified that the canopy was more open in sites of initial regrowth, localized over areas with low soil fertility and a history of more intensive use, compared to those with higher soil fertility and a less intensive land use, both areas with the same age of natural succession. During the intermediate phase of secondary succession, the canopy remains more open, and the mortality of some pioneer species which are not tolerant to shadow is higher, but presenting a larger vertical stratification. With the increase of age from the natural regrowth, the horizontal and vertical structures become more complex, presenting biophysical characteristics close to those of primary forest. Such biophysical arrangements, which include the structure/density of the canopy, the occurrence of a larger or smaller number of arboreal/bush individuals (scatterers) in these types of vegetation formations, can lead to distinct polarimetric intra-class and inter-class configurations, with pedestal height values which are also different, as observed on Figure 7. Figure 8 shows the distribution of pixels extracted from the eleven ROIs in the bi-dimensional classification space $(H, \overline{\alpha})$, whose numerical results (i.e., percent of pixels by scattering zone) are presented in Table 1. At Figure 8 one observes that all 7 ROIs representative of primary forest, forest with timber exploitation and advanced secondary succession, show a concentrated pixel distribution (~70%) at zones Z4, Z5 and Z9. This indicates that the polarimetric response of such classes from already defined forest structure (containing 3 or 4 strata) is based mainly on multiple and volumetric scattering (the backscattering signal is mainly a function of geometry of scatters elements from dense cover) at zones of medium entropy. Additionally there is also some influence of a surface scattering but derived from a zone of low entropy. The two representative plots of classes from intermediate recovery have a higher distribution of pixels (>60%) at the zone of medium entropy (Z5), due to dipole scattering objects and in zones of low entropy (Z9, Z7), with scattering objects regularly distributed in space, and also as a consequence of the double-bounce effect caused by scattering objects such as twigs-trunk-soil-trunk. The initial secondary succession is positioned preferentially in the space of attributes of low entropy (Z9) and also configured by double-bounce interaction mechanisms (medium entropy multiple scattering – Z4), affected by the propagation of the canopy, as a result of the growth uniformity of pioneer species where there is a predominance of only an upper homogeneous stratum, and a lower dispersed one, formed by small plants which germinate and develop in a late succession process, in the shadow of the pioneers.

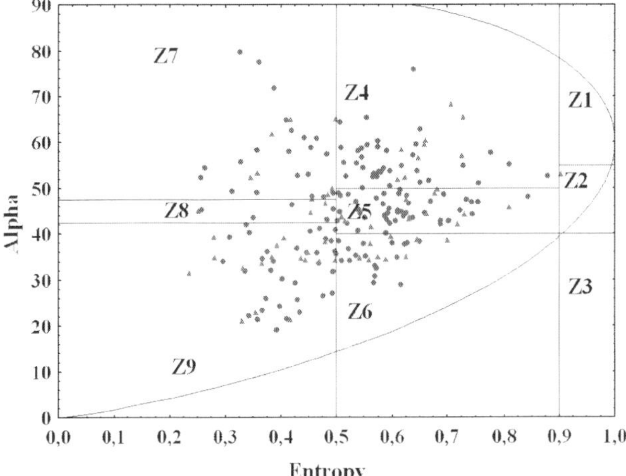

Fig. 8. L-band PALSAR α-H distribution for scattering mechanism: green colour are pixels of primary forest, forest with old selective logging and advanced secondary succession; red colour are initial and intermediate secondary succession. Source: Santos et al. (2009).

van der Sanden, J. J. (1997). Radar remote sensing to support tropical forest management. Tropenbos-Guyana Series 5: Georgetown, Guyana, PhD Thesis Wageningen Agricultural University. ISBN: 90-5485-778-1. 330p.

van Zyl, J. J., Zebker H. A., Elachi C. (1987). Imaging radar polarimetric signatures: theory and observation. Radio Science, 22(4): 529-543.

Woodhouse, I. H. (2006). Introduction to microwave remote sensing. Boca Raton: CRC Press Taylor & Francis Group, 370p.

Wulder, M.; Niemann, K.; Goodenough, D. (2000). Local maximum filtering for the extraction of tree locations and basal area from high spatial resolution imagery. Remote Sensing of the Environment, 73(1): 103-114.

Zebker, H. A.; Norikane, L. (1987). Radar polarimeter measures orientation of calibration corner reflectors. In: Proceedings of the IEEE, vol. 75, n° 12, pp. 1686-1688.

Zhang, L.; Zhang, J.; Zou, B.; Zhang, Y. (2008). Comparison of methods for target detection and applications using polarimetric SAR image. PIERS Online, 4(1): 140-145.

9

Discovery of user-generated geographic data using web search engines

Carlos Abargues, Carlos Granell, Laura Díaz, Joaquín Huerta
and Arturo Beltran
Universidad Jaume I de Castellón
Spain

1. Introduction

Traditionally the discovery of geographic data has been performed locally, from a centralized, desktop-based perspective. In the last decade, though, the notion of Spatial Data Infrastructure (SDI) (Masser, 2005) appeared as a net of interconnected SDI nodes to build spatial information infrastructures at different scales and levels (local, regional, national, European and even global) (INSPIRE, 2007). These SDI nodes represent an effort to link, share, and exchange geographic data from different and disperse sources. Discovery geographic data through this immense network of SDI nodes has been usually delegated to the use of catalogues services (http://www.opengeospatial.org/standards/cat). These specialized services base their functionality on the use of metadata that describes the geographic content they register to offer users and client applications effective interfaces for searching and publishing that content.

With the advent of the popular Web 2.0 the traditional Web moved towards a more social platform where the roles of content producer and content consumer have been diluted and often played by the same actor. The appearance of new technologies and applications (AJAX, Web APIs, social networks, etc.) offered new possibilities that facilitated and accelerated the content creation and sharing using the Web as the basic environment. Some of them deal specifically with geographic content and provide a new way to create and publish geographic content by people independently of their technical skills or knowledge on Geographic Information System (GIS). This new trend is currently known as Neogeography (Turner 2006). Millions of people are currently potential data providers and are able to create and share fresh geographic content for a given area acting as truly sensors (Goodchild 2007). All these quantities of geographic data on the Web have made the concept of the Geospatial Web (Scharl & Tochtermann, 2007) to gain relevancy. This term defines the use of the Web in such a way that the content has a geographic component defining its geospatial location in the world that is used to access to it and also linked it with the resources. In deep the Geospatial Web could be seen as the product of organizing the Web based in the location of its resources.

The openness, simplicity and easy of share of the new Web 2.0 tools and services also facilitate the content creation what results in an increase of geographic data in a proportion never seen till the date. The rate and speed at which the quantity of content increases and the nature of most of the content creators, most of them without deep technical skills, make the use of catalogues with its strict metadata requirements and its user-directed process of content submission an ineffective solution for its management. For these reasons new methods for discovering and managing this user-generated content should be used. During the evolution of the Web its size started to reach a dimension that made the task of searching for information along directories with linked documents a problem. This problem could happen again with the almost unmanageable quantity of content. In the first case the web search engines (Google, Yahoo, AltaVista) appeared with successful results becoming an essential tool for almost any Web user and for innovative and distinct uses (Al-Masri & Mahmoud, 2008). Maybe these applications can represent again a solution for the discovery of geographic data in the actual Web.

2. Background

2.1 Web 2.0 and the Geospatial Web.

Despite the fuzziness that use to follow its definition the term Web 2.0 does not refer to any new and different version of the traditional Web but a change in the technology and design to improve its functionality, communications, information sharing, creativity and collaboration along it. This new vision of the Web could be considered as the evolution and merge of different streams such as the technological improvement, the appearance of new applications and also the socialization of the Web, which comprises the participation and contribution of the users in it (Vossen & Hagemann 2007).

Still a few years ago the model of content consumption on the Web was represented by a commonly used client server model. In this model the content provider acted as server, producing content that was directly used by the client or content consumer defining clearly each role. With the spreading of the Web 2.0 philosophy this model completely changed dissolving the limits of each role. Nowadays the Web is full of services and tools with one major objective, sharing content (photos, videos, bookmarks, knowledge, etc.) among users.

The Geospatial Web represents the merging of different types of information that are already present on the Web (i.e. HTML pages, images, etc.) with geographic content. This facilitates discovering and searching any type of content based on its geographical component. In other words, the Geospatial Web structures each piece of Web content (photos, videos, web pages, 3D models, etc) according to its geospatial location, what is called georeferencing. This linkage between data and its geographic location enables their discovery and use by location approaching, as defined by the Geospatial Web.

2.2 Geographic services and user-generated data

Since the appearance of the first web-based GIS systems a lot of tools and services have emerged. It is especially in these last years when more and more services appeared probably thanks to the entering of big companies in the business of the geographic information (i.e. Google, Yahoo or Microsoft), the increase and improvement of the Internet communications and the spread use of positioning devices such as GPS receivers. Probably the most common of these tools are the web mapping services (Mitchell 2005). There exist many web mapping service implementations available, both from private companies such as Google

Maps, Yahoo! Maps, Bing Maps (formerly Microsoft Live Maps), MapQuest and much others, and integrated in Geoportals (Bernard et al., 2005) present usually in SDI nodes.

Not only the web mapping services have become popular but also a new type of desktop applications have recently emerged to visualize geographic content. Geobrowsers (virtual globes or digital earths) offer a three-dimensional view of the earth where the geographic content is displayed over that virtual representation of the planet's surface. These tools offer totally new ways for visualizing data and they have started a process of change from the two-dimensional to a more realistic method of visualizing data. Users can find geobrowsers such as Google Earth, NASA World Wind and ESRI's ArcGIS Explorer. These services are not limited to show to end users simple cartography and are also enriched with different sources of information. For instance users can find addresses and directions, business, traffic conditions or even content created by other users. These applications do not offer just the visualization of content but also provide a searching mechanism over the geographic data they manage.

The use of the web-based and desktop applications is extended with the release of Application Programming Interfaces (API) and Software Development Kits (SDK). In the first case APIs enable third parties developments to use geographic services and data through calls to the API in their own projects. For instance, anybody can create geographic data and visualize it using either Google Maps or Google Earth in their own website. The use of these APIs popularized the term mashups that denotes the creation of new web applications for specific purposes by combining and integrating remote data from other distributed sources. The SDK offers the opportunity of freely reuse or extend desktop applications such as NASA World Wind or ESRI's ArcGIS Explorer enabling the users to create their own applications with this capabilities.

The emerge of all these new brand services and tools along with the vision of content creation and sharing in Web 2.0, have led to new ways to create and distribute geographic content. Probably one of the pioneering and most important movements nowadays is the OpenStreetMap project whose objective is the collaborative creation of freely available cartography all around the world (Haklay & Weber 2008). This cartographic data is collected, edited and uploaded by its own users. Another example is the concept of Public Participation GIS (PPGIS) (Nyerges et al., 1997) that enables participation of a given community in the future urbanism development of their own area. Not only projects such as the OpenStreetMap promote this type of geographic content creation but also private companies offer their own tools to do so. This is the case of Google My Maps and Google Map Maker, this last one intended also to facilitate the creation of cartography for those countries that experience a lack of it.

2.3 Keyhole Markup Language: The Geospatial Web's HTML.

The Keyhole Markup Language (KML) is a XML-based language designed to express geographic annotation and visualization. The geographic visualization includes not only the representation of the graphical data but also establishes order and control over the data navigation. KML is used in a broad range of applications such as web mapping services, two-dimensional maps including those used in mobile devices and geobrowsers. Most of these applications or services also make use of KMZ files. These ones are basically compressed files containing a KML file and other resources such as images or icons that could be referenced on it. KMZ files facilitate the sharing and distribution of content

Craglia, M.; Kanellopoulos, I. & Smits, P. (2007). Metadata: where we are now, and where we should be going, Proceedings of 10th AGILE International Conference on Geographic Information Science 2007, Aalborg University, Denmark.

Foerster, T.; Schaeffer, B.; Brauner, J. & Jirka, S. (2009). Integrating OGC Web Processing Services into Geospatial Mass-Market Applications, Proceeding of the Intl. Conference on Advanced Geographic Information Systems & Web Services, 98-103, Cancun, Mexico

Goodchild, M (2007). Citizens as Voluntary Sensors: Spatial Data Infrastructure in the World of Web 2.0. International Journal of Spatial Data Infrastructures Research, 2, 24-32

Haklay, M. & Weber, P. (2008). OpenStreetMap: User-generated Street Maps. IEEE Pervasive Computing, 7, 4, 12-18

INSPIRE EU Directive (2007). Directive 2007/2/EC of the European Parliament for establishing an Infrastructure for Spatial Information in the European Community (INSPIRE). Official Journal of the European Union, L 108/1, Vol 50, 25 April 2007.

ISO 19115 (2003). Geographic information - Metadata. ISO/TS 19115:2003, International Organization for Standardization (ISO).

ISO 19139 (2007). Geographic information – Metadata – XML schema Implementation. ISO/TS 19139:2007, International Organization for Standardization (ISO).

Masser, I. (2005). GIS Worlds: Creating Spatial Data Infrastructures, ESRI Press, ISBN: 1-58948-122-4, Redlands, California

Mitchell, T. (2005). Web Mapping Illustrated: Using Open Source GIS Toolkits, O'Reilly Media, ISBN 9780596008659

Nyerges, T; Barndt, M. & Brooks, K. (1997). Public participation geographic information systems, Proceedings of Auto-Carto 13 - American congress on surveying and mapping, pp. 224–233, Seattle, WA

Rajabifard, A.; Feeney, M-E. & Williamson, I. P. (2002). Future directions for SDI development. Intl. Journal of Applied Earth Observation and Geoinformation, 4, 11–22

Scharl, A. & Tochtermann, K. (2007). The geospatial web: how geobrowsers, social software and the Web 2.0 are zapping the network society. Springer Verlag, ISBN: 978-1-84628-826-5

Turner, A. (2006). Introduction to Neogeography, O'Reilly Media, ISBN: 9780596529956

Vossen, G. & Hagemann, S. (2007). Unleashing Web 2.0: From Concepts to Creativity. Morgan Kaufmann, Burlington, MA.

10

Estimation of soil properties using observations and the crop model STICS. Interest of global sensitivity analysis and impact on the prediction of agro-environmental variables

Hubert Varella, Martine Guérif and Samuel Buis
INRA-UAPV EMMAH
France

1. Introduction

Dynamic crop models are very useful to predict the behaviour of crops in their environment and are widely used in a lot of agro-environmental work such as crop monitoring, yield prediction or decision making for cultural practices (Batchelor et al., 2002; Gabrielle et al., 2002; Houlès et al., 2004). These models usually have many parameters and their estimation is a major problem for agro-environmental prediction (Tremblay and Wallach, 2004; Makowski et al., 2006). For spatial application, the knowledge of soil parameters is crucial since they are responsible for a major part of the variability of the crop model output variables of interest (Irmak et al., 2001; Launay and Guérif, 2003; Ferreyra et al., 2006). These parameters may be estimated from different techniques: either with soil analysis at the different points of the study area, from a soil map and the application of pedotransfer functions (Reynolds et al., 2000; Murphy et al., 2003), from remote sensing images (Lagacherie et al., 2008) or by using electrical resistivity measurements (Golovko and Pozdnyakov, 2007). The choice of the first method is difficult because of practical limitations, as well as time and financial constraints. Detailed soil maps adapted to the scale of precision agriculture and even to the scale of catchment are scarcely available (King et al., 1994), while the use of remote sensing images or electrical resistivity is still hampered by a lack of robust interpretation of the signal (Lagacherie et al., 2008). Moreover, these techniques do not permit to access the values of all the soil parameters required to apply a complex crop model. Fortunately, techniques derived from remote sensing images (Weiss and Baret, 1999; Houborg and Boegh, 2008) or yield monitoring (Blackmore and Moore, 1999; Pierce et al., 1999) allow soil parameters being estimated through the inversion of crop models.
Estimating parameters of complex models such as crop models may be not so easy (Tremblay and Wallach, 2004; Launay and Guérif, 2005). One of the reasons for the difficulties encountered may be a lack of sensitivity of the observed variables to the parameters, making the estimation process inefficient. Another reason may be that the influence of the parameters on the observed variables takes place mainly through

interactions, making it difficult to identify the relevant factors (Saltelli et al., 2000). For complex non-linear models such as crop models, global sensitivity analysis (GSA) methods are able to give relevant information on the sensitivity of model outputs to the whole range of variation of model inputs. Many studies have focused on this subject, namely, how to choose the main parameters to be estimated for the model calibration (Campolongo and Saltelli, 1997; Ruget et al., 2002; Gomez-Delgado and Tarantola, 2006; Makowski et al., 2006) and ranked the importance of the parameters by calculating global sensitivity indices: first-order indices (the main effect of the parameter on the output) and total indices (sum of all effects involving the parameter, including the interactions with other parameters). The common practice is consistent with the principles expressed by Ratto et al. (2007). Small total sensitivity indices indicate a negligible effect of the parameter on the model output concerned. These parameters can be fixed at a nominal value ("Factor Fixing setting"). High first-order indices reveal a clearly identifiable influence of the parameter on the model output concerned, and therefore the parameters need to be determined accurately ("Factor Prioritization setting"). Small first-order indices combined with large interaction indices result in a lack of identification. In practice, the two first rules are commonly used to select the set of parameters to be estimated in a calibration problem. GSA can also be used to evaluate the quantity of information contained in a given set of observations for estimating parameters and thus to determine which is the best observation set for estimating the parameters (Kontoravdi et al., 2005). Although the results of GSA are often used to design the estimation process, the link between GSA indices and the quality of parameter estimation has never been quantified.

Our objectives in this study are twofold. Firstly, we propose to use GSA results in order to measure the quantity of information contained in different sets of observations and to illustrate the link between this measurement and the quality of parameter estimates. Secondly, we propose to study the impact of the quality of parameter estimates on the prediction of variables of interest for agro-environmental work. As the performance of the estimation process is supposed to depend on several conditions such as soil type, cropping conditions (preceding crop and climate) or available observations, we chose to conduct the study on synthetic observations in order to be able to generate variability in parameter retrieval performance as well as in sensitivity structure of the observed model outputs to soil parameters and in the prediction performance. This choice also allows eliminating the impact of model errors, which may complicate the interpretation of the results. Finally, we considered in this study the STICS-wheat crop model and various synthetic observations on wheat crops: derived from remote sensing images (LAI and absorbed nitrogen) as well as grain yield.

2. Material and methods

2.1 The crop model, output variables and soil parameters

2.1.1 The STICS model

The STICS model (Brisson et al., 2002) is a nonlinear dynamic crop model simulating the growth of various crops. For a given crop, STICS takes into account the climate, type of soil and cropping techniques to simulate the carbon, water and nitrogen balances of the crop-soil system on a daily time scale. In this study, a wheat crop is simulated. The crop is essentially characterized by its above-ground biomass carbon and nitrogen, and leaf area index. The main outputs are agronomic variables (yield, grain protein content) as well as environmental variables (water and nitrate leaching). Yield, grain protein content and nitrogen balance in the soil at harvest are of particular interest for decision making, especially for monitoring nitrogen fertilization (Houlès et al., 2004). Nitrogen absorbed by the plant and leaf area index are also important to analyze the health and growth of the plant during the crop's growing season.

The STICS model includes more than 200 parameters arranged in three main groups: those related to the soil, those related to the characteristics of the plant or to the genotype, and those describing the cropping techniques. The values of the last group of parameters are usually known as they correspond to the farmer's decisions. The parameters related to the plant are generally determined either from literature, from experiments conducted on specific processes included in the model (e.g. mineralization rate, critical nitrogen dilution curve etc.) or from calibrations based on large experimental database, as is the case for the STICS model (Hadria et al., 2007). The soil parameters are difficult to determine at each point of interest and are responsible for a large part of the spatial variability of the output variable. That is why the sensitivity analysis and parameter estimation processes described in this study only concern soil parameters.

2.1.2 Output variables considered

In this study, we focus on two types of STICS output variables. First, those corresponding with observations that may be done on wheat canopy by automated measurements. They consist in:

- the leaf area index (LAI_t) and the nitrogen absorbed by the plant (QN_t) at various dates t during the crop season - as potentially derived from remote sensing image inversion (Weiss and Baret, 1999; Houborg and Boegh, 2008),
- the yield at harvest (Yld) as potentially provided by yield monitoring systems.

These output variables, hereafter referred to as "observable variables" can be observed at different dates during the growing season.

Second, a main objective of this study, beyond the estimation of soil parameter values, lies in the prediction of some output variables of interest, and its improvement as compared to the prediction obtained with a lack of precise values on soil parameters. They consist in:

- yield at harvest (Yld),
- protein in the grain at harvest ($Prot$),
- nitrogen contain in the soil at harvest (Nit).

Yield, grain protein content and nitrogen balance in the soil at harvest are of particular interest for decision making, especially for monitoring nitrogen fertilization (Houlès et al., 2004). Nitrogen absorbed by the plant are also important to analyze the health and growth of the plant during the crop's growing season (Baret et al., 2006).

2.1.3 The soil parameters estimated

The STICS model contains about 60 soil parameters. In our case, in order to limit the problems of identifiability, the number of soil parameters to be estimated has been reduced. First, among the available options for simulating the soil system, the simplest was chosen, by ignoring capillary rise and nitrification. These assumptions define the domain of validity of the model considered and hence, of the results that are found. We then considered the soil as a succession of two horizontal layers, each characterized by a specific thickness parameter. From the observation of the tillage practices in the region around our experimental site of Chambry (49.35°N, 3.37°E) (Guérif et al., 2001), the thickness of the first layer was set at 0.30 m. We performed a first sensitivity analysis on the 13 resulting soil parameters. This allowed us to fix those whose effects on the observed variables were negligible: for each parameter we computed the values of its effects on all the observed variables considered for a lot of soil, climate and agronomic conditions, and dropped the parameters for which all these values were less than 10% of the total effects generated by the 13 parameters. We thus restricted the study to 7 parameters.

The 7 soil parameters considered (Table 1) characterize both water and nitrogen processes. They refer to permanent characteristics and initial conditions. Among the permanent characteristics, clay and organic nitrogen content of the top layer are involved mainly in organic matter decomposition processes and nitrogen cycle in the soil. Water content at field capacity of both layers affects the water (and nitrogen) movements and storage in the soil reservoir. Finally, the thickness of the second layer defines the volume of the reservoir. The initial conditions correspond to the water and nitrogen content, *Hinit* and *NO3init*, at the beginning of the simulation, in this case the sowing date.

Parameter	Definition	Range	Unit
argi	Clay content of the 1rst layer	14-37	%
Norg	Organic nitrogen content of the 1rst layer	0.049-0.131	%
epc(2)	Thickness of the 2nd layer	0-70 or 50-130*	cm
HCC(1)	Water content at field capacity (1rst layer)	14-30	g g-1
HCC(2)	Water content at field capacity (2nd layer)	14-30	g g-1
Hinit	Initial water content (both layers)	4-29	% of weight
NO3init	Initial mineral nitrogen content (1rst layer)	4-21.5 or 25-86**	kg N ha-1

* the first range is for a shallow soil and the second for a deep soil; ** the first range is for a wheat cultivated after sugar beet and the second for a wheat cultivated after pea

Table 1. The 7 soil parameters and ranges of variation.

2.2 Global sensitivity analysis

In this study, we chose a variance-based method of global sensitivity analysis which allows calculating the sensitivity indices for a non-linear model such as STICS. More precisely, the method we chose is Extended FAST.

2.2.1 Variance decomposition and sensitivity indices

We denote a given output variable of the STICS model as Y. The total variance of Y, $V(Y)$, caused by variation in the 7 selected soil parameters θ, can be partitioned as follows (Chan et al., 2000):

$$V(Y) = \sum_{i=1}^{7} V_i + \sum_{1 \leq i < j \leq 7} V_{ij} + \cdots + V_{1,2,\cdots,7} \tag{1}$$

where $V_i = V\left[E\left(Y|\theta_i\right)\right]$ measures the main effect of the parameter θ_i, $i = 1,\cdots,7$, and the other terms measure the interaction effects. Decomposition (2) is used to derive two types of sensitivity indices defined by:

$$S_i = \frac{V_i}{V(Y)} \tag{2}$$

$$ST_i = \frac{V(Y) - V_{-i}}{V(Y)} \tag{3}$$

where V_{-i} is the sum of all the variance terms that do not include the index i.

S_i is the first-order (or main) sensitivity index for the ith parameter. It computes the fraction of Y variance explained by the uncertainty of parameter θ_i and represents the main effect of this parameter on the output variable Y.

ST_i is the total sensitivity index for the ith parameter and is the sum of all effects (first and higher order) involving the parameter θ_i.

S_i and ST_i are both in the range (0, 1), low values indicating negligible effects, and values close to 1 huge effects. ST_i takes into account both S_i and the interactions between the ith parameter and the 6 other parameters, interactions which can therefore be assessed by the difference between ST_i and S_i. The interaction terms of a set of parameters represent the fraction of Y variance induced by the variance of these parameters but that cannot be explained by the sum of their main effects. The two sensitivity indices S_i and ST_i are equal if the effect of the ith parameter on the model output is independent of the values of the other parameters: in this case, there is no interaction between this parameter and the others and the model is said to be additive with respect to θ_i.

2.2.2 Extended FAST

Sobol's method and Fourier Amplitude Sensitivity Test (FAST) are two of the most widely used methods to compute S_i and ST_i (Chan et al., 2000). We have chosen here to use the extended FAST (EFAST) method, which has been proved, in several studies (Saltelli and Bolado, 1998; Saltelli et al., 1999; Makowski et al., 2006), to be more efficient in terms of number of model evaluations than Sobol's method. The main difficulty in evaluating the first-order and total sensitivity indices is that they require the computation of high dimensional integrals. The EFAST algorithm performs a judicious deterministic sampling to explore the parameter space which makes it possible to reduce these integrals to one-dimensional ones using Fourier decompositions. The reader interested in a detailed description of EFAST can refer to (Saltelli et al., 1999).

We have implemented the EFAST method in the Matlab® software, as well as a specific tool for computing and easily handling numerous STICS simulations. The uncertainties

Saltelli, A., Tarantola, S.,Campolongo, F., 2000. Sensitivity analysis as an ingredient of modeling. Statistical Science 15 377-395.

Saltelli, A., Tarantola, S.,Chan, K.P.S., 1999. A quantitative model-independent method for global sensitivity analysis of model output. Technometrics 41 39-56.

Spearman, C., 1904. The proof and measurement of association between two things. Amer. J. Psychol. 15 72-101.

Tremblay, M.,Wallach, D., 2004. Comparison of parameter estimation methods for crop models. Agronomie 24 351-365.

Weiss, M.,Baret, F., 1999. Evaluation of canopy biophysical variable retrieval performances from the accumulation of large swath satellite data. Remote Sensing of Environment 70 293-306.

11

Modeling of wake-vortex detection by a ground-based fiber LIDAR system

S. Brousmiche, L. Bricteux,
G. Winckelmans, B. Macq and P. Sobieski
Ecole Polytechnique de Louvain, Université catholique de Louvain
Belgium

1. Introduction

Due to its lift force, an aircraft releases large scale swirling flows in its wake. As these vortices can impact significantly the trajectory of a following aircraft, their study is of great importance for practical applications concerning safety in aircraft traffic management. A well-adapted system to detect them is the heterodyne Doppler LIDAR (Light Detection And Ranging). Moreover, fiber-based LIDAR are known for their higher flexibiliy and compacity despite that they are limited in pulse energy. However, to reach optimal performances, it is often needed to precisely model the whole measurement process including atmospheric effects such as the refractive turbulence. Some parameters have to be tuned depending on the application as, for example, the pulse energy and duration, the pulse repetition frequency or the telescope dimensions. Wake vortices detection and intensity estimation also require a good compromise between spatial resolution and velocity precision which also depends on the scanning configuration.

The aim of this chapter is twofold. On the one hand, it describes a complete simulation method to evaluate the performance of a Doppler LIDAR for aircraft wake vortices detection near the ground. The principal interest of the simulation presented here is to combine a LASER beam propagation method with fluid dynamics simulations. On the other hand, it proposes effective wind velocity map reconstruction algorithms. The performance of spectral and correlogram accumulation algorithms in the vicinity of wake vortices is analyzed. The effect of velocity gradients on the estimation error is thus addressed.

After introducing the wake vortex phenomenon and its detection using a heterodyne Doppler LIDAR , the section 2 will discuss the measurement simulation. The numerical simulation of the wake in ground effect is then detailed in section 3. The processing algorithms are finally explained in section 4.

1.1 Wake-vortex fundamentals

The lift force acting on an aircraft is due to the pressure difference between the lower side (pressure side) and the upper side (suction side) of its wing. As a result of this pressure difference, there is a spanwise flow at the edges of the wing from the pressure side to the suction side. This pressure difference forces the suction side streamlines to converge toward the center of the wing and the pressure side streamlines to diverge from it. This spanwise flow combined with the free-stream velocity produces a swirling motion of the air trailing downstream of the wing. Just behind the trailing edge, a vortex sheet is shed which rolls-up rapidly within a few span lengths to form a pair of counter rotating vortices of equal strength. The circulation, Γ [m^2/s], of the velocity field [m/s], $\mathbf{u}$, for each vortex is a quantity of primary importance as it gives an image of the vortex strength. For a closed contour $\mathcal{C} = \partial\Omega$ enclosing a patch of vorticity defined by the field $\boldsymbol{\omega} = \nabla \times \mathbf{u}$,

$$\Gamma = \int_\Omega \boldsymbol{\omega} \cdot \mathbf{n}\, dS = \oint_\mathcal{C} \mathbf{u} \cdot \mathbf{dl}. \tag{1}$$

In the particular case of a vortex alone, it is natural to choose a circular integration contour $\mathcal{C}$ and since the velocity field has the form $\mathbf{u} = u_\theta(r)\,\widehat{\mathbf{e}}_\theta$ it follows that:

$$\Gamma(r) = 2\pi r\, u_\theta(r) \tag{2}$$

The velocity profile is thus closely related to the circulation profile. The circulation and the velocity tangential distribution are given by the following relations:

$$\Gamma(r) \;=\; \Gamma_0 \frac{r^2}{\left(r^2 + r_c^2\right)}\,, \tag{3}$$

$$u_\theta(r) \;=\; \frac{\Gamma(r)}{2\pi r}\,. \tag{4}$$

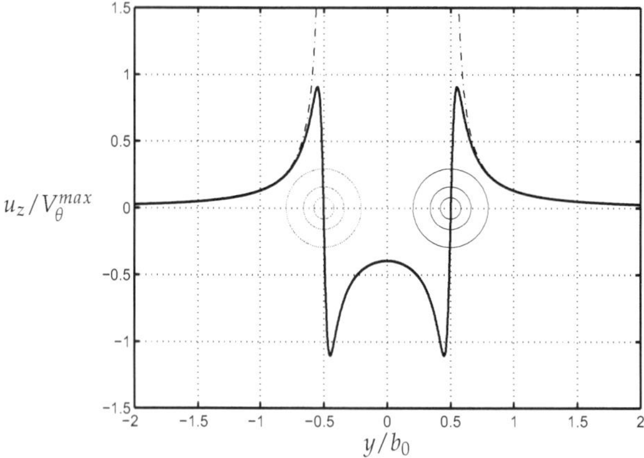

Fig. 1. Vertical velocity profile in the far wake of an aircraft. The dashed-dotted lines represent the potential velocity profile $V_\theta = \frac{\Gamma_0}{2\pi r}$. The curve is normalized by $V_\theta^{max} = \frac{\Gamma_0}{4\pi r_c}$ in the case of a low order algebraic model.

The parameters are Γ_0 [m^{-2}/s], the total circulation of the vortex and r_c [m], the radius of maximum induced tangential velocity. It is an indicator of the core size of a vortex; typically it amounts, after the initial roll-up phase, to $3 - 5\%$ of the wing span b. A typical velocity profile for a pair of vortices is illustrated in Fig. 1. A downwash velocity will alter the trajectory of an airplane passing between the two vortices. If it passes on one of the sides of the vortex pair, it will feel an upwash velocity. If the airplane enters in a vortex zone, a strong rolling moment

Leader aircraft max take-off weight	Follower aircraft (metric tons)	Separation Nautical miles	Time delay [s] Approach speed: $70m/s$
Heavy	Heavy ($\geq 136\,T$)	4	106
Heavy	Medium($< 136\,T$)	5	132
Heavy	Light ($\leq 7\,T$)	6	159
Medium	Light	5	132

Table 1. International Civil Aviation Organization (ICAO) aircraft separation distances to avoid wake vortex encounter, in nautical miles ($1\ NM = 1.852\ km$). For all other combinations, the separation is $3\ NM$. Table reproduced from Gerz et al. (2001).

will be applied to it. The momentum conservation allows to bind the initial circulation Γ_0 of an aircraft vortex pair to the lift force itself, equal to the weight of the aircraft:

$$L = Mg = \rho U b_0 \Gamma_0, \tag{5}$$

where U [m/s] is the flight velocity and ρ is the fluid density. The circulation is thus proportional to the mass M of the aircraft and inversely proportional to its flight speed U. Typical values for the circulation range in $\Gamma_0 \approx 400 - 600$ m^2/s for heavy aircrafts. The length b_0 is the distance between the vortex cores; its value is approximately $b_0 = \frac{\pi}{4}b$ (assuming elliptical loading) with b the wingspan. The vortex spacing b_0 can be deduced following the lifting line theory (Prandtl, 1957). Wake vortices generated at take-off are very strong since the airplane evolves at low speed and is fully loaded. Aircraft wake vortices are also the source of the drag induced by the lift force acting on the aircraft: the so-called *induced drag*.

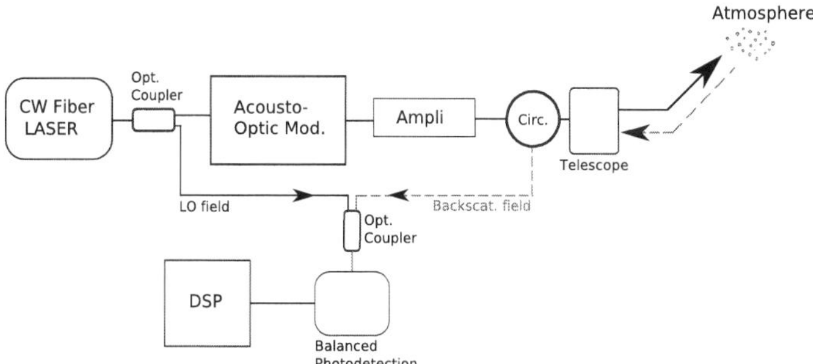

Fig. 2. Classical heterodyne Doppler LIDAR architecture based on a fiber LASER source.

1.2 Detection using a heterodyne Doppler LIDAR

A general architecture for a heterodyne Doppler LIDAR is given at Fig. 2. The heart of the system is a continuous LASER source producing a quasi-monochromatic optical wave. The 1.55 μm fiber-based LASER sources are used as an alternative to classical sources due to their higher compacity and flexibility. Moreover, although the pulse energy is rather low (i.e. a few hundreds of microjoules), they are able to produce a higher pulse repetition frequency. When passing through the acousto-optic modulator, the wave is shifted in frequency and modulated

Simulation parameters	
Laser wavelength	$1.55\ \mu m$
Pulse energy	$150\ \mu J$
Detector bandwidth	50 MHz
Pulse repetition frequency	400 Hz
Intermediate frequency	80 MHz
Sampling frequency	250 MHz
Angular resolution	1.25 mrad
1/e radius of the emitted beam	30mm
Radius of the transmitter aperture	50mm
Aerosol backscattering coefficient	$10^{-7}\ \mathrm{m}^{-1}\ \mathrm{sr}^{-1}$
Linear extinction coefficient	$7.10^{-5}\ \mathrm{m}^{-1}$

Table 2. LIDAR parameters used for the simulations

We use a numerical grid for the screen of 512×512 with 2 mm resolution. We simulate a continuous random medium by using 10 phase-screens and the statistics have been obtained by running 400 samples. Fig. 4. gives the SNR as well as the system efficiency at low and medium refractive turbulence level for both monostatic and bistatic configurations. We see on the one hand that an increase of the turbulence level degrades significantly the system performances with range in the bistatic configuration. On the other hand, performance enhancement is observed in the monostatic case which is due to the fact that the emitted and backpropagated beams propagate through the same turbulence. This effect is limited in range and the SNR eventually becomes lower than the one obtained at low C_n^2.

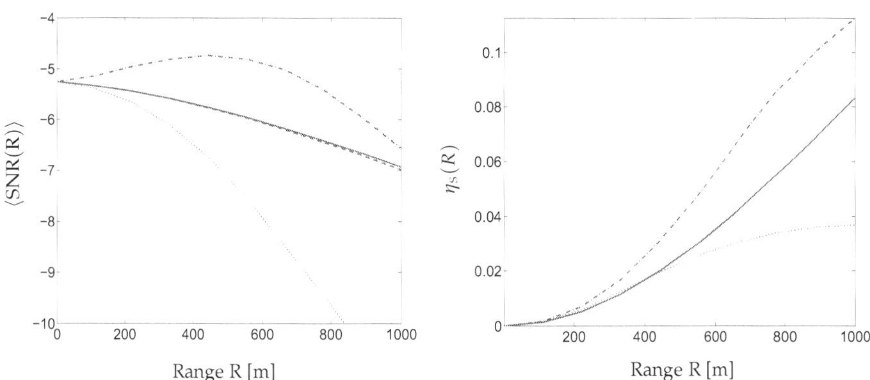

Fig. 4. Signal-to-noise ratio (left) and system efficiency (right) with range obtained for different measurement conditions : free-space (solid); monostatic configuration, $C_n^2 = 10^{-15}$ $\mathrm{m}^{-2/3}$ (dash); bistatic configuration, $C_n^2 = 10^{-13}$ $\mathrm{m}^{-2/3}$ (dot) and monostatic configuration, $C_n^2 = 10^{-13}$ $\mathrm{m}^{-2/3}$ (dash-dot).

3. Simulation of wake vortices in ground effect

In this section, we describe the fluid simulation that have been performed to analyze the detection capability of the heterodyne LIDAR in the presence of wake vortices. Since we consider here a ground-based system, simulating the interaction of wake vortices with the ground is of great importance since it produces secondary vortices.

3.1 Description of the code

The governing equations of the flow considered are the Navier-Stokes equations for incompressible flow, and supplemented by a term which models the effect of the small dissipatives scales of turbulence (not captured by the grid) on the big scales:

$$\nabla \cdot \mathbf{u} \;=\; 0 \tag{20}$$

$$\frac{\partial \mathbf{u}}{\partial t} + (\mathbf{u} \cdot \nabla)\,\mathbf{u} \;=\; -\nabla P + \nu \nabla^2 \mathbf{u} + \nabla \cdot \boldsymbol{\tau}^M, \tag{21}$$

where P is the reduced pressure, ν is the kinematic viscosity, and $\boldsymbol{\tau}^M$ is the subgrid scale stress tensor model. These equations are solved using a fractional-step method with the *delta* form for the pressure (Lee et al., 2001). This form allows simple boundary conditions for the pressure and the intermediate velocity field. The convective term is integrated using an Adams-Bashforth 2 scheme and the diffusion term using a Crank-Nicolson scheme. The time-stepping scheme reads

$$\frac{\mathbf{u}^* - \mathbf{u}^n}{\Delta t} \;=\; -\frac{1}{2}\left(3\mathbf{H}^n - \mathbf{H}^{n-1}\right) - \nabla \phi^n$$

$$+\frac{1}{2}\nu \nabla^2 \left(\mathbf{u}^* + \mathbf{u}^n\right)$$

$$+\frac{1}{2}\left(3\left(\nabla \cdot \boldsymbol{\tau}^M\right)^n - \left(\nabla \cdot \boldsymbol{\tau}^M\right)^{n-1}\right) \tag{22}$$

$$\nabla^2 \varphi \;=\; \frac{1}{\Delta t}\nabla \cdot \mathbf{u}^* \tag{23}$$

$$\frac{\mathbf{u}^{n+1} - \mathbf{u}^*}{\Delta t} \;=\; -\nabla \varphi \tag{24}$$

$$\phi^{n+1} \;=\; \phi^n + \varphi, \tag{25}$$

where $\mathbf{H}^n$ is the convective term, $\mathbf{u}^*$ is the intermediate velocity field and ϕ^n is the modified pressure. The model term is integrated explicitly, also using the Adams-Bashforth 2 scheme. The subgrid scale model used here is the WALE regularized variational multiscale model as defined in (Bricteux, 2008). The equations are discretized in space using fourth order finite differences and the scheme of Vasilyev (Vasilyev, 2000) for the discretization of the convective term, which conserves energy on Cartesian stretched meshes and is therefore particularly suited for direct or large-eddy simulations of turbulent flows. The Poisson equation for the pressure is solved using a multigrid solver with a line Gauss-Seidel smoother. The code is implemented to run efficiently in parallel.

3.2 Radial velocity interpolation technique

In order to compute the Doppler term appearing in the equation giving the heterodyne current (Equation 15), the radial velocity in each layer l and for each LOS is mandatory. As the fluid flow velocity field $(v(y,z), w(y,z))$ issued from the Navier-Stokes simulation is computed on cartesian grid, it is necessary to interpolate the data onto the LIDAR line-of-sight and perform a projection of the velocity vector on the LOS in order to have the radial velocity.

The interpolation scheme selected in this work to interpolate information from a fluid simulation grid grid to the LIDAR measurement points is the M'_4 scheme (Winckelmans, 2004). This technique is widely used in vortex particles methods to redistribute vorticity from a distorted set of particles onto a regular cartesian grid. In this case we perform the inverse operation as we transfer information from a cartesian rectangular grid onto a polar one. This scheme is defined as:

$$\phi^I(\vec{x}) = \sum_{i=0}^{M-1} \sum_{j=0}^{N-1} W(\vec{x} - \vec{x}^S_{ij}) \phi^S_{ij}, \tag{26}$$

where $\vec{x}$ is the lidar sensing location while $\vec{x}^S$ is the location at which the information is computed. The kernel for the M'_4 scheme is defined as:

$$W(x) = \begin{cases} 1 - \frac{5}{2}\left|\frac{x}{h}\right|^2 + \frac{3}{2}\left|\frac{x}{h}\right|^3 & \text{if } \left|\frac{x}{h}\right| \leq 1 \\ \frac{1}{2}\left(2 - \left|\frac{x}{h}\right|\right)^2 \left(1 - \left|\frac{x}{h}\right|\right) & \text{if } 1 < \left|\frac{x}{h}\right| \leq 2 \\ 0 & \text{if } \left|\frac{x}{h}\right| > 2 \end{cases} \tag{27}$$

In Fig. 5, one observes the number of points on the cartesian grid which influence the target point on a LOS. The radial velocity profiles are then obtained by computing the projection of the interpolated velocity profile along the different line-of-sight constituting the scanning pattern.

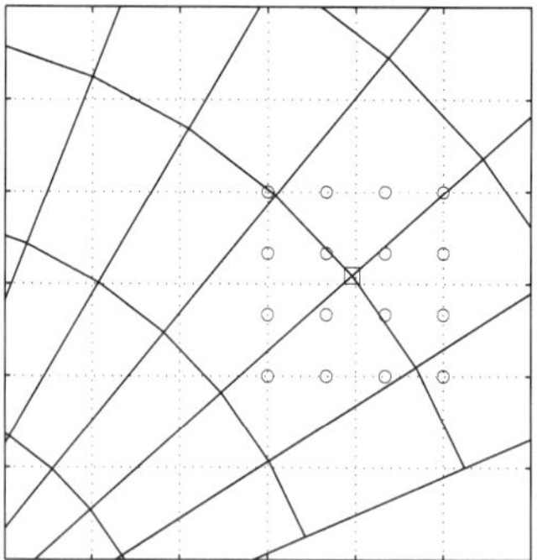

Fig. 5. Interpolation Stencil: the value at a target point (square) located on a given LOS at a given range is obtained by interpolation using the 16 neighbouring points (circles) of the cartesian fluid dynamics simulation grid with a spacing h.

In order to evaluate the estimation error using the fluid simulation velocity field and compare the results with different pulse durations, we have to take into account the radial velocity profile filtering during the measurement process. Hence, the estimated velocity is a filtered version of the exact radial velocity. The weighting function is the square norm of the transmitted pulse defined in Equation 7. A good approximation of the mean radial velocity estimate is given by (Frehlich, 1997):

$$v_{pulse}(R) = \int_0^\infty v_R(\rho) \left| \tilde{f}((R - \rho)2/c) \right|^2 d\rho \tag{28}$$

3.3 Results

The results are given with a dimensionless time τ defined as $\tau = tV_0/b_0$, where $V_0 = \Gamma_0/(2\pi b_0)$ is the intial descent velocity of the vortex pair. The flow is evolving in the following way: due to the mutually induced velocity, the vortex pair moves down. The impermeability condition at the wall can be modeled by a mirror vortex pair, resulting in an hyperbolic trajectory. However, as the induced tangential velocity at the wall is nonzero, the no-slip condition is at the origin of the development of boundary layers beneath the vortices. Those are clearly visible in Fig. 6. Starting from the stagnation point on the symmetry plane between the vortices, the pressure gradient forcing the boundary layers is first favorable and then adverse. This leads to their separation, when the primary vortices are close enough to the wall. The detached vortex-sheet then rolls-up and forms two coherent secondary vortices. They induce an upward velocity on the primary vortices and force the latter to rise; this is the rebound phenomenon. At this time, the flow is still essentially two-dimensional. But the secondary vortices are unstable with respect to short-wavelength modes. These perturbations are growing while the secondary vortex is orbiting around the primary vortex, as can be seen in Figs. 6 at time $\tau = 2$. The instability mechanism can be related to the elliptic instability of Widnall. The flow eventually evolves to a fully turbulent vortex system.

The Γ_{5-15} is an important hazard criterion and is defined as the circulation profile average between 5 and 15 meters. Its evolution is reported in Fig. 7. The initial 2-D decay phase is governed by the molecular viscosity and is due to the slow spreading of the core size, it is however negligible in the case of a high Re number flow. The fast decay phase starts at $\tau \approx 2.8$ and the final circulation level attained is reduced by 40% compared to the initial one. During the first part of the descent, the trajectoriy follows the inviscid theory. As soon as the boundary layer starts to separate, the rebound occurs and the trajectory is significantly different. One observes the loop trajectory induced by the rebound.

The radial velocity maps in the early case (τ=2.5) and the late case (τ=5) are given Fig. 8.

4. Velocity map reconstruction

Assuming that the LIDAR parameters (pulse energy, duration and PRF) have been optimized for the detection problem, the only constraint subsisting on the signal processing algorithms is to guarantee the best spatial resolution while keeping the radial velocity estimation variance as low as possible. This constraint is related to the use of accumulation methods to compensate the low SNR. As circulation estimation method are usually based on a model-fitting method, a poor spatial resolution inside vortices will induce inherently a higher estimation variance on the vortex circulation. The goal of this section is therefore to compare the efficiency of periodogram-based and correlogram-based accumulation methods in terms of spatial resolution in the vicinity of the wake vortex.

4.1 Radial velocity estimation

Each time a pulse is sent, along a particular line-of-sight (LOS) of given elevation angle defined by the current position of the telescope, a signal $s(nT_s)$ is measured where T_s is the sampling period. In order to evaluate the velocity profile along this LOS, we estimate the energy spectral density of $s(nT_s)$ at each position of a sliding window of N time samples. The observation time is then $T_{obs} = NT_s$ which corresponds to a range gate size $\Delta p = T_{obs}c/2$. The time offset of the pth window is t_p and the spatial offset is then $R_p = t_p c/2$. A direct spectral estimate is the periodogram defined by

$$\widehat{P}_s(k\Delta f) = \frac{1}{N}\left|\sum_{n=0}^{N-1} h(nT_s)s(t_p + nT_s)\exp\left(-2\pi ikn/N\right)\right|^2 \tag{29}$$

If no spectral interpolation is performed, the frequency resolution is $\Delta f = F_s/N$. An estimate of mean velocity, $\hat{v}_R$, is given by evaluating

$$\hat{v}_R = \frac{\lambda}{2}\left(f_c - \frac{F_s}{N}\text{argmax}_k\left\{\hat{P}_s[k]\right\}\right) \tag{30}$$

where f_c [Hz] is the intermediate frequency and h is the signal weighting function (usually Gaussian). A parametric spectral estimation method is also possible usually leading to better velocity estimates at moderate noise level. The signal $s(nT_s)$ at a given position of the window can be modeled by a p-order autoregressive model, noted AR(p). The difference equation describing this model is

$$x(n) = -\sum_{k=1}^{p} a_k x(n-k) + w(n) \tag{31}$$

where $\{a_k\}$ are the parameters of the model and $w(n)$ is the input sequence of the model, i.e. a white noise with zero mean and variance σ^2. The corresponding power spectrum estimate can then be expressed in term of the AR coefficients

$$\widehat{P}_s^{AR}(k\Delta f) \equiv \frac{\sigma^2}{|A(k)|^2} = \frac{\sigma^2}{\left|1 + \sum_{k=1}^{p} a_k exp(-j2\pi kf)\right|^2} \tag{32}$$

The $\{a_k\}$ parameters are calculated in this work by the Yule-Walker method and the Levinson-Durbin algorithm for which an estimation of the signal correlogram is needed. The spectrum of an AR(p) process has at most p poles, i.e. frequencies minimizing $A(k)$. Those models are good candidates to represent sharp spectral peaks but not spectral valleys since they have no zeros. However, if p is chosen large enough, AR techniques can approximate adequately any kind of spectral shapes. The model parameters estimation technique is described in (Proakis, 1965). The velocity estimate is then computed using Equation 30. We use a fixed model order of $p=16$.

$\tau = 0.0$

$\tau = 2.0$

$\tau = 5.0$

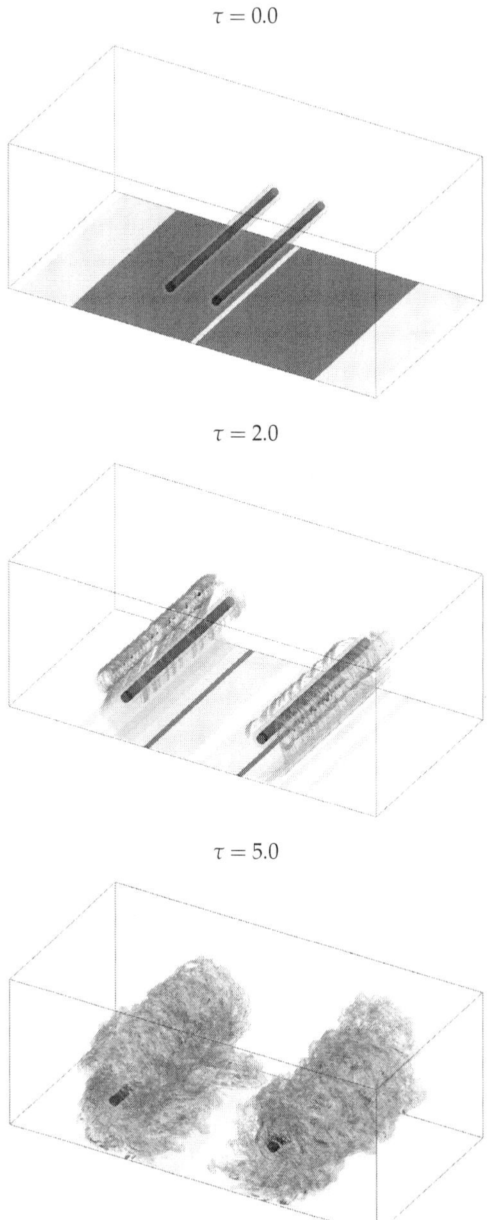

Fig. 6. Visualization of the flow field using isosurfaces of $\|\vec{\nabla} \times \mathbf{u}\| b_0^2 / \Gamma_0 = 1$ and 10.

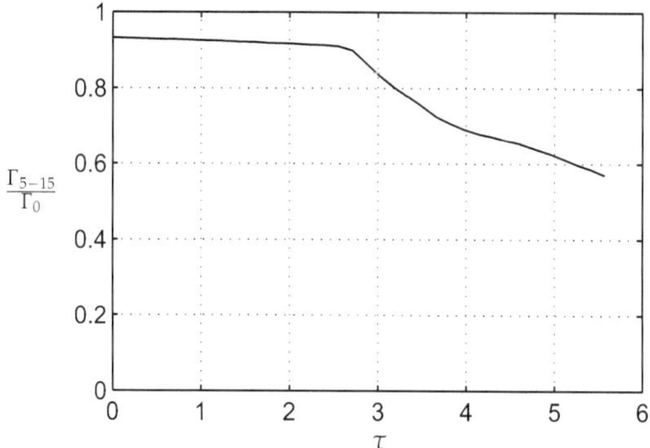

Fig. 7. Temporal evolution of the Γ_{5-15}.

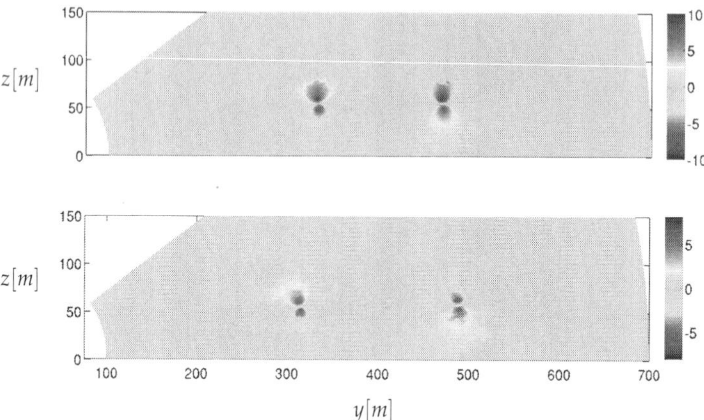

Fig. 8. Radial velocity maps v_R [m/s] of the wake vortices in ground effect at dimensionless time τ=2.5 (top, early case) and τ=5 (bottom, late case).

4.2 Accumulation strategies

Due to the low pulse energy provided by fiber-based LIDAR , the velocity estimates obtained from a single spectral density estimate are rather poor. In order to reduce the estimation variance, two different kinds of accumulation strategies are used: a periodogram accumulation and a correlogram accumulation (Rye & Hardesty, 1993). We apply these concepts here in the case of 2D velocity map reconstruction in order to compare their performances when wake vortices are present in the sensing volume.

Each signal window is defined by a vector $\mathbf{k}$ with polar coordinates $R = \frac{c}{2}(t_p + N_w T_s/2)$ [m] and θ [rad] being the elevation angle defining the line-of-sight. The processing vector set $\{\mathbf{k}\}$ is denoted $\mathcal{K}$. We define I_k as either the periodogram or the correlogram computed from the signal window $s_\theta(t_p + nTs)$ with $n = 0,\ldots,N_w - 1$. The main idea is to compute an averaged estimate I_p at a set of spatial positions $\mathbf{p}$ using the estimations I_k included in a circular region-of-interest $\mathcal{C}$ around this point $\mathbf{p}$. The accumulation algorithm is given in Table 3. Once the radial velocity has been estimated at each point $\mathbf{p}$, the estimation error is computed by the mean square error between the estimated radial velocity $\hat{v}_R$ and the exact filtered radial velocity $v_{pulse}(\mathbf{p})$:

$$\mathrm{MSE}(\mathbf{p}) = |\hat{v}_R(\mathbf{p}) - v_{pulse}(\mathbf{p})|^2 \tag{33}$$

4.3 Results

The accumulation level, defined as the number of estimates I_k used for one radial velocity estimation $\hat{v}_R$, is given at Fig. 9. In order to have the same spatial resolution at each point $\mathbf{p}$ a constant radius ρ with distance R have been chosen. As the measurement is performed on a polar grid, the accumulation level decreases with R. The vortex pair is situated around 400 m from the LIDAR. At this distance, a radius ρ of 1, 2 and 4 m corresponds to an accumulation level per velocity estimate of respectively 20, 60 and 120 k-points. For a given radius, the accumulation level depends also on the pulse repetition frequency.

Using a fixed radius implies the variance of the spectral density estimates to increase with range R due to variable accumulation level and SNR fluctuation. However, it is more interesting to choose a range-dependent radius by determining an optimal trade-off between spatial resolution (accumulation level) and velocity resolution (spectral density estimate variance). Fig. 10 gives the spatial variation of the MSE compared with the exact radial velocity field. We observe that the estimation error increases inside the vortex where the velocity gradients along the propagation direction are the most important. High velocity gradients decrease the temporal coherence of the signal and consequently induce a broadening of the signal spectrum with a decrease of its maximum value. Moreover, the accumulation process combined with the *argmax* estimation tends to promote narrow spectra computed outside the vortex. This effect must be take into account since it distorts the retrieved vortex shape and increases the error made on subsequent model-fitting algorithms dedicated to circulation estimation.

Accumulation algorithm :

$\forall \mathbf{k} \in \mathcal{K},$
    Estimates I_k from $s_\theta(t_p + nTs)$, $n = 0, \ldots, N_w - 1$
$\forall \mathbf{p} \in \mathcal{R},$
    Define $\mathcal{C}$ with radius $\rho = \rho(R)$.
    $n \leftarrow 0;$
    $\forall \mathbf{k} \in \mathcal{K} \mid \mathrm{dist}(\mathbf{p}, \mathbf{k}) \leq \rho(R)$
        $I_p \leftarrow (nI_p + I_k)\frac{1}{1+n};$
        $n \leftarrow n + 1;$
    Estimate spectral density from I_p
    Estimate $\hat{v}(\mathbf{p})$

Table 3. Accumulation algorithm. I_k is either the periodogram or the correlogram of the signal.

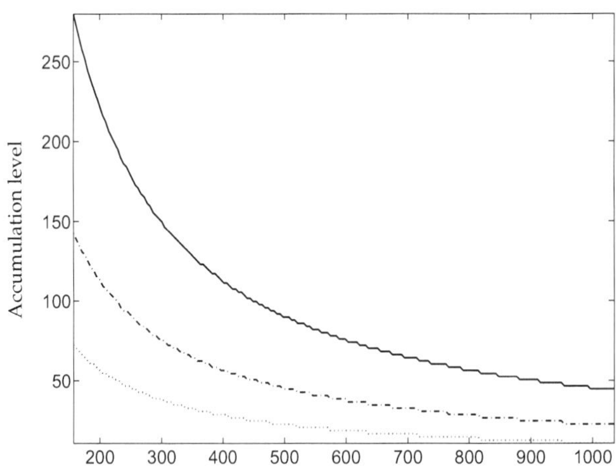

Range [m]

Fig. 9. Accumulation level for different radius ρ of the circular region-of-interest: $\rho = 1$ m (dot), $\rho = 2$ m (dash-dot) and $\rho = 4$ m (solid).

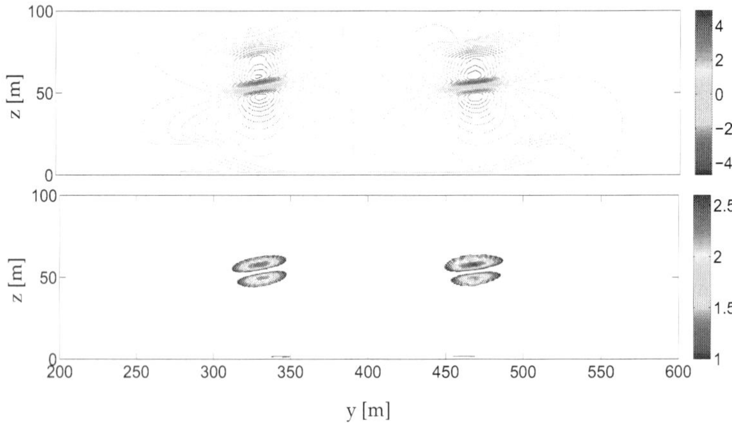

y [m]

Fig. 10. Influence of the accumulation process on the estimation error. Radial velocity map v_R (top) and standard deviation σ_v [m/s] (bottom).

The processing results obtained for wake vortices in the late case with a pulse duration of 400 ns and a refractive index $C_n^2 = 10^{-15}$ m$^{-2/3}$ is given Fig. 11. The window length has been fixed to $N = 64$ samples which gives a range gate of 38.4 m. Comparing this results with Fig. 8

(bottom) allows to illustrate the radial velocity filtering inherent to wind measurement with a pulsed LASER.
Fig. 12. shows the standard deviation of the radial velocity estimates averaged over all the line-of sight $\langle \sigma_v \rangle$ for different values of the pulse energy E_t. It has been obtained in the early case (young vortices) with a pulse duration of 200 ns and strong turbulence $C_n^2 = 10^{-13}$ m$^{-2/3}$. The algorithm used is the correlogram-accumulation method. We see that reducing the pulse

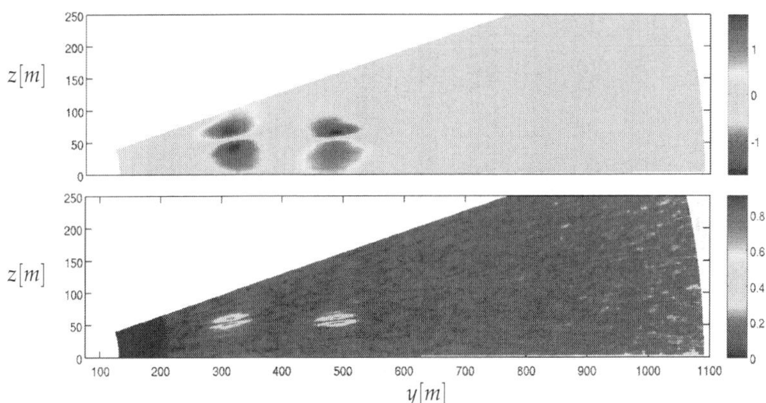

Fig. 11. Signal processing results: estimation of the radial velocity map (top) and MSE (bottom) obtained in the late case. The pulse duration is 400 ns and the refractive index is $C_n^2 = 10^{-15}$ m$^{-2/3}$.

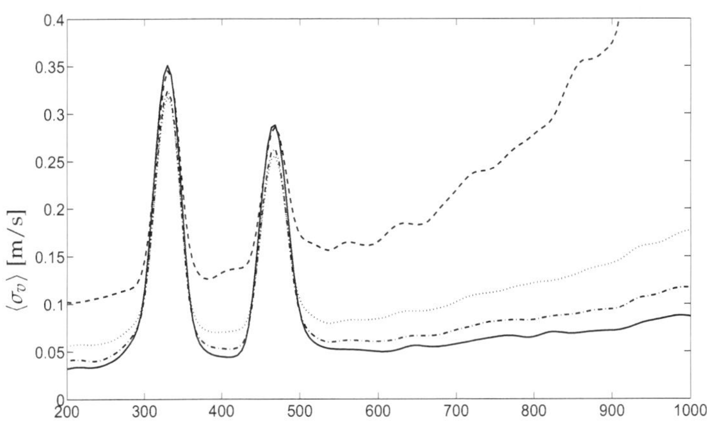

Fig. 12. Influence of the pulse energy on σ_v. $E_t = 50\mu J$ (dash); $E_t = 150\mu J$ (dot-dot); $E_t = 300\mu J$ (dash-dot) and $Et = 600\mu J$ (solid) obtained with the correlogram accumulation method ($\rho = 2$ m). The pulse duration is 200 ns. The refractive index is $C_n^2 = 10^{-13}$ m$^{-2/3}$.

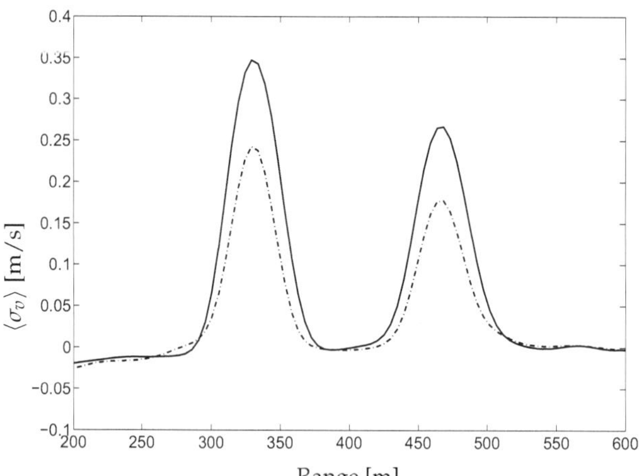

Fig. 13. Comparison of the correlogram (dash-dot) and periodogram (solid) accumulation methods at 200 ns for young vortices. The accumulation radius is 2 m.

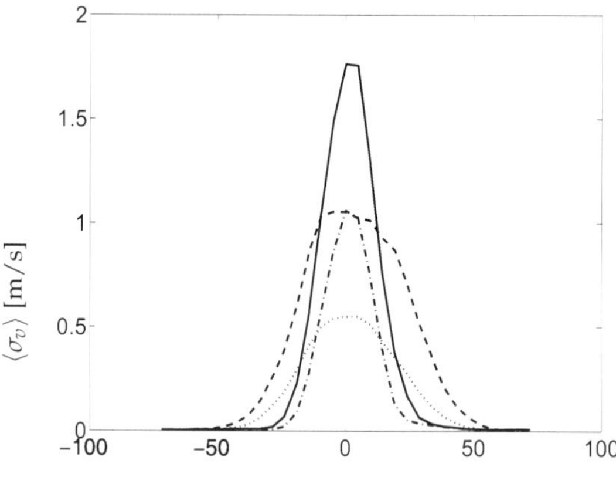

Fig. 14. Comparison of the correlogram (CA) and periodogram (PA) accumulation methods: PA, 200ns (solid); CA, 200ns (dash-dot); PA, 400ns (dash); CA, 400ns (dot). The refractive index is $C_n^2 = 10^{-15}$ m$^{-2/3}$. The range is given such that $R = 0$ m at the vortex center.

energy has no influence inside vortices whereas it monotically increases with range. It must be noted that, as pointed by Equation 14, for the same telescope configuration and beam sizes, the SNR is identical provided that we keep the ratio $\beta E_t / B_w$ constant. Hence, decreasing the pulse energy by a factor of 2 for example, gives the same estimation variance than decreasing by the same factor the aerosol particles density. The comparison between the two accumulation strategies are given at Fig. 13 and Fig. 14. The standard deviation of the velocity estimate are given after having removed the averaged evolution due to SNR variation. The correlogram accumulation gives better results inside the vortices at 200 ns and 400 ns. The error profiles are both narrower (spatial resolution) and smaller (velocity resolution) which induces better results for subsequent circulation estimation algorithm based on model-fitting as described in (Brousmiche et al., 2009).

5. Conclusions

The main originality of the simulation framework proposed here is to link numerical simulation techniques belonging to optical propation and fluid dynamics theory in order to assess the feasibility of wake vortex detection using a fiber-based heterodyne Doppler LIDAR. Moreover, the generation of realistic LIDAR signals with temporal decorrelation due to Speckle and beam propagation through wake vortices allows to develop efficient velocity estimation algorithms. Two accumulation-based algorithms have been tested and their performances in the vicinity of the vortices have been compared. It results that the correlogram accumulation method gives the best compromise between velocity and spatial resolution than the periodogram accumulation method generally used in such application. This feature is important since the main objective is the estimation of wake vortex circulation in order to evaluate their strength (usually by means of model-based algorithms).
Our analysis has principally focused on the detection of wake vortices interacting with the ground. An interesting study would be to consider cross wind as it appears in real atmospheric conditions. The transmission of radial velocity estimation errors on the estimation of vortex strength would also be considered.

6. Acknowledgements

The fluid and LIDAR simulations were done using the facilities of the Calcul Intensif et Stockage de Masse (CISM) on the UCL Lemaître cluster funded by Fond National de Recherche Scientifique (FNRS) under a FRFC Project and by UCL (Fonds Spéciaux de Recherche). Results were obtained in the framework of the LASEF project funded by the Région Wallonne of Belgium.

7. References

Banakh, V. A., Smalikho, I. N. & Werner, C. (2000). Numerical simulation of the effect of refractive turbulence on coherent lidar return statistics in the atmosphere, *Applied Optics*, **Vol.** 39(No. 30): 5403-5414.
Belmonte, A. & Rye, B. J. (2000). Heterodyne lidar returns in the turbulent atmosphere: performance evaluation of simulated systems, *Applied Optics*, **Vol.** 39(No. 15): 2401-2411.
Belmonte, A. (2003). Analyzing the efficiency of a practical heterodyne lidar in the turbulent atmosphere: telescope parameters, *Optics express*, **Vol.** 11(No. 17): 2041-2046.
Bricteux, L. (2008). A new multiscale model with proper behaviour in both vortex flows and wall bounded flows, *Direct and Large-Eddy Simulation VII book, ERCOFTAC Series*, Springer.
Brousmiche, S., Bricteux, L., Sobieski, P., Winckelmans, G., Macq, B. & Craeye, C. (2009). Parameters Estimation of Wake Vortices in Ground Effect, *Proceedings of the 15th Coherent Laser Radar Conference*, Toulouse, pp. 149-152.
Frehlich, R. G. and Kavaya, M. G. (1991). Coherent laser radar performance for general atmospheric refractive turbulence, *Applied Optics*, **Vol.** 30(No. 36): 5325-5351.

Frehlich, R. G. (1993). Effect of refractive turbulence on coherent laser radar, *Applied Optics*, **Vol.** 32(No. 12): 2122-2139.

Frehlich, R. G. (1997). Effect of Wind Turbulence on Coherent Doppler Lidar Performance, *Journal of Atmospheric and Oceanic Technology*, **Vol.** 14.

Frehlich, R. G. (2000). Simulation of laser propagation in a turbulent atmosphere, *Applied Optics*, **Vol.** 39(No. 3).

Gerz, T., Holzapfel, F. & Darracq D. (2001). Aircraft wake vortices: a position paper, Wakenet, the european thematic network on wake vortex position paper.

Levin, M. J. (1965). Power Spectrum Parameter Estimation, *IEEE Trans. on Information theory*, **Vol.** IT-11: pp 100-107

Lee, M. J., Oh, B. D. & Kim, Y. B. (2001). Canonical Fractional-Step Methods and Consistent Boundary Conditions for the Incompressible Navier Stokes Equations, *J. Comp. Phys.*, **Vol.** 168: 73-100.

Martin, J. M. & Flaté, S. M. (1990). Simulation of point-source scintillation through three-dimensional random media, *J. Opt. Soc. Am. A*, **Vol.** 7(No. 5): 838-847.

Prandtl, L. & Tietjens, O.G (1957). *Fundamentals of hydro- and aero-mechanics*, Dover publ.

Proakis, J. G. (1965). *Digital Signal Processing : Principles, Algorithms and Applications*, Prentice Hall.

Rye, B. J. & Hardesty, R. M. (1993). Discrete Spectral Peak Estimation in Incoherent Backscatter Heterodyne Lidar. I: Spectral Accumulation and the Cramer-Rao Lower Bound, *IEEE Trans. on Geoscience and Remote Sensing*, **Vol.** 31(No. 1): 16-27.

Rye, B. J. & Hardesty, R. M. (1993). Discrete Spectral Peak Estimation in Incoherent Backscatter Heterodyne Lidar. II: Correlogram Accumulation, *IEEE Trans. on Geoscience and Remote Sensing*, **Vol.** 31(No. 1): 28-35 .

Salamitou, P., Dabbas, A. & Flamant, P. (1995). Simulation in the time domain for heterodyne coherent laser radar, *Applied Optics*, **Vol.** 34(No. 3): 499-506.

Van Trees, H. (2001). *Detection, Estimation and Modulation Theory. Radar-sonar Processing and Gaussian Signals in Noise*, Wiley, 2001

Vasilyev, O. V. (2000). High order finite difference schemes on non-uniform meshes with good conservation properties, *J. Comp. Phys.*, **Vol.** 157(No. 2): 746-761.

Winckelmans, G. S. (2004). Chapter 5: Vortex Methods, *Encyclopedia of Computational Mechanics*, Erwin Stein, René de Borst, Thomas J.R. Hughes (Eds), John Wiley & Sons.

12

Earthquake damage mapping techniques using SAR and Optical remote sensing satellite data.

Marco Chini
Istituto Nazionale di Geofisica e Vulcanologia
Italy

1. Introduction

The objective of this chapter is to show how it is possible to determine damages caused by seismic events in urban areas using optical and radar data, and automatic or semi-automatic remote sensing techniques. These techniques have revealed themselves a suitable monitoring tool for disaster management since they provide a quick detection of land changes in wide areas, especially in remote areas or where the infrastructures are not well developed to ensure the necessary communication exchanges. Indeed, in the aftermath of these severe disastrous events one of the most urgent needs is to estimate with sufficient reliability and rapidity the amount of population and infrastructures affected for different degrees of damage (Voigt et al., 2007).

The use of these sensors has been addressed to scenario classification procedures, which allow also to recognize objects and changes occurred in the area. Automatic classification procedure concerning changes in urban areas could be used for mapping damage caused by earthquakes as well, being seismic events not so frequents.

Actually, timely and accurate change detection of Earth's surface features is extremely important for understanding relationships and interactions between human and natural phenomena, especially in supporting better decision making. In general, change detection involves the application of multi-temporal datasets to quantitatively analyse the temporal effects of the phenomenon. Because of the advantages such as repetitive data acquisition, synoptic view, and digital format suitable for computer processing, remote sensing data are primary sources which have been using extensively for change detection since the last decades (Lu et al., 2004).

The contribution of space technologies has been demonstrated to be effective for regional/continental damage assessment using low- or medium-resolution remotely sensed data (ranging from 30m to 1 km), and both automatic and manual interpretation approaches have been successfully used for extraction of information at a nominal scale ranging from 1 : 100 000 to 1 : 1 000 000 (Belward et al., 2007).

The space technologies have also been demonstrated their effectiveness for damage assessment at local scale, ranging from 1 : 10 000 to 1 : 25 000 nominal scales. The information extracted at this level was crucial for calibration and estimation of the reliability of low- and medium-resolution assessment, for planning logistics for relief action on the field immediately after the event, and for planning the resources needed for recovery and reconstruction.

Today, local or detailed damage assessment can also be addressed using Very High Resolution (VHR) satellite data with a spatial resolution ranging from 0.6 to 1m. At the beginning, , the operational methodology for extracting the information at this scale of details was based on manual photo-interpretation of the satellite images, which were processed on the screen by the photo-interpreter as for any other aerial imagery. However, the traditional photo-interpretation methodology has some drawbacks firstly linked to the time (and cost) needed for manual processing of the data and, secondly, to the difficulty in maintaining coherent interpretation criteria in case a large number of photo-interpreters, working in parallel in wide areas in a short time, is available. The long required processing time is in conflict with the need for rapid damage estimation, and the solution to involve parallel photo-interpreter teams often leads to an increase of time-consuming organizational problems and additional coherency lack in the information produced (Pesaresi et al., 2007). Accordingly, some automatic procedures for exploiting these data have been developed in order to give information at this scale of detail. The most innovative automatic approaches will be described in this chapter.

For an operational use of this kind of techniques the major limitation is the availability of the images within a short time to manage the crisis. This is a key point for Civil Protections who needs a fast and draft overview of the epicentral area, quick information relative to the extension and distribution of damages, and the evaluation of infrastructure (roads, bridges) conditions. A single satellite can provide access time to a specific site in the order of some days, as a result the necessity to use any kind of satellites data available and an integration of those data is mandatory to increase the chance to collect information on near real time.

The following paragraphs are addressed to the analysis of the different aspects leading to obtain maps representative of damage caused by earthquakes.

2. SAR and Optical satellite sensors and missions

Earth observation satellites have a major mission to observe the entire Earth, so they circle the most suitable Sun-synchronous sub-recurrent orbit. A Sun-synchronous orbit refers to the orbit where the positioning between the satellite and the Sun is always identical . Namely, the incidence angle of sunlight to the land surface is always the same. This orbit characteristic allows to accurately observing radiation and reflection from the land surface. While the sub-recurrent orbit means that after a certain number of days, the satellite repeats its original orbit. Thus, Earth observation satellites utilize the combination of these two orbits, which allow observing the same area at regular intervals under the condition of the same sunlight angle to the land surface.

Furthermore, sensors can be divided into two types: Optical and Microwave sensors.

Optical sensors observe visible lights and infrared rays (near infrared, intermediate infrared, thermal infrared). There are two kinds of observation methods using optical sensors: visible/near infrared remote sensing and thermal infrared remote sensing. The first one consists in acquiring visible light and near infrared rays of sunlight reflected by objects on the ground. By examining the strength of reflection, we can understand conditions of land surface, e.g., distribution of plants, forests and farm fields, rivers, lakes, urban areas. During period of darkness this method cannot be observe. Instead, the latter method acquires thermal infrared rays radiated from land surface heated by sunlight. It can also observe the high temperature areas, such as volcanic activities and fires. By examining the strength of radiation, we can understand surface temperatures of land and sea, the status of volcanic activities and forest fires.

Microwave sensors transmit and/or receive microwaves, longer wavelength than visible light and infrared rays, and the observation is time and weather independent. There are two types of observation methods using microwave sensor: active and passive. In the first one the sensor aboard earth observation satellite emits microwaves and observes microwaves reflected by land surface, as the case of Synthetic Aperture Radar (SAR). The second one observes microwaves naturally radiated from land surface. It is suitable to observe sea surface temperature, snow accumulation and thickness of ice.

While optical sensors are affected by cloud cover limitations, SAR is widely used in environmental studies due to its fairly synoptic view in almost completely weather and time independent conditions. Moreover, like the optical sensors which have reached resolution less than one meter (e.g. QuickBird, Ikonos, WorldView – 1, GoeEye-1) since 2001, SAR has overtaken its own limitations in terms of ground resolution with the launch of the new generation satellite missions TerraSAR-X and COSMO-SkyMed .

Destructive earthquakes challenge satellite remote sensing damage mapping techniques to demonstrate their usefulness in supporting intervention and relief actions. The use of remote sensing data in a disaster context has been widely investigated from a theoretical point of view, but only recently the developed methods seem to be useful for the operational use (Chini et al., 2008d).

Nowadays the implementation of satellite constellations is reducing the access time with the same sensor to 12 hours, as in the case of the Italian COSMO-SkyMed system. One the most important aspects of this new satellite mission is that the COSMO-SkyMed system is a constellation of four satellites (three of which are already in orbit, the first of which was launched in June 2007), developed to provide fast, meter level-resolution, all-weather imagery for disaster management (Stramondo et al., 2008). COSMO-SkyMed is a project in cooperation between the Italian Space Agency (ASI) and the Italian Defence Ministry (MD) and it is the first satellite constellation devoted to the Earth's observation for both civil and military purposes. The system hosts a flexible, multimode X-band synthetic aperture radar (SAR), with right and left looking imaging capabilities, an incidence angle range of 20°–60°, and a minimum revisit time of 12 hours. The fixed antenna has electronic scanning capabilities in both the azimuth and the elevation planes. It has been designed to implement three different operation modes in order to acquire the images at different resolutions: i) Spotlight mode, for metric (1m) resolutions over small images; Stripmap mode for metric resolutions (5 m) over tenth of km images; iii) ScanSAR mode, for medium to coarse (100 m) resolutions over large swaths (Boni et al, 2008).

3. Damage mapping by SAR data

Multi-temporal observations from SAR can be used to detect urban changes either looking at the image intensity changes, as in the case of optical images, but also taking advantage from the information on the phase of the returned signal. This is specific of the radar technique and in particular of the interferometric SAR (InSAR). The intensity correlation and the InSAR complex coherence are two important features for recognizing changes on the surface caused by an earthquake. These two features hold different information concerning changes in the scene (Stramondo et al., 2006). The complex coherence is prevalently influenced by the phase difference between radar returns, a distinctive parameter measured by a coherent sensor. It is particularly related to the spatial arrangement of the scatterers within the pixel and thus to their possible displacements. Conversely, the intensity correlation is more related to change in the magnitude of the radar return (Bignami et al., 2004).

Changes in SAR intensity and phase backscattering have been largely studied by many scientists for earthquake damage mapping purposes. An index to estimate damage level from SAR data by combining image intensity changes and the related correlation coefficient has been achieved and applied to some case studies: the Hyogoken-Nanbu earthquake (Aoki et al., 1998; Matsuoka & Yamazaki, 2004) and the Izmit and Gujarat seismic events (Matsuoka & Yamazaki, 2002; Stramondo el al., 2006). Yonezawa & Takeuchi (2001) have compared changes in the SAR intensity and phase backscatter with damage observed in Kobe. Ito et al. (2000) have assessed different SAR change indicators, derived from L- and C-band sensors, and evaluated the frequency-dependent effects of spatial and temporal decorrelations. Chini et al. (2008a) have detected wide uplift and subsiding areas, as well as large modifications of the coastline associated to the Indonesian earthquake of 2004, using only pre- and post-earthquake SAR backscattering.

It is commonly acknowledged that due to speckle effects (Li & Goldstein, 1990), single-pixel classification of SAR images leads to unsatisfactory results, and this seems to hold true also when damage assessment is concerned. Satisfying results may be achieved if the damage is assessed at a block level, mitigating the speckle noise (Bazi ert al., 2005) by means of averaging in somehow the unreliable results of pixel-wise comparing pre-and post-event images (Gamba et al., 2007). The partition of the image into blocks can be made by visual inspection or using a GIS layer provided by local authorities during crisis phase; otherwise procedures for segment images starting from SAR or optical images are reported in literature. Indeed, city blocks are generally marked by their respective bounding urban roads, which allow to segmenting the urban area.

Nowadays the new very high resolutions SAR sensors, few meters resolution, allow to classifying urban areas, producing reliable land use maps exploiting contextual information from backscattering data. Chini et al., 2009b have investigated the use of contextual information with TerraSAR-X backscattering images for classifying urban land-use. The anisotropic multi-scale morphological filters, coupled with the pruning neural network as a features selection tool, was able to provide urban land-use maps with accuracies of about 0.90 in terms of K-coefficient. This kind of techniques will allow producing damage map at building scale using very high resolution SAR data as it is already done using Very High Resolution (VHR) optical images.

4. Damage mapping by Very High resolution Optical data

The presence of shadows, variations in solar illumination, and geometric distortions may prevent the use of automatic damage detection procedures in optical images. Because of these problems, visual image inspection is still the most widely used method to produce a realistic and reliable inventory of damage (Saito et al., 2004; Yamakaki et al., 2004). Sakamoto et al. (2004) compared with an automatic technique combined with visual interpretation results using QuickBird data. Matsuoka et al. (2004) proposed to detect damage by analyzing edges in high-resolution images. More in general, automatic change-detection algorithms using either QuickBird (Pacifici et al., 2007; Chini et al., 2008) or SAR images (Inglada & Mercier, 2007) are also present in the literature, even if using SAR data have been only exploited the medium resolution.

One of the most important issue when we set a change detection classifier for mapping damages is the rapidity for producing damage maps, especially if we are supporting rescue teams during crisis phase. This characteristic is related to the number of inputs, which are the bands used for classifying, the number of samples for training the supervised classifier and the time spent for selecting training samples representative of the classes we want to identify.

Usually, the unsupervised classifiers are used for overcoming the time consuming step for selecting training pixels, since the post processing task for labelling the classes of interest is faster. The use of an unsupervised classifier, with QuickBird panchromatic images, has provided a damage map at a pixel scale of 0.6 m which allow to detect the complete or partial collapse of individual buildings in the Bam city (Iran) caused by the earthquake occurred on 2003 (Chini et al., 2009a). Moreover, the mathematical morphology (Benediktsson et al., 2003; Benediktsson et al., 2005; Soille, 2003) has proved to be a powerful tool for automatically analyze the panchromatic images. Furthermore, the unsupervised classifier, the mathematical morphology and very high resolution images have permitted to give damage level at building scale which is a more realistic damage degree and more useful for civil protection activity. The damage level at building scale is closer to damage index provided by ground survey team (Chini et al. 2008c).

When the number of training pixels is fixed or limited by whatever factor (including the cost of gathering them by in situ campaigns), experience has shown an initial increase in the classification accuracy by adding features, followed by a subsequent decay in classification skill as more features are considered (Hughes, 1968).

The increase in dimensionality (i.e., number of inputs) of the data worsens the parameter estimates, overcoming the expected increase in class separability associated with the inclusion of additional features. Therefore, one needs to use a robust method to estimate parameters or to reduce the number of inputs. For instance, principal component analysis is used to diminish the number of features (Landgrebe, 2003). Regularization methods have also been proposed in the literature, as discussed in (Berge & Solberg, 2004), attempting to stabilize the estimated class covariance matrix by replacing it with a weighted sum of the class sample covariance matrix or the common (pooled) covariance matrix. Additional methods include regularizing discriminant analysis (Friedman, 1989), leave-one-out covariance estimation (Hoffbeck & Landgrebe, 1996), and pruning methods when we deal with neural networks (Pacifici et al., 2009).

A robust method for estimating parameters has been proposed by Chini et al., (2008b), which is different respect to method presented previously. The number of inputs is not reduced, but they are composed differently in the new architecture to overcome the difficulty of having poor training samples for some classes. This method speeds up the overall analysis time (computation plus error correction) and increases the change detection accuracies. It consists on using a parallel approach to exploit the multi-spectral and multi-temporal characteristics of two scenes, thus combining two approaches which are typically used for supervised change detection, the post classification comparison (Del Frate et al.,

2004; Serra et al., 2004; Sunar Erbek et al., 2004) and the direct multi data classification (Sunar Erbek et al., 2004; Castelli et al., 1998).

The post classification comparison detects changes by comparing the classification maps obtained by independently classifying the two sequential remote sensing images of the same scene. This method does not require normalization to compensate for atmospheric conditions and sensor differences between the acquisition dates. At the same time, this method does not exploit the correlation between multi-temporal data sets, and its accuracy depends on the product of the accuracy of single classification maps (Yuan & Elvidge, 1998). The change detection in direct multi data classification is directly performed by considering each transition as a class in a unique vector obtained by stacking the features related to the individual images; this method is more sensitive to solar and atmospheric conditions.

An efficient method for reducing the number of inputs and selecting the most important features for classifying a certain typology of urban landscape has been implemented by Pacifici et al., (2009), using QuickBird and Worldview-1 panchromatic images. This work has investigated on the potential of these very high resolution panchromatic imageries (the most resolute optic satellite sensors) to classify the land-use of urban environments. Usually spectral-based classification methods may fail with the increased geometrical resolution of the data available. Indeed, improved spatial resolution data increases within-class variances, which results in high interclass spectral confusion. In many cases, several pixels are representative of objects, which are not part of land-use classes defined. This problem is intrinsically related to the sensor resolution and it cannot be solved by increasing the number of spectral channels. To overcome the spectral information deficit of panchromatic imagery, it is necessary to extract additional information to recognize objects within the scene. The multi-scale textural (Haralick, 1979; Haralick et al., 1973) approach exploited the contextual information of panchromatic images. Moreover, neural network pruning (Kavzoglu & Mather 1999; Del Frate et al., 2005) and saliency (Yacoub & Bennani, 1997; Tarr, 1991) measurements have allowed to give an indication of the most important textural features for sub-metric spatial resolution imagery and different urban scenarios.

This has to be a preparatory work, it should be done before the emergency phase, because it gives very useful information concerning the minimum number of inputs necessary for obtain reliable classification maps. Another important aspect of this study is that it uses a very small dataset, only one panchromatic image, all others inputs are contextual information, second order spatial statistic, extracted from the original panchromatic image. The latter aspect is crucial during crisis, where it is necessary processing data as soon as they are acquired from satellites.

5. Conclusions

In this chapter the sensitivity to the urban damage level of different parameters extracted from different types of remotely sensed images has been analyze. It has been highlighted that the parameters to be extracted are strictly related to the type of exploited data (i.e., their resolution and spectral channels). Additionally, this chapter aimed to demonstrate how very high resolution (VHR) images can detect damage at the pixel scale (≤ 1 m) with automatic techniques. Associated problems have been pointed out and ways to overcome them have been proposed.

The chapter has provided a quite realistic view of what remote sensing can offer for this application and which methods and sensors worth to be further developed in the future. From our experience in this field it is possible to affirm that damage products can be obtained at different ground scales and with different satellite data. Those products can be valuable for managing different phases of the crises, and especially mitigation of the effect in the course of the event and precise inventory of the damage in the post-event phase.

It has been pointed out that an operational use of remote sensing for earthquake damage assessment strongly depends on the number of available images, their type (SAR, optical, or both), and quality and timeliness of the data sets (i.e., time delay of the post-seismic images with respect to the destructive event). In particular, the availability of images within a short time to manage the crisis is a key point for Civil Protections who needs a fast and draft overview of the epicentral area, quick information relative to the extension and distribution of damages, and the evaluation of infrastructure (roads, bridges) conditions. Since, a single satellite can provide access time to a specific site in the order of some days, the use of any type of satellites data available, an integration of those data and the exploitation of data provided by constellation missions is mandatory to increase the chance to collect information on time.

The conclusive indication on the effectiveness of remote sensing for earthquake damage detection requires the accumulation of many analyses by different scientists and the work done here will hopefully contribute to this end. For all of these reasons, research of change detection techniques is still an active and very challenging topic and new techniques are needed to effectively use the increasingly diverse and complex remotely sensed data available.

6. References

Aoki, H.; Matsuoka, M. & Yamazaki, F. (1998). Characteristics of satellite SAR images in the damaged areas due to the Hyogoken-Nanbu earthquake, Proceedings of 19th Asian Conference on Remote Sensing 1998, Vol. C7, pp. 1–6.

Bazi, Y.; Bruzzone, L. & Melgani, F. (2005). An unsupervised approach based on the generalized Gaussian model to automatic change detection in multitemporal SAR images, IEEE Transaction of Geoscience and Remote Sensing, Vol.43, No.4, pp. 874-887.

Belward, A. S.; Stibig H. J.; Eva, H.; Rembold, F.; Bucha, T.; Hartley, A.; Beuchle, R.; Khudhairy, D.; Michielon, M. & Mollicone, D. (2007). Mapping severe damage to land cover following the 2004 Indian Ocean tsunami using moderate spatial resolution satellite imagery, International Journal of Remote Sensing, Vol. 28, No. 13-14, pp. 2977–2994.

Benediktsson, J. A.; Pesaresi, M. & Arnason, K. (2003). Classification and feature extraction for remote sensing images from urban areas based on morphological transformations, IEEE Transactions on Geoscience and Remote Sensing, Vol. 41, No. 9, pp. 1940–1949.

Benediktsson, J. A.; Palmason, J. A. & Sveinsson, J. R. (2005). Classification of hyperspectral data from urban areas based on extended morphological profiles, IEEE Transactions on Geoscience and Remote Sensing, Vol. 43, No. 5, pp. 480–491.

Berge, A. & Solberg, A. S. (). A comparison of methods for improving classification of hyperspectral data, Proceedings of IEEE IGARSS 2004, Anchorage (Alaska, USA), Vol. 2, pp. 945–948, September 20-24, 2004.

Bignami, C.; Chini, M.; Pierdicca, N. & Stramondo, S. (2004). Comparing and combining the capability of the detecting earthquake damages in urban areas using SAR and optical data, Proceedings of IEEE IGARSS 2004, Vol. 1, pp.55-58, Anchorage (Alaska, USA), September 20-24, 2004.

Boni, G.; Castelli, F.; Ferraris, L.; Pierdicca, N.; Serpico, S. & Siccardi, F. (2007). High resolution COSMO/SkyMed SAR data analysis for civil protection from flooding events, Proceeding of IEEE IGARSS 2007, pp. 6 – 9, Denver (CO - USA), July 23-28, 2007.

Castelli, V.; Elvidge, C. D.; Li, C. S. & Turek, J. J. (1998). Classificationbased change detection: Theory and application to the NALC data set, In: Remote Sensing Change Detection: Environmental monitoring methods and applications, pp. 53-73, R. S. Lunetta and C. D. Elvidge, Eds. Chelsea, MI: Ann Arbor Press, 1998.

Chini, M.; Bignami, C.; Stramondo S. & Pierdicca, N. (2008a). Uplift and subsidence due to the 26 December 2004 Indonesian earthquake detected by SAR data, International Journal of Remote Sensing, Vol. 29, No. 13, pp. 3891–3910.

Chini, M.; Pacifici, F.; Emery, W. J.; Pierdicca, N. & Del Frate, F. (2008b). Comparing statistical and neural network methods applied to very high resolution satellite images showing changes in man-made structures at Rocky Flats, IEEE Transactions on Geoscience and Remote Sensing, Vol. 46, No. 6, pp. 1812–1821.

Chini, M.; Bignami, C.; Stramondo, S.; Emery, W. J. & Pierdicca, N. (2008c). Quickbird Panchromatic Images for Mapping Damage at Building Scale Caused by the 2003 Bam Earthquake, Proceedings of IEEE IGARSS 2008, Vol. 2, pp. 1029-1031, Boston (USA), July 7 – 11, 2008.

Chini, M.; Bignami, C.; Brunori, C.; Atzori, S.; Tolomei, C.; Trasatti, E.; Kyriakopoulos, C.; Salvi, S.; Stramondo, S. & Zoffoli, S. (2008d). The SIGRIS project: a remote sensing system for seismic risk management, Proceedings of IEEE IGARSS 2008, Vol. 3, pp. 624-627, Boston (Massachusetts, USA), July 7 – 11, 2008.

Chini, M.; Pierdicca, N. & Emery, W. J. (2009a). Exploiting SAR and VHR optical images to quantify damage caused by the 2003 Bam earthquake. IEEE Transaction on Geosciences and Remote Sensing, Vol. 47, No. 1, pp. 145 - 152.

Chini, M.; Pacifici, F. & Emery, W. J. (2009b). Morphological operators applied to X - band SAR for urban land use classification, Proceedings of IEEE IGARSS 2009, Cape Town (South Africa), July 13 – 18, 2009.

Del Frate, F.; Schiavon, G. & Solimini, C. (2004). Application of neural networks algorithms to QuickBird imagery for classification and change detection of urban areas, Proceedings of IEEE IGARSS 2004, Anchorage (Alaska, USA), pp. 1091–1094, September 20-24, 2004.

Del Frate, F.; Iapaolo, M.; Casadio, S.; Godin-Beekmann, S. & Petitdidier, M. (2005). Neural networks for the dimensionality reduction of GOME measurement vector in the estimation of ozone profiles, Journal of Quantitative Spectroscopy & Radiative Transfer, Vol. 92, No. 3, pp. 275–291.

Friedman, J. (1989). Regularized discriminant analysis, Journal of the American Statistical Association, Vol. 84, No. 405, pp. 165–175.

Gamba, P.; Dell'Acqua, F. & Trianni, G. (2007). Rapid Damage Detection in the Bam Area Using Multitemporal SAR and Exploiting Ancillary Data, IEEE Transaction of Geoscience and Remote Sensing, Vol.45, No.6, pp.1582-1589.

Haralick, R. M.; Shanmuga, K. & Dinstein, I. H. (1973). Textural features for image classification, IEEE Transactions on Systems, Man, and Cybernetics, Vol. SMC-3, No. 6, pp. 610–621.

Haralick, R. M. (1979). Statistical and structural approaches to texture. Proceedings of the IEEE, Vol. 67, No.5, pp. 786–804.

Hoffbeck, J. & Landgrebe, D. (1996). Covariance matrix estimation and classification with limited training data, IEEE Transactions on Pattern Analysis and Machine Intelligence, Vol. 18, no. 7, pp. 763–767.

Hughes, G. F. (1968). On the mean accuracy of statistical pattern recognizers, IEEE Transactions on Information Theory, Vol. IT-14, No. 1, pp. 55–63.

Inglada, J. & Mercier, G. (2007). A new statistical similarity measure for change detection in multitemporal SAR images and its extension to multiscale change analysis, IEEE Transactions on Geoscience and Remote Sensing, Vol. 45, No. 5, pp. 1432–1445.

Ito, Y.; Hosokawa, M.; Lee, H. & Liu, J. G. (2000). Extraction of damaged regions using SAR data and neural networks, Proceedings of 19th ISPRS Congress 2000, vol. 33, pp. 156–163, Amsterdam, (NETHERLANDS), July 16–22, 2000.

Kavzoglu, T. & Mather, P. M. (1999). Pruning artificial neural networks: An example using land cover classification of multi-sensor images. International Journal of Remote Sensing, Vol. 20, N. 14, 2787–2803.

Landgrebe, D. A. (2003). Signal Theory Methods in Multispectral Remote Sensing, Hoboken, NJ: Wiley, 2003.

Li, F. K. & Goldstein, R. M. (1990). Studies of multibaseline spaceborne interferometric synthetic aperture radar," IEEE Transaction of Geoscience and Remote Sensing, Vol. 28, No. 1, pp. 88-97.

Lu D.; Mausel, P.; Brondizio, E. & Moran, E. (2004). Change detection techniques. International Journal of Remote Sensing, Vol. 25, No. 12, pp. 2365–240720.

Matsuoka, M. & Yamazaki, F. (2002). Application of the damage detection method using SAR intensity images to recent earthquakes, Proceedings of IEEE IGARSS 2002, Vol. 4, pp. 2042-2044, Toronto (CANADA), June 24-28, 2002.

Matsuoka, M. & Yamazaki, F. (2004). Use of satellite SAR intensity imagery for detecting building areas damaged due to earthquakes, Earthquake Spectra, Vol. 20, No. 3, pp. 975–994.

Matsuoka, M.; Vu, T. T. & Yamazaki, F. (2004). Automated damage detection and visualization of the 2003 Bam, Iran, earthquake using high resolution satellite images, Proceedings of 25th Asian Conference on Remote Sensing 2004, pp. 841–845, Chiang Mai (THAILAND), November 22-26, 2004.

Pesaresi, M.; Gerhardinger, A. & Haag, F. (2007). Rapid damage assessment of built-up structures using VHR satellite data in tsunami-affected areas, International Journal of Remote Sensing, Vol. 28, No. 13–14, pp. 3013–3036.

Pacifici, F.; Del Frate, F.; Solimini, C. & Emery, W. J. (2007). An innovative neural-net method to detect temporal changes in high-resolution optical satellite imagery, IEEE Transactions on Geoscience and Remote Sensing, Vol. 45, No. 9, pp. 2940-2952.

Pacifici, F.; Chini, M. & Emery, W. J. (2009). A neural network approach using multi-scale textural metrics from very high resolution panchromatic imagery for urban land-use classification, Remote Sensing of Environment, Vol. 113, No. 6, pp. 1276 – 1292.

Saito, K.; Spence, R. J. S.; Going, C. & Markus, M. (2004). Using high resolution satellite images for post-earthquake building damage assessment: A study following the 26 January 2001 Gujarat earthquake, Earthquake Spectra, vol. 20, No. 1, pp. 145–169.

Sakamoto, M.; Takasago, Y.; Uto, K.; Kakumoto, S. & Kosugi, Y. (2004). Automatic detection of damaged area of Iran earthquake by high-resolution satellite imagery," Proceedings of IEEE IGARSS 2004, vol. 2, pp. 1418–1421, Anchorage (Alaska, USA), September 20-24, 2004.

Serra, P.; Pons, X. & Saurì, D. (2004). Post-classification change detection with data from different sensors: Some accuracy considerations, International Journal of Remote Sensing, Vol. 24, No. 16, pp. 3311–3340.

Soille, P. (2003). Morphological Image Analysis—Principles and Applications, 2nd ed. Berlin, Germany: Springer-Verlag, 2003.

Stramondo, S.; Bignami, C.; Chini, M.; Pierdicca, N. & Tertulliani, A. (2006). The radar and optical remote sensing for damage detection: results from different case studies, International Journal of Remote Sensing, Vol. 27, No. 20, pp. 4433 – 4447.

Stramondo, S.; Chini, M.; Salvi, S.; Bignami, C.; Zoffoli, S. & Boschi, E. (2008). Ground Deformation Imagery of the 12 May 2008 Sichuan Earthquake, EOS Transaction American Geophysical Union, Vol. 89, No. 37, pp. 341-342.

Sunar Erbek, F.; Ozkan, C. & Taberner, M. (2004). Comparison of maximum likelihood classification method with supervised artificial neural network algorithms for land use activities, International Journal of Remote Sensing, Vol. 25, No. 9, pp. 1733–1748.

Tarr, G. (1991). Multi-layered feedforward neural networks for image segmentation, Ph. D. dissertation, School of Engineering, Air Force Institute of Technology. Wright-Patterson AFB OH.

Voigt, S.; Kemper, T.; Riedlinger, T.; Kiefl, R.; Scholte, K. & Mehl, H. (2007). Satellite image analysis for disaster and crisis-management support, IEEE Transaction on Geoscience and Remote Sensing, Vol. 45, No. 6, pp. 1520–1528.

Yacoub, M. & Bennani, Y. (1997). HVS: A heuristic for variable selection in multilayer artificial neural network classifier. In: Intelligent engineering systems through artificial neural networks, pp. 527-532, Dagli, C.; Akay, M.; Ersoy, O.; Fernandez, B. & Smith, A. (Eds.), Missouri, USA: St Louis.

Yamakaki, F.; Matsuoka, M.; Kouchi, K.; Kohiyama, M. & Muraoka, N. (2004). Earthquake damage detection using high-resolution satellite images, Proceedings of IEEE IGARSS 2004, Vol. 4, pp. 2280–2283, Anchorage (Alaska, USA), September 20-24, 2004.

Yonezawa, C. & Takeuchi, S. (2001). Decorrelation of SAR data by urban damage caused by the 1995 Hoyogoken-Nanbu earthquake, International Journal of Remote Sensing, Vol. 22, No. 8, pp. 1585–1600.

Yuan, d. & Elvidge, C. (1998). NALC land cover change detection pilot study: Washington D.C. area experiments, Remote Sensing of Environment, Vol. 66, No. 2, pp. 168–178.

13

Spectroscopic Microwave Dielectric Model of Moist Soils

Valery Mironov and Pavel Bobrov
Kirensky Institute of Physics Siberian Branch of the Russian Academy of Sciences
Russian Federation

1. Introduction

Microwave dielectric models (MDM) of moist soils are an essential part of the algorithms used for data processing in the radar and radio thermal remote sensing (Ulaby et al., 1986). So far, the MDMs for moist soils developed in (Wang & Schmugge, 1980; Dobson et al., 1985; Peplinski et al., 1995, a, b) have become a routine mean for creating processing algorithms, which concern radar and radio thermal remote sensing of the land. In our recent publications (Mironov et al., 2002; Mironov, 2004; Mironov et al., 2004; Mironov et al. 2009 a, b) the error of dielectric predictions for moist soils has been noticeably decreased due to taking into account, on a physical bases, the impact of bound soil water (BSW). This is an essential distinction of the dielectric models developed in (Mironov, et al. 2002; Mironov, 2004; Mironov et al., 2004; Mironov et al., 2009 a, b) from the previous dielectric models, which are mainly of regression origin when it concerns impact of BSW. In these works, the MDM for moist soil was worked out using an extensive continuum of experimental dielectric data, both measured by the authors and borrowed from the literature. As a result, the newly developed MDM for moist soils attained features of a physical law and ensured more accurate dielectric predictions, with the soil clay content being an intrinsic texture parameter of the soil. Further research in this area has to be aimed at broadening the ensemble of soil types, temperatures, and frequency ranges over which the new MDM can be effectively used for the radar and radio thermal remote sensing of the land, including the ground penetrating radars and TDR applications.

In this chapter, a methodology to develop physically based MDM for moist soils thawed is presented. The moist soil consists of a matrix of mineral and organic particles, air, and aqueous solutions, the latter being further referred to as soil water. It is evident that, when developing a physically based MDM, the law of wave propagation through such a mixture is the first problem to be solved.

Let us consider the propagation of plane harmonic wave in a lossy, nonmagnetic, and unbounded mixture. In this case, the wave amplitude as a function of time, t, and propagation distance, x, may be written in the form:

$$exp[i(kx-\omega t)] \tag{1}$$

where $i = \sqrt{-1}$ is imaginary unit, $k = k_0\, n^*$ − wave propagation constant in a mixture, $k_0 = \omega/c$ − wave propagation constant in a vacuum, $\omega = 2\pi f$ − circular wave frequency, f − wave frequency, c − velocity of light in a vacuum, $n^* = n + i\kappa$ − complex index of refraction (CIR), $n = Rek/k_0$ − index of refraction (IR), $\kappa = Imk/k_0$ − normalized attenuation coefficient (NAC), with n^*, n, and κ being related to a mixture. The CIR is a square root of complex dielectric constant (CDC), $\varepsilon = \varepsilon' + i\varepsilon''$:

$$n^* = \sqrt{\varepsilon} \qquad\qquad (2)$$

For the real, ε', and imaginary, ε'', parts of the CDC, there will be used terms dielectric constant (DC) and loss factor (LF), respectively. According to (2), the DC and LF can be expressed through the IR and NAC and vise versa:

$$\varepsilon' = n^2 - \kappa^2, \qquad \varepsilon'' = 2n\,\kappa \qquad\qquad (3)$$

$$n = \sqrt{\sqrt{\varepsilon'^2 + \varepsilon''^2} + \varepsilon'}\Big/\sqrt{2}, \quad \kappa = \sqrt{\sqrt{\varepsilon'^2 + \varepsilon''^2} - \varepsilon'}\Big/\sqrt{2}, \qquad\qquad (4)$$

In general, a frequency and temperature, T, dependent CDC for a mixture containing N components can be written in the form

$$\varepsilon_s = \varepsilon_{sm}(m_1,\ldots,m_N,\ \varepsilon_1,\ldots,\ \varepsilon_N\,,f,T) = \varepsilon_{sv}(v_1,\ldots,v_N,\ ,\varepsilon_1,\ldots,\ \varepsilon_N,\ f,T) \qquad\qquad (5)$$

where ε_p , m_p and v_p $(p=1,\ldots,N)$ are respectively CDC, gravimetric and volumetric percentages of individual components. We consider the following individual components mixed in moist soil (Ulaby et al., 1986):
1. Mineral matrix consisting of particles of various minerals, which is characterized by gravimetric percentages of clay, C, sand, S, and silt, $L = 100\% - C - S$;
2. Air contained in soil pores;
3. Bound soil water (BSW) consisting of conglomerates of water molecules bound on the surface of the soil mineral matrix;
4. Free soil water (FSW), which is located in soil pores and exists in a form of capillary aggregations;
The CDC of air is considered to be equal to one, $\varepsilon_a = 1$. It has been shown (Ulaby et al., 1986) that the CDC of mineral matrix, $\varepsilon_m(C, S)$, depends on gravimetric percentages of clay, C, sand, S, and silt, $100\% - C - S$, being practically independent on the temperature and wave frequency. On the contrary, due to water molecules relaxation (Hasted J., 1973) the CDCs of BSW, $\varepsilon_b(f, T, C, S)$, and FSW, $\varepsilon_u(f, T, C, S)$, are dependent on frequency and temperature. In addition, these values are considered to be dependent on the soil texture parameters. The latter is justified by bound water molecules interaction with the soil mineral matrix surface and the mineral matrix impact on geometric form and size of capillary water aggregations. At the same time, we assume the CDCs of BSW and FSW to be mutually independent on their percentages, m_b and m_u. With these suggestions, the CDC of moist soil can be expressed in the form

$$\varepsilon_s(v_m,\ v_a,\ v_b,\ v_u,\ f,T)\ =\varepsilon_s(v_m,\ v_a,\ v_b,\ v_u,\ \varepsilon_m(C,\ S),\ \varepsilon_b(f,\ T,\ C,\ S),\ \varepsilon_u(f,\ T,\ C,\ S)). \qquad\qquad (6)$$

As seen from (6), the assumptions made allowed to decompose the variables in (5) representing the percentages of moist soil components and consider the CDCs of individual components as parameters of a mixture dielectric model, which are in their turn dependent on frequency, temperature, and soil texture characteristics. The equation (6) suggests the CDCs of individual components to be independent on their percentages. Such mixtures can be identified as conservative, once their individual components do not affect each other. This assumption may not be true, for instance, in the case of saline soils, with the values of $\varepsilon_b(v_b, f, T, C, S)$, $\varepsilon_u(v_u, f, T, C, S)$ varying due to possible change of soil solute concentration, with the percentages v_b and v_u being changed. There have been proposed plenty of mixture models, which are discussed in (Wang & Schmugge, 1980). From all of those, the refractive mixing dielectric model (RMDM) of (Birchak, 1973)

$$(\varepsilon_s)^{\alpha} = \sum v_p(\varepsilon_p)^{\alpha}, \quad p=m, a, b, u \tag{7}$$

was used as a basis for the moist soil dielectric models suggested in (Wang & Schmugge,1980) and (Dobson et al., 1985). In formula (7), a is a constant shape factor, and the subscripts $m, a, b,$ and u designate mineral matrix, air, bound water, and free water, respectively. As seen from (7), the RMDM meets a conservative mixture condition. In the following section, there are discussed some specific equations for the RMDMs contemporarily used.

2. Refractive mixing dielectric models

Let us consider the propagation of plane wave through a mixture using the geometrical optics approximation. Given the plane wave ray is straight line, a statistically averaged complex phase of the wave that propagated the distance L in a mixture can be written in the form

$$\left\langle n_s^* \right\rangle k_0 L = \left\langle \sum_{i=1}^{N_m} n_m^* k_0 L_{mi} + \sum_{i=1}^{N_a} n_a^* k_0 L_{ai} + \sum_{i=1}^{N_b} n_b^* k_0 L_{bi} + \sum_{i=1}^{N_u} n_u^* k_0 L_{ui} \right\rangle \tag{8}$$

In equation (8), the symbol <> designates statistical averaging over an ensemble of rays, N_m, N_a, N_b, N_u and L_{mi}, L_{ai}, L_{bi}, L_{ui} are respectively the numbers of wave paths and the distances propagated through the media of soil matrix, air, BSW, and FSW, as related to an individual cell of the soil. After statistical averaging and dividing both parts of (8) by $k_0 L$, the moist soil CIR can be represented as follows:

$$n_s^* = n_m^* L_m^{av} + n_a^* L_a^{av} + n_b^* L_b^{av} + n_u^* L_u^{av} \tag{9}$$

where the symbols L_m^{av}, L_a^{av}, L_b^{av}, and L_u^{av} stand for the normalized by L mean wave paths in the media of soil mineral matrix, air, BSW and FSW, respectively. Assuming the wave beam to be non-divergent when propagating in the mixture, one can link the normalized mean wave paths to the volumetric contents of respective substances contained in a mixture:

$$L_p^{av} = V_p / V \tag{10}$$

where V and V_p are respectively the volumes of the mixture as a whole and the ones related to specific components contained in the mixture ($p=m$, a, b, and u). Finally, substituting (10) in (9) yields a refractive mixing dielectric model in the form of (7), with the shape factor, a, being equal to:

$$a = 1/2. \tag{11}$$

Following the paper (Mironov et al., 1995), we re-arrange equation (7) as follows;

$$n_s = \begin{cases} n_d + (n_b - 1)W, & W \leq W_t \\ n_d + (n_b - 1)W_t + (n_u - 1)(W - W_t), & W \geq W_t \end{cases}, \tag{12}$$

$$\kappa_s = \begin{cases} \kappa_d + \kappa_b W, & W \leq W_t \\ \kappa_d + \kappa_b W_t + \kappa_u (W - W_t), & W \geq W_t \end{cases} \tag{13}$$

where the volumetric soil moisture, $W= V_w/V$, is a ratio of the total soil water content, V_w, to the given volume, V, of dry soil sample, W_t is the maximum bound water fraction (MBWF) in a specific type of the soil. The MBWF is such an amount of water in soil that any additional water added to the soil in access of this amount behaves as free soil water. As seen from (1), the CIR is a piecewise linear function of soil moisture, with the MBWF being a transition point in terms of slope angle between the two linear legs relating to the bound, $W \leq W_t$, and free, $W > W_t$, moisture ranges. In formulas (12) and (13), the variables n_d and κ_d are respectively the IR and NAC for a dry soil. With the use of formula (7), these values can be derived in the form

$$n_d = 1 + \frac{n_m - 1}{\rho_m} \cdot \rho_d = 1 + (n_m - 1) \cdot (1 - P) \tag{14}$$

$$\kappa_d = \frac{\kappa_m}{\rho_m} \cdot \rho_d = \kappa_m \cdot (1 - P) \tag{15}$$

where ρ_m and ρ_d are the specific and bulk soil densities, respectively, $P=V_a/V_d$ is the soil porosity, which is equal to the ratio, of the air volume in a dry soil sample, V_a, to a total volume of dry soil sample, V_d. The RMDM was initially proposed in (Birchak, 1973), with no distinction being made between the bound and free soil water. Later (Mironov et al., 1995), it was generalized in the form given by the equation (12), (13), which made possible to distinguish between contributions in the moist soil CIR, generated by the bound and free types of soil water.

The term $(n_m^*-1)/\rho_m$ can be considered collectively as a single mineral parameter of the soil CIR. In the formulas (12)-(15), the values $n_p^*=n_p+i\kappa_p$, ($p= s$, m, d, b, u), W_t are understood as functions of the temperature, soil texture, and frequency: $n_d^*=n_d^*(T, C, S)$, $n_m^*=n_m^*(T, C, S)$, $n_p^*=n_p^*(T, C, S, f)$; $p=s$, b, u; $W_t=W_t(T, C, S)$.

3. Parameters of the RMDM

As seen from (12)-(13), the major advantage of the RMDM is that this model decomposed the soil moisture and soil density variables, thus reducing the problem of developing the moist soil dielectric model, as a function of temperature, soil texture, and frequency, to the one pertaining to the bound. free soil water. and to the dry soil . To obtain building blocks for developing frequency dependent dielectric models for the bound and free soil water, one has to measure, as a function of frequency, the CIRs for these types of water, as parameters in the formulas (12), (13), using the data for the moist soil CIRs measured at varying moistures and frequencies (Mironov et al., 2004).

When fitting the IRs and NACs measured as a function of soil moisture, W, with the formulas (12), (13), the values of n_d and κ_d can be obtained as the interceptions of the respective fits with the axis of ordinates at $W=0$. While the values of MBWF, W_t, can be identified and measured as a point at which the legs of the piecewise dependences (12), (13) intercept each other. Also, the bound and free soil water RIs and NACs can be derived from the slopes of the linear segments in the fits made with (12) and (13), respectively. As a result, the RMDM parameters n^*_d, W_t, n^*_b, and n^*_u can be obtained for a specific soil, with the texture parameters, C and S, temperature, T, and wave frequency, f, being fixed. Fitting the CIR data for moist samples measured at varying frequencies, one can obtain the spectra $n^*_b(C, S, T, f)$, and $n^*_u(C, S, T, f)$. The latter are further applied to derive the Debye relaxation parameters for both the bound and free soil water (Mironov et al., 2004). In our publications, the CIRs measured for moist soils were shown to be well fitted with the equations (12) and (13).

To quantitatively demonstrate a degree of correlation of the RMDM model with the CIRs measured for different types of soil at varying frequencies and temperatures, let us transform the equations (12) and (13) into a universal piecewise linear moisture dependences: which are able to accommodate all the data measured:

$$n_s = \begin{cases} (n_b - 1)W_b, & W_b \le W_t \\ (n_b - 1)W_t + (n_u - 1)W_u, & W_u \ge W_t \end{cases} \tag{16}$$

$$\kappa_s = \begin{cases} \kappa_b\, \bar{W}_b, & W_b \le W_t \\ \kappa_b\, \bar{W}_t + \kappa_u\, \bar{W}_u, & W_u \ge \bar{W}_t \end{cases} \tag{17}$$

where n_b, n_u and κ_b, κ_u are the bound and free soil water IR and NAC reduced values, respectively, $\bar{W}_t$ is the reduced value of the MBWF assigned for the universal piecewise linear RMDM. In the further analysis, we use the following numerical values for these parameters: $W_t = 0.3$; $n_b - 1 = 10/3$; $n_u - 1 = 9$, $k_b = 2/3$; $k_u = 8/3$.

This universal RMDM gives the values of reduced RI, $\bar{n}_s$, and NAC, $\bar{\kappa}_s$:

$$n_s(W) = (n_s(W) - n_d)(n_b - 1)W /(n_b - 1), \quad W \le W_t;$$

$$\tag{18}$$

(24) have real values provided the combined parameter, s, lies in the range $0 \le s \le (1/8)$. Hence, in the case of DC and LF spectra for soil water, which dielectric parameter s exceeds the value of 1/8, neither the relaxation nor the transition frequencies are observed, with the LF spectra being a function monotonically increasing with decreasing frequency.

From the equations of (24), the ranges for the relaxation and transition frequency variations as a function of the combined parameter s, can be determined as follows:

$$1\!\!\left/\sqrt{3}\tau_p\right. \le \omega_{rp} < 1/\tau;\,; \qquad\qquad 0 \le \omega_{np} < 1\!\!\left/\sqrt{3}\tau_p\right. . \tag{25}$$

The least and largest limits of these ranges are calculated from (24), with the values of combined characteristic being equal to $s=1/8$ (at this value, the least limit for relaxation loss range occur and largest limit for ohmic loss range) and to $s=0$ (at this value, least limit for the ohmic loss range occur and largest limit for the relaxation range). In addition, according to (24), the relaxation and transition frequencies are related to each other by the equations given below in two alternative forms:

$$\omega_{np} = (1/\tau_p)\sqrt{\left[1-(\omega_{rp}\tau_p)^2\right]\!\!\left/\!\left[1+3(\omega_{rp}\tau_p)^2\right]\right.} \quad \text{and}$$

$$\omega_{rp} = (1/\tau_p)\sqrt{\left[1-(\omega_{np}\tau_p)^2\right]\!\!\left/\!\left[1+3(\omega_{np}\tau_p)^2\right]\right.} \quad \text{or}$$

$$\left\{\left[\omega_{np}\tau_p\right]^2\!\!\left/\!\left[1+(\omega_{rp}\tau_p)^2\right]\right.\right\} + \left\{\left[\omega_{rp}\tau_p\right]^2\!\!\left/\!\left[1+(\omega_{np}\tau_p)^2\right]\right.\right\} = 1/2 \tag{26}$$

The Debye equations (22), (23) can be recast as a pair of equations which are linear in the Debye parameters. These equations can be then solved via linear regression to yield least squares estimates of the Debye parameters. The first of the two linear equations is

$$\varepsilon'_p(f) = z_{p\sigma} - \tau_p z_{p\tau}(f) \tag{27}$$

where

$$z_{p\sigma} = \varepsilon_{p0} + \sigma_p \tau_p / \varepsilon_0 \tag{28}$$

is a constant and

$$z_{p\tau}(f) = 2\pi f \varepsilon''_p(f) \tag{29}$$

is a reduced variable, which is supposed to be calculated from dielectric measurements. A linear regression with regard to the measured pairs $(\varepsilon'_p, z_{p\tau})$ of (27) yield the estimates of the parameters τ_p and z_{p0} from the measured pairs $(\varepsilon'_p, \varepsilon''_p)$.

The second equation is

$$\varepsilon'_p(f) = \varepsilon_{p0} - \left(\varepsilon_{p0} - \varepsilon_{p\infty}\right) z_{p\varepsilon}(f) \tag{30}$$

where the variable

$$\frac{1}{z_{p\varepsilon}(f)} = 1 + \frac{\varepsilon''_p(f)/2\pi f \tau_p}{z_{p\sigma} - \varepsilon'_p(f)} \tag{31}$$

is thought to be known from measurements, with the values of τ_p and $z_{p\sigma}$ having been obtained from the previous regression with the use of (27). At this point, all four Debye parameters can be found using the values of τ_p, $z_{p\sigma}$, ε_{p0}, and $(\varepsilon_{p0}\text{-}\varepsilon_{p\infty})$ attained from the regressions. The linear fitting of the data measured for $\varepsilon'(f)$ and $\varepsilon''(f)$ with the use of (27) and (30) also provide correlation coefficients and standard deviations to estimate compliance between the measured data and the fitting ones calculated with the Debye equations (22), (23).

To universally represent various dielectric spectra of soil water measured for bound or free soil water, different types of soils and temperatures, The linear functions (27) and (30) representing the Debye equation (22), (23) can be can be modified in a universal linear function:

$$\varepsilon'_{pj}(f) = 1 - z_{pj}(f) \tag{31}$$

where $\varepsilon'_{pj}(f)$ and $z_{pj}(f)$ are the reduced DCs and frequencies which can be obtained after some algebraic manipulations performed with the pairs (27), (29) and (30), (31):

$$z_{p1}(\omega) = \left[\tau_p \omega \varepsilon''_p(\omega) - \left(\varepsilon_{0p} + \sigma_p\right)\right]\left[\varepsilon_{0p} - \varepsilon_{\infty p}\right]$$

$$\varepsilon'_{p1}(\omega) = \left[\varepsilon'_p\left(z_{p1}(\omega)\right) - \varepsilon_{\infty p}\right]\left[\varepsilon_{0p} - \varepsilon_{\infty p}\right] \tag{32}$$

$$z_{p2}(\omega) = \left[\varepsilon_{0p} + \sigma_p - \varepsilon'_p(\omega)\right]\left[\varepsilon_{0p} + \sigma_p + \left(\varepsilon''_p(\omega)\,\tau_p\omega\right) - \varepsilon'_p(\omega)\right]$$

$$\varepsilon'_{p2}(\omega) = \left[\varepsilon'_p\left(z_{p2}(\omega)\right) - \varepsilon_{\infty p}\right]\left[\varepsilon_{0p} - \varepsilon_{\infty p}\right] \tag{33}$$

The values j =1 and j=2 in (32), (33) correspond to the pairs (26), (29) and (30), (31), respectively.

As a result, the equation (30) in conjunction with the expressions (32), (33) can universally represent any DC, $\varepsilon'_p(f)$, and LF, $\varepsilon''_p(f)$, spectra for both the bound, $p=b$, and free, $p=u$, soil water measured for a specific type of soil, at a given temperature, provided these spectra follow the Debye equation (21). From equations (22), (23), (32), (33) follow that the reduced functions $\varepsilon'_{p1}(f)$, $\varepsilon'_{p2}(f)$ and their arguments $z_{p1}(f)$, $z_{p2}(f)$ are defined across the same segment from zero to unit, [0,1].

Substituting the values of $\varepsilon'_p(\omega)$ and $\varepsilon''_p(\omega)$ given by the formulas (22) and (23) in the expressions for the reduced frequencies $z_{p1}(f)$ and $z_{p2}(f)$ of (31), (32), we come up with the following simple formula:

$$z_{p1}(f) = z_{p2}(f) = z_p(f) = (\omega\tau_p)^2 \big/ \left(1 + (\omega\tau_p)^2\right) \tag{34}$$

which can be applied only in the case of formal transforming the equations (27), (28) and (30), (31) to the equations (31), (32), and (33), but not for fitting the equations (31), (32), and (33) to the DC and LF data measured. In addition, using the expressions (25) and (34) and the eqution $\tau\omega_f = 1$, we find for the reduced maximal dispersion frequency, $z_p(\omega_f)$, and the ranges for relaxation, $z_p(\omega_r)$, and ohmic, $z_p(\omega_n)$, frequencies to be as follows:

$$z_p(\omega_f) = 1/2 ; \quad 0 \le z_p(\omega_n) \le 1/4 ; \quad 1/4 \le z_p(\omega_r) \le 1/2 ; \tag{35}$$

When presenting the data measured in the form of reduced frequencies, one can easily identify using the formula (35) specific frequency ranges which these data belong to. These ranges appear to be dimensionless due to the respective transformation conducted with (31)-(33).

An example for the DC and LF spectra following the Debye formulas (22), (23) are shown in Figure 3. They were calculated for a sample of saline water, having solution concentration and temperature of 1.88 g/liter and 20°C, respectively, with the use of the empirical formulas obtained in (Stogryn, 1971). In Figure 3b, the same spectra for the DC and LF reduced according to (32), (33) are plotted as a function of reduced frequency, $z_p(f)$, calculated with (34). The following parameters of the Debye spectra (22), (23) were used (τ_u=9.27 ps ; ε_{0u}=79.44; σ_u=0.3; $\varepsilon_{\infty u}$=4,9), regarding the liquid saline water, to perform transition to reduced spectrum and frequency, as given by (32), (33). The spectroscopic parameters for saline water listed above were determined with empirical formulas of (Stogryn, 1971). Here and henceforth, we understand the low and high frequency limits as the values related exclusively to a specific Debye relaxation frequency range, within which the measured data are located and the low and high frequency limits have been determined using these data. These limits can not be understood as the values of real DCs which may occur beyond the frequencies measured. Such a situation may take place if some other relaxation processes considerably contribute in the values of real DCs, making, for instance due to the Maxwell-Wagner relaxation (Hasted, 1973), real values of ε' a lot greater then the low frequency limit, ε_0, determined via data measured for its deriving. In the range of frequencies exceeding those measured, the same understanding with regard to the high frequency limit, ε_∞, has to be used, as many other relaxation processes may also occur when moving towards the optical range of frequencies.

The vertical lines in Figures. 3a and 3b mark the maximal dispersion frequency, f_f, maximal relaxation loss frequency, f_r, and the frequency of transition from relaxation to ohmic losses, f_n, as these are given with the formulas (24) and (32), (33), to display the intermediate range of frequencies in which both the relaxation and ohmic losses are observed, as well as to show the position of maximal dispersion frequency.

As any Debye spectra can be universally presented like in Figure 3, for the spectra measured a possibility arises to align those along one and the same straight line, which makes possible not only to test whether the measured spectra belong to the Debye class, but also to monitor the position of measured spectra with regard to the Debye spectra characteristic frequencies, that is, f_f, f_r, and f_n, pictured for all the data measured in one and the same graph.

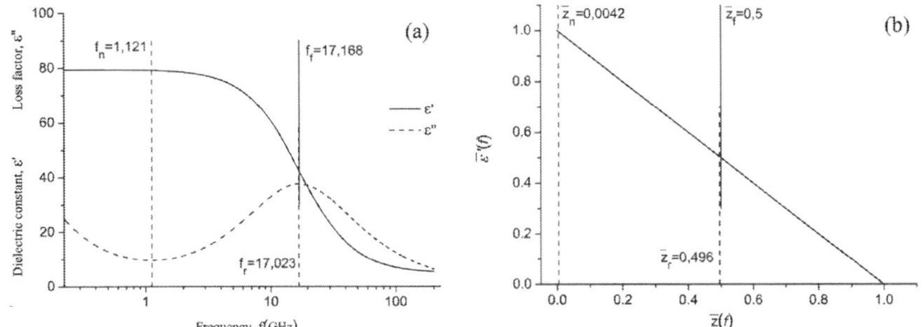

Fig. 3. Relaxation frequency spectra for the DC and LF in the case of saline liquid water, having the following values of Debye parameters: τ=9.27 ps ε_0=79.44 σ=0.3 ε_∞=4,9. a) Initial values of DC, LF and frequency; b) Reduced values of DC, LF, and frequency. With vertical lines, are marked both the real (in Fig. 3a) and reduced (in Fig. 3b) positions of the maximal dispersion frequency (f_f, and z_f , solid lines), maximal relaxation loss frequency, (f_r and z_r , dotted lines), and the frequency of transition to ohmic losses (f_n and z_n, dotted lines). While their numerical values are placed next to the respective lines.

Fitting formulas (12) and (13) to the data for DCs and LFs as a function of moisture for the soils listed in Table 1, first, the values of DCs and LF were obtained with the technique outlined in Section I. Second, the DC and LF spectra for both the bound and free soil water obtained were fitted with the formulas (27)-(31) in order to derive the parameters present in the Debye equation (22), (23), that is the low and high frequency limits, ε_∞ and ε_0, relaxation time, τ, and conductivity, σ, regarding both types of soil water. The values thus obtained are given in Table 2.

Soil number in Table 1	7		9		19		20	
Wave frequencies measured	4 to 18 GHz		4 to 18 GHz		0.5 to 10 GHz		3.01 to 15 GHz	
Soil water type	p=b	p=u	p=b	p=u	p=b	p=u	p=b	p=u
Low frequency limit of DC, ε_{0p}	58.5	103	38.60	97.90	65.08	73.15	43.45	79.52
High frequency limit of DC, $\varepsilon_{\infty p}$	4.9	4.9	4.9	4.9	4.9	4.9	14.2±1.2	14.1±1.2
Relaxation time, τ_p(ps)	1.14	8.3	11.30	8.00	21.41	9.23	65.71	11.57
Conductivity, σ_p(S_p/m)	0.7	1.03	1.24	2.15	3.48	1.29	0,061	3,31

Table 2. Debye relaxation parameters and conductivities for some soils listed in Table 1.

We did not have the dielectric data measured for the frequencies larger then the maximal dispersion frequency, f_f, which, according to Figure 3, could make possible to derive the high frequency limit with reliable accuracy, except for the soil 20. Therefore, the value of 4.9 was preset before fitting with the formulas (27)-(31), in accordance with previously estimated data for this parameter (Mironov et al., 2004).

As follows from Table 2, the Debye relaxation parameters and conductivities relating to the bound and free water can be clearly distinguished via their magnitudes. The most important feature, seen in Table 2, is that not only the low frequency limits but also the relaxation times, as well as conductivities are found to noticeably differ from each other. It implies that, in accordance with the analysis conducted above the maximum relaxation frequencies and transition to ohmic losses frequencies pertaining to the different types of soil water must be also different.

At the same time, in the dielectric model developed in (Dobson et al., 1985), both the bound and free water spectra were assumed to be similar to that of free liquid water existing out of soil. The mixing formula for the moist soil CDC used in (Dobson et al., 1985) can be presented as follows:

$$\varepsilon_s'(f,v_a,v_m,v_w,C,S)^\alpha = v_a(\varepsilon_a)^\alpha + v_m\varepsilon_m(C,S)^\alpha + v_w(\varepsilon_w(f)^\alpha \cdot v_w^{(\beta_1(C,S)-1)} - 1)$$

$$\varepsilon_s''(f,v_w,C,S)^\alpha = v_w^{\beta_2(C,S)} \cdot \varepsilon_w''(f,\sigma(\rho_d,C,S))^\alpha$$

were subscrip w stands for the water out of soil content and $\varepsilon_w(f)$ is a preset function characterizing the water out of soil. At that, only the power index a and the values $\beta_1(C, S)$, $\beta_2(C,S)$ and $\sigma(\rho_d,C,S)$ were adjusted to the dielectric data measured via regression analysis. While the mixing formula applied in (Wang&Schmugge, 1980) suggested the following expression for the CDC of moist soil:

$$\varepsilon_s(f,v_a,v_m,v_w,C,S) = v_a\varepsilon_a + v_m\varepsilon_m(C,S) + v_w \cdot Y \cdot \varepsilon_w(f) \quad if \quad v_w \le v_t ;$$

$$\varepsilon_s(f,v_a,v_m,v_w,C,S) = v_a\varepsilon_a + v_m\varepsilon_m(C,S) + v_t \cdot Y \cdot \varepsilon_w(f)$$
$$+ (v_w - v_t)\varepsilon_w(f) \qquad if \quad v_w \ge v_t ;$$

with the values of v_t and Y being adjusted to the dielectric data measured only at 1.4 and 5 GHz. From the expressions above pertaining to the dielectric models of (Dobson et al.,1985) and (Wang&Schmugge, 1980) follows that the intrinsic properties of soil bound water spectra in terms of all the Debye relaxation parameters and conductivity could neither been measured nor taken into account on a physical ground.

The detailed consideration of both the bound and free soil water spectra measured for the soils listed in Table 2 is conducted in Figure 4 with regard to reduced spectra.

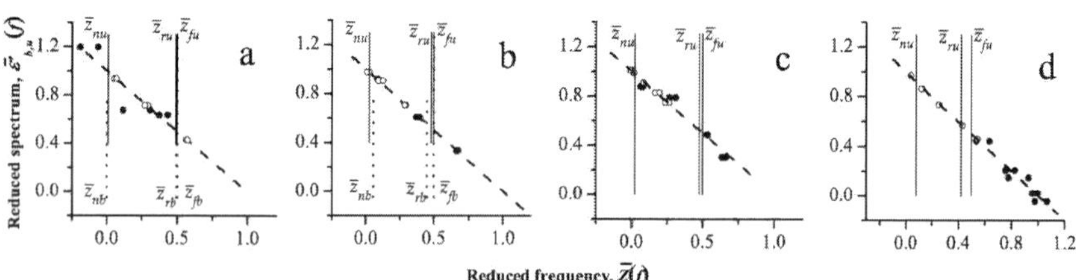

Fig. 4. Reduced spectra (slanting dotted line) for the bound and free water pertaining to some soils listed in Table 1. Position of the maximal dispersion frequency $z_{fu} = z_{fu} = 0.5$ is marked with vertical soilid lines. While the positions of maximal relaxation loss frequency, z_{rp}, and the frequency of transition to ohmic losses range, z_{np}, for the BSW, p=b, and FSW, p=u, are shown with vertical dotted and solid lines, respectively. a) Soil 7, $z_{rb} = 0.498$, $z_{nb} = 0.002$, $z_{ru} = 0.49$, $z_{nu} = 0.01$; b) Soil 9, $z_{rb} = 0.448$, $z_{nb} = 0.05$, $z_{ru} = 0.478$, $z_{nu} = 0.02$; c) Soil 19, $z_{ru} = 0.477$, $z_{nu} = 0.02$; d) Soil 20, $z_{ru} = 0.421$, $z_{nu} = 0.08$.

In contrast to the ideal reduced spectra shown in Figure 3b, some of measured data given in Figure 4 , for reduced DC, $\varepsilon'(f)$ and frequency, $z(f)$, fall beyond the range $0\leq \varepsilon'(f),\ z(f)$ ≤ 1. This occurred not only due to error in the spectroscopic parameters derived, which are taken from Table 2 to be used for calculating the reduced value with the formulas (32), but also because the frequencies $z_1(f)$ and $z_2(f)$ were calculated using in formulas (27)-(31) the data for $\varepsilon''(f)$ measured with some error. As seen from Figure 4, for both the bound and free soil water, the measured spectra for the soils 7, 9,and 19 appeared to be mainly located in the frequency range from the transition to ohmic losses frequency, f_n, to the relaxation loss frequency, f_r, with only a few frequencies exceeding the maximal dispersion frequency, f_f. For this group of soils, there is only a little difference between the characteristic frequencies related to the bound or free soil water types. Though, as to the soil 19, the intermediate frequency range, from f_n to f_r, did not appeared to be observed at all, because , in accordance with the formula (24) and data given in Table 2, the condition $0\leq s\leq(1/8)$ was not met. Finally, we found the spectral properties of the bound and free soil water contained in the arctic shrub tundra soil to be quite different in terms of their frequency ranges As seen from Figure 4 d, the frequencies measured appeared to be located below and higher the respective maximal dispersion frequencies of the bound and free soil water types, respectively. In addition, there was no interim frequency interval for the bound water spectrum observed. Once in the case of this soil, the frequencies measured happened to be higher then the maximal relaxation one, the high frequency limit was measured with an acceptable error. The numerical value of this parameter is given in Table 2. As seen from Table 2, for the organic rich (95%) soil 20, the high frequency limits appeared to be about three times as large compared to the value of 4.9, which is a conventional estimate for the mineral soils (Mironov et al., 2004).

Similar to Figure 4, all the spectra measured for the soils 7, 9, 19, and 20, are presented in the reduced form in Figure 5.

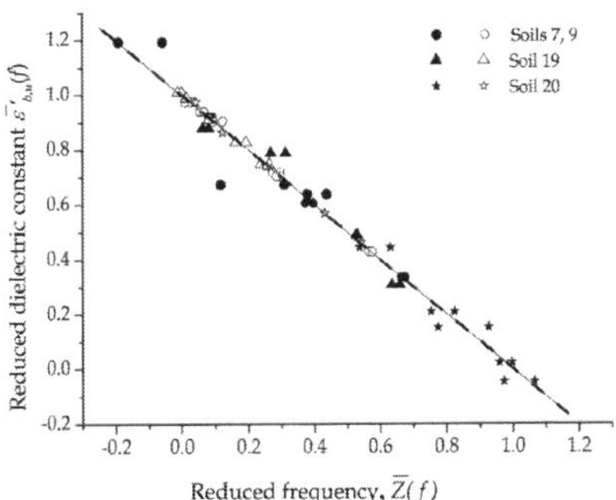

Fig. 5. Reduced spectra for both the bound (filled symbols) and free (empty symbols) soil water regarding the soils 7, 9, 19, and 20. The linear fits related to the bound and free water are drawn with dotted and solid lines, respectively. The correlation coefficient and standard deviation regarding the linear fit are equal to R= 0.988 and SD= 0.0471, respectively.

As seen from Figure 5, the data measured follow quite well the Debye relaxation formula (21) for both the bound and free soil water spectra regarding rather a broad variety of soils and frequencies measured.

The results of the analysis conducted in this Section not only substantiates the use of Debye spectra for the soil water dielectric spectra, but also propose the methodology for deriving soil water Debye parameters and conductivity. Previously, the Debye spectrum model was used in the soil dielectric models (Wang&Schmugge, 1980; Dobson et al., 1985; Mironov et al., 2004), though the applicability of this spectrum class with regard to the soil water types, especially in the case of bound water, has never been substantiated. Summing up the result of Sections 2 and 3, let us summarize the FD RMDM parameters that have to be derived from dielectric measurements to ensure the DC and LF predictions for a specific type of moist soil. As can be seen from the equations (3), (4), (12), (13), (22), (23), a certain type of moist soil, in terms of its dielectric spectra, can be completely determined via a set of the following FD RMDM parameters:

1. DC, $\varepsilon'_d(T,C,S)$, for dry soil;
2. LF, $\varepsilon''_d(T,C,S)$, for dry soil;
3. Value of MBWF, $Wt(T,C,S)$;
4. Low frequency limit dielectric constants, $\varepsilon_{0b}(T,C,S)$ and $\varepsilon_{0u}(T,C,S)$, for bound and free soil water;
5. High frequency limit dielectric constants, $\varepsilon_{\infty b}(T,C,S)$ and $\varepsilon_{\infty u}(T,C,S)$, for bound and free soil water;
6. Relaxation times, $\tau_b(T,C,S)$ and $\tau_u(T,C,S)$, for bound and free soil water;
7. Conductivities, $\sigma_b(T,C,S)$ and $\sigma_u(T,C,S)$, for bound and free soil water.

As shown in Sections 2 and 3, for a specific type of soil, all of these parameters can be derived with the use of conventional dielectric measurements of moist soils. Therefore, to be employed in the microwave remote sensing processing algorithms, the FD RMDM requires prior dielectric measurements to be carried out for a set of the individual soils involved in remote sensing data processing, and the error of dielectric predictions for each individual soil be tested. Some tests to prove the predictive power of the FD RMDM have been already conducted in (Mironov et al., 2004). Further, we use the dielectric data of (Hallikainen et al., 1985; Dobson et al., 1985, Curtis et al. 1995) to analyze prediction errors of the FD RMDM over a wide ensemble of soil types in terms of texture, as well as extended ranges of frequencies measured. Based on this analysis, the FD RMDM can be generalized to predict DCs and LFs of moist soils, with their textures being an input parameter. That frequency and texture dependent refractive mixing dielectric model (FTD RMDM) is outlined in the next section, following the results of (Mironov et al., 2009a).

5. Frequency and texture dependent RMDM

The FD RMDM was tested in (Mironov et al., 2004 a,b; Mironov et al., 2009a) with the dielectric data measured in (Hallikainen et al., 1985; Dobson et al, 1985; Curtis et al., 1995). Given the numerical values for the FD RMDM parameters (see Table 2), the ensemble of the FD RMDM formulas (3), (4), (12), (13), (22), (23) can be applied to calculate the DCs and LFs predicted with this model. Texture characteristics of the combined set of soils measured in (Hallikainen et al., 1985; Dobson et al, 1985; Curtis et al., 1995) are given in Table 3.

The ensemble of dielectric data in (Curtis et al., 1995) appeared to be incomplete, in terms of a number of moistures measured, to be applied for deriving the FD RMDM spectroscopic parameters as it was done in Sections 2 and 3. Based on the results of these sections, we suggested the FD RMDM to be applicable to the dielectric data of (Curtis et al., 1995) measured in the domain of moistures (from nearly dry soils to field capacity values), and frequencies (from 0.45 to 26.5 GHz).

Order number	Sample code	Texture class	Clay %	Sand %	MBWF
1.	B	Sand (SP), Light Gray	0	98	0.005
2.	L	Sand (SP), Brown	0	99	0.001
3.	I	Sand (SP), White	0	100	0.001
4.	C	Silty Sand (SM), Brown	4	88	0.02
5.	E	Clayey Silt (ML), Brown	7	0	0.1
6.	K	Clayey Silt (ML), Brown; Trace of Sand	7	4	0.11
7.	F2	Loam	8.53	41.96	0.075
8.	G	Clayey Sand (SC), Dark Brown	13	55	0.07
9.	F1	Sandy Loam	13.43	51.51	0.07
10.	F3	Silt Loam	13.48	30.63	0.073
11.	D	Silty Sand, (SM), Reddish Brown	14	77	0.05
12.	H	Clay (CH), Gray	34	2	0.15
13.	F5	Silty Clay	47.4	5.02	0.17
14.	J	Silt (ML), White	54	0	0.15
15.	A	Clay (CH), Light Gray	76	2	0.28

Table 3. Texture characteristics for the soils measured in (Hallikainen et al., 1985; Dobson et al, 1985; Curtis et al., 1995).

Therefore, there was applied a procedure of fitting simultaneously the FD RMDM formulas to all the DC and LF spectra corresponding to the whole set of moistures available for each particular type of soil presented in Table 3, as was done in (Mironov et al., 2006; Mironov et al. 2009a). As an example of fitting effectiveness, in the case of soil with code D, the fits and the experimental spectra are shown in Figure 6. The results presented in this figure proves the FD RMDM ability to accurately fit the measured CDs and LFs.

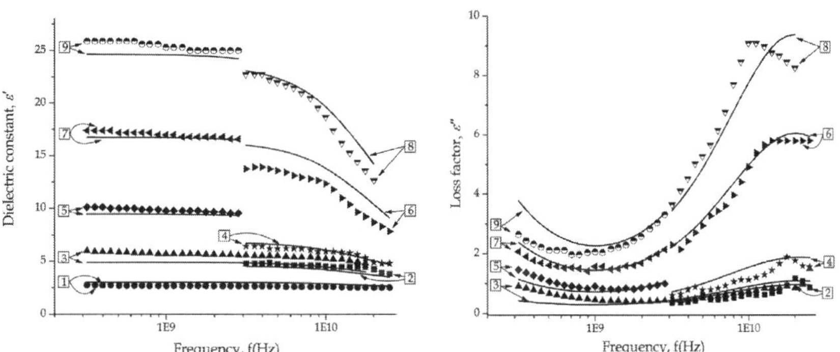

Fig. 6. DC, ε', and LF, ε'', spectra for soil D measured (dots) and fitted (solid line) with the use of the GRMDM model.. Shown data correspond to the following soil volumetric moistures, $W(\%)$: 1) 3.2; 2) 8; 3) 8.8; 4) 13.2; 5) 18.4; 6) 29.1; 7) 29.7; 8) 38.2; 9) 39.4

With this technique of fitting, the FD RMDM spectroscopic parameters were attained for the whole set of soils listed in Table 3. Thus obtained, the FD RMDM parameters are shown with symbols in Figure 7 as functions of clay content.

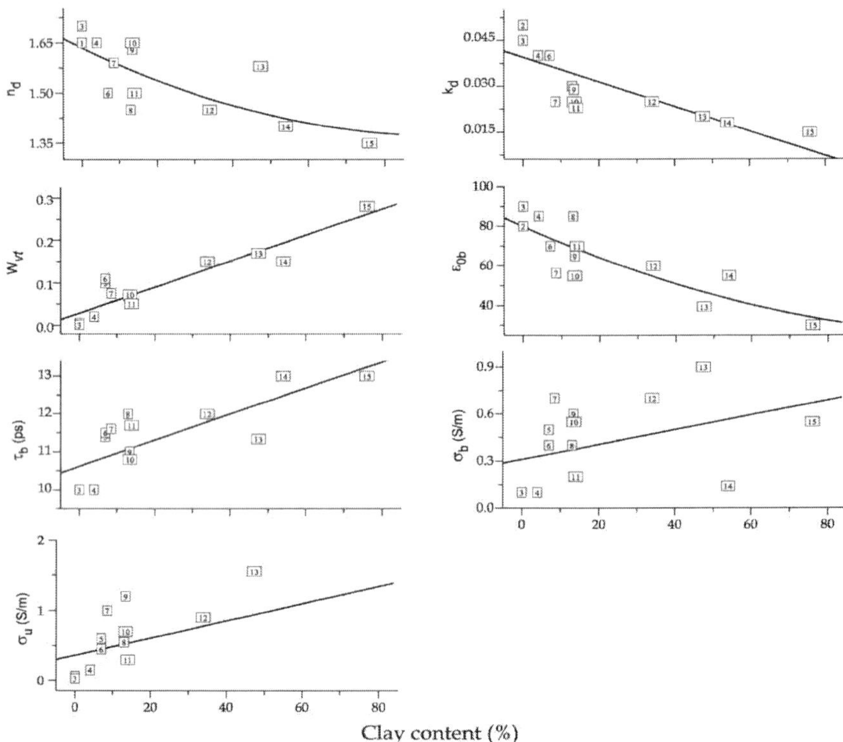

Fig. 7. The GRMDM spectroscopic parameters for the soils presented in Table 1 as a function of gravimetric clay content. The squares represent values derived through fitting the dielectric spectra. The numbers shown in squares correspond to those given in Table 3.

To estimate the error of the FD RMDM predictions for the DCs and LFs over this ensemble of soils, there were applied equations (3), (4), (12), (13), (22), (23) in conjunction with the FD RMDM spectroscopic parameters given in Figure 7. The measured DCs and LFs versus the predicted ones are demonstrated in Figure 8, with the error of prediction being given in the figure caption. According to Figure 8, the error in terms of standard deviation, correlation coefficient and squint of the linear regression with regard to 1 to 1 line appeared to be quit acceptable, given a wide domain of frequencies, soil textures, and moistures.

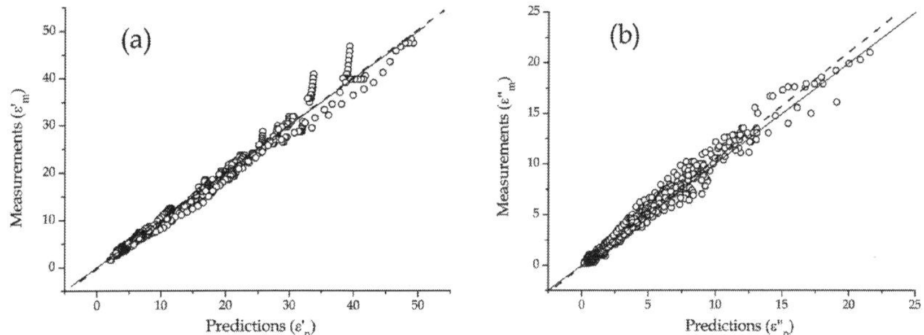

Fig.8. Correlation of the FD RMDM predictions, ε'_p, ε''_p, for DCs (a) and LFs (b) with the measured ones, ε'_m, ε''_m, in the case of soils measured in (Curtis et al., 1995). Solid and dotted lines represent bisectors and linear fits, respectively. Correlation coefficients, R_{DC} and R_{LF}, and standard deviations, SD_{DC} and SD_{LF}, are equal to: R_{DC}=0. 995, R_{LF}=0.991, SD_{DC}=1.023, SD_{LF}=0.4899.The linear fits are expressed as follows: ε'_m=-0.2753 + 1.013 ε'_p, ε''_m=-0.0943 + 1.058 ε''_p.

Generalizing the above analysis, it is possible to assert that the FD RMDM, having been adjusted to each specific soil, as in Figure 6, can ensure acceptable accuracy of dielectric predictions. This result makes possible to use the FD RMDM as a building block to create a frequency and texture dependable dielectric model (FTD RMDM) for moist soils. For this purpose, the whole set of FD RMDM parameters in Figure 7 was fitted with the functions of clay percentage (Mironov et al., 2009a), yielding the equations given in Table 4.

n_d=1.634-0.539·10⁻²C+0.2748·10⁻⁴C² ε_{0b} =79.8-85.4C+32.7·10⁻⁴C² ε_{0u} =100

k_d=0.03952-0.04038·10⁻²C τ_b =1.062·10⁻¹¹+3.450·10⁻¹²·10⁻²C τ_u =8.5·10⁻¹²

W_t=0.02863+0.30673·10⁻²C σ_b =0.3112+0.467·10⁻²C σ_u =0.3631+1.217·10⁻²C

Table 4. Fits for the FD RMDM as a function of clay percentage

The clay content, C, and relaxation times, τ_b and τ_u, in the formulas of Table 4 are expressed in percentages and seconds, respectively. Further on, the equations in Table 4 are referred to as the FTD RMDM fits, while the values calculated with these equations will be identified as the FTD RMDM spectroscopic parameters. Finally, the DCs and LFs calculated with formulas (3), (4), (12), (13), (22), (23) in conjunction with the spectroscopic parameters given in Figure 7 are identified as the FTD RMDM dielectric predictions. Clay percentage, C, is the only input parameter of the FTD RMDM in terms of soil texture to account for a specific type of soil. It should be especially noticed that the equations in Table 4 generalize all the information presented in Figure 7 on the FD RMDM spectroscopic parameters regarding the 15 soils of Table 3.

The same correlation analysis, as shown in Figure 8 in the case of the FD RMDM, was conducted with respect to the FTD RMDM. It proved the FTD RMDM prediction errors to be on the same order as those given in Figure 8. This result signifies that, over the whole variety of soil types, moistures, and frequencies measured in (Curtis et al., 1995) and (Hallikainen et al., 1985; Dobson et al. 1985) the FTD RMDM can predict the DCs and LFs with the same accuracy, in terms of correlation coefficient and standard deviation, as the FD RMDM does.

Nevertheless, it is worth mentioning that the FTD RMSD prediction error has been tested only over the dielectric data, which were used for its developing. In case the test is carried out regarding the dielectric data with which a respective regression model is created, it can reveal only an error of regression analysis itself, as was done in (Wang&Schmugge, 1980; Dobson et al. 1985). To verify an intrinsic predictive power of the FTD RMDM, it has to be tested over the dielectric data set other than the one used for deriving formulas like in Table 4. To perform such a validation of the FTD RMDM, the formulas in Table 4 were obtained, first, with the use of spectroscopic parameters relating to all the soils excluding the ones coded as F1, F2, F3, and F5 and, second, only with the use of soils F1, F2, F3, and F5. As a result, these two new versions of the FTD RMDM could be treated as independent with regard to the soil sets which were not used in their developing. Then the tests like that in Figure 8, were conducted for the both FTD RMDM versions with respect to the DC and LF data independently measured. With these tests, the errors for both FTD RMDM versions was found (Mironov et al., 2009a) to be on the same order as the ones seen in Figure 8. In addition, we have tested the FTD RMDM against the independently measured dielectric data, which are available in (Wang&Schmugge, 1980; Sabburg et al. 1997; Mironov et al., 2004; 2005; 2006), finding the error of the FTD RMDM predictions to be on the same order as that in Figure 8 is.

As seen in Figure 9 borrowed from (Mironov et al., 2009a], in contrast to the physically based FTD RMDM, the regression model of (Dobson et al., 1985) provided for DC and LF predictions with considerably greater error than that observed in Figure 8, pertaining to the FTD RMDM.

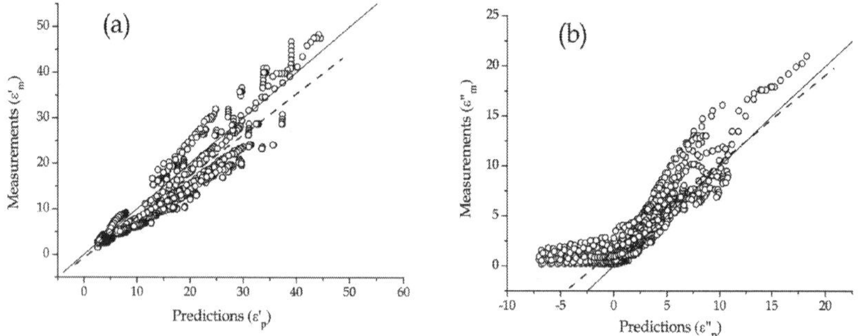

Fig.9. Correlation of the semiempirical model (Dobson et al. 1985) predictions, ε'_p, ε''_p, for DCs (a) and LFs (b) with the measured ones, ε'_m, ε''_m, in the case of soils measured in (Curtis et al.,1995). Solid and dotted lines represent bisectors and linear fits, respectively. Correlation coefficients, R_{DC} and R_{LF}, and standard deviations, SD_{DC} and SD_{LF}, are equal to: R_{DC}=0.942, R_{LF}=0.882, SD_{DC}=3.391, SD_{LF}=1.695. The linear fits are expressed as follows: ε'_m=-0.753 + 0.902ε'_p, ε''_m=1.483 + 0.881ε''_p.

It is worth noting here that the methodology employed for developing the FTD RMDM is based on two key elements. The first one is the FD RMDM, which ensures accurate dielectric predictions using a cluster of the FD RMDM, parameters pertaining to the respective cluster of soil types with specific textures. The second key element is the completeness of such soil clusters in terms of texture diversity. An example of such a cluster is shown in Figure 7. Within the FTD RMDM methodology, the cluster of the FTD RMDM parameters pertaining to a specific ensemble of soils is turned into a specific collection of the regression equations given in Table 4, which represent the example of a specific version of the FTD RMDM related to the ensemble of soils listed in Table 4. The more complete ensemble of soils is available the less error of prediction with the respective version of the FTD RMDM can be attained, especially in regard with the soils other than those used to develop the particular version of the FTD RMDM. Therefore, the FTD RMDM prediction error, regarding the whole variety of natural soils, must depend on both the predictive capability of the FD RMDM, which is a physically based building block of the FTD RMDM, and the completeness of the ensemble of soils employed, in terms of diversity of their texture, which is a regression analysis building block of the respective version of the FTD RMDM. The more complete such a data set is the less regression analysis error in the equations similar to those given in Table 4 can be obtained, thus ensuring less error of predictions related to the respective FTD RMDM version.

One more problem regarding the FTD RMDM prediction error needs to be discussed. It concerns the question whether the FTD RMDM trained on a specific variety of soils is able to provide for dielectric spectra (22), (23) to the data measured predictions in the case of an individual soil, which belong to that variety. Let us consider such a situation. In Figure 10, experimental DCs and LFs are plotted versus the FTD RMDM predictions in the case of specific soil D measured in (Curtis et al., 1995). At that, in Figure 10, there was used the version of FTD RMDM based on the formulas from Table 4, which correspond to the whole variety of soils listed in Table 1. The data predicted with the FTD RMDM instead of the initial FD RMDM parameters as shown in Figure 6, appeared to follow the measured ones quite well for DCs and good enough in the most cases for LFs, excluding the data for greater moistures (graphs 7 and 9) in the range of frequencies below 2 GHz. As seen in Figure 7, this deviation arises from a noticeable difference between the free water conductivities derived with the FD RMDM fitting technique and calculated with the FTD RMDM formulas of Table 4.

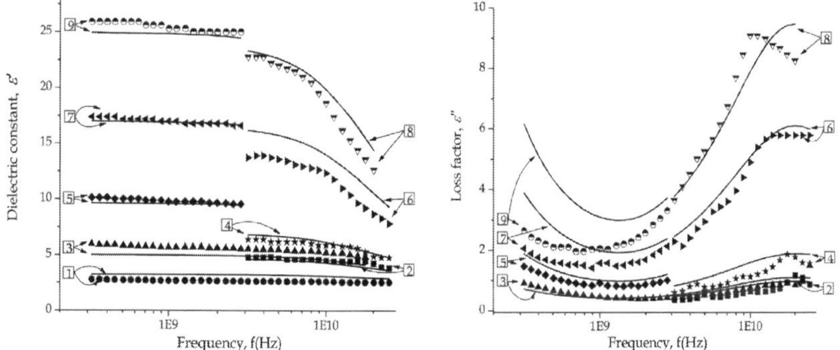

Fig. 10. DC, ε', and LF, ε'', spectra for soil D measured (dots) and calculated with the use of FTD RMDM (solid lines). Shown data correspond to the following soil volumetric moistures, W(%): 1) 3.2; 2) 8; 3) 8.8; 4) 13.2; 5) 18.4; 6) 29.1; 7) 29.7; 8) 38.2; 9) 39.4

pixel test. The variance between adjacent pixels along the scan line in the difference image is compared to a threshold value to detect a cloud edge. The threshold value is fixed for all pixels at 7.0 K and was subjectively determined from viewing a large number of difference image fields. This test is most successful in identifying the edges of clouds during the day. The second step in the BCT method attempts to fill-in between the cloud edges by analyzing the one-dimensional spatial variability of the pixels. The difference between two adjacent pixels is calculated. For a cloud to be detected, this calculated difference value must be less than the cloudy threshold value if the preceding image location was cloudy, or it must be either less than the negative of the clear threshold value or greater than two-thirds of the clear threshold if the preceding image location was clear. In this way the spatial variability in the difference image corresponding to a cloud free surface versus a cloud is considered. Threshold values of 3.0 and 0.0 K, for the clear and cloudy regions, respectively, were derived in a similar fashion as for the first spatial test. The threshold values for these first two tests can be adjusted (tuned) to tweak algorithm performance for particular applications and regions. These tests only detect a small percentage of clouds. However, these clouds are usually on the edge of a larger cloud field, and are diffuse or only partially contaminate a pixel in the image, and are not easily detected by other means.

The third step in the BCT method is used to detect clouds in regions where the first two steps do not detect clouds. It utilizes a positive and negative difference image threshold in a robust 11.0 – 3.9 micrometer difference image test. This test is the most important cloud determination component in the BCT technique and is used to detect the majority of the clouds in a satellite image. The positive and negative difference image composites derived for each observation time (described below) represent the smallest positive and negative observed difference image values, respectively, for the preceding twenty day period. This test compares the current difference image value to these composite images. A pixel is deemed cloudy if the difference image value is smaller than the negative composite value or if the difference image value, is larger than the positive composite value. The fourth and final test in the BCT cloud detection method involves using the longwave 11.0 micrometer channel information. This infrared threshold test uses a twenty day composite of the second warmest thermal infrared channel values at each pixel location. This product is essentially a "warm" cloud free thermal infrared image. A pixel in the observed infrared image is deemed cloudy if its infrared temperature is colder than the composite infrared threshold value corresponding to its location and time period.

In the bispectral composite threshold technique, the 11.0 and 3.9 micrometer channels are used to produce a difference image (11.0-3.9 micrometer) used in the cloud-free generation of the composites. Both positive differences, which mainly occur at low sun angles and at night, and negative differences that occur at all times, are preserved in the difference image. From this difference image information, two composite images are created for each observation time, which represent the smallest negative (values closest to zero) and the smallest positive difference image values, from the preceding twenty day period. The premise here is that difference image values close to zero have the highest probability of representing cloud-free pixels. These composite images serve to provide spatially and temporally varying thresholds for the BCT method. An additional twenty day composite image is also generated for each time using the warmest longwave (11.0 micrometer) brightness temperature for each pixel from the previous twenty day period. These warm

11.0 micrometer brightness temperature composite images are assumed to represent warm cloud-free thermal images, one for each observation time of day.

The twenty day composite images (positive and negative difference images and the warm 11.0 micrometer images for each hour) used by the BCT algorithm, represent the innovative aspect of the method and provide both spatially and temporally varying clear-sky threshold values for comparison to the observed data. By producing these composites, the BCT approach uses a different threshold value from one pixel to the next and therefore the pixel location, underlying terrain features, present sun angle and other surface conditions (e.g. snow cover) are all implicitly taken into account. These composites capture a large variation in the 11.0 – 3.9 micrometer differences across the region which range from close to zero to above -15 K. The smaller differences (closest to zero) generally occur over water, and the larger values are over land. Rivers can be distinguished from the surrounding land in the composite imagery. The warm infrared brightness temperature composite better represents the thermal structure of the surface than a single threshold value in the absence of clouds. These infrared composite images (from the preceding twenty days) contrast the cold surface of the ocean with warmer land regions during the warm season. The opposite contrast occurs in the cold season. Temperature variations across the domain are considerable (270 - 310 K) and using the composite allows for a spatially representative surface temperature value to be used in the detection approach.

3. Validation of cloud detection techniques

The performance of automated cloud detection algorithms is usually evaluated by comparison to other satellite imagery. However, Schreiner et al. (2001) and Schreiner et al. (1993) used ground based observations from Automated Surface Observing Station (ASOS) and aircraft pilot reports (PIREP) to construct a validation data set. Although limited in coverage and cloud altitude, the use of these data in their evaluation showed that the NESDIS method correctly detected clouds 71% of the time in a 14 month comparison over the continental U.S. region. In a more limited study of a dataset from March 2000, Hawkinson et al. (2001) reported that the method determined the correct sky conditions (cloudy and clear) 75% of the time.

A few studies have used the subjective determination of the presence of clouds to generate a "truth" data set for the quantitative evaluation of the accuracy of a particular scheme. Saunders and Kriebel (1988) determined the presence of a cloud over a region by visually identifying clouds in the satellite imagery for 250 points over a limited region over Europe. They used this data set to validate the performance of their algorithm on AVHRR data. For a day-time scene from April 14, 1985, the algorithm had a hit rate for cloud detection of 98% but with a false alarm rate of 34% and a skill score of around 65%. These statistics indicate that the application of the algorithm to this data produced a very conservative cloud mask for this AVHRR data set. Merchant et al. (2005) developed a quasi-objective "tailored" mask, using a "widget" in the Interactive Data Language (IDL), which was used as "truth" data to validate their Bayesian approach. Using a 99%clear sky probability as a threshold for a binary mask, they reported a hit rate of 97.2%, and a false alarm rate of 23.7% with a skill score of 73.5 for 43 night-time cases detecting clouds over the western Pacific. This approach was a significant improvement over the Saunders and Kriebel (1988) method which had a hit rate of 96.3%, a false alarm rate of 32.3% with a skill score of 64.1% for the same data.

Jedlovec et al. (2008) took a subjective approach to generating a "truth" data set for validation of the bispectral composite threshold approach to cloud detection with GOES imagery. In their validation work, satellite meteorologists manually determined sky conditions for 30 points per time period, per day, during four different months throughout 2004-2005 using GOES visible and infrared imagery. Subjectively determined sky conditions varied by just a few percent between meteorologists and the individual "truth" values were averaged between the meteorologists for each point. Their combined results for night and day applications of the bispectral composite threshold technique applied to GOES data in 4 different seasons produced a hit rate of 84.5%, with a false alarm rate of only 8.1%, with a skill score of 76.4% for 16,675 validation data points. These combined results indicated good performance of the algorithm with only minimal over determination of clouds in the comprehensive data set.

It is difficult to inter-compare the performance of various cloud detection techniques because they are often used in different settings. Each cloud detection technique may use different satellite sensors, focus on different geographical regions or even times of the day. To better understand an algorithm's performance relative to another, the algorithms need to be applied to the same data sets allowing for a direct comparison of the resulting cloud masks. The example below presents such a comparison for August 2004 over the eastern two-thirds of the continential U. S. The two algorithms used were the the NASA EOS science team MODIS cloud mask algorithm and the bispectral composite threshold technique (BCT) adapted for MODIS data (Haines et al. 2005). The MODIS collection 5 cloud algorithm (EOS-5; Frey et al. 2008) is similar to that described above with improvements in performance at night and in polar regions by the inclusion of several additional tests. The MODIS collection 5 data set was obtained from the EOS Data Active Archive Center (DAAC). The application of the BCT technique to MODIS data uses five tests applied to the shortwave and longwave infrared window channels and a similar compositing approach. The algorithms were applied to all Terra MODIS passes over the eastern two-thirds of the United States (both day and night) during this case study period. An example of the cloud masks produced by these algorithms is shown in Figure 4 for a nighttime pass on August 5, 2004. Note that the EOS-5 algorithm assigns confidence levels (i.e., cloudy, possibly cloudy, probably clear and clear) to each pixel in the cloud mask with corresponding grey shades in Figure 4 (from white to dark grey, to light grey to transparent, respectively). An appropriate land type color is assigned to clear regions in both products for display. A qualitative comparison of these cloud mask products to the corresponding infrared window channel image indicates that both cloud masks do quite well in capturing the gross cloud features on this day and time. They both capture the obvious cold clouds observed in the infrared image, but some differences exist in cloud detection for low (warm) clouds over the ocean and land regions.

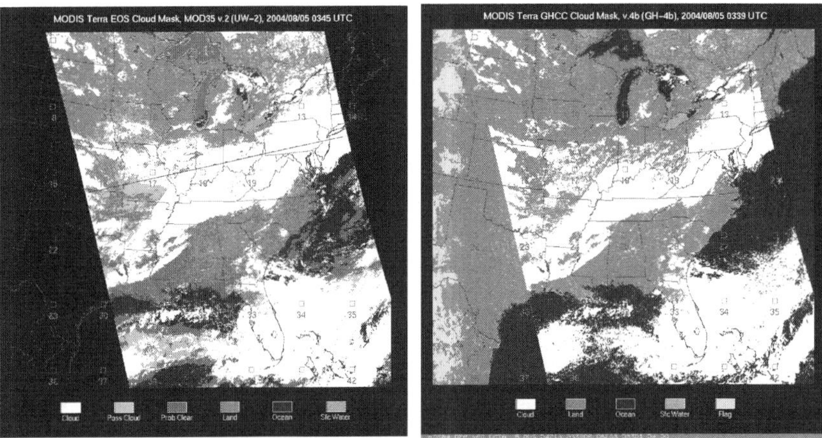

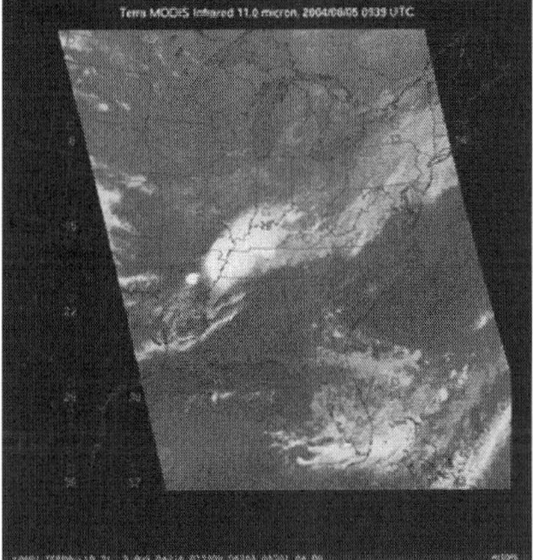

Fig 4. MODIS cloud masks and infrared imagery for August 5, 2004 for the EOS science team collection 5 cloud detection algorithm (upper left) and the bispectral composite threshold technique (upper right). The corresponding infrared MODIS image is presented in the lower panel. Each image is overlaid with grid boxes used in the validation scheme.

To better quantify the performance of these algorithms in detecting clouds over this region, the determination of the presence of clouds by each algorithm was compared to a set of "truth" data subjectively obtained by satellite meteorologists viewing high resolution visible and infrared imagery from MODIS corresponding to the cloud mask times. A fixed set of 42 comparison points were strategically selected over the region as shown on the images in Figure 4. The locations were selected based on a fixed grid with adjustments to include a variety of ocean and land regions (including coastal areas) with a variety of cloud distributions. This validation approach is similar to that used by McNally and Watts (2003) and Alliss et al. (2000)

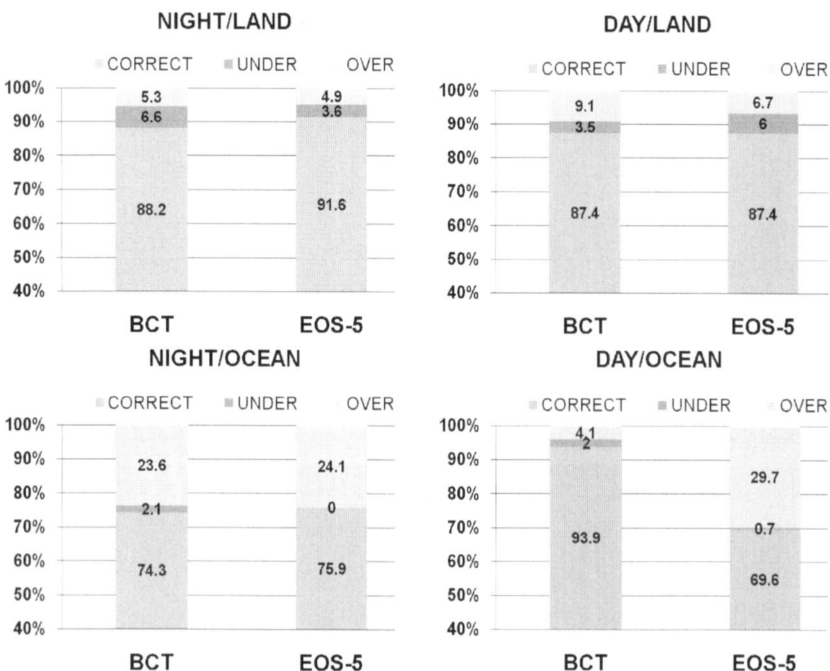

Fig 5. Performance statistics of the bispectral composite threshold (BCT) and the EOS science team collection 5 (EOS-5) MODIS cloud mask algorithm for August 2004 over the eastern half of the United States for night over ocean (upper left), night over land (lower left), day over ocean (upper right), and day over land (lower right). The numbers correspond to the accuracy (%) of the cloud detection algorithm – either correctly determining the correct sky conditions (clear or cloudy – blue bar), or the over- (yellow bar) or under- (pink bar) determination of the cloud conditions.

and nearly identical to that of Jedlovec et al. (2008). For the EOS-5 algorithm, only points having a probability determination of "cloudy" were considered to contain clouds. Other probablistic determinations (possibly cloudy or probably clear) were considered clear. The statistical results of this comparison were categorized by day and night and land and ocean locations and are presented in Figure 5. Both algorithms perform nearly as well at night, correctly detecting clear or cloudy conditons about 75% of the time over the ocean and around 90% of the time over land. Over the ocean, both schemes over determined the presence of clouds by nearly 25%. Over land, the EOS-5 algorithm correctly determined clear or cloud conditions 91.6 percent of the time, with the remaining misses split between over and under determination. The BCT scheme did nearly as well with 88.2% of the points correct.

The performance of the two algorithms varys significantly during the day over the ocean, with significant over determination of clouds in the EOS-5 scheme contributing to its poor performance. During the day over the ocean, the BCT scheme does extremely well, correctly determining 93.9% of the points. The performance of the two algorithms are similar over land during the day with 87.4% of the points being determined correctly as being either cloudy or clear. The BCT cloud mask has a slightly greater over determination value than the EOS-5 approach for the points that were incorrectly identified. One should note that the use of only cloud points having the highest confidence were considered cloudy in with the EOS-5 data. If a more conservative cloud choice was used (by including "possibly cloudy" points), the performance statistics for the EOS-5 algorithm would have been worse as a greater number of points would be considered cloudy increasing the over determination of clouds at all times and regions (not shown). This qualitative comparison and statistical analysis indicates that with the proper determination of thresholds, a simple two channel cloud detection technique, such as the bispectral composite threshold approach (Jedlovec et al. 2008), can detect clouds in a variety of situations quite well.

4. Summary

Many different approaches have been used to automatically detect clouds in satellite imagery. Most approaches are deterministic and provide a binary cloud – no cloud product used in a variety of applications. Some of these applications require the identification of cloudy pixels for cloud parameter retrieval, while others require only an ability to mask out clouds for the retrieval of surface or atmospheric parameters in the absence of clouds. A few approaches estimate a probability of the presence of a cloud at each point in an image. These probabilities allow a user to select cloud information based on the tolerance of the application to uncertainty in the estimate. Many automated cloud detection techniques develop sophisticated tests using a combination of visible and infrared channels to determine the presence of clouds in both day and night imagery. Visible channels are quite effective in detecting clouds during the day, as long as test thresholds properly account for variations in surface features and atmospheric scattering. Cloud detection at night is more challenging, since only courser resolution infrared measurements are available. A few schemes use just two infrared channels for day and night cloud detection. The most influential factor in the success of a particular technique is the determination of the thresholds for each cloud test. The techniques which perform the best usually have thresholds that are varied based on the geographic region, time of year, time of day and solar angle.

5. References

Ackerman, S. A., et al., (1998). Discriminating clear-sky from clouds with MODIS. J. Geophys. Res. Vol. 103, pp. 32,141-32,158.

Alliss, R. J., et al. (2000). The development of cloud retrieval algorithms applied to GOES digital data. 10th Conference on Satellite Meteorology and Oceanography, Amer. Meteor. Soc., Long Beach, 2000, pp. 330-333.

Coakley, J. A., and F. P. Bretherton, (1982). Cloud cover from high-resolution scanner data: Detecting and allowing for partially filled fields of view, J. Geophys. Res., Vol. 87, pp. 4917-4932.

Frey, R. A., S. A. Ackerman, Y. Liu, K. I. Strabala, H. Zhang, J. R. Key, and X. Wang, (2008). Cloud detection with MODIS. Part I. Improvements in the MODIS cloud mask for collection 5. J. Tech., 25, 1057-1072.

to the user, though the cost of priority commissioned data is significantly greater than that of archived imagery (RADARSAT-2, TerraSAR-X and Cosmo-Skymed).

4. Remote sensing applications

4.1 Reduction

Disasters are social constructs in that social drivers such as migration (forced and voluntary), conflict, modification of natural buffer systems, reliance on shrinking resources, private property rights, urban intensification, artificial protection structures, and economic and political vulnerability are all contributors to people living in hazardous locations or at levels of vulnerability that make a disaster more likely. Remote sensing technology can assist with addressing some of these "disaster drivers", through providing the data required to assist land use planners, emergency managers, and others tasked with disaster management. Reduction of risk, and therefore reduction in the probability of a disaster occurring, is an important part of the disaster management cycle. Remote sensing can be applied in disaster reduction initiatives through identification and understanding of hazards (Table 1). This knowledge is then applied to mitigation activities such as land use planning, engineering structures, building codes and hazard consequences modelling to determine methods for reducing vulnerability (Gregg & Houghton 2006). Note that the sensor examples given in Table 1 and subsequent tables are indicative of current or potential instrument use. Many alternative sensors with similar characteristics could also be used.

Understanding of hazards, their magnitude, frequency, duration, location, range and manifestation (e.g. heavy rainfall, tephra, strong winds) has long been accepted as essential to disaster management. Although it is primarily social factors that amplify a hazard event into a disaster (Quarantelli 1985, Wisner 2004), improved knowledge of hazards and their potential consequences is essential for decision making about modifying hazard characteristics, or modifying vulnerability of people and assets. Remote sensing can be used directly for hazard identification (e.g. flood plain modelling, slope stability and landslide susceptibility), but can also be used to derive hazard-independent information that can be used for disaster reduction (e.g. baseline building, infrastructure, and topographic mapping). An excellent example of the use of remote sensing for hazard identification is provided with LiDAR mapping of active fault location (Begg & Mouslpoulou 2009 in press). Traditionally fault location is conducted using stereo aerial photography interpretation followed by intensive field survey. However the horizontal and vertical resolution provided by airborne LiDAR imagery provides the capability for identifying fault traces and extracting elevation offsets with digital data in an objective manner. The identification of many previously unknown faults in northern New Zealand is shown in Figure 2.

Type of information	Data required	Sensor example	Application example
Location of fault traces and rupture zones	High resolution DEM	Airborne LiDAR, SAR	Use for land use planning around active faults to reduce risk from future development in fault hazard locations
Fault displacement	Interferometric SAR	ERS1/2, ENVISAT ASAR, ALOS PALSAR	Knowledge of fault displacement rates are used in numerical models in order to forecast the magnitude of possible earthquakes
Flood plain mapping	DEM	Airborne LiDAR, ERS1/2, ENVISAT ASAR, ALOS PALSAR	Identification of flood plains can help inform changes in land use, and identify areas developing protective measures (e.g. stopbanks)
Land cover / land use	Optical and polarimetric SAR	SPOT, ASTER RADARSAT-2	Used for catchment management planning to reduce flood and landslide risk
Vegetation change	Consistent time series of data	SPOT, ASTER RADARSAT-2	Determine drought zones, inform fire hazard mapping
Determining lahar and lava flow paths	DEM, high resolution optical imagery	SAR, Airborne LiDAR, SPOT, AVNIR-2, ASTER	Hazard zonation, public awareness, determining location of safety shelters
Locating potential and actual unstable slopes	DEM, Interferometric SAR, high resolution stereo optical imagery	Airborne LiDAR, ERS1/2, ENVISAT ASAR, ALOS PALSAR, aerial photography	Hazard mapping for infrastructure planning
Baseline infrastructure maps	Very high resolution optical imagery	Aerial photography, Quickbird, Ikonos, Worldview	Assist with hazard mapping to identify key infrastructure at risk – the risk can then be addressed through mitigation or built in redundancy. Can also be used for later damage assessment post-disaster
Baseline topographic data	Moderate to high resolution optical imagery	SPOT, AVNIR-2, Aerial photography, Quickbird, Ikonos, Worldview	Hazard modelling

Table 1. Examples of information and data requirements during the reduction phase

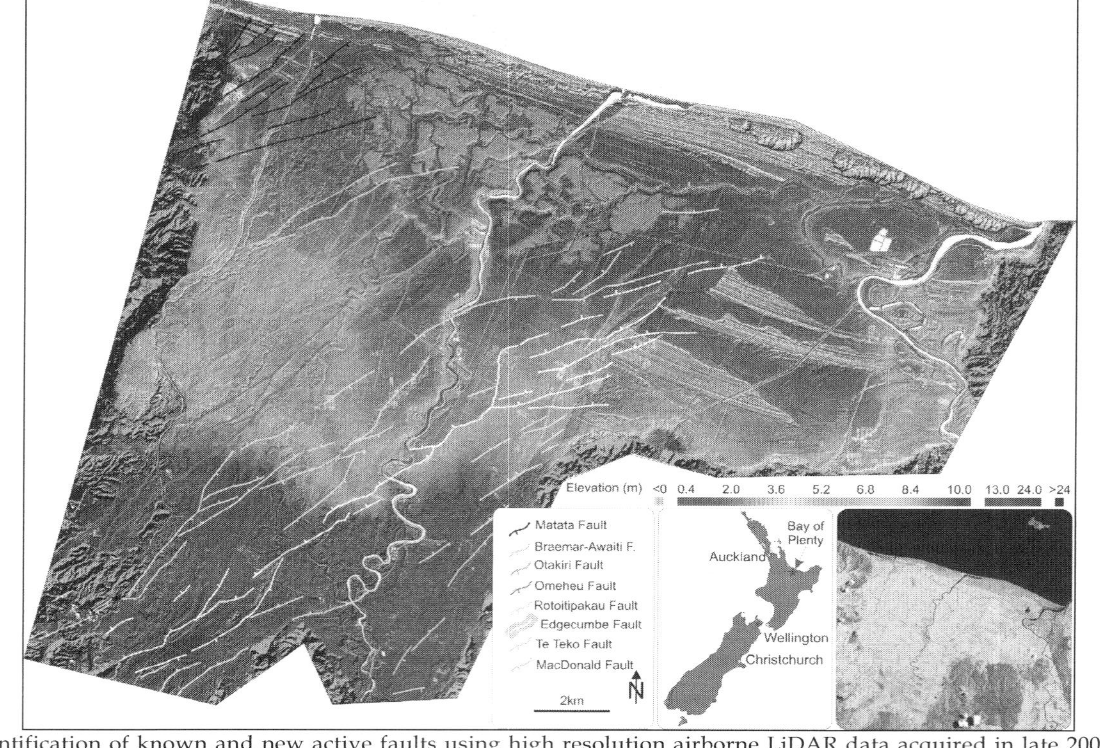

Fig. 2. Identification of known and new active faults using high resolution airborne LiDAR data acquired in late 2006 (Begg & Mouslpoulou 2009 in press). Landsat ETM+ false colour composite (5,4,2) acquired in 2001 is inset for a contextual overview of the site. Of the active fault traces shown here, approximately 85% were unknown before undertaking this study. Also of note also is the discovery of a large inland area that is below sea level (elevation <0m) and is a potentially hazardous region for tsunami related inundation

Remotely sensed data acquisitions can be used to inform land use planning, a key tool that authorities and communities employ to avoid or mitigate hazard risk (Burby 1998). By identifying the location and characteristics of hazards, land use planning methods can be applied to address the risk these hazards pose. Planning methods include mapping hazard zones (location and range of hazard impact) and identifying the probability of occurrence. Hazard maps are applied to developed and green field (undeveloped) land and options for risk treatment determined. Treatment options can include measures such as setback zones (no development within the hazard zone, e.g. proximal to active faults or within coastal erosion or inundation zones), or special building codes (e.g. minimum floor heights above base flood level) can be introduced to reduce the risk to assets and people (Godschalk et al. 1998). Understanding of hazard information is one of a number of critical factors influencing individual and group decision making for risk management (Paton & Johnston 2001). Where hazard information is readily available to the public in a variety of forms, including maps, there is a greater likelihood of public support for risk reduction initiatives introduced through land use planning (Burby 2001).

Other methods for land use planning based on remote sensing data include identifying changes in land use on flood plains to assist with flood hazard modelling. In the city of London, Canada, Landsat images taken over a 25 year period have been used to determine the spread of urban development (Nirupama & Simonovic 2007). The consequent increase in impermeable surface cover facilitated more rapid runoff and less natural absorption of rainfall. When compared with flood hydrographs, the rate of land use change correlates with smaller rainfall events producing flooding. The benefits to future land use planning are that it can be determined how land use changes affect the flood hazard risk, and this will guide future development in a way that mitigates the effects of continued urban sprawl.

Collecting asset data via high resolution remote sensing allows for identification of infrastructure and buildings in hazardous locations, which can then be targeted for strengthening or re-location. Asset data is also essential for hazard consequence modelling, whereby hazard data is combined with asset data and fragility (vulnerability) information to determine potential losses. Building fragility to hazards is based on such factors as construction materials (earthquake, volcanic ash fall, tsunami), engineering design (tsunami, landslide, earthquake), building height (wind), floor areas (earthquake), proximity of other structures and vegetation (fire) and roof pitch angle (ash fall, snow), and floor height (flood, tsunami). Remote sensing methods for collecting building and infrastructure data require high to very high resolution satellite or airborne imagery and is often completed using manual digitizing or more recently, segmentation and object oriented classification. Optical imagery is often complemented by LiDAR data, which can not only aid in detecting building edges, but is also used for calculating building heights. Incorporation of remotely sensed data into a GIS is vital during this phase for recording spatial attributes and combining with other data sets.

Remote sensing technology can also be applied to measure the success of risk reduction initiatives. A common method for addressing flood risk is the construction of stopbanks to contain flood waters for an event of a given magnitude. Aerial reconnaissance during major flooding events can identify whether stopbanks are performing to design standard and identify areas of weakness, overtopping or failure. Monitoring of non-structural risk reduction initiates is also possible. To address coastal hazard erosion and inundation risk, many communities choose non-structural options such as beach renourishment and dune restoration. In Florida, airborne LiDAR captured over time has been applied to measure coastal erosion from hazards, alongside the success of non-structural beach restoration methods through determining changes to beach morphology (Shrestha et al. 2005). Another example of measuring the effects of risk reduction initiatives is analysing post-disaster images of rainfall induced landslides on land under different vegetation covers for large events. From analysis of aerial photographs (oblique and vertical) of an event in 2004 which impacted the lower North Island of New Zealand, it was determined that vegetation cover played an important role in reducing loss of productive soil, and reducing landslide hazard to assets (Hancox & Wright 2005).

4.2 Readiness

Readiness planning and activities are undertaken in the realisation that residual risk from hazards has the potential to create emergencies, and in some cases, disasters for affected populations. Readiness is the identification and development of necessary systems, skills and resources before hazard events occur. The desired outcome of readiness planning and activities is that response to hazards is more coordinated and efficient, communities experience less trauma, and recovery times are reduced (Quarantelli 1997). Examples of readiness activities include public education, preparedness activities, trainng and exercising, evacuation planning, developing hazard monitoring and public alerting systems, and putting in place state, national and international plans and agreements for assistance and aid. Readiness activities and planning are undertaken at a number of levels to increase resilience and response capability for individuals, households, organisations, and states or nations. The provision of good hazard and asset information to assist these activities is essential and examples where remote sensing can assist this phase are given in Table 2. It is important in this phase to prepare an archive of and gain familiarity with the most up to date spatial information including (but not limited to) imagery, DEMs, and vector data. This information is required to assist with damage assessment during the response and recovery phases.

At the individual and household level there are identified factors that contribute to whether people will take actions to prepare for disasters. Personalisation of risk is essential (Barnes 2002, Slovic et al. 2000), e.g. "Will it affect me?", "Do I need to do something about it", and "What can I do about it?". Other factors include belief in the benefits of hazard mitigation (outcome expectancy) and their belief that what they personally can do will make a difference (reduce negative outcome expectancy) (Paton 2006). At a community level, participation in community affairs and projects, and individual's ability to influence what happens in their community (empowerment) and the level of trust they have in different organisations (trust) have also be shown to be key predicators of resilience. Therefore, communication of risk in a meaningful way is an essential part of preparedness planning. Remotely sensed data such as LiDAR are used to produce high resolution hazard and risk maps, which are used by authorities to communicate information about location and range of hazards to their communities. If individuals believe that a hazard is likely to affect them detrimentally within an understandable and pertinent timeframe, they are more likely to take actions to prepare. These actions might include having emergency supplies in the home, an action plan for evacuation and emergency contact with other household members, first aid training or training as a civil defence volunteer. The principle of risk perception aiding preparedness applies to both static and dynamic hazards, e.g. fault trace or flood plain mapping vs. cyclone or bushfire progression. Remotely sensed images showing the progression of a bushfire front or the track of a cyclone are commonly used by the media to inform the public of where hazards are occurring and where they are likely to impact as they evolve. As community resilience research has shown, awareness of hazards is not the only factor in triggering actual preparedness actions; however it is one significant driver (Paton 2006, Paton & Johnston 2001, Ronan & Johnston 2005).

Type of information	Data required	Sensor example	Application example
Severe weather warnings	RADAR, broadscale visible and infra red imagery	GOES, NOAA, Meteosat	Provide valuable advanced warning of severe events to the public and emergency planners via meteorologists
Movement and ground deformation	InSAR and PS-InSAR	ERS-1/2, ENVISAT ASAR, ALOS PALSAR	Rate of movement for slow moving landslides. Often acceleration of deformation rates means that a large event is about to follow. Early detection of deformation in volcanic regions is used for forecasting of possible eruptions
Soil moisture	Long wavelength SAR	SMAP	Water shortage leading to drought and agricultural productivity decline, ability of soils to retain water to indicate flood and landslide potential
Ground temperature variability	Thermal imagery, or SWIR in the case of very hot features	ASTER, MODIS, AVHRR	Monitoring heating and cooling cycles of volcanoes to understand pre-eruptive characteristics for forecasting purposes
Coastal and bathymetric mapping	SONAR, Laser depth ranging	LADS, Topex Poseidon / Jason	Tsunami hazard modelling
Display and advertisement of potential hazards	Moderate to high resolution optical imagery, often overlaying a DEM	Aerial photography, Quickbird, Ikonos - usually using black and white or true colour composites for ease of understanding	For use in public education about hazards and risks to foster greater readiness of individuals, households and organisations Use in civil defence emergency management exercises to provide realistic scenarios that will assist with staff professional development and planning
Detecting sea temperature or atmospheric pressure change in cyclone/hurricane/typhoon generating latitudes	Broad scale thermal imagery, geostationary	MODIS, GOES, AVHRR	Advance warning of severe weather approaching to commence

Table 2. Examples of information and data requirements during the readiness phase

At the institutional level, a strong focus is placed on the development of plans and relationships. A primary way to test the effectiveness of these preparedness plans and relationship functions is through civil defence emergency management exercises. In order

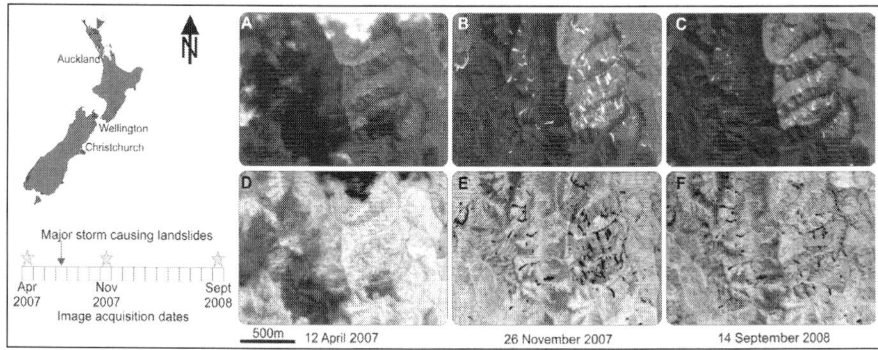

Fig. 8. Recovery of vegetation after a widespread landsliding event in northern New Zealand, July 2007. (a) SPOT-5 CIR obtained before the event; (b) SPOT-5 CIR obtained shortly after the event; (c) ALOS AVNIR-2 CIR imagery obtained one year later; and (d-e) NDVI images of the aforementioned data.

Analysis of time series imagery could also help to monitor the effectiveness of different recovery strategies. By extracting recovery rates from data acquired at appropriate time intervals, this assessment could help guide recovery plans for future events of a similar nature. This would also help identify areas of residual risk that require ongoing monitoring until the physical recovery process completed.

5. Conclusions

Remote sensing can be used to inform many aspects of the disaster management cycle. An exhaustive coverage of all potential applications would be impossible in a single book chapter, however we have shown several good examples from which inspiration can be sought for future use. It is important to consider all aspects of disaster management, rather than focussing on emergency response. By incorporating remote sensing into reduction and readiness activities, this can also educate both emergency management staff and the community about this type of information so that they are familiar with its use under a response and inherently pressured situation.

The key elements to facilitate the usefulness of remote sensing data in support of the disaster management community are being able to provide the appropriate information in a spectrally, temporally, and spatially relevant context. Additionally, one must be aware of the information requirements of that disaster management community, and tailor the remote sensing information to meet those needs. That can only come through close collaborations between the disaster management community and the remote sensing / geospatial community.

6. Acknowledgements

All SPOT 5 imagery used in this chapter is © CNES. This manuscript incorporates data which is © Japan Aerospace Exploration Agency ("JAXA") (2008). The data has been used in this manuscript with the permission of JAXA and the Commonwealth of Australia

(Geoscience Australia) ("the Commonwealth"). JAXA and the Commonwealth have not evaluated the data as altered and incorporated within the manuscript, and therefore give no warranty regarding its accuracy, completeness, currency or suitability for any particular purpose. Environment Bay of Plenty provided the licence to use the LiDAR data. Thank you to Andy Gray for assistance with graphics and to Phil Glassey and David Johnston for chapter review.

7. References

Ambrosia V. G. & Wegener S. S. (2009). Unmanned airborne platforms for disaster remote sensing support. In: *Advances in Geoscience and Remote Sensing*, 317-346 - this book, IN-Tech Publishing, Vienna

Barnes P. (2002). Approaches to community safety: risk perception and social meaning. *Australian Journal of Emergency Management*, 17, 1, 15 - 23

Becker J., Saunders W., Hopkins L., Wright K., & Kerr J. (2008). Pre-event recovery planning for use in New Zealand: An updated methodology. GNS Science Report 2008/11. GNS Science

Begg J. G. & Mouslpoulou V. (2009 in press). Analysis of late Holocene faulting within an active rift using lidar, Taupo Rift, New Zealand. *Journal of Volcanology and Geothermal Research*

Board on Natural Disasters (1999). Mitigation emerges as major strategy for reducing losses caused by natural disasters. *Science*, 284, 1943-1947

Burby R. J. (2001). Involving citizens in hazard mitigation planning: Making the right choices. *Australian Journal of Emergency Management*, 16, 45 - 51

Burby R. J. (1998). Natural hazards and land use: An introduction. In: *Cooperating With Nature: Confronting Natural Hazards with Land Use Planning for Sustainable Communities*, Burby RJ (Ed.), 1 - 26, Joseph Henry Press, Washington D.C.

Carrivick J. L., Manville V., & Cronin S. (2009). A fluid dynamics approach to modelling the 18th March 2007 lahar at Mt. Ruapehu, New Zealand. *Bulletin of Volcanology*, 71, 153-169

Cloude S. R. & Papathanassiou K. P. (1988). Polarimetric SAR Interferometry. *IEEE Transactions on Geoscience and Remote Sensing*, GE36, 1551 - 1565

Cutter S. L. & Finch C. (2008). Temporal and spatial changes in social vulnerability to natural hazards. *Proceedings of the National Academy of Sciences of the United States of America*, 105, 7, 2301-2306, 00278424

Davies A. G., Chien S., Baker V., Doggett T., Dohm J., Greeley R., Ip F., Castano R., Cichy B., Rabideau G., Tran D., & Sherwood R. (2006). Monitoring active volcanism with the Autonomous Sciencecraft Experiment on EO-1. *Remote Sensing of Environment*, 101, 4, 427

Donner W. & Rodriguez H. (2008). Population Composition, Migration anti Inequality: The Influence of Demographic Changes on Disaster Risk and Vulnerability. *Social Forces*, 87, 2, 1089-1114, 00377732

Elachi C. (1987). *Introduction to the Physics and Techniques of Remote Sensing*, John Wiley & Sons, New York

Franceschetti G. & Lanari R. (1999). *Synthetic Aperture Radar Processing*, CRC Press, New York

Galley I., Leonard G. S., Johnston D., Balm R., & Paton D. (2004). The Ruapehu lahar emergency response plan development process: An analysis. *Australian Journal of Disaster and Trauma Studies*, 1

Giglio L., Csiszar I., Restás Á., Morisette J. T., Schroeder W., Morton D., & Justice C. O. (2008). Active fire detection and characterization with the advanced spaceborne thermal emission and reflection radiometer (ASTER). *Remote Sensing of Environment*, 112, 6, 3055

Godschalk D. R., Kaiser E. J., & Berke P. R. (1998). Integrating hazard mitigation and local land use planning. In: *Cooperating With Nature: Confronting Natural Hazards with Land Use Planning for Sustainable Communities*, Burby RJ (Ed.), pp 85-118, Joseph Henry Press, Washington D.C.

Granger K. (2000). An information infrastructure for disaster management in Pacific Island Countries. *Australian Journal of Emergency Management*, 15, 1, 20 - 32

Gregg C. E. & Houghton B. F. (2006). Natural Hazards. In: *Disaster Resilience: An Integrated Approach*, Paton D, Johnston D (Eds.), pp 19-39, Charles C. Thomas Publishers Ltd, Springfield

Hancox G. T. H. & Wright K. (2005). Analysis of landsliding caused by the 15-17 February 2004 rainstorm in the Wanganui-Manawatu hill country, southern North Island, New Zealand GNS Science report. Institute of Geological and Nuclear Sciences (64)

Hanssen R. F. (2001). *Radar interferometry: data interpretation and error analysis*, Kluwer Academic Publishers

Henson R. (2008). *Satellite observations to benefit science and society: recommended missions for the next decade*, National Research Council, National Academies Press, National Acadamy of Science, 0-309-10904-3

Hill A. A., Keys-Mathews L. D., Adams B. J., & Podolsky D. (2006). Remote sensing and recovery: A case study on the Gulf Coasts of the United States, *Proceedings of the Fourth International Workshop on Remote Sensing for Post-Disaster Response*, Cambridge, UK

Ito A. (2005). Issues in the implementation of the International Charter on Space and Major Disasters. *Space Policy*, 21, 2, 141

Joyce K. E., Belliss S., Samsonov S., McNeill S., & Glassey P. J. (2009a). A review of the status of satellite remote sensing image processing techniques for mapping natural hazards and disasters. *Progress in Physical Geography*, 33, 2, 1 - 25

Joyce K. E., Dellow G. D., & Glassey P. J. (2008a). Assessing image processing techniques for mapping landslides, *Proceedings of the IEEE International Geoscience and Remote Sensing Symposium*, Boston, MA

Joyce K. E., Samonsov S., & Jolly G. (2008b). Satellite remote sensing of volcanic activity in New Zealand, *Proceedings of the 2nd Workshop on the use of remote sensing techniques for monitoring volcanoes and seismogenic areas*, Naples, Italy

Joyce K. E., Samsonov S., Manville V., Jongens R., Graettinger A., & Cronin S. (2009b). Remote sensing data types and techniques for lahar path detection: A case study at Mt Ruapehu, New Zealand. *Remote Sensing of Environment*, 113, 1778 - 1786

Kaku K., Held A. A., Fukui H., & Arakida M. (2006). Sentinel Asia initiative for disaster management support in the Asia-Pacific region, *Proceedings of the SPIE*,

Laska S. & Morrow B. H. (2006/7). Social vulnerabilities and Hurricane Katrina: An unnatural disaster in New Orleans. *Marine Technology Society Journal*, 40, 4, 16-26

Leonard G. S., Johnston D. M., & Paton D. (2005). Developing effective lahar warning systems for Ruapehu. *Planning Quarterly*, 158, 6-9

Massonnet D. & Feigl K. (1998). Radar interferometry and its application to changes in the Earth surface. *Reviews of Geophysics*, 36, 4, 441 - 500

Mileti D. S. (1999). *Disasters By Design: A reassessment of Natural hazards in the United States*, Joseph Henry Press, Washington D.C.

Nirupama N. & Simonovic S. (2007). Increase of Flood Risk due to Urbanisation; A Canadian Example. *Natural Hazards*, 40, 25 - 41

Norman S. (2004). Focus on Recovery: A Holistic Framework for Recovery. In: *NZ Recovery Symposium Proceedings 12-13 July, 2004.*, Norman S (Ed.), 31-46, Ministry of Civil Defence and Emergency Management, Wellington

Oppenheimer C. (1993). Infrared surveillance of crater lakes using satellite data. *Journal of Volcanology and Geothermal Research*, 55, 117 - 128

Oppenheimer C. (1997). Remote sensing of the colour and temperature of volcanic lakes. *International Journal of Remote Sensing*, 18, 1, 5-37

Paton D. (2006). Disaster resilience: Integrating individual, community, institutional and environmental perspectives. In: *Disaster resilience: An integrated approach*, Paton D, Johnston D (Eds.), 320, Charles C Thomas Publisher, 0-398-07663-4, Springfield

Paton D. & Johnston D. (2001). Disaster and communities: vulnerability, resilience and preparedness. *Disaster Prevention and Management*, 10, 270-277

Pertrow T., Thieken A., Kreibich H., Merz B., & Bahlburg C. H. (2006). Improvements on flood alleviation in Germany: Lessons learned form the Elbe flood in August 2002. *Environmental Management*, 38, 717-732

Petersen T., Beaven J., Denys P., Denham M., Field B., Francois-Holden C., McCaffrey R., Palmer N., Reyners M., Ristau T., Samonsov S., & Team T. G. (2009 in review). The Mw 6.7 George Sound earthquake of October 16, 2007: response and preliminary results. *Bulletin of New Zealand society for earthquake engineering*

Phinn S. R. (1998). A framework for selecting appropriate remotely sensed data dimensions for environmental monitoring and management. *International Journal of Remote Sensing*, 19, 17, 3457 - 3463

Phinn S. R., Joyce K. E., Scarth P. F., & Roelfsema C. M. (2006). The role of integrated information acquisition and management in the analysis of coastal ecosystem change. In: *Remote sensing of aquatic coastal ecosystem processes: Science and management applications*, Richardson LL, LeDrew EF (Eds.), 325, Springer-Verlag

Pottier E. & Ferro-Famil L. (2008). Advances in SAR polarimetry applications exploiting polarimetric spaceborne sensors, *Proceedings of the 2008 Radar Conference*, pp. 1 - 6, Rome, Italy

Quarantelli E. L. (1997). Ten criteria for evaluating the management of community disasters. *Disasters*, 21, 1, 39-56

Quarantelli E. L. (1985). What is Disaster? The Need for Clarification in Definition and Conceptualisation in Research. In: *Disasters and Mental Health Selected Contemporary Perspectives*, Sowder B (Ed.), 41-73, Government Printing Office, Washington

Ronan K. R. & Johnston D. M. (2005). *Promoting community resilience in disasters : the role for schools, youth, and families*, Springer, 0387238204, New York

Rosen P., Hensley P., Joughin I., Li F., Madsen S., Rodriguez E., & Goldstein R. (2000). Synthetic aperture radar interferometry. *Proceedings of the IEEE*, 88, 3, 333-382

Samsonov S. V., Tiampo K., Gonzalez P., Manville V., & Jolly G. (2009 in review). Modelling deformation occurring in the city of Auckland, New Zealand mapped by differential synthetic aperture radar. *Journal of Geophysical Research*

Shrestha R. L., Carter W. E., Satori M., Luzum B. J., & Slatton K. C. (2005). Airborne Laser Swath Mapping: Quantifying changes in sandy beaches over time scales of weeks to years. *ISPRS Journal of Photogrammetry & Remote Sensing*, 59, 222 - 232

Slovic P., Fischhoff B., & Lichtenstein S. (2000). Cognitive Processes and Societal Risk Taking. In: *The perception of risk*, Slovic P (Ed.), 32-50, Earthscan Publications Ltd, London

Trunk L. & Bernard A. (2008). Investigating crater lake warming using ASTER thermal imagery: Case studies at Ruapehu, Poás, Kawah Ijen, and Copahué Volcanoes. *Journal of Volcanology and Geothermal Research*, 178, 2, 259 - 270

van Zyl J. J., Zebker H. A., & Elachi C. (1990). Polarimetric SAR applications. In: *Radar Polarimetry for Geoscience Applications*, 315 - 360, Artech House, Nomood

Wisner B. (2004). Assessment of capability and vulnerability. In: *Mapping Vulnerability: Disasters, Development and People*, Bankoff G, Frerks G, Hilhorsdt D (Eds.), 183 - 193, Earthscan, London

Wisner B., Blaikie P., Cannon T., & Davis I. (2004). The challenge of disasters and our approach. In: *At Risk: Natural Hazards, people's vulnerability and disasters*, Wisner B, Blaikie P, Cannon T, Davis I (Eds.), 3-48, Routledge, London & New York

Wright R., Flynn L., Garbeil H., Harris A., & Pilger E. (2002). Automated volcanic eruption detection using MODIS. *Remote Sensing of Environment*, 82, 1, 135

Wright R., Flynn L. P., Garbeil H., Harris A. J. L., & Pilger E. (2004). MODVOLC: near-real-time thermal monitoring of global volcanism. *Journal of Volcanology and Geothermal Research*, 135, 1-2, 29-49

8. Glossary

Note that the satellite sensors listed here and within the text are simply examples of the types of instruments that can be used, rather than being a complete listing of all possibilities.

ALOS AVNIR-2	Japanese Space Agency (JAXA) Advanced Land Observing Satellite Advanced Visible and Near Infrared sensor. Useful for local to regional scale mapping and monitoring
ALOS PALSAR	Japanese Space Agency (JAXA) Advanced Land Observing Satellite L Band SAR satellite. Useful for deformation monitoring in regions of dense vegetation
ASAR	C Band Advanced Synthetic Aperture RADAR
ASTER	Advanced Spaceborne Thermal Emission and Reflection Radiometer on board NASA's Terra satellite. Useful for monitoring volcanic activity
AVHRR	Advanced Very High Resolution Radiometer (NOAA – National Oceanic and Atmospheric Administration). Useful for regional to national scale applications
CIR	Colour Infrared – three band standard display of green, red and near infrared light displayed as blue, green and red respectively

DEM	Digital elevation model
DMC	Disaster Monitoring Constellation. International collaboration between space agencies in Algeria, China, Nigeria, Turkey and the UK for regional scale mapping (optical)
ERS	European Space Agency Satellite with a suite of SAR and optical sensors
ENVISAT	European Space Agency Satellite with a suite of SAR and optical sensors
GOES	Geostationary Operational Environmental Satellites – used for metrological applications
Ikonos	Very high spatial resolution commercial satellite (GeoEye). Useful for local scale mapping and monitoring (e.g. buildings and assets)
InSAR	Interferometric Synthetic Aperture RADAR – technique used for measuring surface deformation
KML	Keyhole Markup Language – native language for Google Earth files
LADS	Laser Airborne Depth Sounder
Landsat ETM+	Enhanced Thematic Mapper plus. Useful for long term regional scale mapping and monitoring, though technical malfunctioning limits data coverage
Landsat TM	Thematic Mapper. Useful for long term regional scale mapping and monitoring
LiDAR	Light Detection and Ranging. Used for creating very high spatial resolution DEMs
Meteosat	European geostationary meteorological satellite
MODIS	Moderate Resolution Imaging Spectrometer. Used for hotspot monitoring of fires and volcanic activity on a regional to continental scale
NASA	National Aeronautics and Space Administration
OMI	Ozone Monitoring Instrument – used for monitoring volcanic gas emissions
POLInSAR	Polarimetric Interferometric Synthetic Aperture RADAR
PS-InSAR	Permanent Scatterers Interferometric Synthetic Aperture RADAR
Quickbird	Very high spatial resolution commercial satellite (Digital Globe). Useful for local scale mapping and monitoring (e.g. buildings and assets)
RADARSAT	Canadian Space Agency C Band SAR satellite
RapidEye	Constellation of five high resolution optical satellites, the combination of which provides a daily revisit capability
SAR	Synthetic Aperture RADAR. Active sensor capable of capturing data through clouds, smoke and haze
SMAP	Soil Moisture Active Passive – scheduled for launch in 2012
SPOT	Satellite Pour l'Observation de la Terre – French Space Agency (CNES – Centre National d'Etudes Spatiales) colour infrared and panchromatic earth observation satellite

SWIR	Short Wave Infrared. Used for volcanic ash and gas monitoring and also vegetation applications
Terra SAR-X	X-band SAR satellite
TIR	Thermal Infrared, used for fire and volcanic activity monitoring
Topex Poseidon	Joint CNES / NASA satellite altimetry mission, used for studying sea level, ocean bathymetry, tides and ocean currents (now succeeded by Jason)
UAS	Unmanned Airborne System
UAV	Unmanned Aerial Vehicle
UV	Ultra Violet – non-visible short wavelength radiation, useful for volcanic gas estimation
Worldview	Very high spatial resolution commercial satellite (Digital Globe). Useful for local scale mapping and monitoring (e.g. buildings and assets). Currently only panchromatic.

16

Prediction of Volumetric Shrinkage in Expansive Soils (Role of Remote Sensing)

Fekerte Arega Yitagesu, Freek van der Meer and Harald van der Werff
International Institute for Geo-information Science and Earth Observation, ITC
The Netherlands

1. Introduction

Life time, performance and environmental compatibility of civil engineering infrastructure largely depend on the quality of geotechnical investigations. Apart from being the basis for much of the costing (in terms of time and money both at the construction and maintenance stages) of engineering works, safety of structures lies on information from a geotechnical survey. Existence of expansive soils in construction sites is a great problem that need due consideration in geotechnical investigations.

Expansive soils are weak and unstable when subjected to moisture content fluctuations either due to seasonal climatic variations (cyclic dry and wet periods) or artificial causes. Their presence is one of the crucial factors that can significantly impact engineering costs for it may cause major deterioration and distresses on lightweight and shallowly founded structures. Primary characteristics of expansive soils are their potential to swelling and shrinkage in response to moisture content increase and decrease respectively. These properties are adverse in civil construction works for they pose a huge damage especially on lightweight infrastructures (small buildings, roads and airport runway pavements, pipelines and sewerage systems etc). An increase in volume (swelling) from expansive soils can exceed the downward pressure exerted from lightweight structures and hence cause deformations and development of cracks. Substantial decrease in volume (shrinkage) on the other hand is responsible for uneven settlement. Expansive soils awe their characteristics to their mineralogical assemblage; that is presence of active clay minerals and their amount. Damages due to volume changes of expansive soils in form of swelling and shrinkage that cost billions of dollars are reported in various parts of the world (Al-Rawas, 1999; Erguler and Ulusay, 2003; Gourley et al., 1993; Nelson and Miller, 1992; Ramana, 1993; Shi et al., 2002). Hence expansive soils should be identified and sufficiently characterized at early stages of geotechnical applications in order to guide a detailed design survey so as to avoid or minimize unnecessary expenses and delays in the construction and maintenance of structures.

There are a number of direct and indirect in-situ and laboratory testing procedures available to identify expansive soils and quantify the magnitude of volume change expected. Expansive soils can be identified by surface manifestations in the field, for they show cracks of polygonal pattern in dry seasons (Figure 1) as a result of appreciable volume decrease.

models will perform in the future and the degree of certainty that one might expect while using the models to solve practical problems. Different types of validation methods were discussed and presented in various literature (Martens and Naes, 1989; Wold et al., 2001). Of which, cross validation method has found its application in cases where the data set that one is working on is small and hence a separate or independent and representative validation data set is unavailable (Martens and Naes, 1989). Kooistra (2005) used a full cross validation method, which is based on a leave one out principle where one sample will be left out at a time and the model is calibrated on the remaining samples (CAMO, 2005). This will be repeated N times until every sample is left out once and the model is computed on the remaining samples and the left out sample is predicted.

To avoid erroneous calibration and deterioration of models prediction ability, it is important to detect outliers and remove or replace them by accurate values (Hocking, 2003; Martens and Naes, 1989). Outliers are abnormal observations that show significant deviation from the rest of the dataset in the population. They might arise both due to error in the experiment or instrument, or represent different information other than the material of interest and hence irrelevant. Presence of outliers in the data set is known to influence both the calibration and validation of PLSR models. Different methods are developed to detect sample outliers in PLSR modeling. Martens and Naes (1989) presented outlier detection criteria based on the analysis of residuals and leverages. In PLSR modeling, residuals are of diagnostic interest. It is possible to examine the residual variances (the variation that is left unexplained) in the X as well as the Y-spaces. Wold et al. (2001) demonstrated that large Y-residuals indicate that the model is poor. Normal probability plots of the residuals of a single Y-variable are also useful for identifying outliers in the relationship between T and Y. The X-residuals (part of X that is not used in modeling Y) are also useful for identifying outliers in the X-space, i.e., observations that do not fit the model. In addition, uncertainty tests e.g. Martens uncertainty limit tests (CAMO, 2005; Martens and Naes, 1989) can be used for testing which variables are causing perturbations in the model.

3. Results and Discussions

3.1 Spectral analysis

In the spectral interpretations, spectral libraries of different sources (e.g. TSG (the spectral geologist), ENVI and PIMA view built-in mineral libraries, USGS and JPL mineral libraries) that are developed upon experimental investigations on minerals and verified with variety of conventional testing methods (e.g. X-ray diffraction (XRD), Transmission Electron Micrograph (TEM), Scanning Electron Micrograph (SEM) etc) are used.

Differences in spectral characteristics among spectra of different soil samples were used for differentiating various clay mineral types that are present in the soil samples. Position of absorption features, their shapes, types and number, depth intensity and asymmetry; shape of spectral curves, differences in slopes of spectral curves and variations in reflectance intensity of spectra were some of the important qualitative parameters that helped in identifying spectrally dominant clay mineral from the soil reflectance spectra (Figure 3). Some spectra show a sharp rise in slopes and variable reflectance intensity throughout the whole wavelength region of the electromagnetic spectrum. Others show lower reflectance intensity throughout the whole wavelength range and were on overall dark. The later also exhibited monotonously rising convex slopes in the VNIR (visible near infrared) wavelength region and less variable reflectance intensity in the SWIR. Some show moderate rise in slopes and also moderate increase in reflectance intensity from the VNIR to the SWIR wavelength regions.

In the visible near infrared portion of the spectra changes in slopes were the prominent features that were recognized coupled with changes in reflectance intensities. The absorption features that are apparent on the VNIR region are relatively few, broad, wider and less intense. Whereas, in the SWIR region the main absorption features of clay minerals were observed with variable intensity being the prominent ones at ~ 1400 nanometer, ~ 1900 nanometer and ~ 2200 nanometer (Clark, 1999; Van der Meer, 1999).

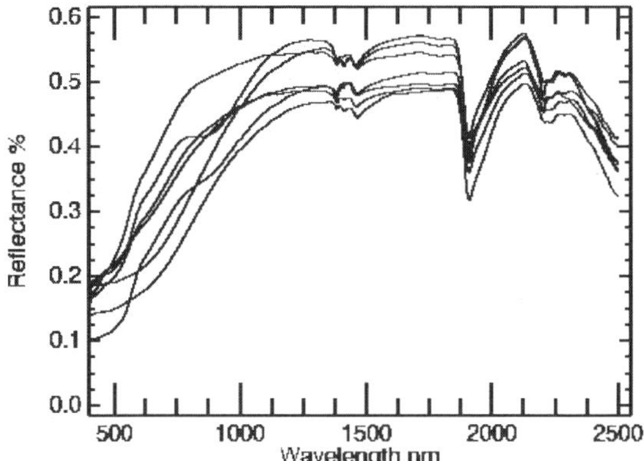

Fig. 3. Spectral reflectance curves of some soil samples (no offset). Note spectral characteristic differences, reflectance and slope variability in the VNIR wavelength region (350 nanometer – 1000 nanometer) and absorption features in the SWIR (1000 nanometer – 2500 nanometer).

3.2 Geotechnical Characteristics

As presented in the plasticity chart (Figure 4) and particle size distribution of soil samples (Figure 6), the soil samples exhibit high plasticity and are finer grained with high percentages of clay fraction. Generally the more plastic and finer grained soils are the greater swell-shrink potential that they are susceptible to, though swell-shrink potential is dictated by mineralogy and other factors as well. Majority of the soil samples are plotted above the A-line spanning from CI to CE zones. The soils that fall in the CI zone are of intermediate plasticity behavior, while those falling in the CH, CV and CE are of high, very high and extremely high plasticity nature. The higher the plasticity the larger will be the susceptibility of the soil to significant volume change characteristics. Accordingly swell-shrink potential of soil samples in the CH, CV, and CE zones are of high, very high and extremely high. Few samples fall below the A-line in the MV zone; these soils have high inherent expansion potential. The 'A' line is an empirical boundary separating inorganic clays from silty and organic soils. Soils of the same geological origin usually plot on the plasticity chart as straight lines parallel to the 'A' line. "Fat" or plastic clays plot above the line. Organic soils, silts and clays containing a large portion of "rock flour" (finely ground non-clay minerals) plot below it.

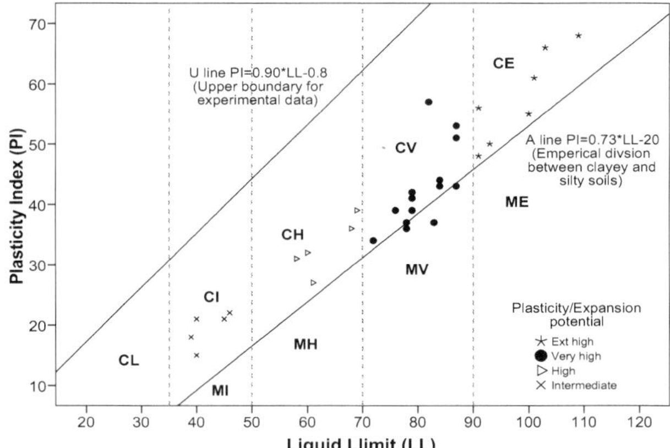

Fig. 4. Plasticity chart showing classes of expansion potential of soil samples.

The scattering of soil specimen in the plasticity chart show a wide range of variability in their swell-shrink characteristics (low to extreme cases). This is important in ensuring representativeness of samples for the intended PLSR modeling since it affects model prediction ability (Martens and Naes, 1989). Calibration in narrow range can bring about a risk of inability to extrapolate model into observations spanning a wider range. That is for instance bad prediction ability in case of failing to cover a total range of variability in a new dataset.

Proctor test results of some soil samples depicting the maximum dry density (MDD) and optimum moisture content (OMC) are shown in Figure 5. As presented in the graph the soil samples are also labeled with mineralogical groups obtained upon interpretation of spectral reflectance curves of respective soil samples.

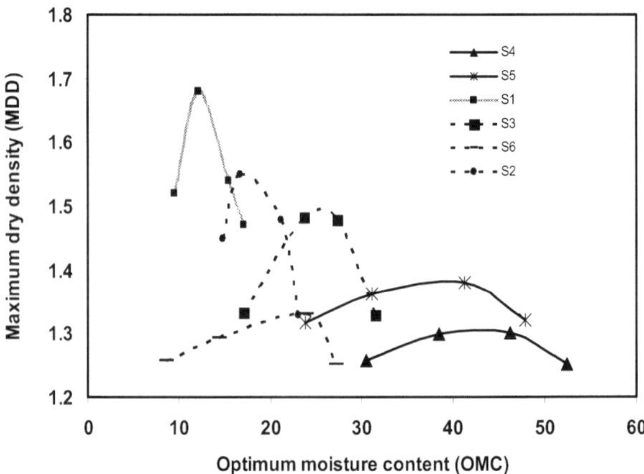

Fig. 5. Compaction curves of selected soil samples from proctor test labeled with mineralogical classes of soil from spectral interpretation.

Smectite clay species dominated soils (S4 and S5) exhibited low dry densities (MDD) with large moisture intake as indicated by their optimum moisture contents (Figure 5). As the grading curves (Figure 6) depicted S4 is finer than S5 with higher clay content. Accordingly S4 shows higher magnitudes of volumetric shrinkage and plasticity character (LL, PI) than its similar species S5 (Table 1).

Halloysite dominated soils (S2, S3 and S6) are characterized by dry densities and moisture intakes that seem to vary with their clay contents. While S6 has highest clay content of the three soils followed by S3 and S2 (Figure 6), the magnitude of volumetric shrinkage and plasticity that it exhibited is also high followed by S3 and S2 (Table 1). Dry density of S2 is the highest followed by S3 and S6 respectively (Figure 5) in the halloysite clay dominated sols.

The quartz dominated soil (S1) attained the highest dry density with lowest moisture intake as compared with soils dominated by halloysite and smectite clay varieties (Figure 5). This soil is also coarser than the other soils (Figure 6) and is non-expansive (Table 1).

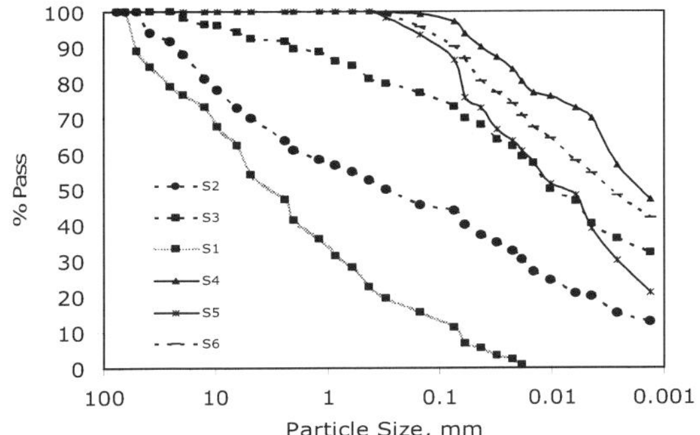

Fig. 6. Particle size distributions of selected soil samples labeled with mineralogical classes of the soil specimen obtained from spectral interpretation.

Even though the clay content of halloysite dominated soil sample S6 is higher than the clay content of smectite dominated soil sample S5, the later is characterized by higher magnitudes of volumetric shrinkage and plasticity, and exhibited low dry density (MDD) coupled with high moisture intake (OMC). This observation can be attributed to the fact that clay mineralogy dominantly controls swell-shrink characteristics of expansive soils. On the other hand within similar clay mineralogy, amount of clay fraction seems to govern the magnitude of volumetric shrinkage, plasticity (LL and PI) and compaction characteristics (MDD and OMC).

Sample ID	Spectrally dominating mineral	Volumetric shrinkage	Liquid limit	Plasticity index
S4	smectite	137	103	58
S5	smectite	118	84	43
S6	halloysite	99	72	34
S3	halloysite	48	48	22
S2	halloysite	32.8	46	22
S1	quartz	-	Np	Np

Table 1. summary of geotechnical properties of selected soil samples. NP: non plastic material

Variations in CBR and CBR swell of the six soil samples from different mineralogical groups are presented in Figure 7. The CBR value is an indicator of soils strength or bearing capacity, which is directly used in the design of subgrade, subbase and base material for pavement. Highest CBR and lowest CBR swell values are attained by the quartz dominated soil sample, S1. For the smectite dominated soils (S4 and S5) lowest CBR and highest CBR swell values are recorded. CBR of S4 and S5 are below standard (e.g. to lay an embankment directly over or to be used as subgrade material in road construction) and their CBR swell is much higher than allowable CBR swell values. Halloysite dominated soils (S2, S3 and S6) exhibited CBR and CBR swell values which seems to vary according to their clay fractions. Among the halloysitic soils S6 show low CBR which is below standard and higher CBR swell which is higher than the allowable CBR swell value. S3 on the other hand exhibited marginal CBR value with high though within the allowable range in different standard specifications.

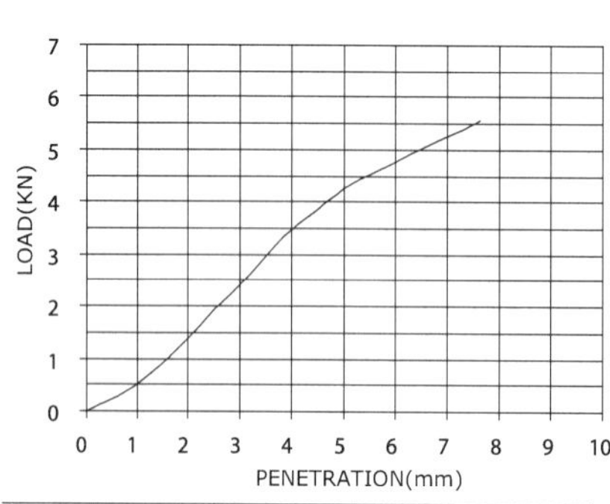

ID	Penetration (mm)	load (KN)	Standard load (KN)	CBR (%)	CBR Swell (%)
S1	2.5	1.98	13.24	14.95	0.31
	5	4.36	20	21.80	

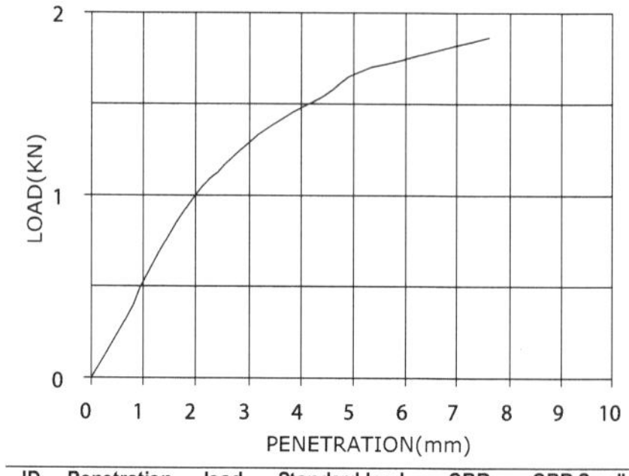

ID	Penetration (mm)	load (KN)	Standard load (KN)	CBR (%)	CBR Swell (%)
S2	2.5	1.16	13.24	8.76	0.94
	5	1.67	20	8.35	

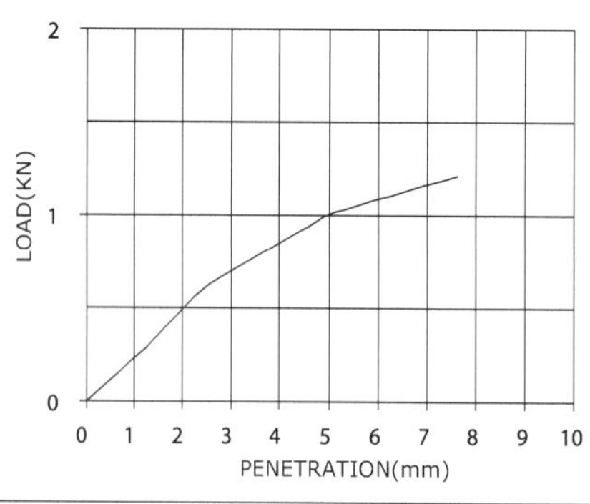

ID	Penetration (mm)	load (KN)	Standard load (KN)	CBR (%)	CBR Swell (%)
S3	2.5	0.63	13.24	4.76	1.89
	5	1.02	20	5.10	

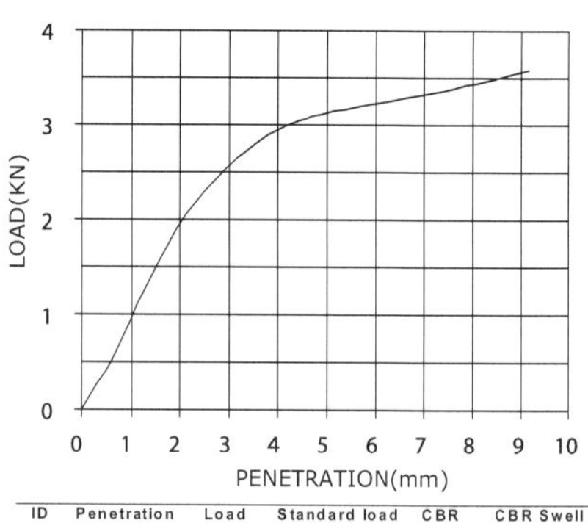

ID	Penetration (mm)	Load (KN)	Standard load (KN)	CBR (%)	CBR Swell (%)
S4	2.5	0.14	13.24	1.06	14
	5	0.19	20	0.95	

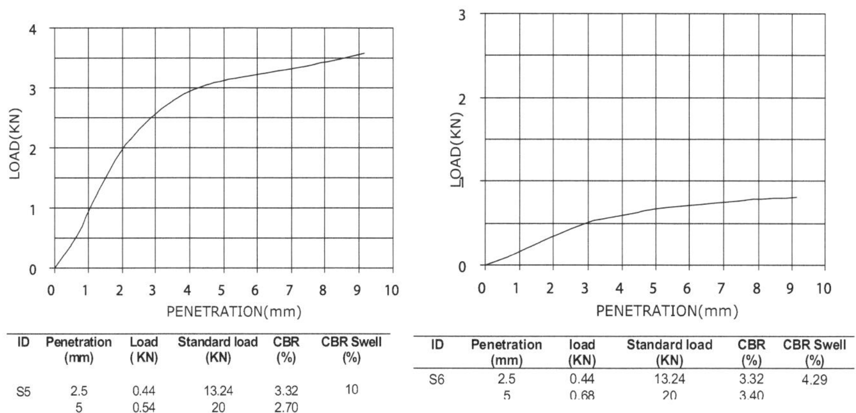

ID	Penetration (mm)	Load (KN)	Standard load (KN)	CBR (%)	CBR Swell (%)
S5	2.5	0.44	13.24	3.32	10
	5	0.54	20	2.70	

ID	Penetration (mm)	load (KN)	Standard load (KN)	CBR (%)	CBR Swell (%)
S6	2.5	0.44	13.24	3.32	4.29
	5	0.68	20	3.40	

Fig. 7. CBR and CBR swell results, graphs with accompanying tables.

XRD test results show that the soil samples contain clay minerals (Figure 8) such as smectite (mainly montmorillonite and nontronite), illite and kaolinite (halloysite and kaolinite) which significantly influence engineering behavior of expansive soil due to their high activity; and original minerals such as quartz, feldspar and mica are which are common constituents of expansive soils but do not contribute to the expansiveness of soils due to their low activity. Qualitative XRD analysis results are summarized in Table 2 and chemical analysis results are resented in Table 3.

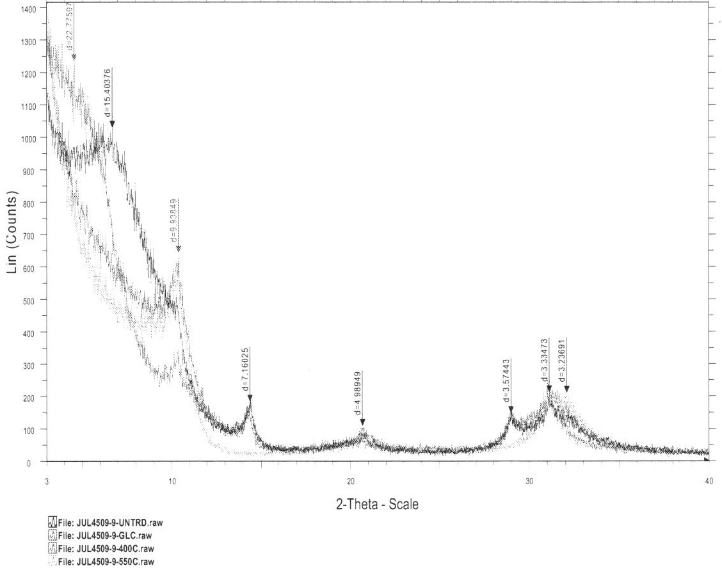

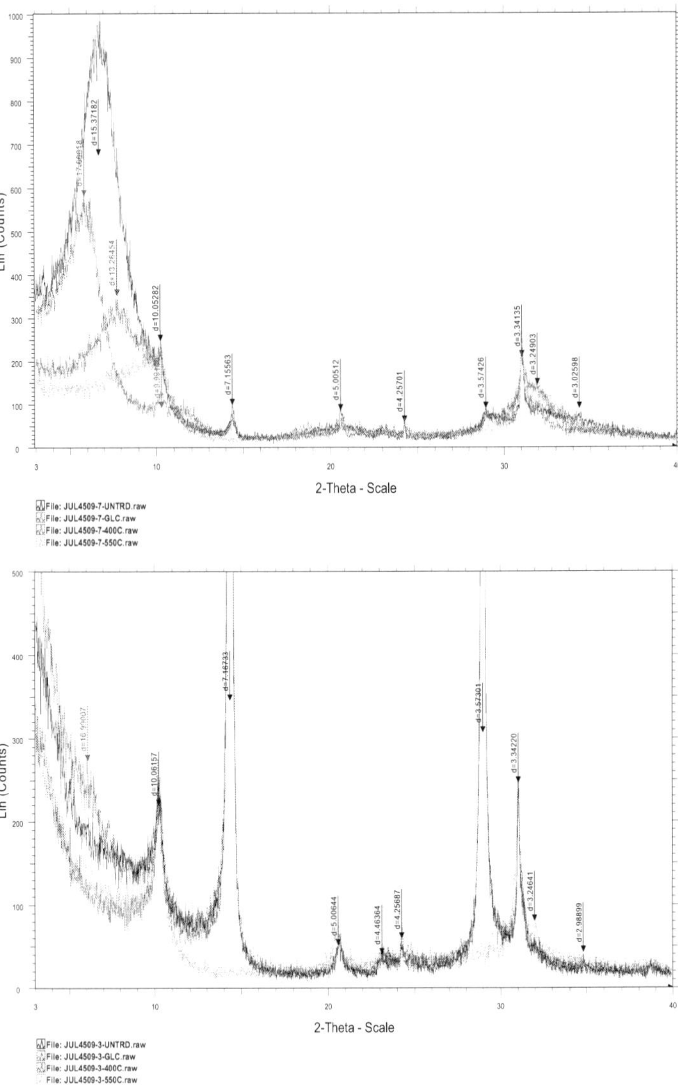

Fig. 8. XRD patterns of clay fractions (< 2µ) of some soil samples (Table 2 and 3 present the mineralogical assemblage and chemical results of these three patterns in descending order).

ID		Major	Moderate	Minor	Trace	Spectrally dominant mineralogy
			XRD mineralogy assignment			
1	Bulk	Montmorillonite	quartz, calcite, Nontronite	kaolinite, plagiodase, potassium feldspar, Illite	*brookite, *rutile, *goethite	Smectite
	Clay Fraction	Montmorillonite	-	illite, kaolinite, (quartz), (calcite), (potassium feldspar)	-	
2	Bulk	Montmorillonite	quartz, Nontronite	kaolinite, plagiodase, potassium feldspar, Illite	*brookite, *rutile, *goethite	Smectite
	Clay Fraction	I/M mixture	Nontronite	illite, kaolinite, (quartz), (calcite), (potassium feldspar)	-	
3	Bulk	-	quartz, kaolinite, halloysite	potassium felspar, illite, plagioclase, montmorillonite, nontronite, I/M mixed	*brookite, *pyrite, *goethite	Halloysite
	Clay Fraction	Halloysite	-	(quartz), illite, I/M mixed	*montmorillonite, *(potassium feldspar)	

Table 2. Summary of Qualitative XRD results showing abundance of major (>30%), moderate (10 -30 %), minor (2 – 10 %) and trace (< 2 %) mineral constituent of soil samples.

ID	Sio2 %	Al2O3 %	Fe2O3 %	MgO %	CaO %	Na2O %	K2O %	TiO2 %	P2O5 %	MnO %	Cr2O3 %	V2O5 %	LOI %	Sum %
1	46	13.8	7.14	2.29	5.66	0.79	1.46	1.14	0.08	0.18	0.02	0.02	21.4	99.98
2	50.8	16.9	8.46	1.5	1.33	2.16	2.53	1.03	0.04	0.24	0.01	0.02	14.5	99.52
3	52.9	19	7.95	1.63	0.3	0.77	1.86	1.22	0.15	0.17	0.01	0.02	13.2	99.18

Table 3. Chemical analysis results.

The information obtained on the clay mineralogical assemblage of the soil samples from X-ray diffraction analysis is in conformity with spectrally dominant mineralogical group assignments from interpreting reflectance spectra of respective soil samples.

Major clay mineral that is responsible for swelling and shrinkage of soils in the study area is smectite (montmorillonite and nontronite) as identified from soil spectral reflectance and confirmed by the X-ray diffraction analysis; and mixed layer combination of montmorillonite and illite. The hydrous variety of kaolinite group (halloysite) shows variable, that is low to appreciable swell-shrink character. Halloysite show low bulk density than kaolinite coupled with high porosity; its hydraulic conductivity is also reported to be higher (West et al., 2004). Geotechnical character of halloysite dominated soils seems to vary according to their particle size distribution (clay to coarser fraction proportions).

Montmorillonite, (Ca, Na) 0.67Al 4(Si, Al) 8 O20(OH)4 nH2O, is a product of weathering of iron and magnesium rich parent materials and is one of the most common smectite minerals (AusSpec International, 2005; Fitzpatrick, 1980) found in soils. It also form from the weathering of volcanic ash or primary silicate minerals such as feldspars, pyroxenes, or amphiboles under conditions of insufficient leaching of soil profile due to low permeability and excessive evaporation (Snethen, 1975). As indicated in the brief summary of the Geology of the study which is covered by rocks of volcanic origin where volcanic debris and

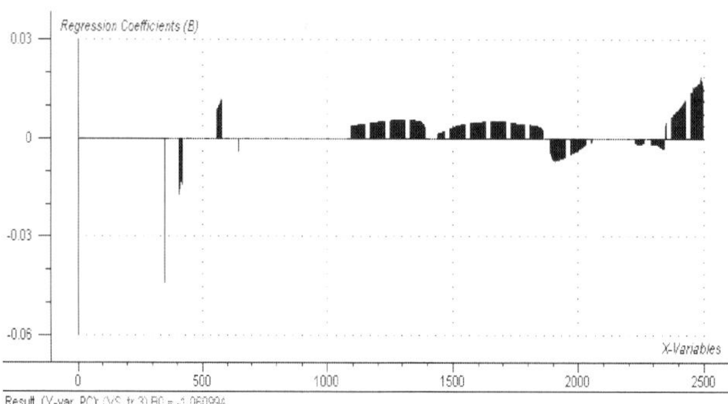

Fig. 12. Result of PLSR modeling for volumetric shrinkage showing: Regression coefficients or statistically significant wavelength regions in predicting volumetric shrinkage from the laboratory soil reflectance spectra.

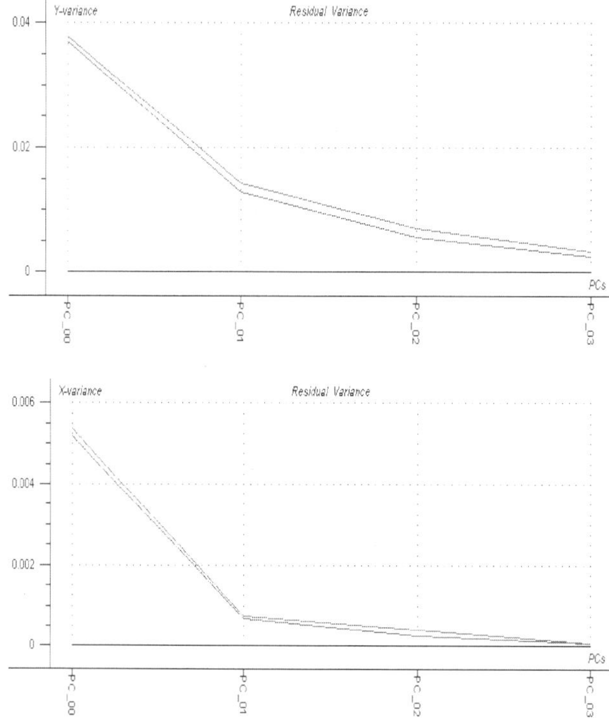

Fig. 13. Result of PLSR modeling for volumetric shrinkage showing: Residual variances (blue during calibration and red during prediction) exhibiting the remaining variation that is not taken into account by the model is minimum after fitting three PLS components suggesting much of the variability in volumetric shrinkage is explained by the model. The upper graph shows Y-variance and the lower graph shows the X-variances.

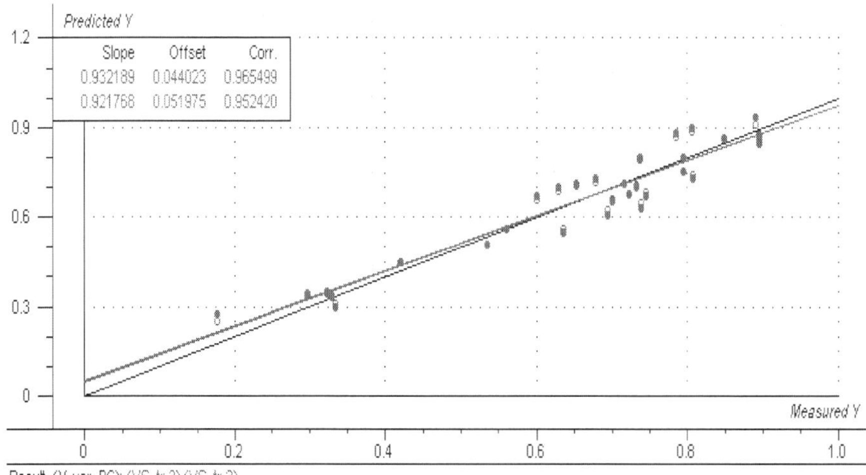

Fig. 14. Result of PLSR modeling for volumetric shrinkage showing: Regression overview showing predicted and measured volumetric shrinkage.

Note that there are no outlying samples lying outside the rest of the cloud (Figure 11). Only one sample lies outside the ellipse; under normal situation it is expected that about 5 % of the samples to lie outside the ellipse.

As shown in Figure 12 Relevant X variables (wavelengths from soil spectra) in predicting volumetric shrinkage fall in the VNIR and SWIR regions. Wavelengths in the SWIR bands are related with clay mineral diagnostic features (Clark, 1999). Wavelengths from the VNIR can be organic matter and iron oxides related (Ben-Dor and Banin, 1994; Wan et al., 2002) and sand or quartz related spectral features (Viscarra et al., 2006).

Figure 13 shows that most of the Y and X variances are accounted for by the three PLS factors, reaching a minimum level (close to zero) of unexplained variances at the third PLS factor. The blue and the red lines represent residual variances during the calibration and validation stages respectively. Note that the two lines do not significantly differ indicating that the calibration data are well fitted and that the model describes the validation data equally well.

In the regression overview (Figure 14) comparing laboratory measured and corresponding predicted (from spectral reflectance of soil samples using PLSR analysis) soil volumetric shrinkage magnitudes, calibration and prediction points lie very close to each other suggesting the model fitted to the calibration data set described the prediction data set as good as possible. Note that an internal calibration, full cross validation leave one out at a time method was used.

Apart from high coefficient of determination (~0.91), the model gives low estimation errors during the calibration (RMSEC = 0.05) as well as the validation (RMSEP = 0.06) stages. Standard errors of calibration (SEC = 0.05) and prediction (SEP = 0.06) which are indicators of precision of the calibration and prediction respectively are also small. The bias which is the average value of the difference between the predicted and measured values, is also small (both in the calibration and prediction stages) for the given number of PLS factors indicating that the effect of bias is negligible (Martens and Naes, 1989). The values of offsets in the calibration and validation stages are small showing minor deviation from an ideal line where the measured and predicted volumetric shrinkage magnitudes are expected to be equal. Model performance indices are presented in Table 4.

Calibration					Validation				
Correlation coefficient	RMSEC	SEC	Bias	Offset	Correlation coefficient	RMSEP	SEP	Bias	Offset
0.965	0.05	0.05	0.005	0.044	0.952	0.586	0.0596	0.001	0.051

Table 4. Summary of model performance indices resulted from PLSR modeling of soil volumetric shrinkage.

4. Conclusions

In this chapter we demonstrated that it is possible to estimate soil volumetric shrinkage from spectral reflectance of soil. The approach can be of potential geotechnical utility contributing to the quality of geotechnical investigations of expansive soil, playing prominent role in planning and designing of infrastructure. This is particularly in minimizing uncertainties through identifying geotechnically problematic areas and estimation of critical parameters (in this case volumetric shrinkage), hence providing better basis for decision making. This in turn present relevant information that can be useful in site selection, route planning and search for construction materials (borrow, subbase etc) especially in the reconnaissance and preliminary design stages of construction projects. In summary;

- The expansive soils in this study are identified by their constituent dominant clay mineral type and accordingly classified into two clusters. Those soils comprised of active clay mineral smectite exhibit high swell-shrink potential as suggested by the magnitude of volumetric shrinkage, plasticity and other related geotechnical characteristics. They also exhibit low maximum dry densities while their optimal moisture contents are high. Halloysite clay mineral dominated soils on the other hand show less swell-shrink susceptibility, and variable maximum dry densities with optimum moisture contents lower than those exhibited by the smectite dominated soil samples. Some halloysite bearing soils can be weak and show high swell-shrink susceptibility as indicated by their geotechnical properties. Note their CBR strengths presented in Figure 7 for the wettest likely condition likely and the associated CBR swell that they exhibited coupled with values of volumetric shrinkage and plasticity characters that they show. Quartz dominated soil samples on the other hand show higher maximum dry density with low moisture intake. These soils also attain high CBR strength and show negligible CBR swell.

- Differences in geotechnical characteristics show dependency on clay mineralogy among different species (smectite, halloysite), and on clay fraction among similar varieties in the studied soils. The fact that clay mineralogy is a crucial factor dictating a number of soil geotechnical behaviors laid emphasis on the importance of qualitative clay mineralogical assemblage analysis of expansive soils. Since spectroscopy is a cheap, rapid and non-destructive way of analyzing soil mineralogy, it can be used for such kind of routine analysis in order to complement quantitative analyses.

- Coefficient of determination obtained in the PLSR analysis show that soil volumetric shrinkage is highly correlated with spectral parameters. The low estimation and interference errors, negligible bias and small offsets indicate spectroscopic techniques potential to be used in routine quantitative analyses of soil shrinkage potential. The presented empirical relations with the added valuable information on the content of major clay minerals (from spectral interpretations) establish a simple way of characterizing expansive soils. Despite the complex nature (comprised of various constituents other than clay materials which are responsible for their expansive characteristics) of soil the results proved that a remarkable amount of information about soil properties can be obtained from their reflectance spectra.
- The current study gave an outlook for future application of optical remote sensing to map soils susceptible to swell-shrink and variation in the magnitude of expansion and shrinkage, provided with the availability of sufficiently high resolution (both spatially and spectrally in order to resolve vital spectral details) data.

5. References

Abebe, Franscesco, Fabrizio and Piero, 1999. Geological Map of Debreziet Area (Ethiopia). GSE, Addis Ababa.

Al-Rawas, A.A., 1999. The factors controlling the expansive nature of the soils and rocks of northern Oman. Engineering Geology, 53: 327-350.

ASD, 1995. Analytical Spectral Devices (ASD): Technical Guide, Portable Spectroradiometers, Analytical Spectral Devices, Inc.

AusSpec International, 2005. The Spectral Geologist software. AusSpec Interbational, Sydney.

Ben-Dor, E. and Banin, A., 1994. Visible and Near-Infrared (0.4-1.1µm) analysis of Arid and Semiarid Soils Remote Sensing of Environment, 48: 261 - 274.

CAMO, 2005. The unscarmbler 9.2.

Chabrillat, S., Goetz, A.F.H., Krosley, L. and Olson, H.W., 2002. Use of hyperspectral images in the identification and mapping of expansive clay soils and the role of spatial resolution. Remote Sensing of Environment, 82: 431 - 445.

Clark, R.N., 1999. Spectroscopy of Rocks and Minerals, and Principles of Spectroscopy. In: A. Rencz (Editor), Chapter 1 in: Manual of Remote Sensing. John Wiley and Sons, Inc, New York.

Crowley, J.K. et al., 2003. Spectral reflectance properties (0.4-2.5 µm) of secondary Fe-oxides, Fe-hydroxides, and Fe-sulphate-hydrate minerals associated with sulphide -bearing mine wastes. Geochemistry; Exploration, Environment, Analysis, 3: 219-228.

Erguler, Z.A. and Ulusay, R., 2003. A simple test and predictive models for assessing swell potential of Ankara (Turkey) clay. Engineering Geology, 67: 331 - 352.

FAO, 1998. World reference base for soil resources. International Society of Soil Science, Rome.

Fitzpatrick, E.A., 1980. SOILS; their formation, classification and distribution. Longman Inc., London, NY, 354 pp.

Goetz, A.F.H., Chabrillat, S. and Lu, Z., 2001. Field Reflectance Spectrometery for Detection of Swelling Clays at Construction Sites. Field Analytical Chemistry and Technology, 5(3): 143-155.

effect were calculated by the so-called dense media radiative transfer theory (DMRT) under Quasi Crystalline Approximation with Coherent Potential (QCA-CP) (Wen et al., 1990; Tsang & Kong, 2001). Dense Media radiative transfer theory was derived from Dyson's equation under the quasi-crystalline approximation with coherent potential (QCA-CP) and the Bethe-Salpeter equation under the ladder approximation of correlated scatterers.

By using the DMRT, the extinction coefficient K_e and albedo ω used in equation (5) were calculated. And then the radiance of each soil slab was calculated by the 4-stream fast model. The radiance just below the soil/air interface was obtained by integrating the radiance from bottom layer to the top layer. Finally, the apparent emission of soil media, T_{bs} in equation (1) and (2), was obtained.

2.3 Surface roughness effects

When an electromagnetic wave reaches the air/soil interface, it suffers the reflection and refraction due to the dielectric constant changing in the two sides of the interface. The roughness of the interface divides the reflected wave into two parts, one is reflected in the specular direction and another is scattered in all directions. Generally, the specular component is often referred to as the coherent scattering component. And the scattered component is known as the diffuse or noncoherent component, which consists of power scattered in all directions but with a smaller magnitude than that of the coherent component. Qualitatively, surface roughness increases the apparent emissivity of natural surfaces, which is caused by increased scattering due to the increase in surface area of the emitting surfaces. And it was demonstrated by many researches that the surface roughness has a nonnegligible effects on the accuracy of soil moisture retrieval by spaceborne microwave sensors (Oh & Key, 1998; Singh et al., 2003). In general, the surface roughness effects are simulated by two ways: semi-empirical models and fully physical-based models.

(1) Semi-empirical models

The semi-empirical models are simply and do not cost too much computation efforts. The parameters used in semi-empirical models are often derived from field observations. Depending on the parameters involved, there are three different semi-empirical models: Q-H model (Choudhury et al., 1979; Wang & Choudhury, 1981), Hp model (Mo & Schmugge, 1987; Wegmuller & Matzler, 1999; Wigneron et al., 2001) and Qp model (Shi et al., 2005).

(2) Fully physical-based model

In our algorithm, we simulated the land surface roughness effect using the Advanced Integral Equation Model (AIEM) (Chen et al., 2003). AIEM is a physically-based model with only two parameters: standard deviation of the height variations s (or rms height) and surface correlation length l. AIEM is an extension of the integral Equation Model (IEM) (Fung et al., 1992). It has been demonstrated that IEM has a much wider application range for surface roughness conditions than other models such as the Small Perturbation Model (SPM), Physical Optics Model (POM) and Geometric Optics Model (GOM). AIEM improves the calculation accuracy of the scattering coefficient compared with IEM by retaining the absolute phase term in the Green's function.

By coupling AIEM with DMRT (DMRT-AIEM), this radiative transfer model for soil media is fully physically-based. As a result, the parameters of DMRT-AIEM, such as the rms height, correlation length and soil particle size, have clear physical meanings and their values can be obtained either from field measurement or theoretical calculation.

2.4 Vegetation masking effects

The existence of canopy layers complicates the electromagnetic radiation which is originally emitted solely by soil layers. The vegetation may absorb or scatter the radiation, but it will also emit its own radiation. The effects of a vegetation layer depend on the vegetation opacity τ_c and the single scattering albedo of vegetation ω_c (Schmugge & Jackson, 1992). The vegetation opacity in turn is strongly affected by the vegetation columnar water content W_c. The relationship can be expressed as (Jackson & Schmugge, 1991):

$$\tau_c = \frac{b' \lambda^\chi W_c}{\cos\theta} \tag{6}$$

where λ is the wavelength, θ the incident angle, W_c the vegetation water content. χ and b' are parameters determined by vegetation type.

The single scattering albedo, ω_c, describes the scattering of the emitted radiation by the vegetation. It is a function of plant geometry, and consequently varies according to plant species and associations. The value of it is small in the low frequency microwave region (Palosica & Pampaloni, 1988; Jackson & Oneill, 1990). In our algorithm, ω_c is calculated by

$$\omega_c = \omega_0 \cdot \sqrt{W_c} \tag{7}$$

The value of albedo parameter ω_0 is decided empirically in current researches.Experimental data for this parameter are limited, and values for selected crops have been found to vary from 0.04 to about 0.12.

By combing the T_{bs} solved by equation (5), the surface reflectivity calculated by AIEM, the vegetation opacity τ_c calculated by equation (6), and the vegetation single scattering albedo ω_c estimated by equation (7), a physical-based radiative transfer model was developed.

3. The Algorithm

The basis of our algorithm is a data base of brightness temperature and/or some indexes calculated from brightness temperature. By searching the data base (or look up table) with the satellite observation as the input, soil moisture and other related variables of interest can be estimated quickly. Such high searching speed is the main reason why we adopt the look up table method for soil moisture retrieval. The implementation of our algorithm consists of three steps: (1) fixing the parameters used in the forward model; (2) generating a look up table by running forward model; and (3) retrieving soil moisture by searching the look up table.

3.1 Parameterization

As in other physically-based algorithms, such as that developed by Njoku et al. (2003) and the SCA developed by Jackson (1993), the parameters used in our algorithm have clear physical meanings. This advantage derives from the strength of the forward radiative transfer model. Before running the forward RTM to generate look up table, the parameters should be confirmed at first. The parameters to be confirmed include *rms* height (s), correlation length (l), soil particle sizes (r) and vegetation parameters such as χ and b'. Currently, we can obtain these parameters through two methods: best-fitting method and a

Since the one-to-one relationship in our lookup table is very clear, it becomes simple to reverse the lookup table, so that the soil moisture can easily be estimated from the AMSR-E data set.

4.4 Retrieval soil moisture from AMSR-E

In this study, we retrieved soil moisture data for the period from July to August, 2003. The estimation is shown in **Figure 3** for (a) time variation and (b) accuracy comparison.

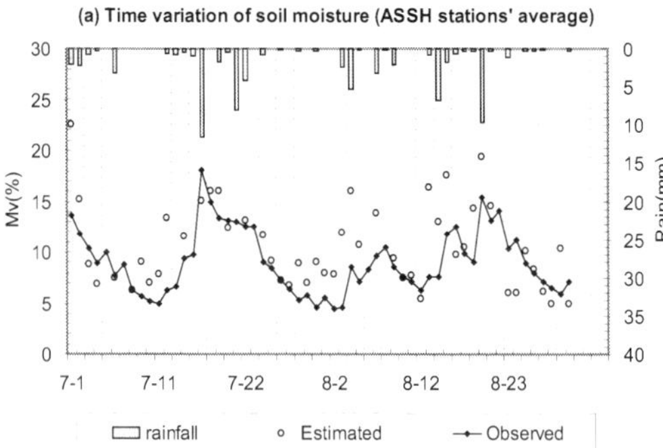

Fig. 3(a). Time series of retrieved soil moisture, observed soil moisture and precipitation.

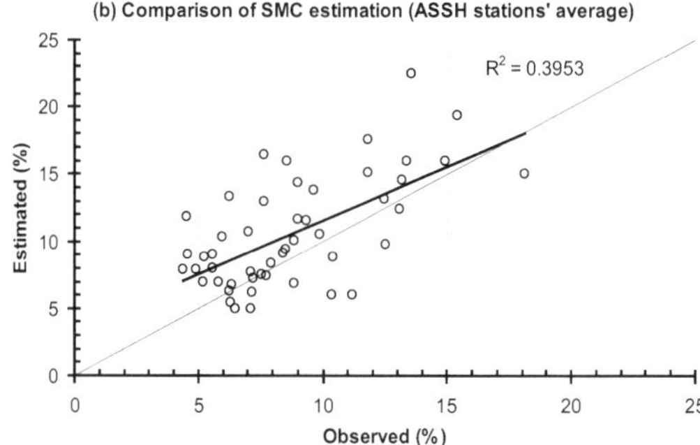

Fig. 3(b). Comparison of retrieval results with in situ observation

It is clear from **figure 3** that the algorithm gives a reliable soil moisture content estimate in both tendency and amplitude. The value of R-square is 0.3953, and the Standard Error of the Estimate (SEE) is 3.8%. From **figure 3(a)**, we find some overestimation around Aug. 4, 14 and 20, when moderate rainfall (5~10 mm) occurred. Such errors can be attributed partly to the difference between the TDR sensor depth and the penetration depth of the X band and Ku band. Moderate rainfall makes the soil surface much wetter than the soil 3cm below the surface where the TDR sensors were located. Such vertical heterogeneity of soil moisture in the first 3cm of soil was not considered in our algorithm. On the other hand, the wet surface situation decreases the penetration depth dramatically. The combination of these reasons makes our algorithm estimate higher soil moisture content than the in situ observations for moderate rainfall periods.

One advantage of our proposed algorithm is that it estimates soil physical temperature and soil moisture simultaneously. This is important for studies involving energy and water budget, such as studies of land surface processes and of weather forecasting.

The average retrieved physical temperature for ASSH stations is shown in **figure 4(a)** and **figure 4(b)**. As in soil moisture comparisons, the algorithm effectively retrieved physical temperature on average for 10 stations. The value of R-square is 0.5458, and the value of SEE is 4.4K. As with our soil moisture analysis, the overestimation of daily temperature variation can also be explained partly as the results of different observation depths.

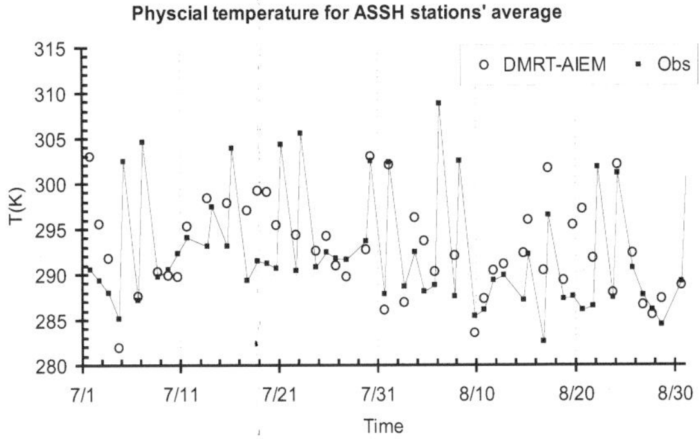

Fig. 4(a). Time series of retrieved and in situ observations of soil physical temperature.

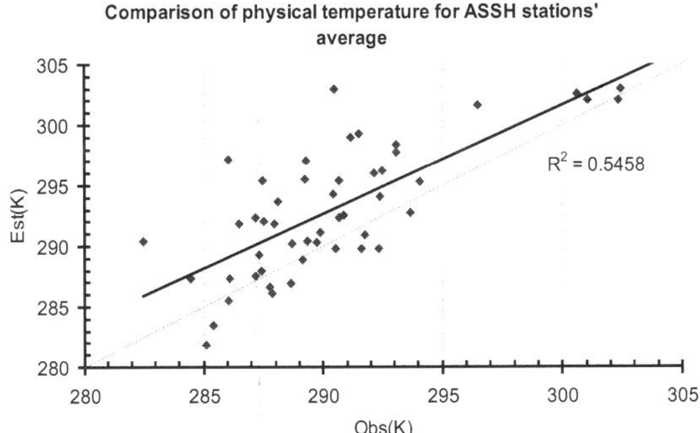

Fig. 4(b) Comparison of algorithm-estimated soil physical temperature with in situ observation

5. Application to SSM/I

Since the parameters used in our RTM have clear physical meaning, they are independent on the configuration of radiometers. The parameters used in AMSR-E soil moisture retrieval therefore can be directly used to the SSM/I data set, in the same region. In this test, we first checked the accuracy of TB simulation of our DMRT-AIEM model. And then a look up table for SSM/I data was generated and soil moisture was retrieved.

5.1 SSM/I TB Validation

Using the parameters optimized by AMSR-E math-up data set, with the in-situ observed soil moisture and temperature as input, we run the DMRT-AIEM model to generate TB at 19.35 and 37.0 GHz, two frequencies operated by SSM/I. The SSM/I TB validation results were shown in figure 5, for the A3 station, during the period from Jul. 1st to Jul. 30th, 2003.

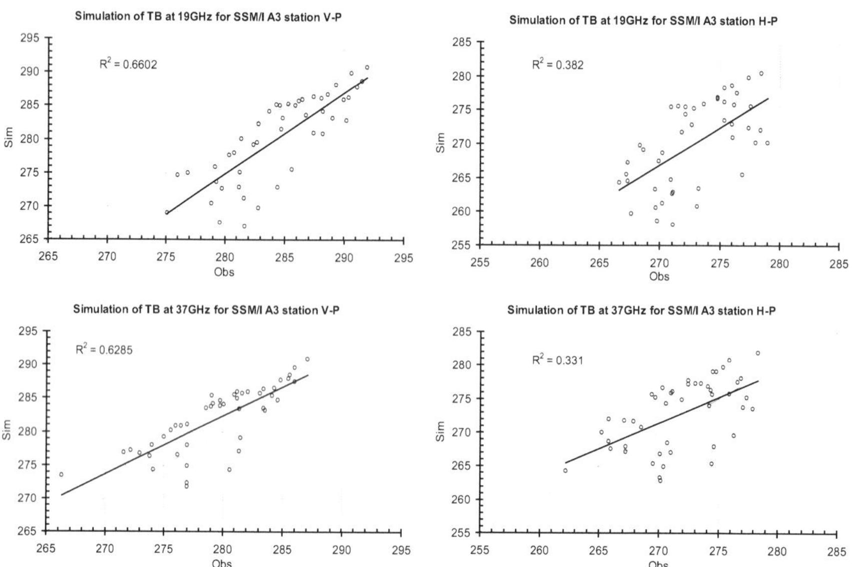

Fig. 5. Comparison of simulated brightness temperature with the one observed by SSM/I

From figure 5, it is clear that, for the vertical polarization, the TB simulated by DMRT-AIEM is in a good correlation with the SSM/I observation, with slight underestimation for 19 GHz and overestimation for 37 GHz. For the horizontal polarization, the performance of our RTM is not so good. Statistically, the Average Absolute Error (AAE, see equation (10)) and the square of correlation coefficient between observed brightness temperature and simulated one are listed in table 3.

Channel	19V	19H	37V	37H
AAE (K)	4.34	4.41	3.45	3.85
R^2	0.66	0.38	0.63	0.33

Table 3. AAE and correlation coefficient of DMRT-AIEM model for SSM/I data

$$AAE = \{\sum_{i=1}^{n}[ABS(TBS_{i,V} - TBO_{i,V}) + ABS(TBS_{i,H} - TBO_{i,H})]\}/(n*2) \tag{10}$$

where, TBS is simulated brightness temperature, TBO is observed brightness temperature by spaceborne sensor; n is number of samples.

5.2 Look up table of SSM/I

Through the TB validation, it was confirmed that our DMRT-AIEM model was able to produce reasonable TB at vertical polarization channels of SSM/I. But there was some gaps between the simulated TB with the one observed by SSM/I, especially for horizontal polarization channels. Moreover, as we know, the atmosphere effects should be considered for 37 GHz. All of these make it difficult to build a look up table with the same way used for AMSR-E. Authors proposed a simple solution by nudging the SSM/I TB data to fit the simulation and by using PI and ISW indexes to generate a look up table. Detail of the TB adjustments and look up table generation can be found from (Ohta et al., 2007). Figure 6 shows the look up table for SSM/I, in which the PI calculated from 19 GHz and the ISW calculated from the horizontal polarization of 37 and 19 GHz were used. The black points represent the PI and ISW values calculated from corresponding TB data observed by SSM/I.

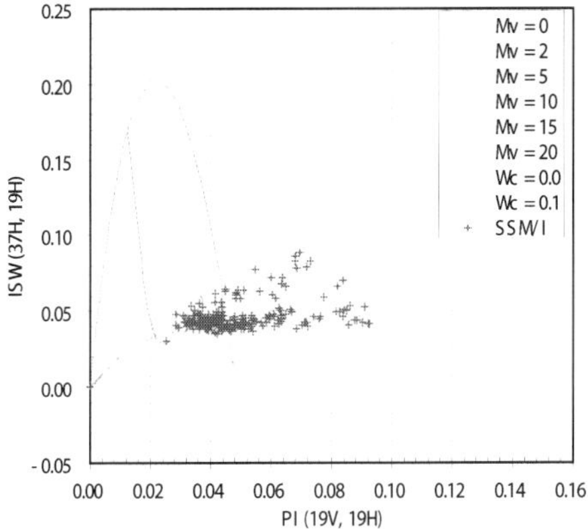

Fig. 6. Diagram of the Look Up Table for SSM/I soil moisture algorithm

5.3 Soil moisture retrieval from SSM/I

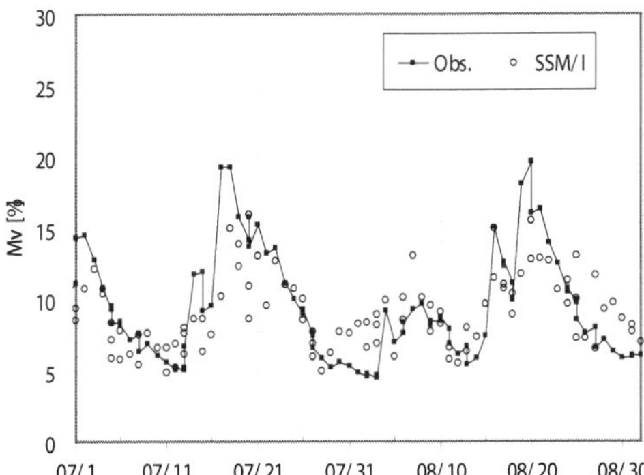

Fig. 7. Comparison of soil moisture retrieved from SSM/I with in-situ observation

By using the PI-ISW look up table, soil moisture was estimated from SSM/I TB data. The results are shown in figure 7, for the period from July to August, 2003. The line represents the in-situ soil moisture observation; the open cycles are SSM/I soil moisture estimate. From this figure, it is clear that the performance of SSM/I soil moisture retrieval algorithm is very good. So, it is feasible to get reasonable soil moisture estimation form SSM/I with the helps from AMSR-E. But we must keep it in mind that the TB adjustment was applied to the SSM/I data. Such good performance of SSM/I algorithm is therefore just for the special cases where the in-situ observation are available and the appropriate TB adjustment are possible.

6. Conclusions

Spatial distributed soil moisture information is an essential parameter for hydrological, meteorological and ecological studies. This paper presents the structure and contents of a soil moisture retrieval algorithm for the spaceborne passive microwave remote sensing. This algorithm was validated by using the AMSR-E match-up data set at CEOP Mongolia reference site. Comparing to the in-situ observation, reliable surface soil moisture was retrieved by the algorithm.

The transferability of our algorithm was also tested by using SSM/I TB data. At first, it was demonstrated that the forward RTM of our algorithm was capable to represent the SSM/I TB data only after the parameters were calibrated by AMSR-E data set. And then, with some adjustments to the SSM/I TB data, reasonable surface soil moisture was also retrieved from SSM/I data by our algorithm.

The results presented in this paper clearly show that we had built a bridge between the parameters retrieved from AMSR-E and those for SSM/I. With some further consideration about the difference between AMSR-E and SSM/I, e.g. the footprint size and the observation patterns, it is believed that our algorithm could provide a possibility to use the long historical global data observed by SSM/I. Moreover, it is possible to extend our algorithm to other available radiometers. And then, we can merge multi-sensor or/and multi-satellite observations to generate a long term global historical soil moisture product. Such a long term historic data set should be much useful for large scale hydrological and climatologic studies.

As mentioned in section 1, the retrieval of surface soil moisture is physically limited by the current satellite instruments which are operating at high frequencies. The low frequency, i.e. L-band, passive microwave soil moisture observation will firstly be available through the launch of Soil Moisture Ocean Salinity (SMOS) mission of ESA (Kerr, et al., 2001). NASA will provide a combined L-band radiometer and L-band radar observation through the Soil Moisture Active and Passive (SMAP) mission (Entekhabi, et al., 2008). Since the configuration of our algorithm was not specified to any sensors, it is also possible to apply our algorithm to the incoming L-band radiometers. We hope this algorithm will be helpful for these future soil moisture missions and for connecting current available C-band and X-band observations to the L-band observations.

7. Acknowledgements

This study was carried out as part of the Coordinated Enhanced Observing Period (CEOP) and Verification Experiment for AMSR/AMSR-E funded by the Japanese Science and

Technology Corporation for Promoting Science and Technology Japan and by the Japan Aerospace Exploration Agency (JAXA). The authors express their great gratitude to them.

8. References

Beljaars, A.C.M.; Viterbo P.; Miller M.j. & Betts A.J. (1996). The anomalous rainfall over the United States during July 1993: sensitivity to land surface parameterization and soil moisture anomalies. *Monthly Weather Review*, 124, 362-383.

Burke, W.J.; Schmugge, T.J. & Paris, J.F. (1979). Comparison of 2.8 and 21 cm microwave radiometer observations over soils with emission model calculation. *Journal of Geophysics Research*, 84, 287-294.

Chen, K.S.; Wu, T.D.; Tsang, L.; Li, Q.; Shi, J.C. & Fung, A.K. (2003). Emission of rough surfaces calculated by the integral equation method with comparison to three-dimensional moment method simulations. *IEEE Transactions on Geoscience and Remote Sensing*, 41, 90-101.

Choudhury, B. J.; Schmugge, T.J.; Chang, ATC & Newton N.R. (1979). Effect of surface roughness on the microwave emission from soils. *Journal of Geophysical Research*, 84, 5699-5706.

Chuluun, Togtohyn & Ojima, D.S. (1996). Simulation studies of grazing in the Mongolian steppe, *Proceedings of the fifth International Rangeland Congress*, Society of Range Management, Denver, CO, USA.

Dobson, M.C.; Ulaby, F.T.; Hallikainen, M.T. & Elrayes, M.A. (1985). Microwave dielectric behavior of wet soil .2. dielectric mixing models. *IEEE Transactions on Geoscience and Remote Sensing* 23, 35-46.

Delworth, T. & Manabe, S. (1988). The influence of potential evaporation on the variability of simulated soil wetness and climate. *Journal of Climate*, 13, 2900–2922.

Entekhabi, D.; Rodrigues-Iturbe I. & Castelli F. (1996). Mutual interaction of soil moisture state and atmospheric processes. *Journal of Hydrology*, 184, 3-17.

Entekhabi, D.; Njoku, E.; Oneill, P. E.; Spencer, M.; Jackson, T.; Entin, J.; Im, E. & Kellogg, K. (2008). The Siol Active Passive Mission (SMAP). *Proceedings of the IEEE 2008 International Geoscience and Remote Sensing Symposium* (IGARSS'08), 3, 1–4, July, 2008, Boston, Doi: 10.1109/IGARSS.2008.4779267.

Fung, A.K.; Li, Z.Q. & Chen, K.S. (1992). Backscattering from a randomly rough dielectric surface. *IEEE Transactions on Geoscience and Remote Sensing*, 30, 356-369.

Gloersen, P. & Barath F. T. (1977). A Scanning Multichannel Microwave Radiometer for Nimbus-G and SeaSat-A. *IEEE Journal of Oceanic Engineering*, 2:172-178.

Henyey, L. C. & Greenstein, J. L. (1941). Diffuse radiation in the galaxy, *The Astrophysics Journal*, 93, 70–83.

Hipp J.E. (1974). Soil electromagnetic parameters as functions of frequency, soil density, and soil moisture, *Proceedings of the IEEE*, 62:98–103

Hollinger, J.P.; Peirce, J.L. & Poe, G.A. (1990). SSM/I Instrument Evaluation. *IEEE Transactions on Geoscience and Remote Sensing*, 28(5), 781-790.

Jackson, T. J. & Oneill, P. E. (1990). Attenuation of Soil Microwave Emission by Corn and Soybeans at 1.4 Ghz and 5 Ghz. *IEEE Transactions on Geoscience and Remote Sensing*, 28(5), 978-980.

Jackson, T.J. & Schmugge, T.J. (1991). Vegetation effects on the microwave emission of soils. *Remote Sensing of Environment*, 36, 203-212.

Jackson, T.J. (1993). Measuring surface soil moisture using passive microwave remote sensing. *Hydrology Processes*, 7, 139-152.

Kawanishi,T.; Sezai, T.;Ito Y.; Imaoka, K.; Takeshima, T.; Ishido, Y.; Shibata, A.; Miura, M.; Inahata, H. & Spencer Roy W. (2003). The advanced microwave scanning radiometer for the earth observing system (AMSR-E), NASDA's contribution to the EOS for global energy and water cycle studies. *IEEE Transactions on Geoscience and Remote Sensing*, 41, 184-194.

Kendra, J. R. & Sarabandi, K. (1999). A hybrid experimental theoretical scattering model for dense random media. *IEEE Transactions on Geoscience and Remote Sensing*, 37, 21-35.

Kerr, Y. H.; Waldteufel, P.; Wigneron, J. P.; Martinuzzi, J. M.; Font, J. & Berger, M. (2001). Soil moisture retrieval from space: The Soil Moisture and Ocean Salinity (SMOS) mission. *IEEE Transactions on Geoscience and Remote Sensing*, **39**, 1729-1735

Koike, T.; Tsukamoto, T.; Kumakura, T. & Lu, M. (1996). Spatial and seasonal distribution of surface wetness derived from satellite data, *Proceedings of International workshop on macro-scale hydrological modeling*, 87-96.

Koike, T. (2004). The Coordinated Enhanced Observing Period – an initial step for integrated global water cycle observation, *WMO Bulletin*, 53(2), 1-8.

Liu, G. (1998). A fast and accurate model for microwave radiance calculations, Journal of Meteorology Society of Japan, 76(2), 335–343.

Lu, H.; Koike, T.; Hirose, N.; Morita, M.; Fujii, H.; Kuria, D. N.; Graf, T. and Tsutsui, H. (2006). A basic study on soil moisture algorithm using ground-based observations under dry condition. *Annual Journal of Hydraulic Engineering, Japan Society of Civil Engineer*, 50, 7-12.

Lu, H.; Koike, T.; Fujii, H.; Ohta, T. & Tamagawa, K. (2009). Development of a Physically-based Soil Moisture Retrival Algorithm for Spaceborne Passive Microwave Radiometers and its Application to AMSR-E, *Journal of The Remote Sensing Society of Japan*, 29(1), 253-261.

Mironov, V.L.; Dobson, C.; Kaupp, V.H.; Komarov, V.A. & Kleshchenko, V.N. (2004). Generalized refractive mixing dielectric model for moist soils. *IEEE Transactions on Geosciences and Remote Sensing*, 42:773–785. doi:10.1109/TGRS.2003.823288

Mo, T. & Schmugge, T. J. (1987). A Parameterization of the Effect of Surface-Roughness on Microwave Emission. *IEEE Transactions on Geoscience and Remote Sensing*, **25**(4), 481-486

Njoku, E.G. & Kong, J.A. (1977). Theory for passive microwave remote sensing of near-surface soil moisture. *Journal of Geophysical Research*, 82, 3108-3118.

Njoku, E.G. & Entekhabi, D. (1996). Passive microwave remote sensing of soil moisture. *Journal of Hydrology*, 184, 101-129.

Njoku, E.G.; Jackson, T.J.; Lakshmi, V.; Chan, T.K. & Nghiem, S.V. (2003). Soil moisture retrieval from AMSR-E. *IEEE Transactions on Geoscience and Remote Sensing*, 41, 215-229.

Oh, Y. & Kay, Y. C. (1998). Condition for precise measurement of soil surface roughness. *IEEE Transactions on Geoscience and Remote Sensing*, **36**(2), 691-695.

Ohta, T; Koike, T.; Lu, H.; Kuria, D.N.; Tsutsui, H.; Graf, T.; Kaihotsu, I; Davaa, G. & Matsuura, N. (2007). Development of a special sensor microwave imager (SSM/I) algorithm for long term monitoring of soil moisture, *Annual Journal of Hydraulic Engineering, Japan Society of Civil Engineer*, 51, 205-210.

Owe, M.; De Jeu, R.A.M. & Holmes, T.R.H. (2008). Multi-sensor historical climatology of satellite-derived global land surface moisture. *Journal of Geophysics Research*, 113, doi:10.1029/2007JF000769.

Paloscia, S. & Pampaloni, P. (1988). Microwave polarization index for monitoring vegetation growth. *IEEE Transactions on Geoscience and Remote Sensing*, 26, 617-621.

Paloscia, S.; Macelloni, G.; Santi, E. & Koike, T. (2001). A multifrequency algorithm for the retrieval of soil moisture on a large scale using microwave data from SMMR and SSM/I satellites, *IEEE Transactions on Geoscience and Remote Sensing*, 39, 1655~1661.

Paloscia, S.; Macelloni, G. & Santi E. (2006). Soil moisture estimates from AMSR-E brightness temperatures by using a dual-frequency algorithm, *IEEE Transactions on Geoscience and Remote Sensing*, 41(11), 3135-3144.

Prigent, C.; Aires, F.; Rossow, W. B. & Robock, A. (2005). Sensitivity of satellite microwave and infrared observations to soil moisture at a global scale: Relationship of satellite observations to in situ soil moisture measurements. *Journal of Geophysical Research*, 110, D07110. doi:10.1029/2004JD005087

Ray, P. S. (1972). Broadband Complex Refractive Indices of Ice and Water. *Applied Optics*, 11 (8), 1836-1844.

Said, S., Kothyari, U.C. & Arora, M.K. (2008). ANN-based soil moisture retrieval over bare and vegetated areas using ERS-2 SAR data. *Journal of Hydrologic Engineering*, 13, 461-475.

Schar, C.; Luthi, D.; Beyerle, U. & Heise, E. (1999). The soil-precipitation feedback: A process study with a regional climate model. *Journal of Climate*, 12, 722–741.

Schmugge, T.J. & Choudhury B.J. (1981). A comparison of radiative transfer models for predicting the microwave emission from soils. *Radio Science*, 16, 927-938.

Schmugge, T.J.; Wang, J.R. & Asrar G. (1988). Results from the Push Broom Microwave Radiometer flights over the Konza Prairie in 1985. *IEEE Transactions on Geoscience and Remote Sensing*, 26(5), 590-597.

Schmugge, T. J. & Jackson, T. J. (1992). A Dielectric Model of the Vegetation Effects on the Microwave Emission from Soils. *IEEE Transactions on Geoscience and Remote Sensing*, **30**(4), 757-760.

Seneviratne, S. I.; Luthi, D.; Litschi, M. & Schar, C. (2006). Land–atmosphere coupling and climate change in Europe. *Nature*, 443(7108), 205–209.

Shi, J. C.; Jiang, L.M.; Zhang, L.X.; Chen, K.S.; Wigneron, J.P. & Chanzy, A. (2005). A parameterized multifrequency-polarization surface emission model, *IEEE Transactions on Geoscience and Remote Sensing*, 43, 2831-2841.

Shibata, A.; Imaoka, K. & Koike, T. (2003). AMSR/AMSR-E level 2 and 3 algorithm developments and data validation plans of NASDA. *IEEE Transactions on Geoscience and Remote Sensing*, 41, 195-203

Singh, D.; Sing, K. P.; Herlin, I. & Sharma, S.K. (2003). Ground-based scatterometer measurements of periodic surface roughness and correlation length for remote sensing. *Advances in Space Research*, **32**(11), 2281-2286.

Singh, D. & Kathpalia, A. (2007). An efficient modeling with GA approach to retrieve soil texture, moisture and roughness from ERS-2 SAR data. *Progress in Electromagnetics Research-Pier*, 77, 121-136.

Tsang, L. & Kong, J. A (1977). Theory for thermal microwave emission from a bounded medium containing spherical scatterers, *Journal of Applied Physics*, 48, 3593-3599.

Tsang, L. & Kong, J. A (2001). *Scattering of Electomagnetic Waves: Advanced Topics*, Wiley, New York.

Ulaby, F.T.; Razani, M. & Dobson, M.C. (1983). Effects of vegetation cover on the microwave radiometric sensitivity to soil moisture. *IEEE Transactions on Geoscience and Remote Sensing*, 21, 51-61.

Ulaby, F.T. ; Moore, R.K & Fung A.K. (1986). *Microwave Remote Sensing: Active and Passive*, Artech House, Norwood, MA.

Verstraeten, W.W.; Veroustraete, F.; van der Sande, C.J.; Grootaers, I. & Feyen, J. (2006). Soil moisture retrieval using thermal inertia, determined with visible and thermal spaceborne data, validated for European forests. *Remote Sensing of Environment*, 101, 299-314.

Wagner, W.; Scipal, K.; Pathe, C.; Gerten, D.; Lucht, W. & Rudolf, B. (2003). Evaluation of the agreement between the first global remotely sensed soil moisture data with model and precipitation data. *Journal of Geophysical Research*, 108(D19), 4611- , doi:10.1029/2003JD003663

Wang, J.R. & Schmugge, T.J. (1980). An empirical model for the complex dielectric permittivity of soils as a function of water content. *IEEE Transactions on Geoscience and Remote Sensing*, 18, 288-295.

Wang, J. R. & Choudhury, B. J. (1981). Remote-Sensing of Soil-Moisture Content over Bare Field at 1.4 Ghz Frequency. *Journal of Geophysical Research-Oceans and Atmospheres*, **86** (Nc6), 5277-5282.

Wegmuller, U. & Matzler, C. (1999). Rough bare soil reflectivity model. *IEEE Transactions on Geoscience and Remote Sensing*, 37, 1391-1395.

Wen, B.; Tsang, L.; Winebrenner, D. P. and Ishimura, A. (1990). Dense media radiative transfer theory: comparison with experiment and application to microwave remote sensing and polarimetry, *IEEE Transactions on Geoscience and Remote Sensing*, 28, 46-59.

Wigneron, J. P.; Laguerre, L. & Kerr, Y. (2001). A simple parameterization of the L-band microwave emission from rough agricultural soils. *IEEE Transactions on Geoscience and Remote Sensing*, **39**(8), 1697-1707.

Yang, K.; Watanabe, T.; Koike, T.; Li, X.; Fujii, H.; Tamagawa, K.; Ma Y. & Ishikawa H. (2007). An auto-calibration system to assimilate AMSR-E data into a land surface model for estimating soil moisture and surface energy budget. *Journal of Meteorology Society of Japan*, 85A, 229-242.

Yang, K.; Koike, T.; Kaihotsu, I. & Qin, J. (2009). Validation of a Dual-pass Microwave Land Data Assimilation System for Estimating Surface Soil Moisture in Semi-arid Regions, *Journal of Hydrometeorology*, 10(3), 780-794.

18

Multiwavelength Polarimetric Lidar for Foliage Obscured Man-Made Target Detection

Songxin Tan[1] and Jason Stoker[2]
[1]*South Dakota State University*
[2]*USGS Earth Resources Observation and Science (EROS)*
USA

1. Introduction

LIDAR is an acronym for LIght Detection And Ranging. It is also called optical radar, laser radar or ladar under different application scenarios. The fundamental principle of lidar is similar to that of the microwave radar. However, lidar operates in optical frequency while radar works in microwave frequency. There are many unique ways that light interacts with matter, hence lidar differs in many respects from radar. The interaction between light and matter has different physical mechanisms, which include Rayleigh scattering, Mie scattering, Raman scattering, resonance scattering, fluorescence, absorption, and differential absorption and scattering, etc. (Measures, 1992). The interaction is detected by lidar, and the information is used to obtain the target characteristics. Lidar can be used to characterize a wide variety of targets, from a hard target such as a vehicle or a tree, to a pure phase object such as the atmosphere or the cloud. It has been used in many remote sensing applications, including vegetation and forest monitoring, land-use and land-cover detection, land management, cloud and aerosol detection, sea shore bathymetry, global ice sheet monitoring, among many others. The information provided by lidar improves our understanding of the environment, the ecological system, and the biocomplexity of the planet Earth.

A typical lidar system consists of a laser source, transmitting optics and receiving optics, photodetector, analog–to-digital (A/D) converter, signal and data processor, and output device. A block diagram of a lidar system is illustrated in Figure 1. The sequence of lidar operation is described as follows. The laser beam is sent out through the transmitter towards the target, and the transmitted light interacts with the media and is reflected/scattered back toward the receiving optics. The photodetector then converts the received light signal into an analog electrical signal, which is further transformed into a digital signal by an A/D converter. The signal is analyzed by the signal and data processor, and the final result is sent to an output device for display.

While most existing commercial lidar systems employ one single laser wavelength and lack the polarimetric measurement capability, a new type of lidar, the multiwavelength polarimetric lidar, was introduced in vegetation remote sensing (Tan & Narayanan, 2004). The multiwavelength lidar is able to provide more spectral information about the target under study. For example, it is well known that both coniferous and deciduous trees have similar reflectivity in visible wavelengths, but deciduous trees have distinctively higher reflectivity at near-infrared compared to coniferous trees. As a result, such spectral information at several laser wavelengths can be used for tree species classification (Jaaskelainen et al., 1994). On the other hand, polarimetric lidar provides polarization state of the backscattered laser which contains information such as target composition, surface roughness, water content, etc. The information is essential in order to understand the target characteristics. For example, a previous study found that deciduous trees had distinct depolarization signature at 1064-nm while coniferous trees had such signature at 532-nm (Kalshoven & Dabley, 1993). At this time, the understanding of polarimetric scattering property of vegetation at optical wavelengths is very limited. Both theoretical and experimental studies (Ma et al., 1990) are needed in order to fill the gap. As a new concept lidar sensor, the potential of multiwavelength polarimetric lidar is yet to be explored.

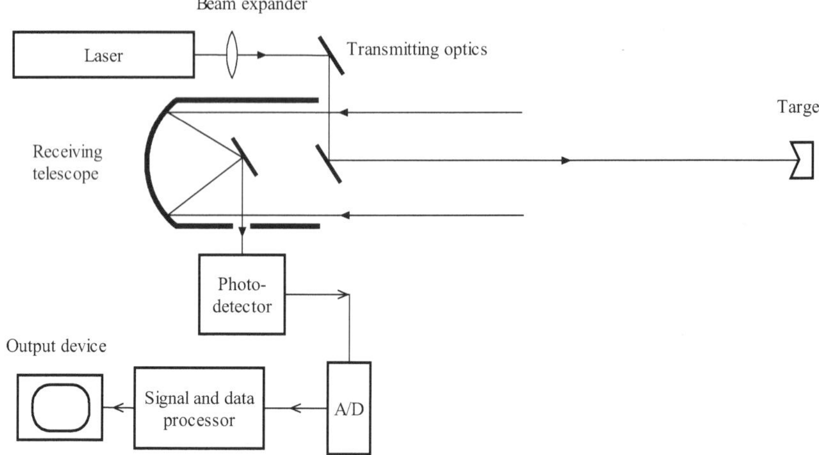

Fig. 1. Illustration of a typical lidar system.

2. Foliage Obscured Target Detection

Many remote sensing applications demand for the foliage obscured man-made objects detection. These applications may range from civilian applications, such as search and rescue missions in densely forested areas, to military applications, such as detecting camouflaged enemy vehicles. Traditional passive remote sensing has very limited capability in foliage obscured target detection. Active sensors such as the foliage penetration (FOPEN) radar has been proven successful in many applications and has become the primary tool when foliage obscured target detection is needed. Studies on foliage obscured target detection using radar technology have been reported in the literature (Nashashibi & Ulaby, 2005; Xu & Narayanan, 2003). The reason is that microwave has longer wavelength and can penetrate the vegetation canopy. The penetration ability of microwave or lightwave is related to its wavelength. Generally speaking, the longer the wavelength, the better it penetrates. Lidar is not traditionally considered when discussing foliage penetration (FOPEN) applications. This is due to the fact that the laser wavelength is much shorter than the microwave wavelength. Therefore, a laser beam is considered as incapable of penetrating through the vegetation canopy. However, multiwavelength polarimetric lidar can be a viable alternative even though it has received little attention.

Nevertheless, gaps do naturally occur inside a forest. These include the gaps inside a single tree crown and gaps in between different tree canopies. Consequently, the target is usually not 100% covered by the foliage. If the laser beam or even a portion of the laser beam is able to go through the gaps and hit the target and return to the detector, then it is still possible to detect the hidden target using a lidar sensor. For example, a previous study demonstrated that a scanning lidar system could be used to detect the hard obstacles on the ground which were non-drivable and the foliage area which was drivable for a robotic vehicle (Castano & Matthies, 2003). It demonstrated that lidar does have certain foliage penetration capability. With its inherent shorter wavelength, lidar provides better range resolution and better spectral purity in comparison with radar. Therefore, it can offer advantage in certain target detection and identification applications.

In this study, a multiwavelength polarimetric lidar was used to detect a man-made target obscured by trees. The multiwavelength polarimetric lidar provides several advantages in target detection. Firstly, the spectral reflectance of a man-made target is usually different from that of the foliage. Even when certain type of camouflage cover or paint is applied, it is extremely difficult to match the natural foliage reflectance pattern at several different wavelengths. Secondly, the polarimetric scattering properties of the man-made target and the foliage are generally quite different. Since the polarimetric scattering property at an optical wavelength is determined by many factors such as the material composition, target shape, surface roughness, etc., it is extremely difficult to produce a camouflage that has the same polarimetric scattering feature as the foliage. Therefore, the detection rate and identification success can be improved using the combined information from spectral and polarimetric features.

3. Lidar System Description

3.1 MAPL lidar system

The lidar system built and used in this study is the Multiwavelength Airborne Polarimetric Lidar (MAPL) system. The MAPL is composed of several sub-systems: the laser source, the optical receiver assembly and the data acquisition and processing hardware and software. The MAPL system block diagram is shown in Figure 2.

The MAPL system functions as follows. A computer controls a digital delay/pulse generator through an RS-232 interface. TTL signals are then generated by the delay generator. One TTL signal is sent to trigger the Q-switch of the laser; the other TTL signal is delayed and sent to trigger the digitizer. The delay time is determined by the range of the target. The laser simultaneously sends out two laser pulses at 1064-nm and 532-nm. On top of the laser head, there is a silicon PIN detector attached to monitor the relative laser output. The laser pulses are backscattered by the target, and received by four photomultiplier tube (PMT) detectors. These detectors enable detection at both co-polarized and cross-polarized direction at the two wavelengths. Outputs from the PMT detectors are digitized by a high-speed analog to digital (A/D) converter and then sent to the computer through a PCI interface card. The ranging ability is achieved by precisely timing the transmission and reception of the laser pulse. Data are stored in a hard drive and post-processed by the computer to extract information.

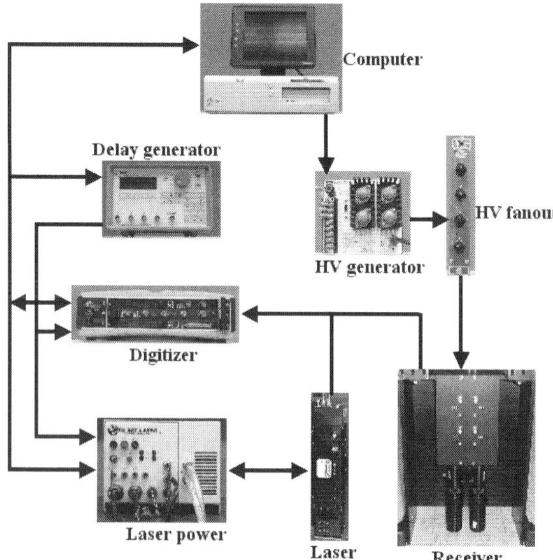

Fig. 2 Block Diagram of the MAPL system.

3.1.1 Laser

The MAPL employs a Big Sky CFR-400 Nd:YAG laser. This is a multiwavelength pulsed laser system, and is designed to be compact, rigid and stable, and is thus suitable for field applications. The laser has two outputs, one at 1064-nm and one at 532-nm, emitting from two separate apertures. Special optical wedge assembly is provided inside the laser module to ensure that the two beams are parallel. The two beams are linearly polarized in the horizontal direction. The polarization purity is greater than 100:1 at both wavelengths. Each beam has a divergence angle of about 4 mrad, which will form a laser footprint with a diameter of about 4 m at a distance of 1000 m. The highest pulse repetition rate is 10 Hz. The laser output energy is adjustable and typically 30 mJ/pulse at each wavelength is used. An attenuator can also be mounted in front of the laser head to decrease the laser output energy for safety considerations. The near field beam diameter is about 6 mm at each wavelength. The pulse width is 10 ns, which yields a range resolution of 1.5 m. To ensure accurate timing, trigger of flash lamp and Q-switch are separately controllable.

A Si PIN detector from Thorlabs is sealed using an O-ring and attached externally on top of the laser head, and is used to monitor the relative output laser energy. This serves as a reference in order to calibrate the received laser return. The power supply unit and cooling group unit are integrated into a Mini-ICE (integrated cooler and electronics) subsystem. It is a compact and rugged sub-system designed to operate in a harsh outdoor environment. A remote control box is provided to run all the controls on the laser. An RS-232 interface is also provided for computer controls.

3.1.2 Receiving Optics

A total number of four receiving channels are employed in the MAPL system. There are two channels, one co-polarized and one cross-polarized, for the 1064-nm light; and the same configuration is used for the 532-nm light. Adding more receiving channels can enable the measurement of the Stokes parameters, which are used to fully describe the polarization state of the backscattered light. However, the circularly polarized light from the vegetation and the ground is very weak under linearly polarized laser illumination (Kalshoven & Dabney, 1993). Useful information is extremely difficult to retrieve from the circular polarization channel because of the low signal to noise ratio (Tan, Narayanan & Kalshoven, 2001). The other limiting factor comes from the system size limit. More receiving channels require a larger system which is impractical for field deployment. Thus, fully polarimetric ability is not currently integrated in the MAPL system.

The optical receiver apertures are 25 mm in diameter. There is a linear polarizer, an optical interference filter, and a single plano-convex lens with a focal length of 65 mm, inside each receiver, as shown in Figure 3. The received light is diffused onto the PMT detection plane, instead of imaging onto it. The advantage of the so-called straight-through structure is to minimize the reflection and refraction optics inside the receivers, thus reducing possible modifications to the polarization state of the backscattered light (Kalshoven & Dabney, 1993). The 1064-nm polarizers inside the receivers are from Corning Polarcor and the 532-nm polarizers are from Polaroid, both of them have polarization extinction ratios higher than 1,000:1. The system also uses narrowband interference filters from Andover. These filters have full-width-at-half-maximum (FWHM) bandwidths of 1 nm. The narrowband filters serve to reduce the background solar radiation. The peak transmission for 1064-nm filter is ~38%, and for 532-nm filter is ~47%. For each receiver there is a field stop at the focal plane of the lens, which defines the detector FOV to be 6 mrad, slightly larger than the laser beam divergence angle.

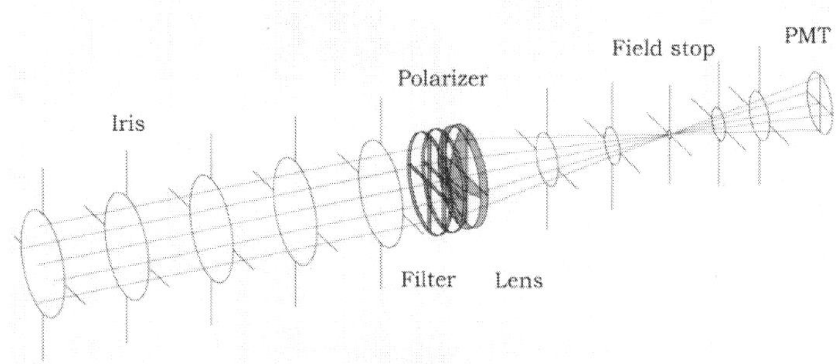

Fig. 3. The receiver optics.

Highly sensitive and ultrafast PMT detectors from Hamamatsu are used as photodetectors. Hamamatsu R632-01 is used to detect 1064-nm light, and R1464 is used for 532-nm, making good use of their spectrum response.

One important parameter for a photodetector is the minimum detectable signal. Usually this is specified by the noise equivalent power (NEP). This quantity refers to the radiant flux in watts to produce a signal to noise ratio of unity, at the output end of the detector under dark current only condition. This is also called the photon limited signal to noise ratio. In dark current limited case, the NEP is defined as (Budde, 1983)

$$NEP = \sqrt{2e \cdot i_{dark} \cdot G \cdot \Delta f} \, / \, S \, , \tag{1}$$

where S is the PMT current sensitivity at peak response wavelength, e is the electron's charge, i_{dark} is the dark current, G is the detector gain, and Δf is the bandwidth of the electronic system. At a bandwidth of 1 Hz, the NEP is 4.2×10^{-13} watts for the R632-01 detector at 1064-nm, and 1.1×10^{-15} watts for the R1464 detector at 532-nm.

From the backscattered laser signal, the cross-polarization ratio is used to quantitatively describe the polarimetric scattering property of the target surface. It is defined as the ratio between the cross-polarized return and the co-polarized return, i.e.,

$$\delta = \frac{I_\perp}{I_{//}} , \tag{2}$$

where $I_{//}$ represents the co-polarized laser return and $I_\perp$ represents the cross-polarized return received at the PMT detectors. The value of the cross-polarization ratio is usually within the range of 0 to 1.0. However, for some targets, it was observed that this ratio can exceed 1.0 (Kalshoven & Dabney, 1993).

3.1.3 Data Acquisition and Processing

A Berkeley Nucleonics model 555 digital delay/pulse generator is integrated in the system as a timing source. It is used to generate the control TTL signals to trigger the laser and the digitizer. The timing accuracy of the delay generator is 1 ns. For A/D conversion, two 8-bit digitizing cards, DC256 and DC110, from Acqiris are selected. The DC256 has four input channels, and a sampling rate of 500 MS/s, and is used to digitize the outputs from the PMT detectors. The DC110 has one channel, also samples at 500 MS/s, and is used to digitize the reference signal. Both cards are housed inside a modular CompactPCI crate, and connected to the computer through a PCI interface card with 100 Mb/s transfer rate. The Acqiris digitizers employ an internal time trigger interpolator with 5 ps resolution, which is used to assist timing calibration and trigger positioning. The accurate timing ability of the digitizer improves the range measurement accuracy, and helps obtain accurate lidar waveforms which record the entire history when the laser propagates and interacts with the target.

The system software is programmed using Labview. Functions of the Labview program include control of the digital delay generator, the laser, the digitizer, data recording and real time waveform display. Data are recorded and post-processed.

The photographs of the MAPL lidar system are shown in Figure 4, where the photo at the left side shows the laser head and the optical receiver assembly and the photo at the right side shows the electronic devices mounted inside a standard 19-inch rack. This is the configuration for ground experiments. The system needs to be repackaged in order to be installed inside an aircraft for future airborne missions.

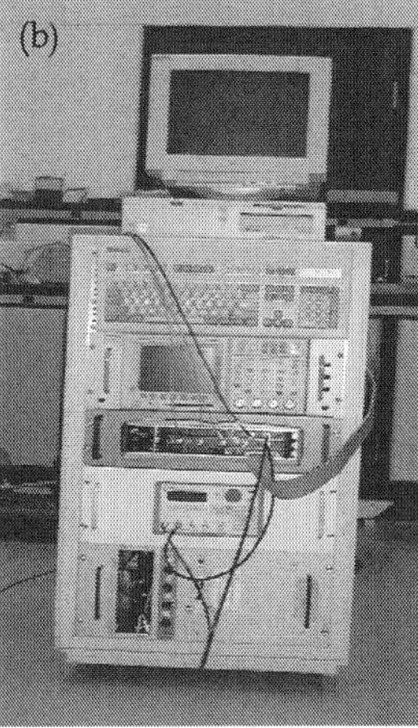

Fig. 4. Photographs of the MAPL system, where (a) is the optical assembly and (b) is the electronic subsystem.

3.2 Multiwavelength polarimetric lidar calibration

To account for the laser power output instability, photodetector gain variation and other potential error sources, the lidar data must be calibrated. Calibration is always a very difficult task. This is especially true for a polarimetric lidar where the polarimetric reflectance data calibration requires special consideration. In general, an ideal Lambertian reflection surface with known reflectivity needs to be used as a calibration standard. But an ideal Lambertian surface is extremely difficult to obtain even for laboratory measurement, not to mention the field calibration.

Measurement of materials with potential to be used as calibration standards was done inside an optical laboratory, in search of a better field calibration standard. Table 1 lists all six materials used. The Stokes parameters of these materials at both 1064-nm and 532-nm were measured to quantify their polarimetric reflectance properties. The measurement results are shown in Figure 5. We can see that the tarp and the concrete (Figure 5 (a) and (d)) resemble an ideal diffusing scatterer at both wavelengths. Their backscattered light intensities do not show strong dependence on the incident angle. Thus, it is concluded that the backscatter properties of the tarp and the concrete are close to an ideal Lambertian surface. Both of them could be used as a calibration standard in field experiments.

Canvas tarpaulin (Gosport Manufacture Inc.)
White paper (Great White® 86700)
Plywood (APA Champion 329)
Concrete (with surface polished)
Machined aluminum plate
Black anodized aluminum plate

Table 1. Materials used for Stokes parameter measurement.

In practice, the canvas tarp is used as a calibration standard due to its portability. Laboratory measurement of the polarimetric reflectance at co-polarized and cross-polarized directions from the canvas tarp is shown in Figure 6, where solid blue lines are for 1064-nm and dashed red lines are for 532-nm. The data shown here are used in the calibration for the MAPL system. The standard calibration procedure for ground experiments is described as follows. Before measuring each target, the calibration standard (i.e., the canvas tarp) is set up side-by-side with the target, and the lidar reflectance data are collected from the tarp. The high voltage supply to each PMT detector is individually adjusted to make sure that the signal has enough amplitude and is not saturated after A/D sampling. The data are compared with the laboratory measurement data as shown in Figure 6. Finally, the ratio between the averaged field calibration data and the laboratory results from the same tarp was used as a calibration constant to correct for any measurement variations. Once the calibration constant is obtained, the calibration procedure is done. Then the MAPL system is directed to the target under study and multiple measurements are recorded as raw data. These raw data from the target are then multiplied with the calibration constant to obtain the calibrated data. All future data analysis should be based on the calibrated data.

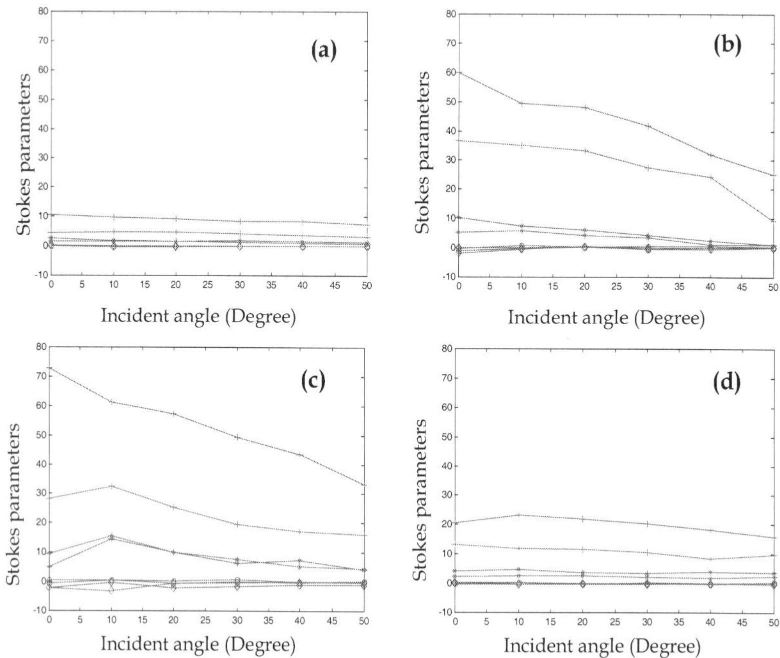

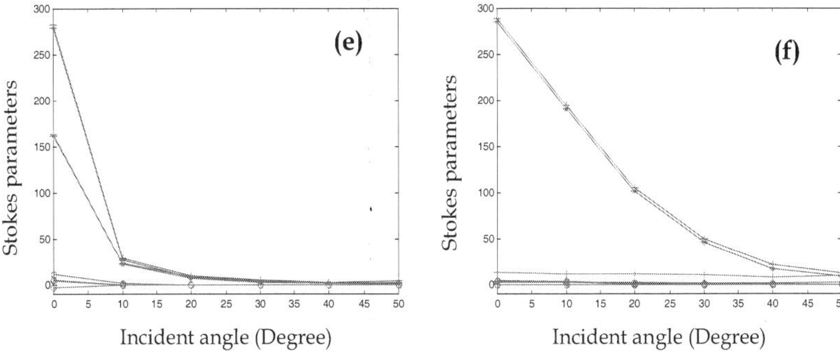

Fig. 5. Stokes parameters measurement of backscattered laser light from various materials. (a). Canvas Tarp.   (b). White paper.   (c). Plywood.   (d). Concrete.   (e). Aluminum plate. (f). Black anodized aluminum plate.   ('+'=I, '*'=Q, '◊'=U and 'o'=V. Solid blue lines are for 1064-nm and dashed red lines are for 532-nm.).

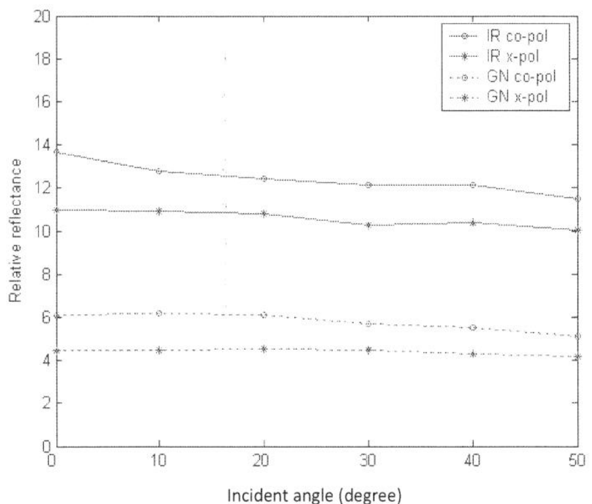

Fig. 6. Laboratory measurement of the polarimetric reflectance of a canvas at 1064-nm (solid blue lines) and 532-nm (dashed red lines) for calibration purpose.

4. Field test and data analysis

4.1 Description of the test site

The field test site was selected at Oakwood Lakes State Park in Eastern South Dakota. There were plenty of trees and shrubs inside this park. The tree species observed at the test site were primarily cottonwood, ash, oak and willow. A silver Dodge minivan was used in this experiment as a man-made target. The vehicle was parked and hidden behind several trees as shown in Figure 7. The MAPL system was set up across the lake at a distance of about 1000 m. The laser beams were shot horizontally across the lake toward the test scene.

Fig. 7. Photograph of the test site, where a vehicle was hidden inside the trees. The vehicle was ~1000 m away from the MAPL sensor. The approximate positions of the laser scans are marked by red arrows here.

The laser scanned over the test area from the left to the right. The scanning pattern is demonstrated in Figure 7. A total number of six separate scans were made during the experiment. Only one of the scans, scan number 4, coincided with the vehicle's location.

4.2 Experimental results

From the lidar data, it is observed that the vehicle paint strongly reflects near-infrared laser at both co-polarized and cross-polarized directions. The co-polarized return at 1064-nm is shown in Figure 8, where a white arrow is pointing at the vehicle position. The reflectivity of the vehicle paint at near-infrared is so strong that the returned signal is saturated after A/D conversion as seen from scan number 4. From scan lines 1, 2, 3 and 5, it is seen that the reflectance of the foliage at 1064-nm is much lower compared with the vehicle paint. The relative return of the vegetation is typically below 55 after A/D conversion. It is also observed that the top of the vegetation (scan number 1) is sparse so the returned signal strength is weaker than that at the bottom of the trees (scan number 5) where the vegetation becomes more dense. In general, it is observed that as the scan line gets lower, the returned signal becomes stronger due to more dense foliage. However, at scan number 6, the laser beam finally reaches the lake water and the returned signal becomes almost invisible. Two combined effects cause the backscattered signal level to be very low in this case. The first one is that water strongly absorbs the near-infrared laser, in comparison with foliage and vehicle paint. In addition to that, the water surface also acts as a specular surface and reflects the laser energy in the forward direction away from the detector due to the very large incident angle. As a result, there is little signal observed from scan number 6 as shown in the bottom of Figure 8. For better clarity, Figure 9 shows the image from scan number 4 separately. An arrow is used to highlight the vehicle inside the trees. All the weak scatterers in this image represent the vegetation; while the strongest reflector represents the hidden vehicle.

The direct return data from the co-polarized and cross-polarized green channels at 532-nm reveal no observable difference between the foliage and the vehicle. This is demonstrated in Figure 10 at the co-polarization channel. The result shows that the vehicle paint has a very similar reflectivity at 532-nm compared with the vegetation. Therefore, using the direct return information from a single wavelength at 532-nm and a single polarization, it is impossible to detect the foliage-obscured vehicle. Conversely, the polarization information can provide enhanced capability in detection and identification. After computing the cross-polarization ratio following Eq. (2), the difference between the vehicle and the vegetation is revealed. The cross-polarization of the vehicle surface is significantly lower than that of the vegetation. This is partially explained by the fact that the surface of the vehicle paint is much smoother compared with the surface of vegetation leaves and barks at optical wavelength. As a result, the vehicle surface can better maintain the original linear polarization state of the incident laser thus leads to low cross-polarization ratio. Moreover, the random orientation and distribution of the tree leaves, i.e., the vegetation canopy structure, also contributes to higher depolarization during the multiple scattering process. As observed in Figure 11, the cross-polarization ratio of the vehicle (as indicated by an arrow) is typical below 0.2. On the other hand, the cross-polarization ratio of the vegetation is consistently above 0.4. At certain vegetated areas, this ratio can reach a value of larger than 0.8. This may be related to trees with very rough leaf surface and trees with very dense canopy structure.

Figure 12 and Figure 13 show the scatter plot of the returned signal at 1064-nm and 532-nm, respectively. It is clear that the classification boundary between the vegetation and the vehicle can be easily obtained in this case. From Figure 12 it is seen that the reflectivity of the vehicle paint at 1064-nm is consistently higher at both co-polarized and cross-polarized directions compared with the vegetation. The vehicle signal is so strong that it is almost always saturated. From Figure 13 it is observed that although the vehicle paint at 532-nm has comparable reflectivity in comparison with the vegetation, the cross-polarization ratio of the vehicle is much lower than that of the vegetation. The cross-polarization ratio at 1064-nm was not computed since the return data from the vehicle are saturated at both the co-polarized and cross-polarized channels. However, a logarithmic amplifier may be used in the lidar system to reduce the dynamic range of the signal. In that case, it is possible to obtain the cross-polarization ratio at 1064-nm. With the additional polarization information at 1064-nm, better detection and identification should be available for more targets.

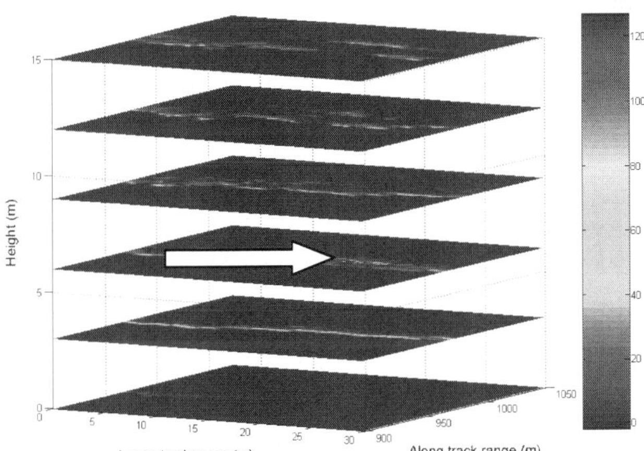

Fig. 8. Scanning result of the relative lidar return from the co-polarization channel at 1064-nm. The arrow is pointing to the vehicle inside the trees.

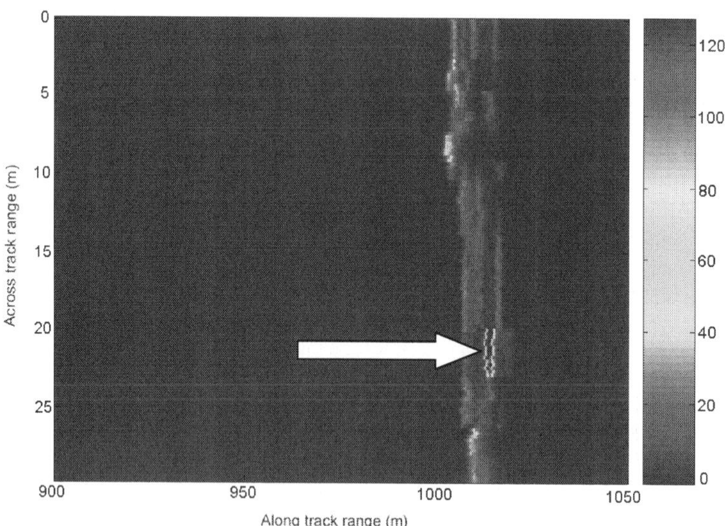

Fig. 9. Scan number 4 of the relative lidar return from 1064-nm at the co-polarization channel. The arrow is pointing to the vehicle (the strong scatterer) inside the trees (the weak scatterers).

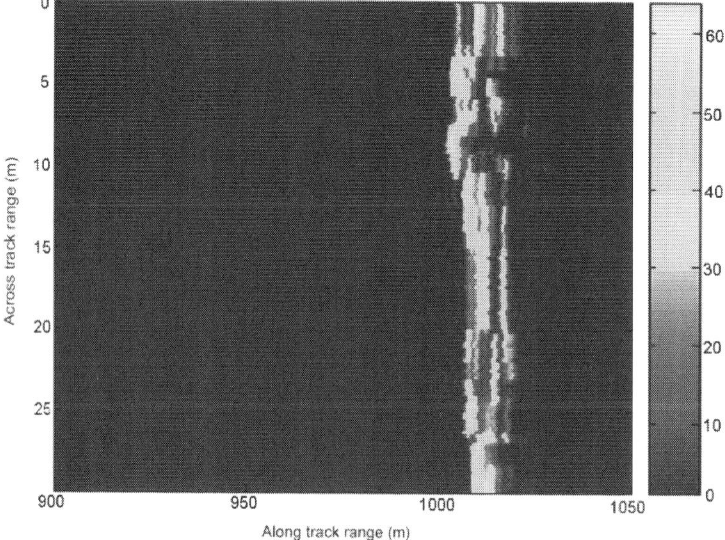

Fig. 10. Scan number 4 of the relative lidar return from 532-nm at the co-polarization channel. The vehicle could not be identified in this image.

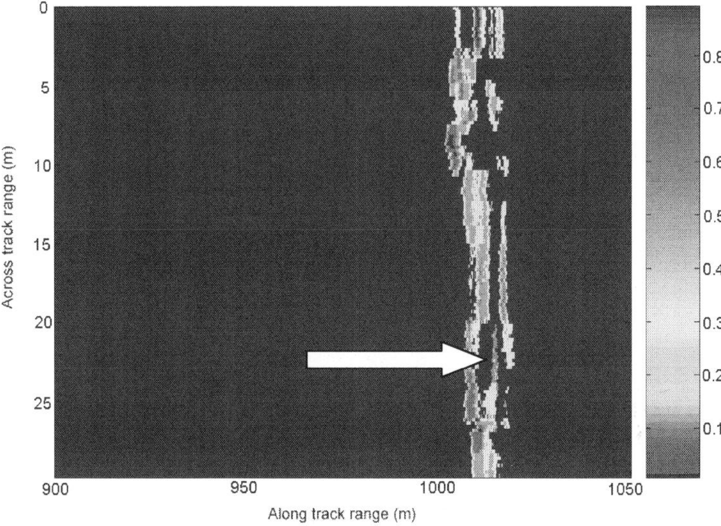

Fig. 11. The cross-polarization ratio at the 532-nm from scan number 4. The arrow is pointing to the vehicle with low cross-polarization ratio inside the trees.

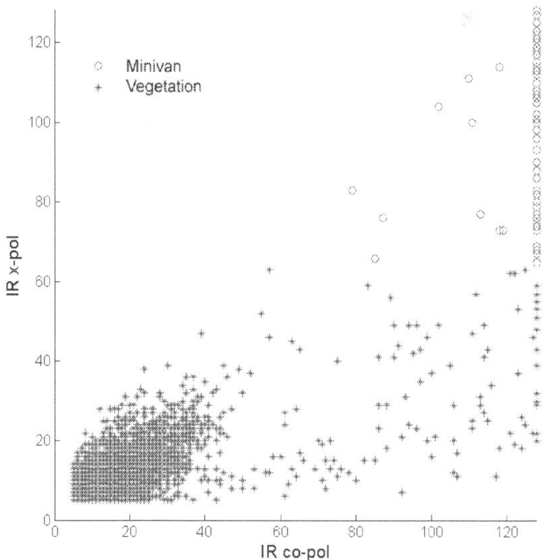

Fig. 12. Scatterplot of the lidar return at 1064-nm, where the x-axis is the co-polarization channel and the y-axis is the cross-polarization channel.

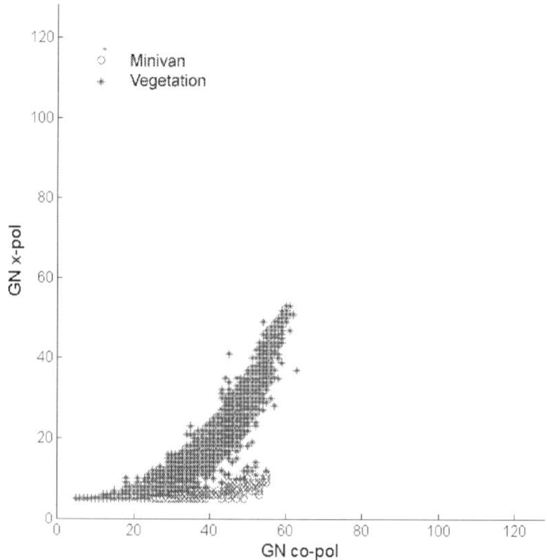

Fig. 13. Scatterplot of the lidar return at 532-nm, where the x-axis is from the co-polarization channel and the y-axis is from the cross-polarization channel.

5. Conclusions and future work

The experimental study demonstrates that the multiwavelength polarimetric lidar provides an enhanced capability in the detection of a foliage-obscured vehicle. It is revealed that the reflectivity of the vehicle and the reflectivity of the foliage are very close at 532-nm. However, at 1064-nm the reflectivity of the vehicle is significantly higher than that of the foliage. Using multiwavelength lidar in this situation demonstrates improvement in target detection. It is also shown that the cross-polarization ratio of the tree canopies is much higher than that of the vehicle at 532-nm. If a logarithmic amplifier was employed in the lidar system, the cross-polarization ratio at 1064-nm could be obtained, which can further improve the detection and identification accuracy. It is evident from the experiment that the polarimetric information provides enhanced target detection and identification capability. The multiwavelength polarimetric lidar certainly has its advantage over the existing commercial lidar systems, which are single wavelength and non-polarimetric, in target detection and identification.

In this study, only ground test data were collected. The airborne lidar data are expensive and were not collected at the time. However, it is known that more openings of the forest canopies are observed looking from above the ground. This means that the laser beam has a better chance to penetrate the tree canopies from airborne sensors. As a result, there is a better chance to detect the foliage obscured target from an airborne lidar sensor than from a ground based sensor.

This study also demonstrates the validity of the technique to detect and identify other foliage-obscured man-made targets using a multiwavelength polarimetric lidar. As long as we can find the difference in the spectral reflectance or polarimetric reflectance between the man-made target and the vegetation, we can always employ this technique. The selection of appropriate laser wavelengths and polarization combinations should be able to maximize the differences between the vegetation and the man-made target, hence enable the detection of many other types of man-made targets.

Further research is needed in order to use this multiwavelength polarimetric lidar detection technique in practical applications. From the lidar hardware side, many improvements could be made. For example, a tunable laser could be used in the system to provide a wide range of laser wavelengths for better detection. Furthermore, we can introduce a logarithm amplifier or a 12-bit (or higher) A/D converter instead of an 8-bit one in the lidar system to make sure that the system can handle signals at a larger dynamic range without signal saturation. Polarimetric lidar calibration is another challenge and better calibration strategy still needs to be explored. A polarimetric calibration standard that has a constant polarimetric reflectance property over repeated use and is easy to deploy in the field is desired. Even more importantly, efforts need to be made for the multiwavelength polarimetric lidar operation and data processing to be simple enough so that even a person with limited engineering background can operate such a sensor. From the software side, an algorithm that can automatically detect and identify the targets needs be developed. Both artificial neural networks (NN) and support vector machine (SVM) seem to be good candidates for automated data processing. The classification accuracy will be improved if we understand the target polarimetric scattering characteristics. Such understanding can greatly facilitate the development of classification algorithm. At this time, however, there is only limited research available. Consequently, more studies on the polarimetric scattering properties of both man-made targets and vegetation are in urgent need (Tan, Narayanan & Helder, 2005). In addition, more ground experiments are also needed. For example, man-made targets under camouflage conditions need to be tested. Camouflage presents a challenge in target detection. However, multiple-wavelength and polarization should provide advantage over existing lidar technology for camouflaged target detection. Camouflage is usually designed to match the spectral reflectance of the background. It is usually not intended to match the polarimetric reflectance of the background. Even if anybody wanted to do so, it would be extremely difficult to control the polarimetric reflectance of the camouflage. Therefore, a polarimetric lidar have a better chance to detect and identify the camouflaged target.

In summary, the multiwavelength polarimetric lidar is a new concept lidar in foliage obscured man-made target detection. Even though more studies are needed in this new technology, the experimental study clearly serves the purpose to demonstrate that the multiwavelength polarimetric lidar can provide better target detection and identification over the traditional single wavelength and non-polarimetric lidar system.

6. References

Budde, W. (1983). Physical detectors of optical radiation, In : *Optical Radiation Measurements*, Grum, F. & Bartleson, C. (Ed.), Academic, New York.

Castano, A. & Matthies, L. (2003). Folidage discrimination using a rotating ladar, *IEEE Internation Conference on Robotics and Automation*, Vol. 1, pp. 1-6.

Jaaskelainen, T.; Silvennoinen, R.; Hiltunen, J.; & Parkkinen, P. (1994). Classification of the reflectance spectra of pine, spruce and birch, *Applied Optics*, Vol. 33, No. 12, pp. 2356-2362.

Kalshoven, J. & Dabley, P. (1993). Remote sensing of the Earth's surface with an airborne polarimetric laser, *IEEE Trans. Geoscience and Remote Sensing*, Vol. 31, pp. 438-446.

Ma, Q.; Ishimazu, A.; Phu P.; & Kuga, Y, (1990). Transmission, reflection, and depolarization of an optical wave for a single leaf, *IEEE Trans. Geoscience and Remote Sensing*, Vol. 28, pp. 865-872.

Measures, R. (1992), *Laser Remote Sensing: Fundamentals and Applications*, Krieder Publishing, Malabar, FL.

Nashashibi, A. & Ulaby, F. (2005). Detection of stationary foliage-obscured targets by polarimetric millimeter-wave radar, *IEEE Trans. Geoscience and Remote Sensing*, Vol. 43, pp. 13-23.

Tan, S.; Narayanan, R. & Kalshoven J. (2001). Measurement of Stokes parameters of materials at 1064-nm and 532-nm wavelengths, *Proc. SPIE, laser radar technology and applications VI*, Vol. 4377, pp. 263-271.

Tan, S. & Narayanan, R. (2004). Design and performance of a multiwavelength airborne polarimertric lidar for vegetation remote sensing, *Applied Optics*, Vol. 43, pp. 2360-2368.

Tan, S.; Narayanan, R. & Helder, D. (2005). Polarimetric reflectance and depolarization ratio from several tree species using a multiwavelength polarimetric lidar, *Proc. SPIE, Polarization Science and Remote Sensing II*, Vol. 5888, 5888-23.

Xu, X. & Narayanan, R. (2003). FOPEN SAR imaging using UWB step-frequency and random noise waveforms, *IEEE Trans. Aerospace and Electronic Systems*, Vol. 37, pp. 1287-1300.

19

Unmixing Based Landsat ETM+ and ASTER Image Fusion for Hybrid Multispectral Image Analysis

Nouha Mezned[1], Sâadi Abdeljaoued[1]
and Mohamed Rached Boussema[2]
[1] Faculté des Sciences de Tunis,
[2] Ecole Nationale d'Ingénieurs de Tunis
TUNISIA

1. Introduction

Monitoring the environmental risks associated with mine tailings in a quick and timely fashion is the first step towards mitigating their impact. Mine tailings may have a widespread geographical distribution; their location and extent may also vary along time, due to reprocessing and disposal activities. For these reasons, the characterization of mine tailings using traditional field work alone is both costly and inefficient. Remote sensing techniques have been proven extremely valuable in the inventory, characterization, and remediation of mine tailings elsewhere (Peters & Hauff, 2000). Determining the location, extent, and geochemistry of mine tailings is the first step towards remediation and hence avoidance of negative health and environmental consequences.

In this chapter we focus on mine tailing cartography and monitoring using multispectral data (Landsat ETM+ and ASTER). These data are well available from much passed years at regular intervals. We are particularly interest in the case study of zinc and lead mines in the north of Tunisia (North Africa). Particularly, the mine of Jebel (Hill) Hallouf-Bouaouane (36°42'N 9°0'5''E), which is forsaken since 1986, is among several types of mines in the Mejerda river watershed (Mansouri, 1980). The Mejerda River is the most important river which is exploited for the agriculture irrigation and drinking water alimentation of the north of Tunisia. In this site, mine tailings cause the environment degradation, so it had polluted soils, vegetation and water quality (Souissi, 2007). Mine tailing cartography becomes fundamental in order to follow the environmental changes and pollution quickly.

The approach was organized in two steps; a coarse cartography based on the Landsat (Enhanced Thematic Mapper Plus) ETM+ spectral unmixing using image derived endmember and a detailed cartography based on a multispectral inter-images fusion using a simplified version of Multisensor Multiresolution Technique (MMT). The analysis of the resulting hybrid image allowed to tailing cartography refinement. Detailed fraction maps were generated.

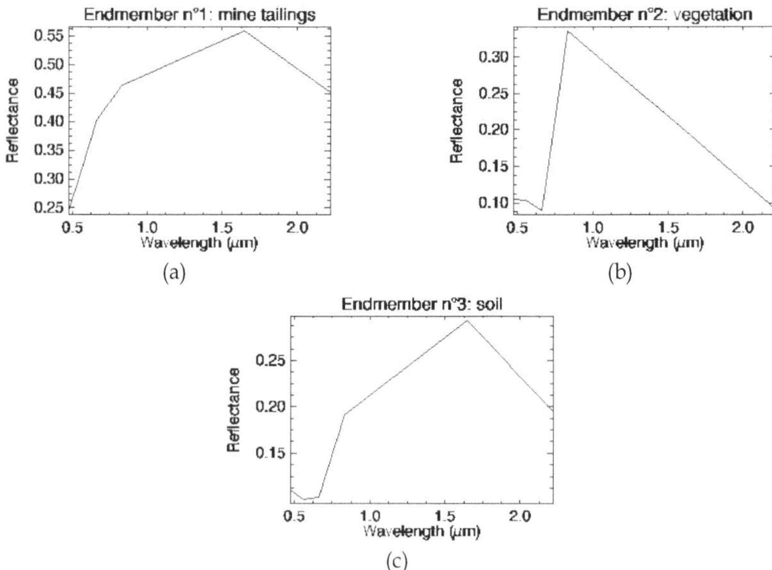

Fig. 1. The three endmember spectra selected from first three PC analyses.

The chapter is structured as follows. Section 2 is a tailing cartography approach description. In section 3, we consider a coarse and refinement cartography of the mine tailings using hybrid Multispectral data and demonstrate that its important special case. In section 4, we present results and discussion. Section 5 concludes the chapter.

2. Mine tailing cartography

The adopted cartography approach includes two main parts: The first one deal with the classification of the ETM+ data which results give the distribution and the abundance image of mine tailing surface cover. The resulting map represents a coarse mine tailing cartography. The second part, focus on the ETM+ and ASTER data fusion and analysis (Mezned et al., 2007). We used The (MS) and (Pan) ETM+ data acquired on May 3, 2000 and Aster Level 2B SWIR reflectance surface product acquired on June 26, 2000. The resulting hybrid image have the high spatial resolution of the (MS)/(Pan) Landsat ETM+ image (15m) and its relative high spectral resolution combined with ASTER SWIR image (10 bands).

2.1 Multispectral image classification
The first classification was performed on Landsat ETM+ MS/Pan fused image (196x196 pixels) with 6 bands using constrained spectral unmixing method. We used the Principal Component (PC) Spectral Sharpening technique to merge (MS) and (PAN) Landsat ETM+ data. Spectral information's are conserved as these two images were acquired by the same sensor. The used images were georeferenced, co-registered, topographically and

atmospherically corrected using the FLAASH (Cooley, 2002) software before performing the sharpening. Indeed, (Pan) data was only topographically corrected.

The first step of the analysis identifies the pure materials that are present in the scene. Principal Component (PC) analysis method was used for optimal endmember extraction (Boardman, 1993); (Carreiras and al., 2002). It allows one to reduce the dimensionality of the data and to know the number of endmembers present in the scene as explained in (Keshava & Mustard, 2002). Endmembers lay correctly at the corners of the triangular shaped diagram, indicating the purest pixel spectra (Boardman, 1993). Figure 1 show the three endmember, which were selected using the first three principal components (contain more than 95 % of the information), corresponding to tailings, vegetation and soil. Their identification was based on their typical spectra and their location.

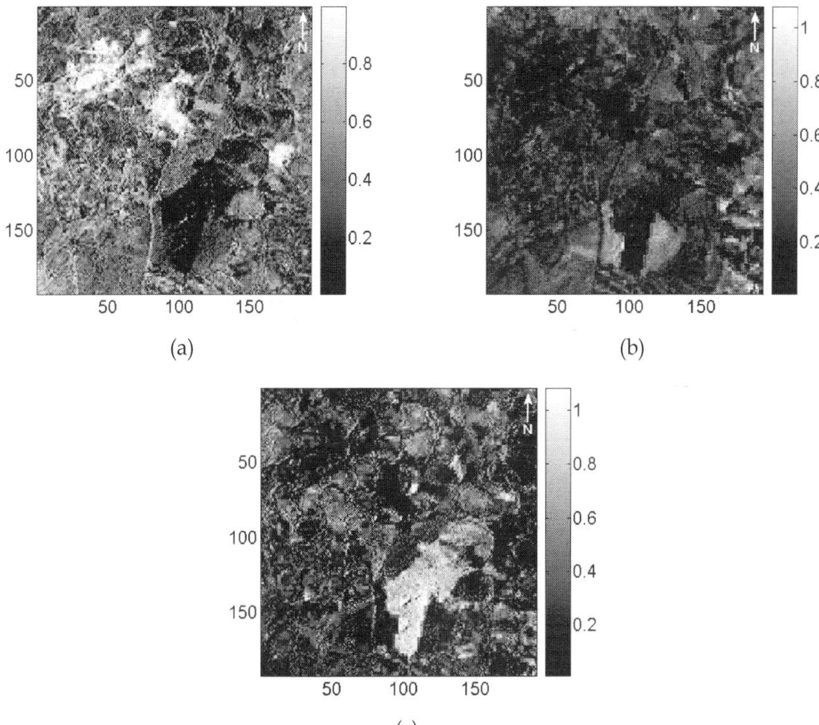

(a)

(b)

(c)

Fig. 2. Fraction maps (196x196 pixels) resulting from Landsat ETM+ coarse classification: (a) tailings, (b) vegetation and (c) soils.

The fraction maps which estimated by the constrained least-squares algorithm (Singer, 1981); (Shimabukuro & Smith, 1991) for the three pure materials are represented in Figure 2. Note that a black (respectively, white) pixel in the map indicates a large (respectively, small) value of the abundance. The accuracy of the endmember fraction maps are evaluated using the Root Mean Square (RMS) error. The most of RMS error (96 %) are lower than 1.35 10^{-5}, which shows a good result.

2.2 Unmixing based multispectral Image Fusion

The proposed unmixing based image fusion method consists on a two data set fusion: the first set has a high spatial resolution (and low spectral resolution), called classifying instrument (CI) and the second one have a high spectral resolution (and low spatial resolution), called measuring instrument (MI). The unmixing method uses the spatial high resolution data classification (CI) to unmix the lower spatial resolution image (MI). Furthermore, the unmixing can be performed only relative to the recognized classes in the (MI) image.

The main contribution consists on considering a fusion of two different multispectral sensors. Thus, we chose the (MS) Landsat ETM+ data combined with the corresponding (PAN) band (Chavez et al., 1991) as the (CI) image (Figure 3 a) and the ASTER SWIR product as the (MI) image (Figure 3 b). These images were co-registered before fusion. The used fusion technique is based on:

-   classification of the high spatial resolution image (CI). In our case, the classification of the CI Landsat ETM+ MS/Pan fused image (only 4 bands were used, the last two bands were removed due to its overlapping with the ASTER SWIR bands which represent better spectral resolution) was performed by the ISODATA unsupervised technique (Richards, 1986) with K_0 = 16 classes (gives minimal error during the constrained unmixing).

-   definition of class contributions to the signal of the low spatial resolution of MI pixels. It is based on the resulting high-resolution classification map k (m, n) which represents different class areas. The contribution of class is given by the following equation:

$$c_i(l, s; k_0) = \sum_{k(m,n)\in k_0} \rho_i(l, s; m, n)$$

    where c_i (l, s; k_0) is the contribution of class k_0 to the signal of low spatial resolution of MI pixel (l, s) in different band i, and ρ_i (l, s; m, n) is a discrete approximation for the sensor PSF (Point spread function) which sum over a high resolution pixel (m, n) is assumed to be normalized to 1. The discrete PSF includes the registration component (MI image is co-registered with CI image) and the atmospheric component (CI image was corrected to eliminate the atmospheric and illumination effects before the classification and thus, these effects will be also removed during the unmixing).

-   we proposed a simplified version of the constrained unmixing algorithm (Hu et al., 1999). In this algorithm, we make the hypothesis that the reflectance of MI pixels equal to the sum of the mean reflectance of each class in the window.

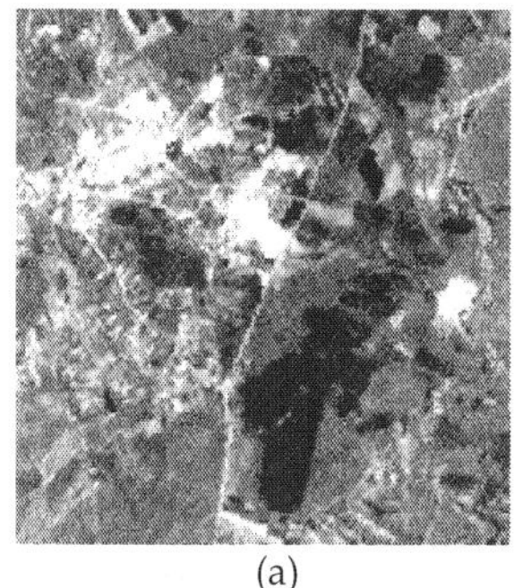

(a)

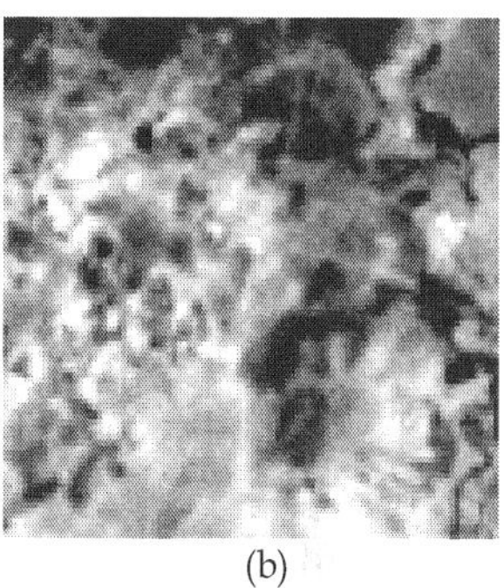

(b)

(c)

Fig. 3. Landsat ETM+ MS/Pan (196x196) with 15 m pixel size acquired in May 2000 shown in gray scale at wavelength λ=0.66 μm (a), ASTER SWIR (196x196) with 30 m pixel size acquired in June 2000 shown in gray scale at wavelength λ=2.4 μm (b) and Hybrid fused image (196x196) with 15 m pixel size shown in gray scale at wavelength λ=0.66 μm (c).

The unmixing of the MI pixels was performed in a 5 x 5 window that was moved with the step of 1 MI-pixel size. The central MI pixel in each window is unmixed by an inversion of a system of linear mixture equations that are written as following for all pixels in the window:

$$R_i(l,s) = \sum_{k=1}^{K} c_i(l,s,k)\, \bar{R}_i(k) + \varepsilon_i(l,s)$$

where R_j (l, s) is the reflectance of MI pixel (l, s) in the window, $\bar{R}_i$ (k) is the mean MI-reflectance for class k in the window and ε_i (l, s) is the model error. The numerical inversion of the linear system given in (Zhukov et al., 1999); (Zurita Milla et al., 2008) was done with the matlab software using the least-square function independently for each MI band (Matlab, 2004) This function returns the vector $\bar{R}$ that minimizes norm (C* $\bar{R}$ - R) subject to $\bar{R} \geq 0$.

- restoration of the unmixed lower resolution channels image was performed by assigning the estimated mean class reflectance to the corresponding high resolution pixels of the classification map.

The resulting hybrid image (Figure 3 c) present the high spatial resolution of the (MS)/(Pan) Landsat ETM+ image (15 m) and its relative high spectral resolution (4 bands) combined with ASTER SWIR bands (6 bands).

3. Tailing cartography refinement

The tailing cartography refinement approach includes the generation of tailing mask needed for unmixing the hybrid image. The mask was generated from the first tailing coarse cartography. It was validated before being used in the hybrid image classification process. The classification was performed using the constrained spectral unmixing method. The resulting maps represent detailed mine tailing cartography. The laboratory analyses were used as field truth for tailing maps validation.

3.1. Generation and validation of the tailing mask

The tailings fraction map represents their abundance and their spatial distribution in the study area. The mean value for tailing proportions is 0.32. This value supposes that the study area is covered by a considerable quantity of tailings. Those for vegetation and soil are estimated by 0.26 and 0.41 respectively. A tailing mask (Figure 4. a) was generated (extracted from fractional abundance map) to highlight the mine tailing area and masking the surrounding zones (vegetation and soils). The spatial distribution of the mine tailing was validated using high resolution data. The validation of tailing mask was based on aerial photo. It consists on the comparison of the tailing distribution (indicated by red lines) detected on the mask as well as aerial photo (Figure 4. c).

Indeed, certain zones are inaccessible or difficult of access. A spatial interpretation will be easier and faster. So, we refer to the interpretation of the aerial photo to analyze the distribution of the tailings in the Jebel Hallouf-Bouaouane region. The comparison reveals the presence of an individual quantity of tailings transported eastward and deposited near the Kassab wad. It is detected on the mask as well as on the aerial photo.

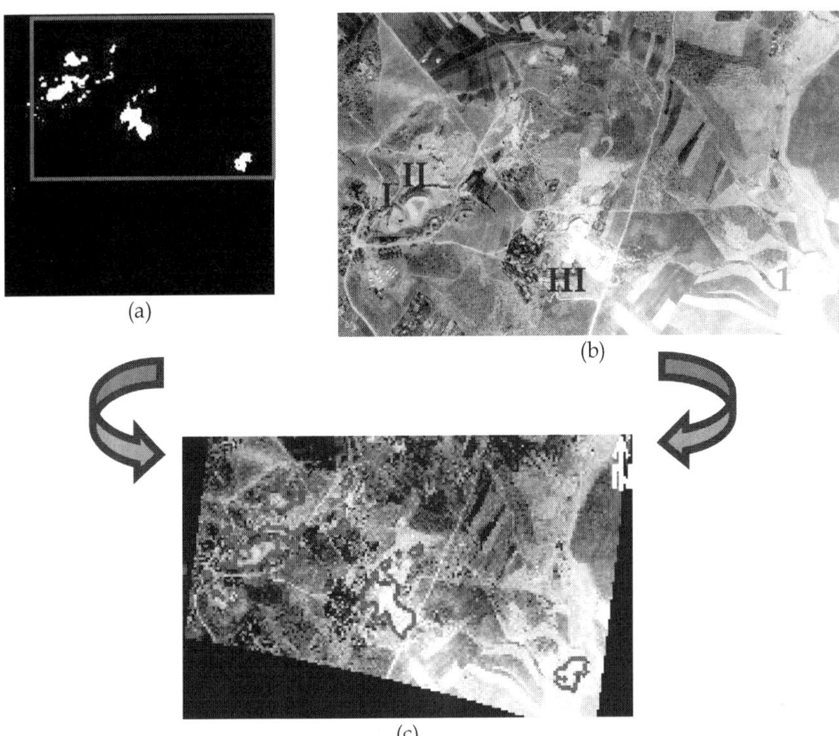

(a)
(b)
(c)

Fig. 4. Figure (a) indicates the generated and filtered tailing mask (196x196 pixels). Figure (b) shows the aerial photo (scale: 1/6250) scanned at 100 dpi of the corresponding Jebel Hallouf-Bouaouane region. Blue roman numbers locate tailing deposits: (I) the first dyke of Jebel Hallouf, (II) the second dyke of Jebel Hallouf and (III) the dykes of Bouaouane. Orange numbers locate the eroded and transported tailings towards: (1) kassab wad and (2) swamps in the north-east. Figure (c) shows the validation of the region of interest with 171 km^2 (indicated with red rectangle). The contours lines (red lines) which correspond to tailing zones were superposed on the georeferenced aerial photo.

3.2. Hybrid Multispectral data analysis

The classification of hybrid multispectral data, which is based on the constrained linear spectral unmixing using JPL mineral library spectra, generates mineral detailed maps. Only mine tailing pixels are used in the analysis. We used the tailing mask generated from the coarse classification to exclude the undesired zones and to analyze minerals in the tailing area.

The used endmembers are: calcite (with additionally cerusite, quartz and barite), gypsum, oxides (hematite and goethite), clays (kaolinite and illite) and pyrite. This mineralogical composition was revealed from samples by

distributed when $D = 0$ and becomes a Dirac delta function when $D = 1$. It is also obvious that multi-look processing improves the phase accuracy, which lead to the decreasing of the phase variance for a larger number of looks (Lee et al., 1994).

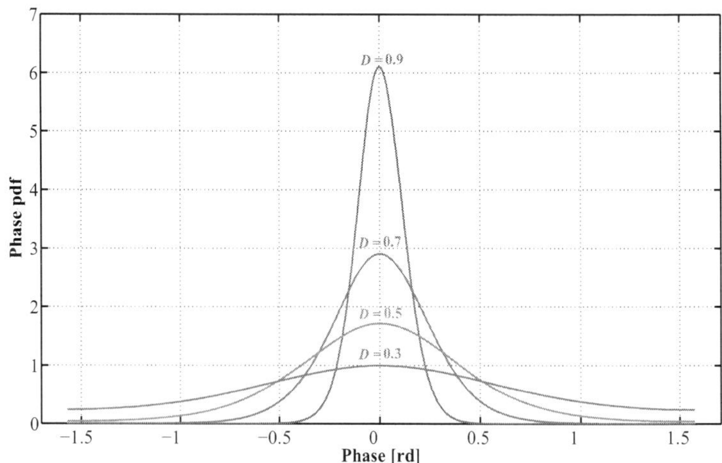

Fig. 1. Theoretical marginal multi-look pdf of the interferometric phase with $N = 9$ and $\beta = 0$.

2.2 Decorrelation effects

The interferometric phase can be affected by mainly three decorrelation factors: thermal ($\hat{\rho}_{thermal}$), temporal ($\hat{\rho}_{temporal}$) and geometrical ($\hat{\rho}_{geom}$) decorrelation (Hanssen, 2001). The Doppler centroid decorrelation and the processing induced decorrelation could be avoided or neglected (Franceschetti & Lanari, 1999). The considered decorrelation effects increase the interferometric phase noise, which will complicate more the unwrapping phase operation. When considered together, the above decorrelation factors are multiplicative so the approximate total decorrelation value $\hat{\rho}_{total}$ can be estimated as (Zebker & Villasenor, 1992):

$$\hat{\rho}_{total} \simeq \hat{\rho}_{thermal} \cdot \hat{\rho}_{temporal} \cdot \hat{\rho}_{geom} \tag{4}$$

where :

- $\hat{\rho}_{thermal}$ is the decorrelation value induced by the temperature of the sensors (thermal noise) on the interferometric phase during acquisition. It can be expressed by the signal to noise (SNR) of the specific sensor by (Zebker & Villasenor, 1992):

$$\hat{\rho}_{thermal} = \frac{1}{1 + SNR^{-1}} \tag{5}$$

The value of the SNR ratio is defined as:

$$SNR = \frac{P_S}{P_N}$$

where P_s is the signal power and P_N the noise power.

- $\hat{\rho}_{temporal}$ is the temporal decorrelation which indicates all the physical changes occurring on the surface between acquisitions, mainly in the case of repeat-pass interferometry. It includes changes of soil moisture content, surface roughness and vegetation. An analytical model of the temporal decorrelation is given in (Zebker & Villasenor, 1992) by:

$$\hat{\rho}_{temporal} = \exp\left(-\frac{1}{2} \left(\frac{4\pi}{\lambda} \right)^2 (\sigma_x^2 \sin^2\theta + \sigma_z^2 \cos^2\theta) \right) \tag{6}$$

where σ_x and σ_z are the average standard deviation, respectively, in azimuth and in range of a random displacement of the reflectors inside a resolution cell. λ is the radar wavelength and θ is its average look angle.

- $\hat{\rho}_{geom}$ represents the effects of the geometric decorrelation on the interferometric phase due to the interferometric acquisition geometry (baseline, registration, target rotation with respect to the sight line). Indeed, the same ground resolution cell is imaged from two slightly different looking angle. This means that the sensors (for both the interferometric acquisitions) don't look at the target with the same incidence angle. The numerical assessments of these effects depend on the length of the baseline, the incidence angle and the spatial resolution. In (Zebker & Villasenor, 1992), the authors provide the geometric baseline decorrelation function as the result of the phase offset due to the difference in the incidence angle between the two InSAR acquisitions:

$$\hat{\rho}_{geom} = 1 - \frac{2B_{\perp} l_g \cos\theta}{\lambda R_0} \tag{7}$$

where $B_{\perp}$ is the orthogonal component of the baseline to the radar look direction, l_g the ground range resolution, and R_0 the slant range. A modified version of (7), given by Lee and Liu (Lee & Liu, 1999), includes the terrain slope α is given by:

$$\hat{\rho}_{geom} = 1 - \frac{cB_{\perp} \mid \cos(\theta_0 - \alpha) \mid}{\lambda R_0 B_w} \tag{8}$$

where c is the speed of light, θ_0 the nominal incidence angle on the ellipsoidal earth ($23°$ for ERS-1 and ERS-2) and B_w the frequency bandwidth of the transmitted signal.

2.3 InSAR phase noise model

The InSAR phase quality is measured with the absolute coherence (Abdelfattah & Nicolas, 2006). Coherence interference due to reflection by random scatterers degrades the complex image. The amplitude is corrupted by multiplicative noise, while the phase is corrupted by additive noise. Indeed, based on (3), Lee et al. proved in (Lee et al., 1998) that: Since φ distribution is symmetrical about β, β is the mean. The standard deviation is independent of β. Consequently, φ can be characterized by an additive noise model in the real domain:

$$\varphi_z = \varphi_x + \nu \tag{9}$$

where φ_z is the measured value, φ_x is the original phase to be estimated and ν is the zero-mean noise with the standard deviation σ_ν. In the real domain, this phase noise model presents the problem of phase jumps. It is due to the complex representation of the interferometric phase

2^{i-1}. Then, the mask growing will depend on the coherence values of the four considered pixels in the band of the scale 2^{i-1}. Figure 5 shows an illustrative example of the mask growing step for a given (m,n) pixel in the band of the scale 2^i and its corresponding four pixels in the band of the scale 2^{i-1}, $\{(k,l);(k,l+1);(k+1,l);(k+1,l+1)\}$. The decision rule will depend on:

1. If the pixel (m,n) in the mask is a signal coefficient, so

$$|\rho(k,l) - \rho_p| \leq \varepsilon_c \Rightarrow p \text{ is a signal coefficient}$$

$$|\rho(k,l) - \rho_p| > \varepsilon_c \Rightarrow p \text{ is a noise coefficient}$$

where $p \in \{(k,l+1);(k+1,l);(k+1,l+1)\}$

2. If the pixel (m,n) in the mask is a noise coefficient, so

$$|\rho(k,l) - \rho_p| \leq \varepsilon_c \Rightarrow p \text{ is a noise coefficient}$$

$$|\rho(k,l) - \rho_p| > \varepsilon_c \Rightarrow p \text{ is a signal coefficient}$$

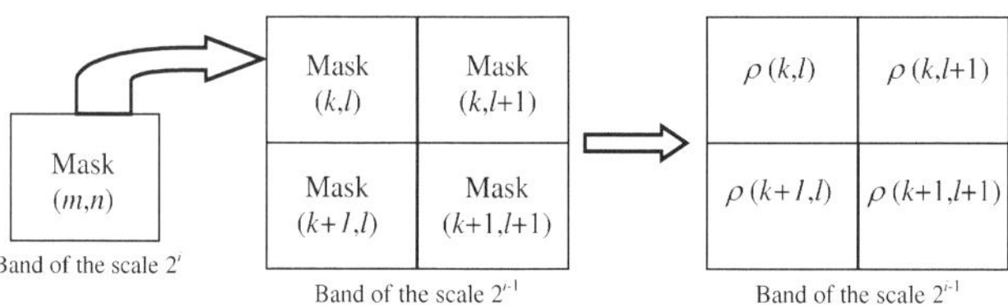

Fig. 5. Mask growing taking into account, the InSAR coherence map.

3.4 Noise reduction on simulated and real interferograms

In order to validate the proposed noise reduction algorithm, synthetic and real interferograms were used. The accuracy of the denoising process was estimated using two different approaches:

- The $PSNR$, computed between the filtered and the original simulated interferograms. It is given by:

$$PSNR = 10log_{10}\left(\frac{(2\pi)^2}{MSE}\right)$$

where MSE is the Mean Square Error between the filtered and the original simulated interferograms.

- The difference image distribution, computed between the filtered and the original real interferograms.

Moreover, a quantitative comparison (using the PSNR ratio) with alternative interferometric phase filters addressed in the literature is reported.

Table 1. Interferometric phase filtering PSNR for the cone and pyramid interferograms, with noise variance of 10^{-2}.

	Interferogram	Gaussian noise (dB)	White noise (dB)	Uniform noise (dB)
Lopez	cone	39.9	39.85	41.5
(López-Martínez & Fabregas, 2002)	pyramid	39.51	39.77	40.89
FAMM	cone	41.36	41.29	42.78
	pyramid	41.13	40.81	41.67
Lee	cone	37.18	37.52	37.22
(Lee & Liu, 1999)	pyramid	36.74	37.03	37.28

The simulated data were a 512x512 pixel interferograms, representing a cone and a pyramid, relatives to single look complexe images, with coherence varying from 0.9 to 0.45. More details could be found in (Abdelfattah & Bouzid, 2008) about the simulation process. The experimental results given in (Abdelfattah & Bouzid, 2008) show that the Daubechies 20-coefficient wavelet transform gives the best PSNR for both cone and pyramid interferograms. Table 1 shows the numerical comparison for these simulated interferograms. The used empirical parameters for the thresholds were $th_{wa} = -0.4$, $th_{wd} = -0.2$ and $\varepsilon_c = 6.10^{-4}$. The results in Table 1 confirm that the FAMM filtering algorithm allows a gain more than 2 dB compared to the one of Lopez and Fabregas (López-Martínez & Fabregas, 2002) and more than 4 dB compared to the Lee one (Lee & Liu, 1999).

In order to test the spatial resolution maintenance properties of the different above mentioned filters, we simulated a pyramid with phase noise coherence equal to 0.45 and a ten-pixel fringe period. The critical point is to maintain the pyramid edges. Fig. 6 shows the noisy phase image and the result obtained after applying the different filters. The pyramid edges are clearly maintained for both the FAMM and the Lopez Filters.

We also tested the developed noise reduction algorithm (FAMM) on a real experimental ERS2 interferogram of Capbon, in the north of Tunisia. This single look complex interferometric phase has a 2048x2048 pixel dimension with an approximate spatial resolution of 25x25 m. This phase image has been filtered by the FAMM filter using the same parameters as those used for the simulated interferograms. From Fig. 9 (c), one can notice that the algorithm, like the Lopez one (López-Martínez & Fabregas, 2002), is able to process, in the same efficient way, areas with smooth or steep slopes at the same time. Figure 9 (d) shows the difference between the original and the filtered phases, which has a mean equal to zero and does not contain any image detail, demonstrating that the proposed filter preserves the topographic information.

4. InSAR phase unwrapping based on MRF model

The aim of this section is to present a general framework for the phase unwrapping problem, based on Markov random fields (MRF(s)) models (Griffeath, 1976). The proposed algorithm exploit Bayesian estimation theory to perform energy minimization in order to reconstruct the original phase field. Our contribution consists in the exploitation of the Interferometric phase distribution (Abdelfattah & Nicolas, 2006) in order to develop an adaptive energy functional for the optimization of the MRF to be retrieved.

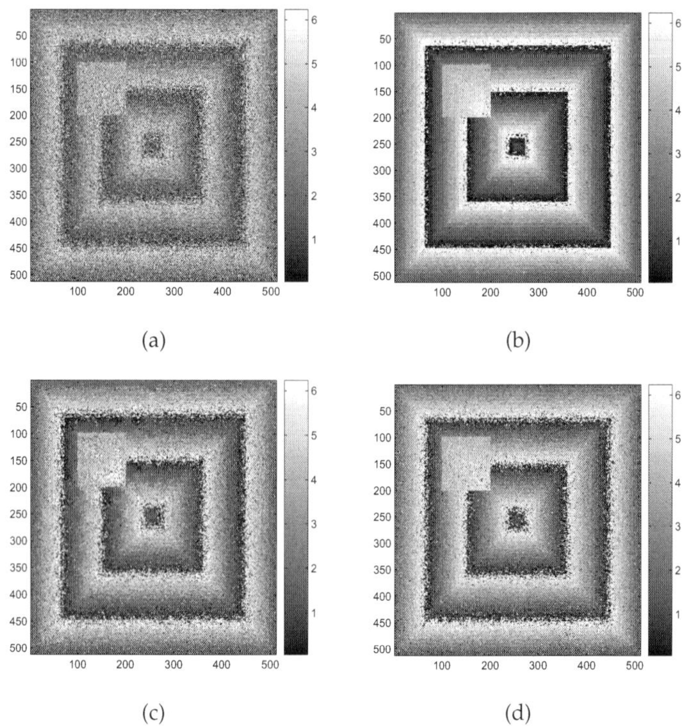

Fig. 6. Interferometric phase filtering results with an interferogram representing a pyramid with $|\rho| = 0.45$ and a variance Gaussian noise 10^{-2}. (a) Original noisy image, (b) FAMM filter, (c) Lopez filter and (d) Lee filter.

4.1 General context

The problem of recovering the absolute phase field from observations is ill-posed and can be solved if one first regularizes it by introducing further information (prior knowledge) about the behavior of the solution. To introduce this additional information, we exploit a statistical framework, which consist in considering that the searched field ϕ is a Markov random one (Griffeath, 1976). Thanks to the Hammersly-Clifford theorem (Griffeath, 1976), which establishes a correspondence between MRF and Gibbs distribution, the distribution of ϕ field is expressed with the following representation (Griffeath, 1976):

$$P(\phi) = \frac{1}{Z}\exp(-U(\phi)) \tag{23}$$

where Z is a normalizing constant, and $U(\phi)$ is a function which encodes the local characteristics of the phase behavior being modelled. The true phase field ϕ is related to the wrapped phase field φ through the equation given by:

$$\phi = \varphi + 2k\pi \tag{24}$$

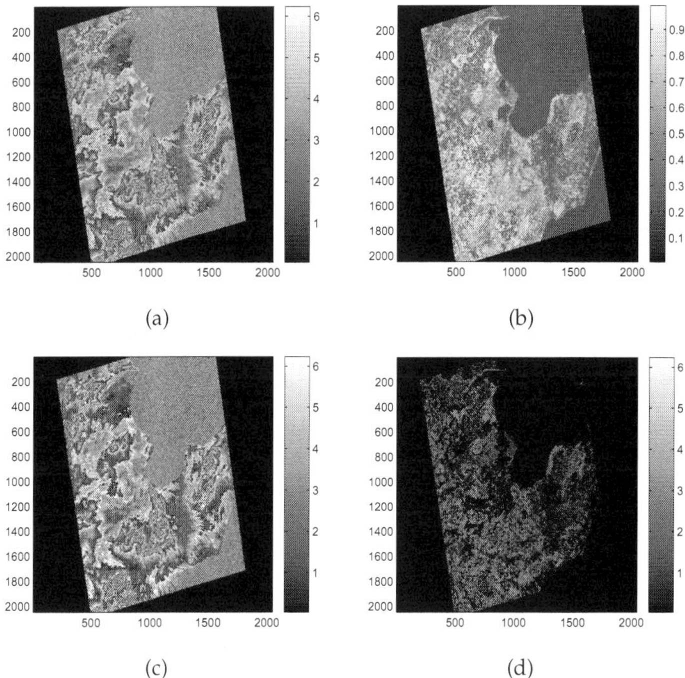

Fig. 7. Filtering result with the FAMM filter of (a) the original interferometric phase of Capbon (ERS2 interferogram), (b) the corresponding coherence map, (c) The filtered interferogram and (d) the difference image (original interferogram - filtered interferogram).

where k is the integer to determine for each (i,j) pixel in the φ field, in order to solve (24) and recover ϕ. Considering the MRF model in two dimensional processing, each pixel (i,j) of an image is defined as a site s in a set of regular lattice of nodes S. Let us assume that the observed data obey to the equation, related to each site s given by (Rodriguez-Vera & Servin, 1995):

$$\phi_s = \varphi_s + 2\pi k_s + q \tag{25}$$

where ϕ_s the phase to be retrieved at the site s, φ_s is the observed phase at the site s and q is a constant. We introduce a new field f as a correction field, defined by the elements f_s, where $s \in S$ and given by (Rodriguez-Vera & Servin, 1995):

$$f_s = k_s + \frac{q}{2\pi} \tag{26}$$

Note that the values of each element f_s of the field f, are not necessarily integers ($f_s \in \mathbb{R}$), but the differences are, so that for any two neighboring sites $\{s,t\} \in S$, we have (Rodriguez-Vera & Servin, 1995):

$$f_s - f_t = r(f_s - f_t) \tag{27}$$

where f_s and f_t are phase values related to two neighboring sites and $r(x)$ is the closet integer to x. Thus, (25) yields (Rodriguez-Vera & Servin, 1995):

$$\phi_s = \varphi_s + 2\pi f_s \tag{28}$$

Assuming (28), that introduce the correction field f, is verified everywhere in the lattice structure of the phase field ϕ, then we obtain the conditional probability, $P(\varphi, f|\phi)$, of the observed phase φ and f given ϕ, that takes the form (Rodriguez-Vera & Servin, 1995):

$$P(\varphi, f|\phi) = \begin{cases} 1 & \text{if} \quad \phi_s = \varphi_s + 2\pi f_s \quad \forall s \in S \\ 0 & \text{otherwise} \end{cases} \tag{29}$$

we note that the likelihood distribution $P(\varphi, f|\phi)$ corresponds to Dirac impulse response, and can be expressed by (Rodriguez-Vera & Servin, 1995):

$$P(\varphi, f|\phi) = \frac{1}{Z_1} \exp\left(-\alpha \sum_{s \in S} \left[\varphi_s - \phi_s + 2\pi f_s\right]^2\right) \tag{30}$$

where Z_1 is a normalizing constant and α is a weight parameter.

4.2 Prior and posterior model

To define a probabilistic model that describes the real interferometric phase information, it is necessary to define an energy functional:

$$U(\phi) = \sum_{c \in C | s \in c} U_c(\phi) \tag{31}$$

where the $U_c(.)$ are called potentials functions, indexed by the clique c in the set of cliques C of the neighborhood system of a given site s. Note that each potential U_c depends only on the values taken on the clique sites $x_s, s \in c$, and, therefore, accounts only for local interactions between neighboring pixels. As consequence, local dependencies in the field ϕ can be modelled by defining suitable potentials. In the literature, a number of models have been proposed for various applications, for further comprehension refer to (Li, 2001). To enforce global smoothness of reconstructed phase, a quadratic potential model could be adopted (Rodriguez-Vera & Servin, 1995):

$$U_c(\phi) = (\phi_s - \phi_t)^2 \tag{32}$$

The equations (31) and (32) gives:

$$U(\phi) = \sum_{c \in C | s \in c} (\phi_s - \phi_t)^2 \tag{33}$$

Probabilistic description of the current estimate of the field (ϕ, f), given the observed data φ, is calculated using Bayes'rule (Rodriguez-Vera & Servin, 1995):

$$P(\phi, f|\varphi) = \frac{P(\varphi|\phi, f)P(\phi, f)}{P(\varphi)} \tag{34}$$

Assuming that $P(\varphi)$ is a constant, and that the correction field f obey to the prior model defined by (23), we obtain:

$$P(\phi, f|\varphi) \quad \propto \quad P(\varphi|\phi, f)P(\phi, f) \tag{35}$$

$$\propto \quad \frac{1}{Z_2} \exp\left(-U(\phi, f)\right) \tag{36}$$

where Z_2 is a normalizing constant, and $U(\phi, f)$ is the posterior energy function, given by merging (23), (30) and (14) (Rodriguez-Vera & Servin, 1995):

$$U(\phi, f) = \sum_{c \in C | s \in c} U_c(\phi) + \alpha \sum_{s \in S} \left(\varphi_s - \phi_s + 2\pi f_s \right)^2 \tag{37}$$

The maximum a posteriori (MAP) estimator for ϕ may now be obtained by the minimization of (37). Using (28), and requiring that $\left[f_s - f_t - r(f_s - f_t) \right]^2$ be small for nearest-neighbor pairs of sites, the minimization problem involving two unknown fields ϕ and f, may be simplified by expressing the energy in term of the field f only. This allows to absorb the noise and interpolate the missing data. The new expression of the posterior energy becomes (Rodriguez-Vera & Servin, 1995):

$$U(f) = \sum_{c \in C | s \in c} U_c(\varphi + 2\pi f) + \alpha \sum_{t \in V_s} \left[f_s - f_t - r(f_s - f_t) \right]^2 \tag{38}$$

where V_s is a set of sites neighbouring the site s and α is a regularization parameter that controls the smoothness of f.

4.3 Minimization algorithm-Gradient descent method

The gradient descent optimization method could be adopted to restor the field f that minimize the energy function denoted by (38), and hence to unwrap φ. this iterative method is useful to prevent the initialization problem related to the application of the MRF models for phase unwrapping procedure (Rodriguez-Vera & Servin, 1995). to start the decent algorithm with the appropriate initial state, we must set the initial value as follow:

$$f_s = -\frac{\varphi_s}{2\pi} \tag{39}$$

The implementation of the minimization algorithm, in which the gradient descent method is performed by updating the value of each variable f_s of the field f, in parallel programming way, according to the following equation (Rodriguez-Vera & Servin, 1995):

$$f_s^{(n+1)} = f_s^{(n)} - h \frac{\partial U[f^{(n)}]}{\partial f_s} \tag{40}$$

where n denotes the iteration number, and h is the step size of the gradient descent. The gradient of U is expressed by (Rodriguez-Vera & Servin, 1995):

$$\begin{aligned}
\frac{\partial U(f^{(n)})}{\partial f_s} &= \sum_{c \in C | s \in c} \frac{\partial U_c(\varphi + 2\pi f^{(n)})}{\partial f_s} \\
&+ 2\lambda \sum_{t \in V_s} \left[f_s^{(n)} - f_t^{(n)} - r(f_s^{(n)} - f_t^{(n)}) \right]
\end{aligned} \tag{41}$$

For the regular lattice S, we consider the first-order system, so we have $C = C_1 \cup C_2$, with $C_1 = \{s | s \in S\}$ and $C_2 = \{\{s,t\} \,|\, t \in V_s, s \in S\}$. Using (32) the developed form of (40) and then the gradient descent automaton is given by (Rodriguez-Vera & Servin, 1995):

$$\begin{aligned}
f_s^{(n+1)} = f_s^{(n)} - 2h \Bigg[&\sum_{t \in V_s} \left\{ 2\pi(g_s - g_t) + 4\pi^2 (f_s^{(n)} - f_t^{(n)}) \right\} \\
&+ \lambda \left(f_s^{(n)} - f_t^{(n)} - r(f_s^{(n)} - f_t^{(n)}) \right) \Bigg]
\end{aligned} \tag{42}$$

where $V_s = t : \|s - t\| = 1$ and where we set $\varphi_s = 0$ and $f_s = 0 \ \forall \, s \notin S$.

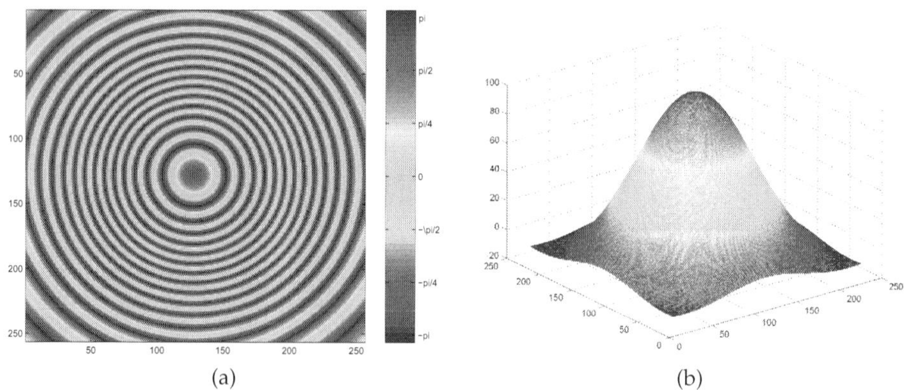

(a)   (b)

Fig. 8. (a) Interferogram generated from a Gaussian DEM with a maximum height of 100 m and standard deviations (in rows (σ_x) and columns (σ_y)) of $\sigma_x = \sigma_y = 2.10^{-4}$. (b) Reconstructed elevation data obtained with the gradient descent algorithm..

Figure 8 displays the interferogram generated from a synthetic Gaussian DEM with a maximum height of 100 m and standard deviations (in rows (σ_x) and columns (σ_y)) of $\sigma_x = \sigma_y = 2.10^{-4}$. The algorithm described above works well with highly coherent interferograms (for data affected by low noise level). However, gradient descent method does not give satisfactory results in the case of noisy interferograms. In fact, the chosen a priori model does not integrate a noise model. Thus, it is necessary to accommodate the potential function to recover the discontinuities (noise) present in the observed data. A modification of the prior potentials is proposed in (Rodriguez-Vera & Servin, 1995), making the phase jumps, that overcomes a certain threshold a, contribute with a fixed amount to the energy, independently of their size. A truncated quadratic potentials, are then proposed and the (32) is replaced by (Rodriguez-Vera & Servin, 1995):

$$U_c(\phi) = \left\{ \begin{array}{lll} (\phi_s - \phi_t)^2 & \text{if} & |\phi_s - \phi_t| < a, \\ a^2 & \text{otherwise} & \end{array} \right. \tag{43}$$

where a is a positive parameter.

Figure 9 (a) illustrates a noisy (mean = 0.6 and standard deviation = 0.9) interferogram generated from a two nearby Gaussians elevation data of height 100 m and 50 m, with local inconsistencies (nul data) marked by blue squares. However, as we can see on Fig. 9 (b), the reconstructed elevation data obtained with the gradient descent algorithm is not perfect, in spite of the correction of the inconsistencies. This have lead us to think for an adaptive energy functional modelling.

4.4 Our contribution

In order to enforce the global smoothness of the algorithm, the above considered quadratic potentials is given by:

$$V_{i,j}(z_i, z_j) = (z_i - z_j)^2. \tag{44}$$

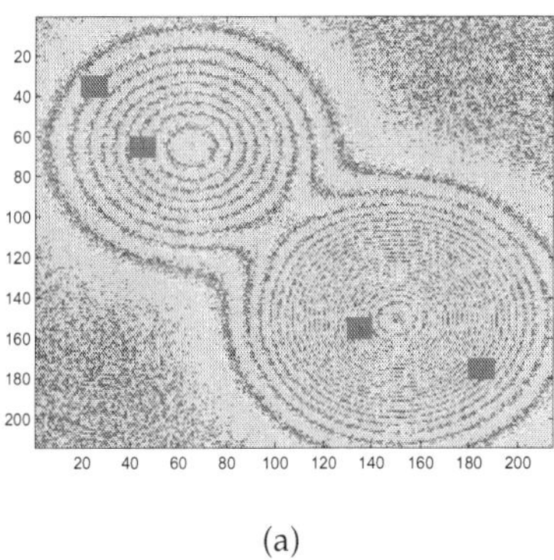

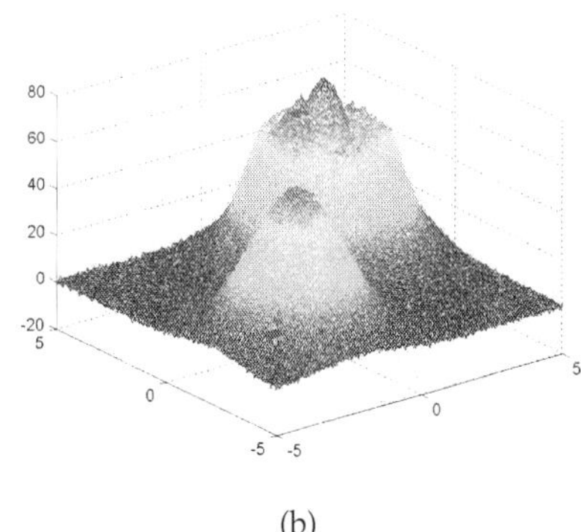

(a)   (b)

Fig. 9. (a) Interferogram generated from a two nearby Gaussians elevation data of height 100 m and 50 m, with local inconsistencies marked by blue squares. (b) Reconstructed elevation data obtained with the gradient descent algorithm..

Such a potential function makes this algorithm work well for smooth surface. However it fails to recover discontinuous surfaces as well. Our main contribution in this work is to propose a modification to the above prior potential so that the jumps contribution to the energy are mitigated by weighting them by the probability of their occurrence which decrease when they increase, the proposed potential function is then equivalent to (Elmzoughi et al., 2008):

$$V_{i,j}(z_i, z_j) = (z_i - z_j)^2 P_Z(z_i - z_j).$$
(45)

In our case, Z corresponds to the absolute phase ϕ which pdf can be written as an expression implying hypergeometric functions. However, A good approximation can be given by a Gaussian model:

$$P_\Phi(\phi) = \frac{1}{\sqrt{2\pi}\sigma_\phi} \exp(-\frac{(\phi - \widehat{\phi})^2}{\sigma_\phi^2}),$$
(46)

where σ_ϕ and $\widehat{\phi}$ are respectively the variance and the mean of the absolute phase ϕ. Theses parameters can be easily estimated from the theoretical hypergeometric pdf. By considering the transfert theorem, the gradient descent automation in equation (38) could be given by:

$$\begin{aligned} f_i^{t+1} &= f_i^t - \frac{\sqrt{2\pi}h}{\sigma_\phi} \times \Sigma_{j \in V_i} \left[[g_i - g_j + 2\pi(f_i^t - f_j^t)] \exp \frac{[g_i - g_j + 2\pi(f_i^t - f_j^t)]^2}{4\sigma_\phi^2} \right] \\ &\quad \times \left[1 - \frac{1}{4\sigma_\phi^2}[g_i - g_j + 2\pi(f_i^t - f_j^t)] \right] + \lambda[f_i^t - f_j^t - r(f_i^t - f_j^t)] \end{aligned}$$
(47)

where V_i is the set of neighbors of the considered site i. In (Elmzoughi et al., 2008), it is shown that both algorithms seems to be equivalent for smooth surfaces. However, using the proposed potential function allowed to ameliorate results, especially in terms of mean square errot and in terms of SNR, for surfaces with discontinuities. This could be explained by a better preservation of these discontinuities in the reconstructed phase.

5. Conclusion

In this chapter, we proposed a modified filtering algorithm to the López and Fabregas (López-Martínez & Fabregas, 2002) noise reduction algorithm for the interferometric phase noise in SAR interferometry using a multiresolution approach. Our contribution to the existing algorithm consists on the exploitation of the InSAR coherence map in order to generate a more adaptive mask for each decomposition level.

Moreover, we presented a general probabilistic framework to phase unwrapping problem based on the work of Rodriguez and Servin (Rodriguez-Vera & Servin, 1995). Both MRF and Bayesian estimator were applied to recover the desired phase, as the optimal field solution of the maximum a posteriori (MAP) estimation criterion. An iterative method that minimizes a general energy function is proposed, and a parallel algorithm based on gradient descent optimization is designed to perform this task. The proposed solution overcomes some important limitations of most of the phase unwrapping procedures, and the results show robustness, and stability.

These results give us new ideas for the applications of the InSAR unwrapping phase MRF algorithm for unwrapping interferometric synthetic Aperture Sonar (InSAS) (Bonifant et al., 2000) phase. Both two phases present similar statistics. However, if the ground truth could be used for the InSAR result validations, it not the case for the InSAS results. In fact, the ability to produce full coverage bathymetric maps and generate accurate measurements of the seafloor height, is limited. So, an adaptive unwrapping procedure could be interesting.

6. References

Abdelfattah, R. & Bouzid, A. (2008). Sar interferogram filtering in the wavelet domain using a coherence map mask, *Proceedings of the 15th IEEE International Conference on Image Processing, ICIP 2008*, San Diego, CA, pp. 1888–1891.

Abdelfattah, R. & Nicolas, J. (2002). Topographic sar interferometry formulation for high-precision dem generation, *IEEE Transaction on Geoscience and Remote Sensing* **Vol. 40**(No. 11): pp. 2415–2426.

Abdelfattah, R. & Nicolas, J. (2006). Interferometric sar coherence magnitude estimation using second kind statistics, *IEEE Transaction on Geoscience and Remote Sensing* **Vol. 44**(No. 7): pp. 1942–1953.

Balzter, H. (2001). Forest mapping and monitoring with interferometric synthetic aperture radar (insar), *Progress in physical geography* **Vol. 25**(No. 2): pp. 159–177.

Bodart, C., Gassani, J., Salmon, M. & Ozer, A. (2009). Contribution of sar interferometry (from ers1/2) in the study of aeolian transport processes: The cases of niger, mauritania and morocco, *in* A. Marini & M. Talbi (eds), *Desertification and Risk Analysis Using High and Medium Resolution Satellite Data Training Workshop on Mapping Desertification*, Springer, pp. 129–136.

Bonifant, W., Richards, M. & McClellan, J. (2000). Interferometric height estimation of the seafloor via synthetic aperture sonar in the presence of motion errors, *IEE Proceedings - Radar, Sonar and Navigation* **Vol. 147**(No. 6): pp. 322–330.

Buccigrossi, R. & Simoncelli, E. (2001). Image compression via joint statistical characterization in the wavelet domain, *IEEE Transactions on Image Processing* **Vol. 8**(No. 12): pp. 1688–1701.

Donoho, D. & Johnstone, I. (1995). The $z\pi m$ algorithm: a method for interferometric image reconstruction in sar/sas, *Journal of the American Statistical Association* **Vol. 90**(No. 432): pp. 1200–1224.

Elmzoughi, A., Maalej, A., Abdelfattah, R. & Belhadj, Z. (2008). Insar phase unwrapping based on a combination of markov random fields and hypergeometric phase pdf models, *Proceedings of the IEEE International Geoscience and Remote Sensing Symposium, IGARSS 2008*, Boston, Massachusetts, pp. 538–541.

Franceschetti, G. & Lanari, R. (1999). *Synthetic Aperture Radar processing*, CRC Press.

Gabriel, A. & Goldstein, R. (1988). Retrieval of vegetation parameters with sar interferometry, *International Journal of Remote Sensing* **Vol. 9**(No. 4): pp. 857–872.

Ghiglia, D. & Pritt, M. (1998). *Two-dimensional phase unwrapping: Theory, algorithms and software*, John Wiley & Sons Inc.

Ghiglia, D. & Romero, L. (1994). Robust two-dimensional weighted and unweighted phase unwrapping that uses fast transforms and iterative methods, *Journal of the Optical Society of America A* **Vol. 11**(No. 1): pp. 107–117.

Goldstein, R. & Werner, C. (1998). Radar interferogram filtering for geophysical applications, *Geophysical Research Letters* **Vol. 25**(No. 21): pp. 4035–4038.

Goldstein, R., Zebker, H. & Werner, C. (1988). Satellite radar interferometry: two-dimensional phase unwrapping, *Radio Science* **Vol. 23**(No. 4): pp. 713–720.

Graham, L. (1974). Synthetic interferometer radar for topographic mapping, *Proceedings of the IEEE* **Vol. 62**(No. 6): pp. 763–768.

Griffeath, D. (1976). Introduction to markov random fields, *in* J. Kemeny, J. Snell & A. Knapp (eds), *Denumerable Markov Chains*, Springer-Verlag, pp. 424–457.

Hanssen, R. (2001). *Radar Interferometry: Data Interpretation and Error Analysis*, Kluwer Academic Publishers.

Hess-Nielsen, N. & Wickerhauser, M. (1996). Wavelets and time frequency analysis, *Proceedings of the IEEE* **Vol. 82**(No. 4): pp. 523–540.

Lee, H. & Liu, J. G. (1999). Spatial decorrelation due to topography in the interferometric sar coherence imagery, *Proceedings of the IEEE International Geoscience and Remote Sensing Symposium IGARSS'99*, Hamburg, Germany, pp. 485–487.

Lee, J., Hoppel, W. & Mango, S. (1994). Intensity and phase statistics of multilook polarimetric and interferometric sar imagery, *IEEE Transaction on Geoscience and Remote Sensing* **Vol. 32**(No. 5): pp. 1017–1027.

Lee, J., Papathanassiou, K., Ainsworth, T., Grunes, M. & Reigber, A. (1998). A new technique for phase noise filtering of sar interferometric phase images, *IEEE Transaction on Geoscience and Remote Sensing* **Vol. 36**(No. 5): pp. 1456–1465.

Li, S. (2001). *Markov Random Field Modeling in Image Analysis*, Springer-Verlag.

López-Martínez, C. & Fabregas, X. (2002). Modeling and reduction of sar interferometric phase noise in the wavelet domain, *IEEE Transaction on Geoscience and Remote Sensing* **Vol. 40**(No. 12): pp. 2553–2566.

Maître, H. (2001). *Traitement des images de RSO*, Hermes.

Mallat, S. (1998). *A wavelet tour of signal processing*, Academic Press.

Massonet, D. & Feigl, K. (1998). Radar interferometry and its application to changes in the earth's surface, *Reviews of Geophysics* **Vol. 36**(No. 4): pp. 441–500.

Mattar, K., Vachon, P., Geudtner, D., Gray, A., Cumming, I. & Brugman, M. (1998). Validation of alpine glacier velocity measurements using ers tandem-mission sar data, *IEEE Transaction on Geoscience and Remote Sensing* **Vol. 36**(No. 3): pp. 974–984.

Odegard, J., Guo, H., Lang, M., Burrus, C., Jr., R. . W., Novak, L. & Hiett, M. (1995). Wavelet based sar speckle reduction and image compression, *Proceedings of the SPIE Symposium on OE/Aerospace Sensing and Dual Use Photonics*, Orlando, Florida, pp. 17–21.

Rodriguez-Vera, R. & Servin, M. (1995). Modeling and reduction of sar interferometric phase noise in the wavelet domain, *Journal of the Optical Society of America A* **Vol. 12**(No. 12): pp. 2578–2585.

Suksmono, A. & Hirose, A. (2002). Adaptive noise reduction of insar images based on a complex-valued mrf model and its application to phase unwrapping problem, *IEEE Transaction on Geoscience and Remote Sensing* **Vol. 40**(No. 3): pp. 699–709.

Touzi, R. & Lopez, A. (1996). Statistics of the stokes parametersand of the complex coherence parameters in one-look and multilook speckle fields, *IEEE Transaction on Geoscience and Remote Sensing* **Vol. 34**(No. 2): pp. 519–531.

Vidakovic, B. (1999). *Statistical Modeling by Wavelets*, Wiley-interscience.

Wegmullerl, U. & Werner, C. (1997). Retrieval of vegetation parameters with sar interferometry, *IEEE Transaction on Geoscience and Remote Sensing* **Vol. 35**(No. 1): pp. 18–24.

Wishart, J. (1928). The generalized product moment distribution in samples from a normal multivariate population, *Biometrika* **Vol. 20A**(No. 1/2): pp. 32–52.

Zebker, H. & Villasenor, J. (1992). Decorrelation in interferometric radar echoes, *IEEE Transaction on Geoscience and Remote Sensing* **Vol. 30**(No. 5): pp. 950–959.

Zhou, Y., Li, D. & Genderen, J. V. (1999). Wavelet based sar speckle reduction and image compression, *Proceedings of the SPIE Symposium Vol. 3723, Wavelet Applications VI*, Orlando, Florida, pp. 208–214.

21

Soil moisture estimation using L-band radiometry

Alessandra Monerris[1] and Thomas Schmugge[2]
[1]*Remote Sensing Laboratory, TSC-UPC, and SMOS-BEC, Barcelona, Spain*
[2]*Physical Sciences Laboratory, New Mexico State University, Las Cruces, NM, USA*

1. Introduction

About 70% of the Earth is covered with water, 97% of which is part of oceans. Heating of oceans by the Sun keeps the Earth's water in a continuous circulation from the atmosphere to the Earth and back to the atmosphere through condensation, evapotranspiration, and precipitation processes. This continuous water motion is called water cycle. The groundwater in the unsaturated or vadose zone of land is usually referred to as soil moisture. The thickness of this zone extends from soil surface to a few metres below the surface in humid regions, and to 300 m or more below surface in arid regions. Although the water held by soils is a small fraction of the Earth's water budget, soil moisture plays an important role in the water cycle since it controls the proportion of rainfall that percolates, runs off, or evaporates from the land, influences plants growth and transpiration, and is related to the precipitation variability within a region (Koster et al., 2004).

Although soil moisture is one of the main parameters used in climate models, because of the large temporal and spatial variations of soil moisture sparse in-situ measurements are inadequate to be of much use in these models. Remote sensing with sufficient accuracy would provide meaningful soil moisture data over large regions. The maximum sensitivity to soil moisture is achieved in the lower range of microwave frequencies, which goes from 1 to 5 GHz (wavelength from 30 cm to 5 cm, respectively). The justification to this behaviour can be found if the dielectric properties of a target are analysed, since they have a large influence on its microwave brightness temperature. Land surfaces can be considered as a mixture of soil, water, and air particles and thus soil moisture might be estimated from measurements if the contrast between the water and soil particles were large enough. At microwave frequencies the real part of dry soil and water dielectric constants are approximately 4 and 80, respectively (Jackson & Schmugge, 1989. The scientific literature concluded that microwave radiometry at L-band (1.4-1.427 GHz) is optimal to estimate soil moisture, not only because it is very sensitive to soil moisture, but also because provides all-weather coverage, since the atmosphere at microwave frequencies may be considered nearly transparent, and vegetation is semi-transparent, which allows observations of the underlying layers (Eagleman, 1976; Wang & Choudhury, 1981; Jackson & Schmugge 1991). This specific band was chosen because it is a radio astronomy quiet band.

1.1. Satellite missions for soil moisture estimation using passive microwave sensors

While L-band microwave sensors provide maximum sensitivity to soil moisture their long wavelength implies the need for large antennae to achieve useful spatial resolution. For example, to obtain a ground resolution of 50 km or less using classical solutions on low-orbit satellites implies an antenna size of up to 20 m. At present, different scientific groups are developing new techniques to face this problem, and two space missions have been proposed to measure soil moisture at global scale: ESA's SMOS and NASA's SMAP.

The Soil Moisture and Ocean Salinity mission

The Soil Moisture and Ocean Salinity (SMOS) mission will observe soil moisture over land and salinity over oceans, and will measure snow and ice areas contributing to studies of the cryosphere (Kerr et al., 2001; Barré et al., 2008). Scheduled for launch in autumn 2009, it is the second Earth Explorer mission and part of ESA's Living Planet Programme. SMOS was thought of as a cost-effective, demonstrator mission with a nominal (extended) lifetime of 3 (5) years. The orbit is quasi-circular, sun-synchronous and dawn-dusk, and will be in the low-Earth range, at 763 km.

The SMOS mission provides a completely new approach in the field of remote sensing at L-band. Its payload MIRAS (Microwave Radiometer by Aperture Synthesis) is the first-ever two-dimensional interferometric radiometer in space. MIRAS consists of a Y-shaped antenna array with three arms, each arm having an approximate length of 4.5 m and 21 dual-polarisation L-band antennae or LICEF (Light Cost Effective Front-end) spaced 0.875 times the wavelength. Each LICEF is a total power radiometer in its own. Nine redundant antenna and three full-polarimetric noise injection radiometers are located in a fixed structure in the centre of the array, while the arms are divided in three segments to be folded during the launch (McMullan et al., 2008). If classical solutions would have been used, with this antenna size and orbit altitude the field of view (FOV) would be of near 3000 km in diameter. However, because of the microwave interferometry technique, the instrument Y-shape and the antenna spacing, the resulting FOV is a hexagonal-like area of less than 1000 km (see Figure 1). The spatial resolution varies from 35 km at the FOV centre to 50 km at the border. MIRAS will provide observations of a single azimuth at various incidence angles (from 0° to 55°) and radiometric resolutions depending on its position within the field of view during a satellite overpass. This fact will much improve the retrieval algorithms because a lot of independent information of each pixel will be registered, and will permit the estimation of soil roughness, vegetation opacity and albedo, etc. together with soil moisture.

To satisfy the scientific requirements, SMOS aims at providing global maps of soil moisture every 3 days with a ground resolution better than 50 km, and a 0.04 m^3/m^3 volumetric humidity. This soil moisture accuracy is referred to pixels outside mountainous, urban, and partially frozen or snow-covered areas. For sea salinity, maps with accuracy better than 1.2 psu and 200 km ground resolution will be acquired every 30 days.

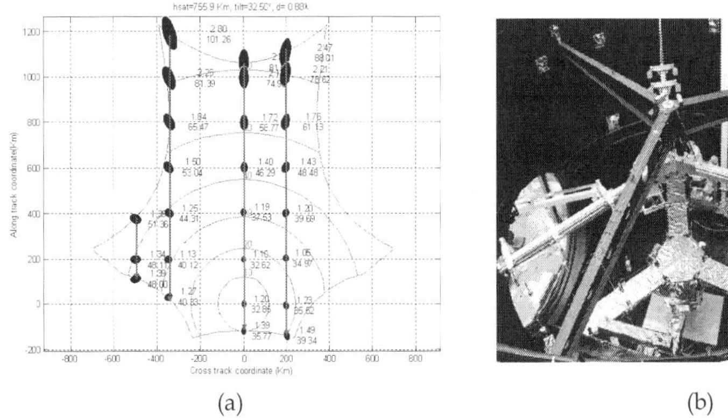

<table>
<tr><td>(a)</td><td>(b)</td></tr>
</table>

Fig. 1. (a) SMOS field of view. Blue lines indicate the observation angle, black ellipses indicate the pixel size depending on the position within the snap-shot, and red lines show how a single pixel can be observed under different viewing positions during a satellite overpass. (b) MIRAS during the IVT campaign at ESA-ESTEC Maxwell anechoic chamber premises (picture courtesy of EADS-CASA Espacio)

The Soil Moisture Active and Passive mission

The Hydrosphere State (HYDROS) mission was proposed by NASA to enhance the understanding of the land hydrosphere state and improve the climate prediction models (Entekhabi et al., 2004). The mission had a circular, polar, sun-synchronous, 6 am/pm equator crossing orbit located at 670 km altitude. Hydros aimed at providing soil moisture estimates with a 4% volumetric accuracy in the top 2-5 cm and capturing freeze/thaw state transitions in integrated vegetation-soil continuum at the spatial scale of landscape variability (3 km). The payload consisted of both active and passive sensors. An L-band radiometer measured the first, second, and third Stokes parameters with a 40 km spatial resolution, and 1 K relative accuracy, while an L-band radar acquired VV, HH, and HV polarisations with a 10 km resolution, and 0.5 dB accuracy for VV and HH. Observations at a constant incidence angle between 35° and 50°. Both the radar and the radiometer share a 6 m diameter reflector antenna which rotates about its nadir axis. The HYDROS mission was cancelled in 2005, but many HYDROS science and technology issues are being reviewed to be implemented in the future Soil Moisture Active and Passive (SMAP) mission.

1.2. Scope of this chapter

Space missions for Earth observation require a huge amount of previous work to validate the existing science and technology. This is accomplished through airborne and ground-based field experiments. This chapter reviews the basics of microwave radiometry and the existing models of soil emissivity and soil dielectric constant. The field experiments over land carried out during the last three decades are summarised, paying special attention to the recent experiments carried out by the Remote Sensing Laboratory from the Universitat Politècnica de Catalunya (UPC) as part of the preparatory activities for the SMOS mission over land.

2. L-band emission of land covers

Microwave remote sensing is based on the measurement of the thermal radiation or brightness temperature of a target, which is determined by its physical temperature and emissivity. The emissivity of land covers depends on soil moisture, but also on soil temperature (Choudhury et al., 1982; Wigneron et al., 2001), soil surface roughness (Mo & Schmugee, 1987; Escorihuela et al., 2007), vegetation canopy (Jackson & Schmugee, 1991; Ferrazzoli et al., 2002), snow cover (Schwank et al, 2004), relief (Mätzler & Standley, 2000; Talone et al., 2007), etc.

2.1. Soil dielectric constant model

The dielectric constant determines the response of the soil to an incident electromagnetic wave. This response is composed of two parts, real and imaginary ($\varepsilon = \varepsilon' + j\varepsilon''$) which determine the wave velocity and energy losses respectively. In a non-homogeneous medium such as soils, the dielectric properties have a strong impact on its microwave emission. However, the relationship between the soil dielectric constant and the soil physical properties is not straightforward. A large number of studies have been performed during the last decades to find out this relationship since it plays an important part in the soil moisture retrieval algorithms from remote sensing data (Hipp, 1974; Wang & Schmugge, 1980; Topp et al., 1980; Hallikainen et al., 1985; Dobson et al., 1985; Roth et al., 1992; Mironov et al., 2004). Some of these models are simple empirical models in which data is fitted by a curve unique for all soils; others propose semi-empirical approaches which take into account some soil physical properties. The dielectric constant of dry soils is almost independent of temperature (Topp et al., 1980) and frequency. On the contrary, wet soils show a complex behaviour depending on the interaction between soil, water, and air particles. Hallikainen et al. (1985) performed a series of dielectric constant measurements of five soils with different texture composition at frequencies between 1.4 and 18 GHz and found out that texture has a strong effect on the dielectric behaviour which is specially pronounced at frequencies below 5 GHz.

In the dielectric-mixing model by Roth et al. (1992), differences in soil texture and bound water and free water are ignored altogether. Other models use a semi-empirical approach that contains a model of the complex dielectric constant and the volume fraction of each of the soil components. This kind of approach was followed by two widely used models presented in Wang & Schmugge (1980) and Dobson et al. (1985). The starting point of both of them is the dielectric mixing model by (Birchak et al., 1974),

$$\varepsilon^{\alpha} = V_s \varepsilon^{\alpha} + V_a \varepsilon_a^{\alpha} + V_{fw} \varepsilon_{fw}^{\alpha} + V_{bw} \varepsilon_{fw}^{\alpha}, \tag{1}$$

where V_s (ε_s), V_a (ε_a), V_{fw} (ε_{fw}) and V_{bw} (ε_{bw}) are the volume fraction (dielectric constant) of solid phase, air, free water, and bound water in the soil, respectively. The expression in (1) can be rewritten as a function of the bulk density ρ_b, the particle density ρ_s, and the volumetric moisture m_v as

$$\varepsilon^{\alpha} = 1 + \frac{\rho_b}{\rho_s}(\varepsilon_s^{\alpha} - 1) + V_{fw} \varepsilon_{fw}^{\alpha} + V_{bw} \varepsilon_{fw}^{\alpha} - m_v. \tag{2}$$

The Wang & Schmugge (1980) model was proposed for 1.4 and 5 GHz frequencies and starts from (2) with α=1. This model provides separate dielectric constant equations for volumetric water content lower than or greater than the transition moisture w_t. The transition moisture is the moisture content at which the free water phase begins to dominate the soil hydraulics, and it is strongly dependent on texture. The soil dielectric constant is then estimated as

$$\varepsilon = \begin{cases} m_v \varepsilon_x + (P - m_v)\varepsilon_a + (1 - P)\varepsilon_r, & m_v < w_t \\ w_t \varepsilon_x + (m_v - w_t)\varepsilon_{fw} + (P - m_v)\varepsilon_a + (1 - P)\varepsilon_r, & m_v > w_t \end{cases}$$ (3)

with

$$\varepsilon_x = \begin{cases} \varepsilon_i + (\varepsilon_{fw} - \varepsilon_i)\dfrac{m_v}{w_t}\gamma, & m_v < w_t \\ \varepsilon_i + (\varepsilon_{fw} - \varepsilon_i)\gamma, & m_v > w_t \end{cases}$$ (4)

and $\gamma = -0.57 w_t + 0.481$, ε_i, ε_a, ε_{fw}, and ε_r the dielectric constants of ice, air, free water, and rock, respectively, ε_x the dielectric constant of the initially adsorbed water, and P the soil porosity. The variable ε_{fw} is estimated using the Debye equation.

On the other hand, the Dobson et al. (1985) model starts from (2) and assumes that there is not distinction between bound and free water. Taking this into account, the new expression for the dielectric constant is

$$\varepsilon = \left(1 + \frac{\rho_b}{\rho_s}(\varepsilon_s{}^\alpha - 1) + m_v^\beta \varepsilon_{fw}{}^\alpha - m_v\right)^{1/\alpha}.$$ (5)

The real and imaginary parts of the dielectric constant of soils are obtained separately using the percentage of sand S and clay C in the soil, and $\beta =(127.48-0.519S-0.152C)/100$ for the real part, and $\beta =(1.33979-0.603S-0.166C)/100$ for the imaginary part. For further information on dielectric constant models refer to Behari (2005) and Chukhlantsev (2006).

2.2. Emission from bare soils

The emissivity of bare soils e_s is given by

$$\Gamma_{s,p} = 1 - e_{s,p}$$ (6)

where $\Gamma_{s,p}$ is the total reflectivity at p polarisation in a chosen observation direction, which can be expressed in terms of the bistatic scattering coefficients σ_{pp} and σ_{pq} as follows (Fung, 1994):

$$\Gamma_{s,p}(\theta_i, \varphi_i) = \frac{1}{4\pi} \int_{4\pi} \left[\sigma_{pp}(\theta_i, \varphi_i, \theta_r, \varphi_r) + \sigma_{pq}(\theta_i, \varphi_i, \theta_r, \varphi_r)\right] d\Omega_r.$$ (7)

Subscripts i and r stand for the incident and scattered radiation, and θ and φ are the incidence and azimuth angles, respectively. The coefficient σ_{pp}, often called coherent component, considers the scattering on the observation direction and polarisation, while σ_{pq}, or incoherent component, considers the scattering in whatever other direction and albedo describes the scattering of the emitted radiation by the vegetation, and is a function of plant geometry.

3. Soil moisture retrieval algorithms

The brightness temperature of land covers is influenced by many variables, the most important being soil moisture and temperature, and vegetation characteristics. The challenge is to reconstruct the environmental parameters from the measured signal by using a minimum of ancillary data. To do this, different soil moisture retrieval algorithms have been developed. The first one is based on the experimental relationship between the geophysical variables and the radiative transfer equation using a regression technique. This approach has limited applicability, since often the regression is valid only for the test sites where they were obtained. The second approach is based on the use of neural networks. These algorithms have been used with satisfactory results in the retrieval of agricultural parameters from radiometric data (Del Frate et al., 2003), but need a training phase that is not always feasible. The third type of algorithms is widely used and is based on the inversion of radiative transfer models. Obviously, this approach has also disadvantages, since errors of the model lead to errors in the retrieval. The soil moisture models are used as forward models, and the geophysical variables are retrieved by minimisation of a cost function of the type (Wigneron et al., 2003; Pardé et al., 2004; Saleh et al., 2006)

$$F = \sum_{\theta} \frac{\left(T_{Bh} - T_{Bh}^{meas}\right)^2 + \left(T_{Bv} - T_{Bv}^{meas}\right)^2}{\sigma_{T_B}^2} + \sum_{N} \frac{\left(P_N^{ini} - P_N\right)^2}{\sigma_{P_N}^2} . \tag{14}$$

The simulated brightness temperature T_{Bp} is computed using (13), T_{Bp}^{meas} is the measured brightness temperature, and P_N is any of the parameters on which T_{Bp} depends. A first-guess value of the parameter P_N (P_N^{ini}) with associated standard deviation σ_{PN} can be also considered in the cost function.

4. Soil moisture field experiments

Space missions for Earth observation require a huge amount of previous work to validate the existing science and technology. This is accomplished through airborne or ground-based field experiments. Table 1 summarises some of the campaigns that have been conducted in the last years over land surfaces using L-band radiometers. Most of these radiometers have been designed within the context of preparatory activities previous to a space mission launch.

In the early 70's NASA was getting ready to launch its first microwave radiometers for Earth observations on the Nimbus 5 spacecraft and it needed to learn more about the capabilities of these sensors for land observations. To do this several field campaigns were conducted in the period 1971 to 1973. For the remote sensing of soil moisture irrigated agriculture test areas were selected in the southwestern part of the United States. These experiments demonstrated the sensitivity of microwave radiometers to surface soil moisture and that the longer wavelength sensors worked best (Schmugge et al. 1974; Burke et al., 1979). These data were used to study the effects of surface roughness (Choudhury et al., 1979), soil temperature (Choudhury et al., 1982) and soil texture (Schmugge, 1980). These experiments defined the basic sensitivity of microwave radiometers for soil moisture sensing and some of the factors affecting this sensitivity. Since that time much has been done to refine and improve this understanding. Some of these later experiments will be decribed in the following sections. A description of the MOUSE 2004, T-REX 2004/2006, and SMOS REFLEX 2003/2006 campaigns over land carried out by UPC and of their results is provided in the next sections. Radiometric measurements were acquired using the UPC L-band Automatic Radiometer (LAURA), which has a working frequency of 1.4135 GHz, the same as the SMOS payload MIRAS (Villarino, 2004). A view of the experiment sites and the LAURA radiometer is shown in Figure 2.

(a) T-REX 2006   (b) SMOS REFLEX 2003   (c) MOUSE 2004

Fig. 2. Picture of the MOUSE 2004, T-REX 2006, and SMOS REFLEX 2003 experiment sites

Year	Campaign	Site	Surface	Instrument
1971/1975		Phoenix,AZ Imperial valley, CA (USA)	Irrigated agriculture	Multi-frequency Microwave Radiometers
1976/1978	Joint US/USSR exchange	Hand County, SD (USA)	Dryland agriculture	PMIS, MFMR
1985/1989	FIFE	Konza Prairie, Kansas (USA)	Short grass, prairie	PBMR
1990/1991	Monsoon 90	Walnut Gulch AZ (USA)	Desert	PBMR, ESTAR
1992	HAPEX-Sahel	Niger, West Africa	Semi-arid savannah	PBMR
1992/1994	Washita	Oklahoma (USA)	Agricultural	ESTAR, AIRSAR, PBMR
1997/1999	SGP	Oklahoma (USA)	Agricultural	ESTAR, PSR, PAL
1999	ESSC	Reading (UK)	Bare soil	SWaMP-L
2001	INRA	Avignon (FR)	Bare, corn	EMIRAD
2001	EuroSTARSS	Toulouse (FR), Utiel (SP)	Various	STARSS
2002	ETH	Eschikon (CH)	Bare	ELBARA
2002	SMEX	Iowa (USA)	Various	ESTAR, PSR, PALS, AIRSAR
2003/2006	SMOS REFLEX	Utiel (SP)	Vines	LAURA
2003	ELBARA 2003	Wageningen (NL)	Bare, grass	ELBARA
2003-2006	SMOSREX	Toulouse (FR)	Bare, fallow	LEWIS
2003	SMEX	OK,AL,GA(USA)	Various	2DSTAR, PSR, AIRSAR
2004	SMEX	AR, NM (USA)	Various	2DSTAR, PSR

2004	MOUSE	Ispra (IT)	Bare soil	LAURA
2004/2006	T-REX	Agramunt (SP)	Bare soil	LAURA
2004	ETH	Eschikon (CH)	Grass	ELBARA
2004	Bray	Les Landes (FR)	Forest	EMIRAD
2005	ETH	Julich (GE)	Forest	ELBARA
2005	SMEX	Iowa (USA)	Various	APMIR
2005	NAFE	Upper Hunter (AU)	Grass, forest	EMIRAD, PLMR
2006	TuRTLE	El Brull (SP)	Mountain	LAURA
2006/2007	MELBEX	Utiel (SP)	Shrub, vines	EMIRAD
2006	NAFE	Yanco (AU)	Various	PSR, PALS
2007	CLASIC	Oklahoma (USA)	Various	PSR, PALS
2008-2010	GRAJO	Vadillo de la Guareña (SP)	Grass, barley, bare soil	LAURA,SMIGOL

Table 1. Summary of L-band field campaigns over land performed since 1970. Acronyms stand for USA (United States of America), UK (United Kingdom), FR (France), SP (Spain), NL (The Netherlands), CH (Switzerland), GE (Germany), IT (Italy), and AU (Australia).

5. Texture effects on the L-band emissivity of bare soils: the MOUSE 2004 experiment

The Monitoring Underground Soil Experiment (MOUSE) 2004 was performed from day of year (DoY) 159 to 184, 2004, at the Joint Research Centre (JRC) complex of the European Union in Ispra, north of Italy (45°48N, 8°37E; 213 m altitude). The test lane consisted of six 6 m long by 6 m wide by 1 m depth bare plots of different soil types which are distributed along a lane with West-East orientation. The LAURA radiometer, an infra-red video-camera, and the control rack were installed on a metallic gantry guided by a rail on one side and freely rolling on the other. A picture of the setup is shown in Figure 2(c). Soil temperature was measured using 107 probes from Campbell Scientifc installed just below the surface, and at 5 cm and 10 cm depth. An infra-red video camera was mounted by LAURA to acquire soil surface temperature images concurrently with the radiometric measurements. Most probably, satellite-borne radiometers will have to rely on this kind of auxiliary sensors to estimate soil moisture from radiometric observations. On the other hand, soil moisture was automatically acquired every 30 min by seven ML2x Thetaprobes per plot installed below the surface (x2), and at 5 cm (x2), 10 cm (x2), and 15 cm (x1). Plots were irrigated three times during the experiment, except RS which was kept dry and protected from water by a sliding tent. Soils at the experiment site were loam (LO), natural sand (SA), sieved sand used in the construction industry (RS), clay from rice fields (CL), a mixture of loam and of commercial products for gardening (OR), and ferromagnetic crushed volcanic soil from Naples (FE).

Figure 3 represents the sensitivity of the H-pol emissivity to the percentage of field capacity, which depends on soil texture. Measurements at 25° (blue), 35° (red), and 45° (green) incidence angles have been plotted. LO, RS, and FE emissivity has a linear dependence on field capacity. The R^2 estimator at 35° for LO is 71%, for RS 70%, and for FE is 85%. Clay has the worst linear fit, with R^2 lower than 20%. Soil samples of each plot were acquired and their dielectric constant measured at laboratory (Vall-llossera et al., 2005a).

Soil moisture was retrieved from radiometric measurements using the cost function in (14) with different constraints. The minimum soil moisture RMSE obtained considering different depths as ground-truth data and three different dielectric constant models (Wang &

Schmugge, 1980; Dobson et al., 1985; Vall-llossera et al., 2005a) is represented in Figure 4. Best results for LO and FE are achieved for the soil moisture in the 0 to 5 cm layer, but few discrepancies between RMSE values at different depths were found out. This is mainly due to their almost linear soil moisture profile. Best results for SA were obtained for the soil moisture in the 0-15 cm layer. As sandy soils dry faster, the penetration depth is higher. On the other hand, as RS plot was kept dry throughout the experiment, no important differences in the RMSE have been found in the comparison. Clay had a high water content at 10 cm (>0.22 m³/m³) and hence the main signal contribution comes from the first 10 cm of soil. Similar RMSE values have been found for OR considering either the 0 to 5 cm layer or the 0 to 10 cm layer as ground-truth. Similarities are because the water content at 5 cm and 10 cm depth is almost the same during the experiment. On the other side, in Figure 4 it can be noted the performance of each dielectric constant models for each soil type. Results for Wang & Schmugge (1980) model are good for loamy and ferromagnetic soils, but their performance for sandy soils is worse than Dobson et al. (1985) and Vall-llossera et al. (2005a). The model in Dobson et al. (1985) offers best results for sandy soils (SA and RS) which registered lower soil moisture values and dried faster than the others. The expressions derived from UPC laboratory experiments show good performance for most of the fields. These facts confirm that the selected dielectric constant model has an impact on soil moisture estimation error and can increase it between 3 to 5% depending on the soil type.

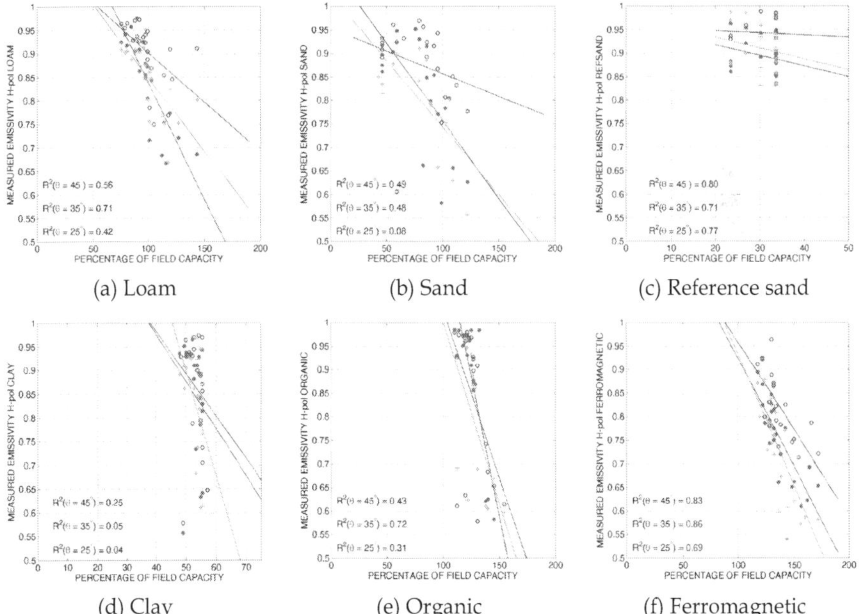

(a) Loam (b) Sand (c) Reference sand

(d) Clay (e) Organic (f) Ferromagnetic

Fig. 3. Variation of the emissivity with the field capacity at H-pol. Colours indicate the incidence angle: 25° (blue), 35° (red), and 45° (green).

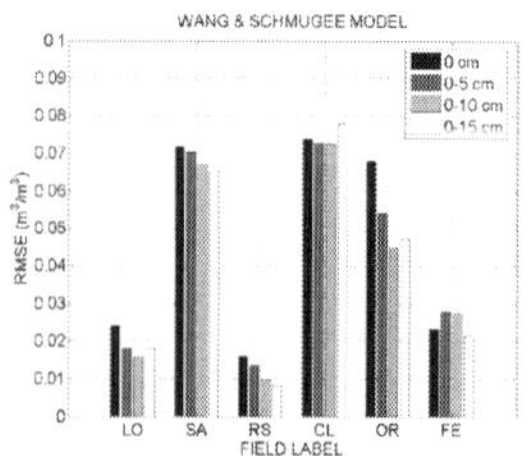
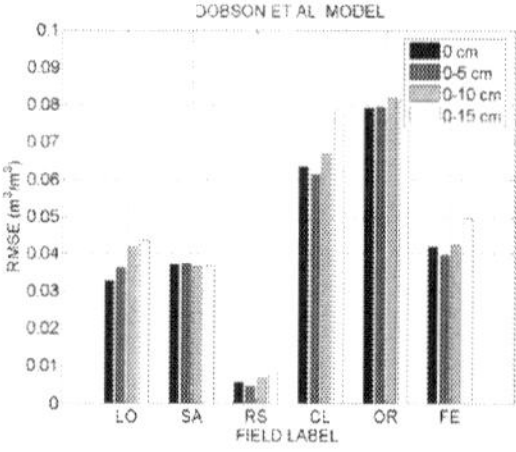
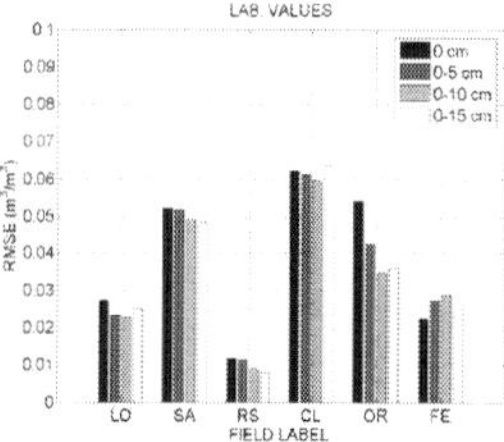

(a) Wang & Schumugge   (b) Dobson et al.   (c) Lab. measurements

Fig. 4. Minimum root mean square error of the soil moisture retrieval. Results for a soil dielectric constant model and different soil moisture ground-truth data are represented.

6. Emission of bare rough soils: the T-REX experiments

Several experiments have been performed in the past years to study the L-band emission from bare soils, both at small and large scales. Although small scale measurements are not representative of onboard sensor data, they can be useful to evaluate the various approaches to the land emission modelling. The Terrain-Roughness EXperiments (T-REX) were carried out in Agramunt, Spain (41°48N, 1°7W; 356 m altitude) from DoY 333 to DoY 337, 2004, and then again from DoY 143 to DoY 160, 2006. The sites consisted of bare soil plots with four different ploughing, but the same textural composition (21% clay, 21.3% sand). This section presents some results from the T-REX 2006 experiment. A picture of the site has been shown in Figure 2(a). Each parcel was 6 m x 15 m and the same ploughing was applied to two different plots to observe spatial diversity effects. LAURA was installed on a trailer towed by a car. Radiometric measurements were acquired at incidence angles from 40° to 65° in T-REX 2006, in 5° steps. Every plough type was measured twice each day of experiment.

Parameter	Plot label			
(mm)	A	B	C	D
l_c	126	100	190	96
σ_s	17	16.6	8	33
σ_λ	7	6.8	2.5	17
$h_{\text{air-soil}}$	31	26	13	46

Table 2. Roughness characteristics of the T-REX 2006 ploughs

Roughness measurements were performed in the along and across directions of LAURA's field of view. The standard deviation of height (σ_s) and the correlation length (l_c) of the ploughs at the T-REX 2006 site are summarised in Table 2. Two additional parameters have also been included: (i) $h_{\text{air-soil}}$, which is the transition layer thickness used to evaluate the small-scale soil roughness in the model proposed in Schneeberger et al. (2004), Mätlzer (2006, section 4.7) and (ii) σ_λ, which is the standard deviation of height assuming a windowing of length λ. Soil temperature was measured at 0, 5, 10, 15, and 20 cm depth using thermometers. On the other hand, soil moisture was measured with ML2x ThetaProbe sensors at various positions within the radiometer's FOV. As the purpose of this experiment was to study the impact of soil roughness, the site was not irrigated to keep as constant a value of soil moisture as possible. The mean soil moisture during T-REX 2006 was of 4% to 6%.

Figure 5 shows the variation of the emissivity measured during the T-REX 2006 as a function of the incidence angle. Colours indicate the polarisation, H-pol (red) and V-pol (blue), while line styles distinguish the ploughing. While soil emission at H-pol decreases almost linearly with increasing incidence angle (and the trend is more significative at incidence angles above 55°), the emission at V-pol slightly increases with incidence angle until an inflexion point at 55°, when it begins decreasing. Ploughs C and D, which correspond to the smoother and rougher plots in the site, have the highest (lowest) range of variation of the emissivity.

The comparison between radiometric measurements and numerical simulations obtained with a numerical approach by means of the integral equation method (IEM; Fung (1994)) are shown in Figure 6. Icons stand for the mean value of LAURA measurements while error bars indicate their standard deviation during the whole T-REX 2006 experiment. Although IEM predicts an increase of the vertical emission with incidence angle, dry soil measurements decrease as incidence angle increases. A good matching between simulations and measurements is observed at horizontal polarisation. The effects of taking into account the incoherent component are only relevant in the case of very rough ploughs (see Figure 6(c), plough D).

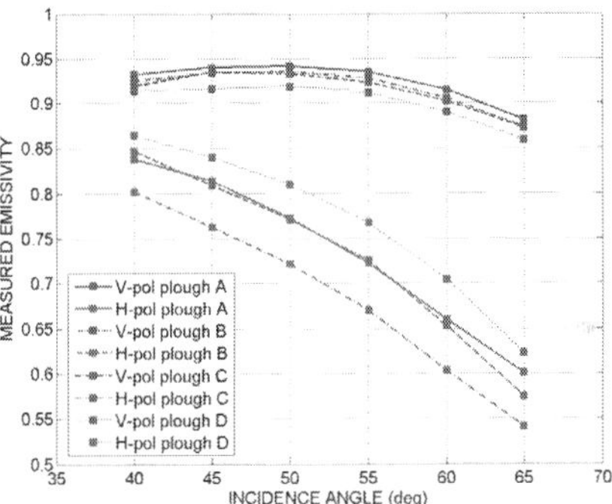

Fig. 5. Variation of the emissivity measured during the T-REX 2006 as a function of the incidence angle for the four ploughs. Colours indicate the polarisation: H-pol (red) and V-pol (blue). Line style and icons distinguish the ploughing.

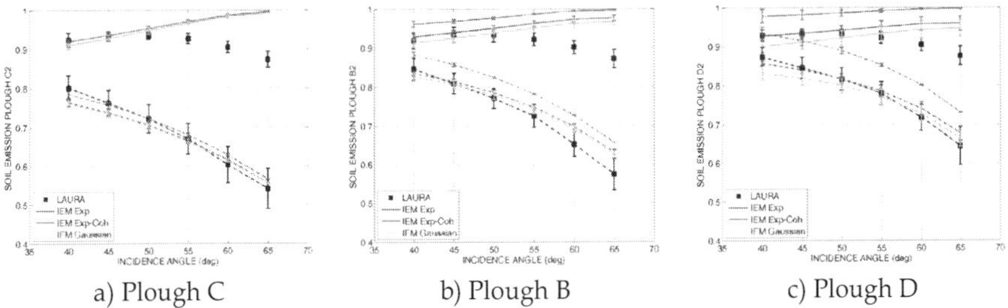

a) Plough C   b) Plough B   c) Plough D

Fig. 6. Variation with time of the emissivity measured during the T-REX 2006 experiment as a function of the incidence angle. The mean value of LAURA measurements (black icons) and their standard deviation during the whole experiment (bars) have been represented together with IEM simulations.

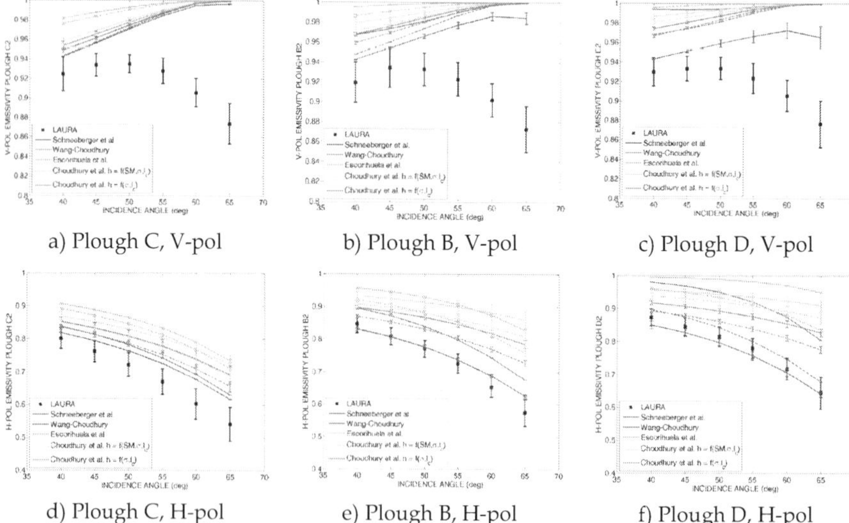

a) Plough C, V-pol   b) Plough B, V-pol   c) Plough D, V-pol

d) Plough C, H-pol   e) Plough B, H-pol   f) Plough D, H-pol

Fig. 7. Comparison between T-REX 2006 measurements (black icons) and the models in Wang & Choudhury (1981), Mätzler (2006, section 4.7), Schneeberger et al. (2004), Mo & Schmugge (1987), and Escorihuela et al. (2007). Solid lines indicate results using the standard deviation of the surface height profile while dashed lines indicates results using the λ-average standard deviation of height

T-REX 2006 measurements were also compared to the semi-empirical soil reflectivity models in Wang & Choudhury (1981), Mo & Schmugge (1987), Escorihuela et al. (2007), and the air-to-soil transition model proposed in Schneeberger et al. (2004). These models propose different expressions for the effective roughness parameter h_s in (12). Results are shown in Figure 7. Upper plots correspond to vertical polarisation while bottom plots correspond to horizontal polarisation. Colours indicate the soil reflectivity model, while line types indicate whether l_c and σ_s of the height profile (solid lines) or their equivalents taking into account the λ-windowing (dashed lines) have been used. Both the model in Wang & Choudhury (1981) using as inputs the λ -windowing soil roughness characteristics and the air-to-soil transition model from Schneeberger et al. (2004) fit data. However, Schneeberger et al. (2004) does not fit T-REX 2006 measurements of wet soil, which suggests that inhomogeneities in soil moisture or the presence of dew make it difficult to properly characterise the upper soil layer, which is needed for the air-to-soil transition model.

7. Radiometric observations of fully developed vines

Several papers are available that present experimental studies of brightness temperatures from vegetation canopies (Jackson et al. 1982; Jackson et al., 1991; Ferrazzoli et al., 2002; Van de Griend & Wigneron, 2004; Saleh et al., 2006). A look at the field experiments conducted during the past decades over vegetated sites shows how they focused their attention on crops such as alfalfa, wheat corn, or soybean, on grass, on forests, and on bushes, but few studies had been conducted over vineyards. In this context, two ground-based field experiments were performed to study the variations on the emissivity due to vines development. The SMOS REFerence pixel L-band EXperiments (REFLEX) were carried out in July 2003 and then again from July to November 2006 at two vineyards within the València Anchor Station (VAS; Figure 2(b)), Spain (Vall-llossera et al., 2005b). The VAS has been selected as a SMOS Cal/Val site because of its almost homogeneous land cover (vineyards) and its size, which is approximately that of a SMOS pixel. The goal of the SMOS REFLEX experiments was to assess the impact of grapevines on the radiometric emission and on the sensitivity of the sensor to soil moisture. Results from the SMOS REFLEX 2003 experiment are presented hereafter.

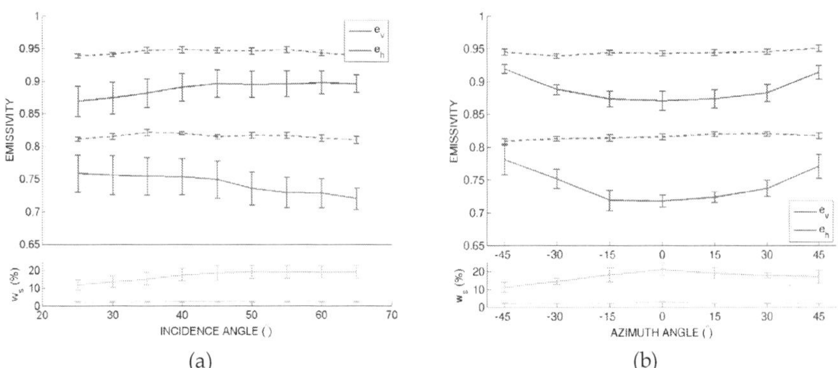

(a) (b)

Fig. 8. Mean value and standard deviation of the emissivity and of the ground-truth soil moisture as a function of (a) the incidence angle and for all azimuth angles, and (b) the azimuth angle and for all incidence angles. DoY 185 (solid line, wet soil) and DoY 181 (dashed-dotted line, dry soil) have been represented. Blue and red indicate vertical and horizontal polarisations, respectively.

During SMOS REFLEX 2003 radiometric measurements were acquired from DoY 181 to 191 at nine incidence angles (from 25° to 65° in 5° steps) and seven azimuth angles. Pictures at every look direction were taken to estimate the percentage of FOV covered by plants. The site was irrigated until saturation on DoY 182 and 185, and then was let to dry out. Even though the field was irrigated until saturation, different soil moisture values were measured depending on the location of the test point within the vineyard. This is most probably because of surface inhomogeneities and variations on the compactness of the terrain due to plough and roots distribution. This large spatial variability in soil moisture, which in this case varies from 3% up to 35% in a 20 m x 20 m area, is one of the main problems soil moisture retrieval algorithms must face since it complicates the comparison between ground-truth and estimated w_s at the plot scale. Volumetric soil moisture, soil temperature, and soil roughness were measured, and plants were fully characterised.

Figure 8(a) represents the mean value and standard deviation of the emissivity and of the ground-truth soil moisture as a function of the incidence angle and for all azimuth angles. Measurements from DoY 185 (solid line) and DoY 181 (dashed-dotted line) have been represented. The mean value of both polarisations seems independent of incidence angle for

Soil roughness has a strong impact on land brightness temperature. This effect is more noticeable in the case of dry soils. Data from bare soils with a standard deviation of height varying from 8 mm to 33 mm was acquired during the T-REX experiments, and was compared to predictions from numerical and semi-empirical soil emission models. The goodness of these models is of key importance for an accurate soil moisture estimation from satellite data. As expected, dry soil emissivity at H-pol decreased almost linearly with increasing incidence angle, being the decrement inversely proportional to soil roughness. Although similar emission was measured at H-pol for all plough at low incidence angles, it was noted that emission decreases slowly for rough soils than for smooth soils. Above an incidence angle of 50°, the decreasing slope of H-pol emission was higher for all plough. The sensitivity of V-pol to roughness is lower than that of H-pol. The trend of V-pol emission was found to be different depending on whether the soil was wet or dry. When the soil is wet the emission increased with the incidence angle, which is in accordance with most model predictions. However, the trend for dry soils is decreasing with increasing incidence angle and roughness. This may suggest the existence of a relationship between soil moisture and the effective roughness.

The integral equation method (IEM) predicts an increase in V-pol emission with incidence angle, which is not in accordance with measurements of dry soils. On the other hand, the predicted descending trend at H-pol with increasing incidence angle is in accordance to measurements, although IEM underestimates the soil emission both for dry and wet soils. In a randomly ploughed field, without a significant tillage direction, the impact of choosing an exponential or gaussian height probability density function in the IEM model is minimum, whereas it was noted that the incoherent term of the reflectivity must be considered for rough soils.

In general, all semi-empirical land emission models follow the trend of dry soils measurements at H-pol, whereas discrepancies exist for wet soils. Neither the semi-empirical models nor the IEM describe the trend of dry soils V-pol measurements, being the error at this polarisation larger as the incidence angle increases. The lowest error between predictions and measurements for H-pol was obtained using the Wang & Choudhury (1981) model, but with roughness statistics averaged as a function of the measurements wavelength. The dependence on the incidence angle proposed by Wang & Choudhury (a squared cosine) had been discussed by other authors who considered it to be too much strong at L-band. However, this simple formulation has been tested with T-REX data with good results if the averaged standard deviation in transects equal to the wavelength is used.

The SMOS REFLEX 2003/2006 experiments site was in the Valencia Anchor Station, a selected area for the SMOS calibration and validation activities. No previous studies over vines were reported in the literature prior to these campaigns. In the first experiment, fully developed vines were characterised during two weeks, while controlled irrigations moistened the field. The second experiment was planned to monitor changes in the L-band emission of vineyards during different stages of plants development.

Since vines do not have a predominant vertical nor horizontal structure, the opacity and the albedo were found to be independent on the polarisation. The Wang & Choudhury (1981) model was used as the soil emission forward model in the retrieval algorithm, while the dielectric constant model was the one from Wang & Schmugge (1980). Good results were obtained for incidence angles up to 55°, but the convergence of the algorithm was rarely achieved above that value. The error between ground-truth and estimated soil moisture was

2.3%, better than the 4% required for SMOS. For incidence angles above 55° the convergence of the algorithm was rarely achieved, probably due to the larger effect of the vegetation at large incidence angles, not accurately described by the model. Higher order models should be accounted for at these angles.

Although the radiometric behaviour varies from one canopy to the other, at a larger scale such as that of SMOS it is very likely that the vegetation types can be averaged over the footprint and that it is not necessary to account for an accurate distinction between canopies.

9. Acknowledgements

A. Monerris would like to thank Spanish Ministry of Science and Education projects MIDAS-2 ESP2002-11648-E, MIDAS-3 ESP2004-00671, and MIDAS-4 ESP2005-06823-C05-02, for funding the SMOS REFLEX, T-REX and MOUSE 2004 field experiments, and the FPU AP2003-1567 grant. Special thanks are also given to M. Vall-llossera and A. Camps, from UPC, for their valuable comments during the analysis of the experimental data.

10. References

Barré, H., Duesmann, B. & Kerr, Y.H. (2008). SMOS: The mission and the system. *IEEE Trans. Geosci. and Remote Sens.*, 46, 587-593

Behari, J. (2005). *Microwave dielectric behaviour of wet soils (Remote Sensing and Digital Image Processing)*, Springer, ISBN-10:1402032714 , The Netherlands

Brady, N.C. & Weil, R.R. (2003). *Elements of the nature and properties of soils*, 2nd ed., Prentice Hall, ISBN-10:013048038X , NJ, USA

Burke, W. J., Schmugge, T. J. & Paris, J.F. (1979). Comparison of 2.8 and 21 cm microwave radiometer observations over soils with emission model calculations, *J. Geophys. Res.*, 84, 287-294.

Chanzy, A., Raju, S. & Wigneron, J.-P. (1997). Estimation of soil microwave effective temperature at L and C bands. *IEEE Trans. Geosci. and Remote Sens.*, 35, 3, 570-580

Choudhury, B.J., Schmugge, T.J., Chang, A.T.C. & Newton, R.W. (1979). Effect of surface roughness on the microwave emission of soils. *J. Geophys. Res.*, 84, 5699-5705

Choudhury, B.J., Schmugge, T.J. & Mo, T. (1982). A simple parameterization of effective soil temperature for microwave emission. *J. Geophys. Res.*, Vol., No., (month and year of the edition) 1301-1304

Chukhlantsev, A.A. (2006). *Microwave radiometry of vegetation canopies*, Springer, ISBN-10 1-4020-4681-2, The Netherlands

Del Frate, F., Ferrazzoli, P. & Schiavon, G. (2003). Retrieving soil moisture and agricultural variables by microwave radiometry using neural networks. *Rem. Sens. Env.*, 84, 2, 174-183

Dobson, M. C., Ulaby, F. T., Hallikainen, M. T. & Elrayes, M. A. (1985). Microwave dielectric behaviour of wet soils 2. Dielectric mixing models. *IEEE Trans. Geosci. and Remote Sens.*, 23, 35-46

Eagleman, J.R. & Lin, W.C. (1976). Remote sensing of soil moisture by a 21 cm passive radiometer. *J. Geophys. Res.*, 81, 3660-3666

Entekhabi, D., Njoku, E., Houser, P., Spencer, M., Doiron, T., Smith, J., Girard, R., Belair, S., Crow, W. ,Jackson, T., Kerr, Y., Kimball, J., Koster, R., McDonald, K., O'Neill, P.,

Wang, J.R. (1983). Passive microwave sensing of soil moisture content: The effects of soil bulk density and surface roughness. *Remote Sens. Environ.*, 13, 329-344

Wang, J.R. & Choudhury, B.J. (1981). Remote sensing of soil moisture content over bare field at 1.4 GHz frequency. *J. Geophys. Res.*, 86, 5277-5282

Wegmüller, U. & Mätzler, C. (1999). Rough bare soil reflectivity model. *IEEE Trans. Geosci. and Remote Sens.*, 37, 1391-1395

Wigneron, J., Laguerre, L. & Kerr, Y.H. (2001). A simple parameterization of the L-band microwave emission from rough agricultural soils. *IEEE Trans. Geosci. and Remote Sens.*, 39, 1697-1707

Wigneron, J.-P., Calvet, J.-C., Pellarin, T., de Griend, A.A. Van, Berger, M. & Ferrazzoli, P. (2003). Retrieving near-surface soil moisture from microwave radiometric observations: current status and future plans. *Remote Sens. Environ.*, 85, 489-506

Wigneron, J.-P., Chanzy, A., de Rosnay, P., Räudiger, C. & Calvet, J.-C. (2008). Estimating the effective soil temperature at L-band as a function of soil properties. *IEEE Trans. Geosci. And Remote Sens.*, 46, 3, 797-807

22

AggieAir: Towards Low-cost Cooperative Multispectral Remote Sensing Using Small Unmanned Aircraft Systems

[a]Haiyang Chao, [a,b]Austin M. Jensen, [a]Yiding Han, [a]YangQuan Chen
and [b]Mac McKee
[a]*Center for Self Organizing and Intelligent Systems (CSOIS),Utah State University*
haiyang.chao@aggiemail.usu.edu, yqchen@ieee.org
[b]*Utah Water Research Laboratory (UWRL), Utah State University*
austin.jensen@aggiemail.usu.edu
United States of America

1. Introduction

This chapter focuses on using small low-cost unmanned aircraft systems (UAS) for remote sensing of meteorological and related conditions over agricultural fields or environmentally important land areas. Small UAS, including unmanned aerial vehicle (UAV) and ground devices, have many advantages in remote sensing applications over traditional aircraft- or satellite-based platforms or ground-based probes for many applications. This is because small UAVs are easy to manipulate, cheap to maintain, and remove the need for human pilots to perform tedious or dangerous jobs. Multiple small UAVs can be flown in a group and complete challenging tasks such as real-time mapping of large-scale agriculture areas.

The purpose of remote sensing is to acquire information about the Earth's surface without coming into contact with it. One objective of remote sensing is to characterize the electromagnetic radiation emitted by objects (James, 2006). Typical divisions of the electromagnetic spectrum include the visible light band ($380 - 720nm$), near infrared (NIR) band ($0.72 - 1.30\mu m$), and mid-infrared (MIR) band ($1.30 - 3.00\mu m$). Band-reconfigurable imagers can generate several images from different bands ranging from visible spectra to infra-red or thermal based for various applications. The advantage of an ability to examine different bands is that different combinations of spectral bands can have different purposes. For example, the combination of red-infrared can be used to detect vegetation and camouflage and the combination of red slope can be used to estimate the percent of vegetation cover (Johnson et al., 2004). Different bands of images acquired remotely through UAS could be used in scenarios like water management and irrigation control. In fact, it is difficult to sense and estimate the state of water systems because most water systems are large-scale and need monitoring of many factors including the quality, quantity, and location of water, soil and vegetations. For the mission of accurate sensing of a water system, ground probe stations are expensive to build and can only provide data with very limited sensing range (at specific positions and second level temporal resolution). Satellite photos can cover a large area, but have a low resolution and a slow update rate (30-250 meter or lower spatial resolution and week level temporal resolution). Small UAVs cost less money but can provide more accurate information (meter or centimeter spatial resolution and hour-level temporal resolution) from low altitudes with less interference from clouds. Small UAVs combined with ground and orbital sensors can even form a multi-scale remote sensing system.

UAVs equipped with imagers have been used in several agricultural remote sensing applications for collecting aerial images. High resolution red-green-blue (RGB) aerial photos can be used to determine the best harvest time of wine grapes [Johnson et al. 2003]. Multispectral images are also shown to be potentially useful for monitoring the ripeness of coffee [Johnson et al. 2004]. Water management is still a new area for UAVs, but it has more exact requirements than other remote sensing applications: real-time management of water systems requires more and more precise information on water, soil and plant conditions, for example, than most surveillance applications. Most current UAV remote sensing applications use large and expensive UAVs with heavy cameras (in the range of a kilogram). Images from reconfigurable bands taken simultaneously can increase the final information content of the imagery and significantly improve the flexibility of the remote sensing process.

Motivated by the above remote sensing problem, AggieAir, a band-configurable small UAS-based remote sensing system has been developed in steps at Center for Self Organizing and Intelligent Systems (CSOIS) together with Utah Water Research Lab (UWRL), Utah State University. The objective of this chapter is to present an overview of the ongoing research on this topic.

The chapter first presents a brief overview of the unmanned aircraft systems focusing on the base of the whole system: autopilots. The common UAS structure is introduced. The hardware and software aspects of the autopilot control system are then explained. Different types of available sensor sets and autopilot control techniques are summarized. Several typical commercial off-the-shelf and open source autopilot packages are compared in detail, including the Kestrel autopilot from Procerus, Piccolo autopilot from CloudCap, and the Paparazzi open source autopilot etc.

The chapter then introduces AggieAir, a small and low-cost UAS for remote sensing. AggieAir comprises of a flying-wing airframe as the test bed, the OSAM-Paparazzi autopilot for autonomous navigation, the Ghost Foto image system for image capture, the Paparazzi ground control station (GCS) for real time monitoring, and the gRAID software for image processing. AggieAir is fully autonomous, easy to manipulate, and independent of a runway. AggieAir can carry embedded cameras with different wavelength bands, which are low-cost but have high spatial resolution. These imagers mounted on UAVs can form a camera array to perform multi-spectral imaging with reconfigurable bands, depending on the objectives of the mission. Developments of essential subsystems, such as the UAV autopilot, imaging payload subsystem, and image processing subsystem, are introduced in detail together with some experimental results to show the orthorectification accuracy.

Several typical example missions together with real UAV flight test results are focused in Sec.3 including land survey, water area survey, riparian applications and remote data collection. Aerial images and stitched maps showed the effectiveness of the whole system. The future direction is more accurate orthorectification method and band-reconfigurable multi-UAV-based cooperative remote sensing for real-time water management and distributed irrigation control.

2. Small UAS Overview

In this paper, the acronym UAV (Unmanned Aerial Vehicle) is used to represent a power-driven, reusable airplane operated without a human pilot on board. The UAS (Unmanned

Aircraft System) is defined as an UAV and its associated elements which may include ground control stations, data communication links, support equipment, payloads, flight termination systems, and launch/recovery equipments (Tarbert et al., 2009). Small UAS (sUAS) could be categorized into five groups based on gross take-off weight by the sUAS aviation rule making committee, as shown in Tab. 1. Group *i* include those constructed in a frangible manner that would minimize injury and damages if there is a collision, compared with group *ii*.

Group	Gross Take-off Weight
i	$\leq$ 4.4 lbs or 2 kgs
ii	$\leq$ 4.4 lbs or 2 kgs
iii	$\leq$ 19.8 lbs or 9 kgs
iv	$\leq$ 55 lbs or 25 kgs
v	lighter than air (LTA) only

Table 1. Small UAS Categories

The topic of small UAS is quite active in the past few years (Chao et al., 2009). A lot of small fixed-wing or rotary-wing UAVs are flying in the air under the guidance from the autopilot systems for different applications like forest first monitoring, coffee field survey, search and rescue, etc. A typical UAS includes:

(1) Autopilot: an autopilot is a MEMS system used to guide the UAV without assistance from human operators, consisting of both hardware and its supporting software. The autopilot is the base for all the other functions of the UAS platform.

(2) Airframe: the airframe is where all the other devices are mounted including the frame body, which could be made from wood, foam or composite materials. The airframe also includes the flight control surfaces, which could be a combination of either aileron/elevator/rudder, or elevator/rudder, or elevons.

(3) Payload: the payload of UAS could be different bands of cameras, or other emission devices like Lidar mostly for intelligence, surveillance, and reconnaissance uses.

(4) Communication system: most UAS have more than one wireless link supported. For example, RC link for safety pilot, wifi link for large data sharing, and wireless serial modem.

(5) Ground control station: ground control station is used for real-time flight status monitoring and flight plan changing.

(6) Launch and recovery devices: some UAS may need special launching devices like a hydraulic launcher or landing devices like a net.

The whole UAS structure is shown in Fig 1. The minimal UAS onboard system requires the airframe for bedding all the devices, the autopilot for sensing and navigation, the basic imaging payload for aerial images, and the communication systems for data link with the ground. The left section concentrates on the autopilot overview since it is the base for the UAS platform and it also needs to provide accurate orientation and position data for each data set the UAS collects.

Autopilot systems are now widely used in modern aircrafts and ships. The objective of UAV autopilot systems is to consistently guide UAVs to follow reference paths, or navigate through some waypoints. A powerful UAV autopilot system can guide UAVs in all stages including

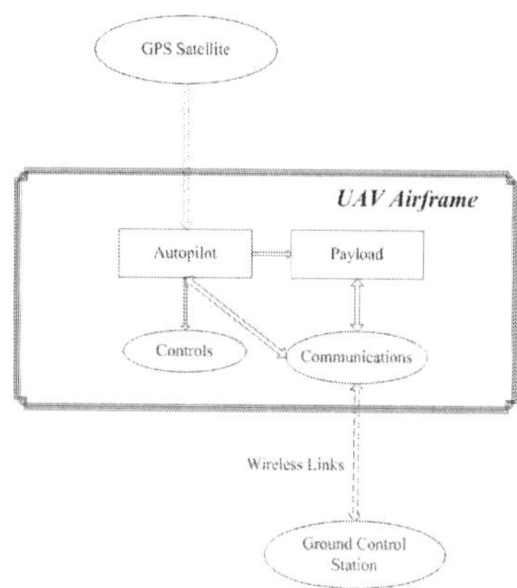

Fig. 1. UAS Structure.

take-off, ascent, descent, trajectory following, and landing. The autopilot needs also to communicate with ground station for control mode switch, to receive broadcast from GPS satellite for position updates and to send out control inputs to the servo motors on UAVs.

An UAV autopilot system is a close-loop control system with two fundamental functions: state estimation and control signal generation based on the reference paths and the current states. The most common state observer is the inertial measurement unit (IMU) including gyros, accelerometers, and magnetic sensors. There are also other attitude determination devices available like infrared or vision based ones. The sensor readings combined with the GPS information can be passed to a filter to generate the estimates of the current states for later control uses. Based on different control strategies, the UAV autopilots can be categorized to PID based autopilots, fuzzy based autopilots, neutral network (NN) based autopilots, etc. A typical off-the-shelf UAV autopilot system comprises of the GPS receiver, the IMU, and the onboard processor (state estimator and flight controller) as illustrated in Fig. 2.

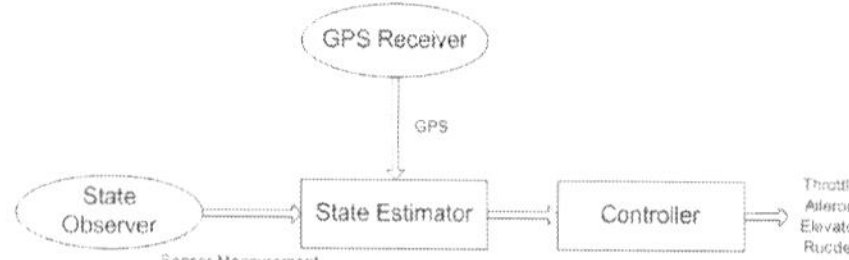

Fig. 2. Functional Structure of the UAV Autopilot.

Due to the limited size and payload of the small UAVs, the physical features like size, weight and power consumption are the primary issues that the autopilot must take into consideration. A good autopilot should be small, light and have a long endurance life. It is not so hard to design the hardware to fulfill the autopilot requirements. The current bottleneck for autopilot systems lies more in the software side.

2.1 Autopilot Hardware

A minimal autopilot system includes sensor packages for state determination and onboard processors for estimation & control uses, and peripheral circuits for servo & modem communications. Due to the physical limitations of small UAVs, the autopilot hardware needs to be of small sizes, light weights and low power consumptions. The accurate flight control of UAVs demands a precise observation of the UAV attitude in the air. Moreover, the sensor packages should also guarantee a good performance, especially in a mobile and temperature-varying environment.

2.1.1 MEMS Inertial Sensors

Inertial sensors are used to measure the 3-D position and attitude information in the inertial frame. The current MEMS technology makes it possible to use tiny and light sensors on small or micro UAVs. Available MEMS inertial sensors include:

(1) GPS receiver: to measure the positions (p_n, p_e, h) and ground velocities (v_n, v_e, v_d).

(2) Rate or gyro: to measure the angular rates (p, q, r).

(3) Acceleration: to measure the accelerations (a_x, a_y, a_z).

(4) Magnetic: to measure the magnetic field for the heading correction (ψ).

(5) Pressure: to measure the air speed (the relative pressure) and the altitude (h).

(6) Ultrasonic sensor or SONAR: to measure the relative height above the ground.

(7) Infrared sensor: to measure the attitude angles (ϕ, θ).

(8) RGB camera or other image sensors: to replace one or several of the above sensors.

GPS plays an indispensable role in the autonomous control of UAVs because it provides an absolute position measurement. A known bounded error between GPS measurement and the real position can be guaranteed as long as there is a valid 3-D lock. For instance, u-blox 5 GPS receiver could achieve a three meter 3-D accuracy (PACC) in the best case for civilian applications in the United States. There are also differential GPS units which could achieve centimeter level accuracy. The disadvantage of GPS is its vulnerability to weather factors and its relatively low updating frequency (commonly 4Hz), which may not be enough for flight control applications.

2.1.2 Possible Sensor Configurations

Given all the above inertial sensors, several sensor combinations could be chosen for different types of UAVs to achieve the basic autonomous waypoints navigation task. Most current outdoor UAVs have GPS receivers onboard to provide the absolute position feedback. The main difference is the attitude measurement solution, which could be inertial measurement unit (IMU), infrared sensor or image sensor etc.

2.1.2.1 Inertial Measurement Unit (IMU)

A typical IMU includes 3-axis gyro rate and acceleration sensors, which could be filtered to generate an estimation of the attitude (ϕ, θ, ψ). A straightforward sensor solution for small UAVs is to use the micro IGS, which can provide a complete set of sensor readings. Microstrain Gx2 is this kind of micro IMU with an update rate up to 100 Hz for inertial sensors. It has 3-axis magnetic, gyro and acceleration sensors (Microstrain Inc., Accessed 2008).

2.1.2.2 Infrared Sensor

Another solution for attitude sensing is using infrared thermopiles. The basic idea of infrared attitude sensor is to measure the heat difference between two sensors on one axis to determine the angle of the UAV because the Earth emits more IR than the sky. Paparazzi Open Source Autopilot group used this kind of infrared sensors as their primary attitude sensor (Paparazzi Forum, Accessed 2008) (Egan, 2006). The infrared sensors can be used for UAV stabilization and RC plane training since it can work as a leveler. However, it is not that accurate for later georeferecing.

2.1.2.3 Vision Sensor

Vision sensor could also be used to estimate the attitude by itself or combined with other inertial measurements (Roberts et al., 2005). The pseudo roll and pitch can be decided from the onboard video or image streams (Damien et al., 2007). Experiments on vision only based navigation and obstacle avoidance have been achieved on small rotary wing UAVs (Calise et al., 2003). In addition, vision based navigation has potentials to replace the GPS in providing position measurements especially in task oriented and feature based applications. Vision based navigation for small fixed wing UAVs is still an undergoing topic and a lot of work are still needed for mature commercial autopilots.

2.2 Autopilot Software

All the inertial measurements from sensors will be sent to the onboard processor for further filter and control processing. Autopilot could subscribe services from the available sensors based on different control objectives.

2.2.1 Autopilot Control Objectives

Most UAVs can be treated as mobile platforms for all kinds of sensors. The basic UAV waypoints tracking task could be decomposed into several subtasks including:

(1) Pitch attitude hold.

(2) Altitude hold.

(3) Speed hold.

(4) Automatic take-off and landing.

(5) Roll-Angle hold.

(6) Turn coordination.

(7) Heading hold.

There are two basic controllers for the UAV flight control: altitude controller, velocity and heading controller. Altitude controller is to drive the UAV to fly at a desired altitude including the landing and take-off stages. The heading and velocity controller is to guide the UAV to fly through the desired waypoints. Most commercial autopilots use PID controllers, and the control parameters could be tuned off-line first and re-tuned during the flight.

2.3 Typical Autopilots for Small UAVs

In this section, several available autopilots, including both commercial and open source ones, are introduced and compared in terms of sensor configurations, state estimations and controller strengths. Most commercial UAV autopilots have sensors, processors and peripheral circuits integrated into one single board to account for size and weight constraints. The advantage of the open source autopilots is its flexibility in both hardware and software. Researchers can easily modify the autopilot based on their own special requirements.

2.3.1 Procerus Kestrel Autopilot

Procerus Kestrel Autopilot is specially designed for small or micro UAVs weighing only 16.7 grams (modem and GPS receiver not included), shown in Fig. 3. The specifications are shown in Table 2. Kestrel 2.2 includes a complete inertial sensor set including: 3-axis accelerometers, 3-axis angular rate sensors, 2-axis magnetometers, one static pressure sensor (altitude) and one dynamic pressure sensor (airspeed). With the special temperature compensations for sensors, it can estimate the UAV attitude (ϕ and θ) and the wind speed pretty accurately (Beard et al., 2005).

Fig. 3. Procerus Kestrel Autopilot (Beard et al., 2005).

Kestrel has a 29MHz Rabbit 3000 onboard processor with 512K RAM for onboard data logging. It has the built-in ability for autonomous take-off and landing, waypoint navigation, speed and altitude hold. The flight control algorithm is based on the traditional PID control. The autopilot has elevator controller, throttle controller and aileron controller separately. Elevator control is used for longitude and airspeed stability of the UAV. Throttle control is for controlling airspeed during level flight. Aileron control is used for lateral stability of the UAV (Beard et al., 2005). Procerus provides in-flight PID gain tuning with real-time performance graph. The preflight sensor checking and failsafe protections are also integrated to the autopilot software package. Multiple UAV functions are also supported by Kestrel.

2.3.2 Cloud Cap Piccolo

Piccolo family of UAV autopilots from Cloud Cap Company provide several packages for different applications. PiccoloPlus is a full featured autopilot for fixed-wing UAVs. Piccolo II is an autopilot with user payload interface added. Piccolo LT is a size optimized one for small electric UAVs as shown in Fig. 4. It includes inertial and air data sensors, GPS, processing, RF data link, and flight termination, all in a shielded enclosure (CloudCap Inc., Accessed 2008). The sensor package includes three gyros and accelerometers, one dynamic pressure sensor and one barometric pressure sensor. Piccolo has special sensor configuration sections to correct errors like IMU to GPS antenna offset, avionics orientation with respect to the UAV body frame.

Piccolo LT has a 40M Hz MPC555 onboard microcontroller. Piccolo provides a universal controller with different user configurations including legacy fixed wing controller, neutral net helicopter controller, fixed wing generation 2 controller, and PID helicopter controller. Fixed

Fig. 4. PICCOLO LT Autopilot (CloudCap Inc., Accessed 2008).

wing generation 2 controller is the most commonly used flight controller for conventional fixed wing UAVs. It includes support for altitude, bank, flaps, heading & vertical rate hold, and auto take-off and landing. Piccolo autopilot supports one ground station controlling multiple autopilots and it also has a hardware-in-the-loop simulation.

2.3.3 Paparazzi Autopilot

Paprazzi autopilot is a pretty popular project first developed by researchers from ENAC university, France. Infrared sensors combined with GPS are used as the default sensing unit. Although Infrared sensors can only provide a rough estimation of the attitude, it is enough for a steady flight control once tuned well. Tiny 13 is the autopilot hardware with the GPS receiver integrated, shown in Fig. 5. Paparazzi also has Tiny Twog autopilot with two open serial ports, which could be used to connect with IMU and modem. One Kalman filter is running on the autopilot to provide a faster position estimation based on GPS updates.

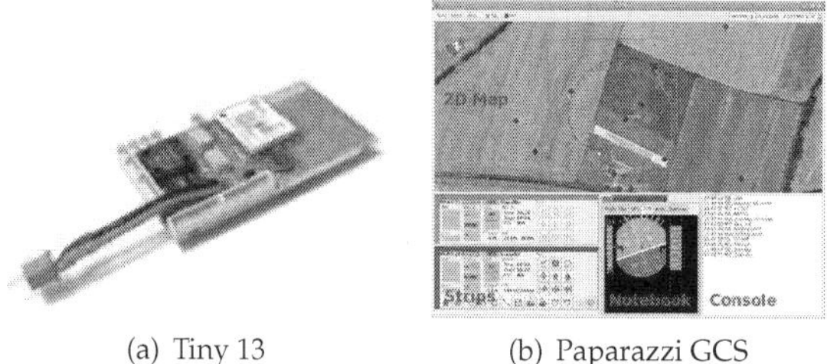

(a) Tiny 13   (b) Paparazzi GCS

Fig. 5. Paparazzi Autopilot System(Brisset et al., 2006).

Paparazzi uses LPC 2148 ARM7 chip as the central processor. For the software, it could achieve waypoints tracking, auto-takeoff & landing, and altitude hold. The flight controller could also be configured if gyro rate is used for roll and pitch tracking control especially for micro UAVs. However, paparazzi doesnt have a good speed hold and changing function currently since no air speed sensor reading is considered in the controller part. Paparazzi is also a truly autonomous autopilot without any rely on the ground control station (GCS). It also has a lot of safety considerations in conditions like RC signal lost, out of predefined range, GPS lost, etc.

2.3.4 Specification Comparisons

The physical specifications of the autopilots are important since small UAVs demand as fewer space, payload and power as possible. The size, weight, and power consumption issues are shown in Table 2. Both the Crossbow MNAV and Procerus Kestrel have a bias compensation to correct the inertial sensor measurement under different temperatures. The functional specifications of these three typical autopilot are listed in detail in Table 3.

	Size (cm)	Weight (g) w/o radio	Power Consumption	Price (k USD)	DC In (V)	CPU	Memory (K)
Kestrel 2.2	5.08*3.5*1.2	16.7	500mA (3.3 or 5V)	5	6-16.5	29MHz	512
Piccolo LT (w.modem)	11.94*5.72*1.78	45	5W	-	4.8-24	40MHz	448
Pprz Twog	4.02*3.05*1	8	N/A	0.125	6.1-18	32 bit ARM7	512

Table 2. Comparison of Physical Specifications of Autopilots (Chao et al., 2009)

	Kestrel	Picocolo LT	Paparazzi
Waypints Navigation	Y	Y	Y
Auto-takeoff & landing	Y	Y	Y
Altitude Hold	Y	Y	Y
Air Speed Hold	Y	Y	N
Multi-UAV Support	Y	Y	Y
Attitude Control Loop	-	-	20/60 Hz
Servo Control Rate	-	-	20/60 Hz
Telemetry Rate	-	25Hz or faster	Configurable
Onboard Log Rate	≤100Hz	-	N

Table 3. Comparison of Autopilot Functions (Chao et al., 2009)

3. AggieAir UAS Platform

Although most current autopilot systems for UAVs have the ability to autonomously navigate through waypoints, it is actually not enough for the real remote sensing applications since the end users need aerial images with certain spatial and temporal resolution requirements from different bands of cameras. More importantly, most civilian remote sensing users want the UAV platform to be inexpensive. AggieAir UAS platform is developed considering all these remote sensing requirements. AggieAir is a small and low-cost UAV remote sensing platform, which includes the flying-wing airframe, the OSAM-Paparazzi autopilot, the GhostFoto image capture subsystem, the Paparazzi ground control station (GCS), and the gRAID software for aerial image processing. All the subsystems are introduced in detail in this section together with a method to help improve the orthorectification accuracy and calibrate the aircraft sensors using ground references.

3.1 Remote Sensing Requirements

Let $\Omega \subset R^2$ be a polytope including the interior, which can be either convex or nonconvex. A series of band density functions $\eta_{rgb}, \eta_{nir}, \eta_{mir} \ldots$ are defined as $\eta_i(q,t) \in [0,\infty) \; \forall q \in \Omega$. η_{rgb} can also be treated as three bands η_r, η_g, η_b, which represent RED, GREEN and BLUE band values of a pixel. The goal of remote sensing is to make a mapping from Ω to $\eta_1, \eta_2, \eta_3 \ldots$ with a preset spatial and temporal resolution for any $q \in \Omega$ and any $t \in [t_1, t_2]$ (Chao et al., 2008).

With the above remote sensing requirements, several specific characteristics need to be considered to get accurate georeferenced aerial images aside from an autonomous flying vehicle:

- Expense: most civilian applications require inexpensive UAS platforms instead of expensive military unmanned vehicles. However, most commercial-off-the-shelf (COTS) autopilots cost more than $6000, let alone the camera and the air frame.

- Orientation Data: the orientation information when the image is taken is critical to the image georeferencing. But many open source UAS autopilot systems don't have good supports to the accurate sensors. For example, Paparazzi uses IR sensors as the main sensing unit by default.

- Image Synchronization: some COTS UAV could send videos down to the base station and record them on the ground computer. But there is a problem that the picture may not match up perfectly with the UAV data on the data log. The images may not synchronize perfectly with the orientation data from the autopilot.

- Band configurable ability: a lot of remote sensing applications require more than one band of aerial images like vegetation mapping and some of them may require RGB, NIR and thermal images simultaneously.

3.2 AggieAir System Structure

AggieAir UAS includes the following subsystems:

(1) The flying-wing airframe: Unicorn wings with optional 48″, 60″ and 72″ wingspans are used as the frame bed to fit in all the electronic parts. The control inputs include elevons and a throttle motor.

(2) The OSAM-Paparazzi autopilot: the open source Paparazzi autopilot is modified by replacing the IR sensors with the IMU as the main sensing unit. Advanced navigation routines like the survey of a random polygon are also added to support image acquisition of an area with a more general shape.

(3) The GhostFoto imaging payload subsystem: a high resolution camera system with both the RGB and NIR band is developed. More importantly, the image system could guarantee an accurate synchronization with the current autopilot software.

(4) The communication subsystem: AggieAir has a 900MHz data link for GCS monitoring, a 72MHz RC link for safety pilot backup, and an optional 2.4GHz wifi link for real time image transmission.

(5) The Paparazzi ground control station (GCS): Paparazzi open source ground station is used for the real-time UAS health monitoring and flight supervising.

(6) The gRAID software: a new World Wind plug-in named gRAID is developed for aerial image processing including correcting, georeferencing and displaying the images on a 3D map of the world.

The physical structure of AggieAir is shown in Fig. 6, with the specifications in Tab 4 and the airborne layout in Fig. 7.

AggieAir has the following advantages over other UAS platforms for remote sensing missions:

(1) Low costs: AggieAir airborne vehicles are built from scratches including the airframes and all the onboard electronics. The total hardware cost is around $3500.

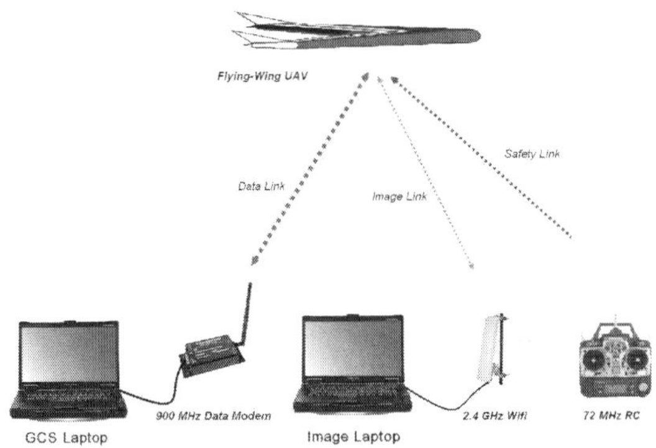

Fig. 6. AggieAir UAS Physical Structure.

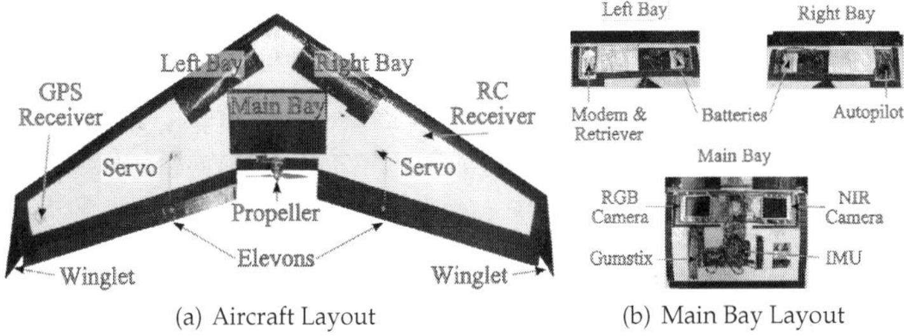

(a) Aircraft Layout   (b) Main Bay Layout

Fig. 7. AggieAir UAS Airborne Layout.

	Specifications
Weight	up to 8 lbs
Wingspan	72″
Flight Time	≤ 1 hour
Cruise Sped	15 m/s
Imaging Payload	RGB/NIR/thermal camera
Operational Range	up to 5 miles

Table 4. AggieAir UAS Specifications

(2) Full autonomy: AggieAir uses the Paparazzi autopilot, which supports the total autonomy of the air vehicle even without the ground station.

(3) Easy manipulation: only two people are required to launch, manipulate and land the vehicle.

(4) Run-way free capability: the bungee launching system supports take-off and landing basically at any soft field with only one launching operator.

- The body frame: the origin is defined at the center of gravity (CG), with the x axis pointing through the nose, the y axis pointing to the right wing and the z axis pointing down.

- The camera frame: the origin is located at the focal point of the camera. The axes of the camera frame are rotated by ϕ_c, θ_c and ψ_c with respect to the body frame.

- The inertial frame: the origin is usually defined on the ground with the x, y, z axes pointing towards the north, east and down, respectively. The orientation of the UAV with respect to the NED frame is given by ϕ, θ and ψ.

- The earth centered earth fixed (ECEF) frame: the z axis passes through the north pole, the x axis passes through the equator at the prime meridian and the y axis passes through the equator at 90° longitude.

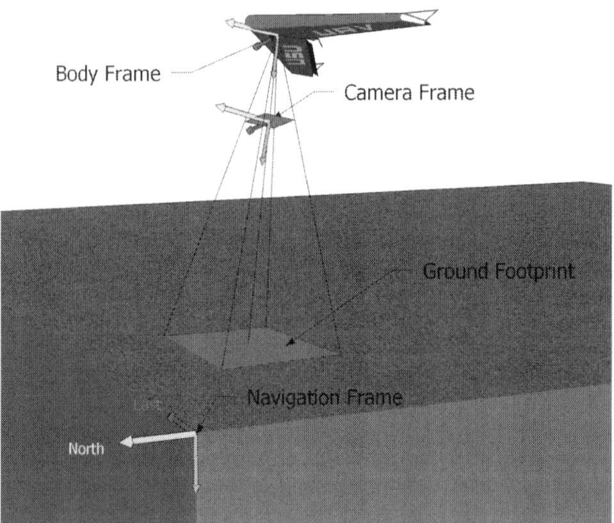

Fig. 10. Aircraft Coordinate Systems

Any point in an image can be rotated from the camera frame to the ECEF coordinate system in order to find where it is located on the earth. However, it is only necessary to find the location of the four corners of the image in order to georeference it. Assuming the origin is at the focal point and the image is on the image plane, equation 1 can be used to find the four corners of the image. As defined in figure 11, FOV_x is the FOV around the x axis, FOV_y is the FOV around the y axis and f is the focal length.

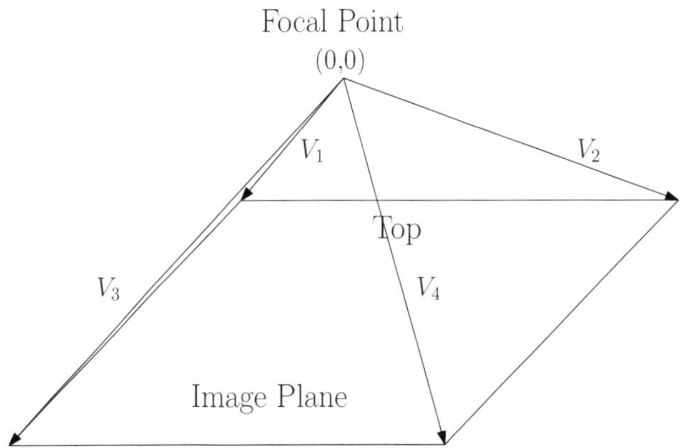

Fig. 11. Definition of Initial Image Corners

$$v_c^1 = \left[\; f\tan(FOV_y/2) \quad -f\tan(FOV_x/2) \quad f \; \right] \tag{1}$$

$$v_c^2 = \left[\; f\tan(FOV_y/2) \quad f\tan(FOV_x/2) \quad f \; \right] \tag{2}$$

$$v_c^3 = \left[\; -f\tan(FOV_y/2) \quad -f\tan(FOV_x/2) \quad f \; \right] \tag{3}$$

$$v_c^4 = \left[\; -f\tan(FOV_y/2) \quad f\tan(FOV_x/2) \quad f \; \right] \tag{4}$$

To rotate the corners into the navigation frame, they first need to be rotated into the body frame. The Euler angles with respect to the body frame are given by ϕ_c, θ_c and ψ_c, and can be used to create a clock-wise rotation matrix R_c^b which rotates a vector in the body frame to the camera frame.

$$R_c^b = R_{xyz}(\phi_c, \theta_c, \psi_c) \tag{5}$$

To rotate from the camera frame to the body frame, the transpose of R_c^b is used.

$$R_b^c = (R_c^b)^T = R_{zyx}(-\theta_c, -\psi_c, -\phi_c) \tag{6}$$

The same rotation matrix is used, with ϕ, θ and ψ, to rotate from the body into the navigation frame.

$$R_n^b = (R_b^n)^T = R_{zyx}(-\theta, -\psi, -\phi) \tag{7}$$

Now each corner is rotated from the camera frame into the navigation frame using equation 8.

$$v_n^i = R_n^b R_b^c v_c^i \tag{8}$$

Now that the corners are in the NED coordinate system, they are scaled to the ground to find their appropriate magnitude (assuming flat earth) where h is the height of the UAV above ground and $v_n^i(z)$ is the z component of v_n^i.

0% to be set up based on experience; it cannot provide an optimal solution. More work on a close-loop real-time solution is needed for an optimal solution.

Preliminary experimental results are shown in this section to demonstrate the effectiveness of the whole UAV remote sensing system on both the hardware and software levels. Three sample applications are introduced in detail. A remote data collection application is offered to demonstrate the feasibility of using UAVs to collect data from ground-based sensors through wireless modems. Acquisition of photographic data over Desert Lake, Utah, illustrates an application involving the use of the RGB and NIR imagers. The data collected by UAVs from ground sensors can also be used for comparison and calibration with information from UAV images. Finally, we present a farmland coverage test with image stitching in our regular test flight site.

4.1 Farmland Coverage

The goal of irrigation control is to minimize the water consumption while sustaining the agriculture production and human needs (Fedro et al., 1993). This optimization problem requires remote sensing to provide real-time feedback from the farmland field including:

- Water: water quantity and quality with temporal and spatial information, for example water level of a canal or lake.

- Soil: soil moisture and type with temporal and spatial information.

- Vegetation: vegetation index, quantity and quality with temporal and spatial information, for example the stage of growth of the crop.

The "real-time" here means daily or weekly temporal resolution based on different applications. AggieAir is currently involved in a large scale agricultural project which will use visual and NIR images from AggieAir to measure the soil moisture of the area to help save water (Jensen, Chen, Hardy & McKee, 2009). The project includes 30 square miles of different types of crops where 87 wireless soil moisture ground probes are placed. The ground probes sample and send the data every hour through wireless to a base station where the data is displayed on the Internet. The ground probes and Landsat data will be used to calibrate the images from AggieAir using the downscaling techniques described in (Kaheil, Gill, McKee, Bastidas & Rosero, 2008) and (Kaheil, Rosero, Gill, McKee & Bastidas, 2008). After calibration, these images should be able to measure soil moisture and evapotranspiration for water managers and farmers whenever it is needed. However, AggieAir has not yet been flown for this project due to the large area and the high altitude AggieAir will need to fly at. Approval from the Federal Aviation Administration (FAA) is currently being sought to ensure the safety of all airborne vehicles before AggieAir is flown for this project. A research farm (one square mile) coverage map is provided in Fig. 15 to show the capability of AggieAir.

4.2 Road Surveying

AggieAir UAS could also provide low-cost aerial images for road and highway construction and maintenance. Figure 16 shows a highway intersection located in Logan Canyon, which was recently rebuilt for better safety to turn onto the main road. The Utah Department of Transportation (UDOT) usually needs to photograph the area to be built or altered before construction with a manned aircraft. However no imagery is available during or after construction due to the cost and availability of the imagery. UDOT is not only interested in AggieAir to lower the cost and increase the availability of imagery for construction, but also to update their inventory of signs, culverts, traffic lines, etc. The aerial images acquired by manned aircraft and by AggieAir are shown in Fig. 16.

Fig. 15. Cache Junction Farm Coverage Map

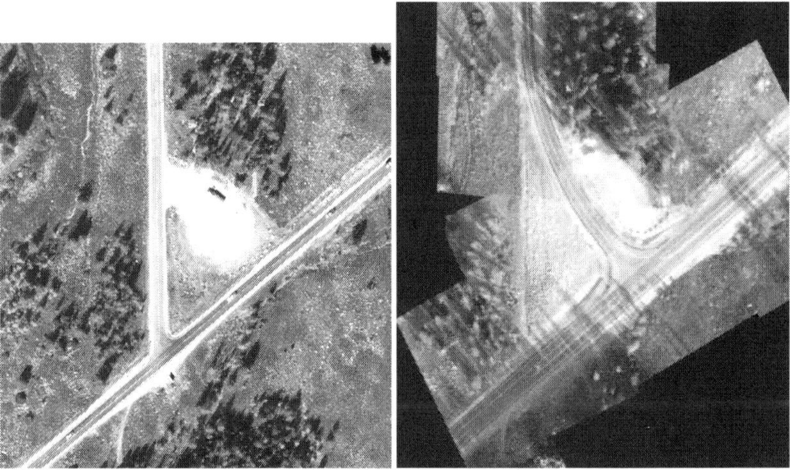

(a) Before Construction   (b) After Construction (AggieAir)

Fig. 16. Beaver Resort Intersection

4.3 Water Area Coverage

Water areas include wetlands, lakes, or ponds, etc. Water areas could provide lots of information to ecological environment changes, flood damage predictions, and water balance management. Desert Lake coverage mission is a typical example, which lies in west-central Utah (latitude: $39°22'5''N$, longitude: $110°46'52''W$). It is formed from return flows from irrigated farms in that area. It is also a waterfowl management area. This proposes a potential problem because the irrigation return flows can cause the lake to have high concentrations of mineral salts, which can affect the waterfowl that utilize the lake. Managers of the Desert Lake resource are interested in the affect of salinity control measures that have been recently constructed by irrigators in the area. This requires estimation of evaporation rates from the Desert Lake area, including differential rates from open water, wetland areas, and dry areas. Estimation of these rates requires data on areas of open water, wetland, and dry lands, which, due to the relatively small size and complicated geometry of the ponds and wetlands of Desert Lake, are not available from satellite images. A UAV can provide a better solution for the problem of acquiring periodic information about areas of open water, etc., since it can be flown more frequently and at little cost.

The whole Desert Lake area is about $2 * 2$ miles. It is comprised of four ponds and some wetland areas. The early version of AggieAir imaging payload, GFDV, is used in this mission together with the Procerus UAV with real-time, simultaneous RGB and NIR videos. Both the RGB and NIR videos are transmitted back to the ground station in real time. The photos are stitched using gRAID, shown in Fig 17.

(a) RGB Image

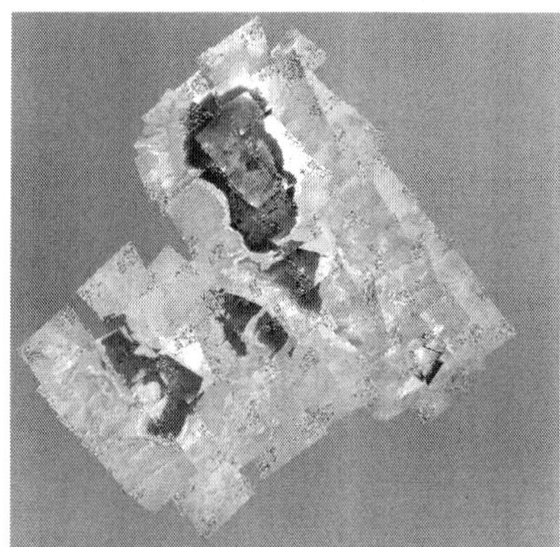
(b) NIR Image

Fig. 17. Desert Lake Coverage Map

4.4 Riparian Surveillance

Riparian buffer surveillance is becoming increasingly more important since it is challenging to maintain stream ecosystem integrity and water quality with the current rapidly changing land use (Goetz, 2006). AggieAir UAS could be used in several applications including river tracking, vegetation mapping and hydraulic modeling, etc.

4.4.1 River Tracking

The path and flow of a river might constantly change due to drought, flood or other natural calamities. Because of this, the aerial images of the river path could be outdated or in low quality, making it difficult to perform studies of the changed river and the variations of its nearby ecological system. AggieAir UAS platform with the high resolution multi-spectral camera system could present a real-time low-cost solution to the river tracking problem Han, Dou & Chen (2009). A flight plan with 3D waypoints could be formed by integrating flow line data from NHDPlus (National Hydrography Dataset Plus) and DEM (Digital Elevation Model) from USGS (U.S. Geological Survey). The images captured by the cameras are processed in real time. Based on the information derived from these images, waypoints are dynamically generated for the autonomous navigation so that the UAV can exactly follow the changed river path and the focus of each image from the camera system is on the center of the river. The actual flight results collected in several flying experiments along a river verify the effectiveness of AggieAir System, shown in Fig 18. Example RGB and NIR pictures acquired by AggieAir is also shown in Fig 19.

Fig. 18. River Tracking Map after Stitching.

4.4.2 Vegetation Mapping & Hydraulic Modeling

Figure 20 shows some imagery taken with AggieAir of a small section of the Oneida Narrows near Preston Idaho. A team of engineers used this imagery to map the substrate and vegetation for 2D hydraulic and habitat modeling. Normally, the team uses low resolution, outdated imagery to map rivers. This can be difficult when the vegetation, the path and the flow of the river are always changing. The imagery from AggieAir, however, was up-to-date (within a week) and had high resolution (5 cm), which made mapping the river quick and easy. Not only could different types of vegetation be distinguished from the imagery, but different types of sediment, like sand piles, could also easily be distinguished.

4.5 Remote Data Collection

Many agricultural and environmental applications require deployment of sensors for measurement of the interested field. However, it is not always easy or inexpensive to collect all

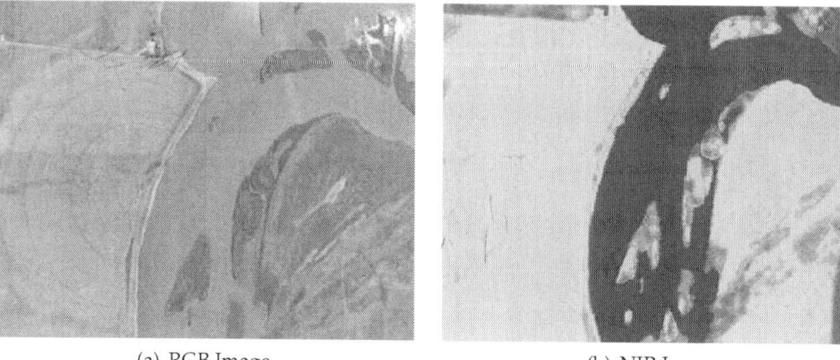

(a) RGB Image   (b) NIR Image

Fig. 19. Sample Picture for River Tracking

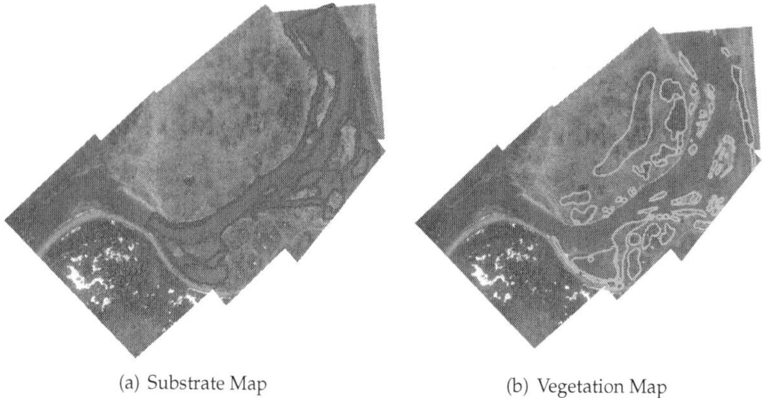

(a) Substrate Map   (b) Vegetation Map

Fig. 20. Oneida Narrows Imagery

the data from remote data loggers for further processing. Many applications still require humans to get close to the ground-based sensors to retrieve the data from their data loggers. Wireless sensor networks and satellite networks are used in environmental data collection applications, but wireless communication can be expensive and vulnerable to changing environmental conditions (such as loss of line-of-sight due to vegetation growth). This problem is especially difficult when the sensors are deployed sparsely over a large geographic area where transportation might be limited by terrain conditions. UAVs can fly into such areas without affecting the vegetation on the ground; they can spare humans from having to enter dangerous or difficult areas; and they might be able to operate at lower costs that might be required for approaches involving direct human access to the data. Moreover, UAVs can achieve better wireless communication since the signal can be transmitted more dependably in the air than near ground level.

One typical example of remote data collection is fish tracking. In order to understand fish habitats, radio transmitters are planted in fish in order to locate and track their movements. Human operators are needed to drive a boat around a lake or down a river following the periodic beacon sent from the transmitters. The beacon is heard through a radio receiver with a directional antenna and its strength is highly dependent on the distance and the direction the antenna is pointed at. AggieAir could be employed here with an onboard self-designed device to catch the signal from the transmitter and record its strength. Thus, the location of the fish can be found and recorded much easier and faster since the wireless signal transmits much better in the air. Hardware developments and real experiments are still undergoing.

5. Towards Band Reconfigurable Multiple UAV based Remote Sensing

A single AggieAir system could have many applications as mentioned above, but some irrigation applications may require remote sensing of a large land area (more than 30 square miles) within a short time (less than one hour). Acquisition of imagery on this geographic scale is difficult for a single UAV. However, groups of UAVs (which we call "covens") can solve this problem because they can provide images from more spectral bands in a shorter time than a single UAV.
The following missions will need multiple UAVs (covens) operating cooperatively for remote sensing:

- Measure $\eta_1, \eta_2, \eta_3 \dots$ simultaneously.

- Measure $\eta_i(q, t)$ within a short time.

To fulfill the above requirements, UAVs equipped with imagers having different wavelength bands must fly in some formation to acquire the largest number of images simultaneously. The reason for this requirement is that electromagnetic radiation may change significantly, even over a period of minutes, which in turn may affect the final product of remote sensing. The "V" or "—" formation, keeping algorithm similar to the axial alignment (Ren et al., 2008), can be used here since the only difference is that the axis is moving:

$$\dot{q}_m^d = - \sum_{n \in \mathcal{J}_m(t)} [(q_m - q_n) - (\delta_m - \delta_n)], \tag{18}$$

where q_m^d is the preset desired waypoints, $\mathcal{J}_m(t)$ represents the UAV group, $\delta_m = [\delta_{mx}, \delta_{my}]^T$ can be chosen to guarantee that the UAVs align on a horizontal line with a certain distance in between.
Based on the theoretical analysis and our preliminary results, more effort is needed to achieve the final band reconfigurable multi-UAV based cooperative remote sensing.

- (1) Multiple UAVs: the preliminary results have shown that the sensing range of the AggieAir UAS is about 2.5×2.5 miles, given the current battery energy density. However, light reflection varies a great deal in one day and accurate NIR images require at most a one-hour acquisition time for capture of the entire composite image. This motivates the use of covens, or multiple UAVs, for this type of application, with each UAV carrying one imager with a certain band.

- (2) More robust control: the current UAV platform requires winds of less than 10 m/s. However, the UAV needs to have the ability to deal with wind gusts, which is problematic for UAV flight.

- (3) Accurate real-time image stitching and registration: the current software requires manual post processing for accurate georeferencing results. For the application to be attractive to managers of real irrigation systems, manual post processing is unacceptable because of its cost and requirement of technically trained individuals.

Acknowledgement

This work is supported in part by the Utah Water Research Laboratory (UWRL) MLF Seed Grant (2006-2009) on "Development of Inexpensive UAV Capability for High-Resolution Remote Sensing of Land Surface Hydrologic Processes: Evapotranspiration and Soil Moisture." This work is also partly supported by NSF Grants #0552758 and #0851709. The authors would like to thank Professor Raymond L. Cartee for providing the Utah State University research farm at Cache Junction as the UAV flight test field, and Calvin Coopmans, Long Di, Daniel Morgan for their technical support and help to the flight tests. The authors would also like to thank Shannon Clemens for georeferencing and generating mosaics.

6. References

Beard, R., Kingston, D., Quigley, M., Snyder, D., Christiansen, R., Johnson, W., Mclain, T. & Goodrich, M. (2005). Autonomous vehicle technologies for small fixed wing UAVs, *AIAA J. Aerospace Computing, Information, and Communication* **5**(1): 92–108.

Brisset, P., Drouin, A., Gorraz, M., Huard, P. S. & Tyler, J. (2006). The paparazzi solution, *MAV 2006*, Sandestin, Florida, USA.

Calise, A. J., Johnson, E. N., Johnson, M. D., & Corban, J. E. (2003). Applications of adaptive neural-network control to unmanned aerial vehicles, *in Proc. AIAA/ICAS Int. Air and Space Symp. and Exposition: The Next 100 Years*, Dayton, Ohio, USA.

Chao, H., Baumann, M., Jensen, A. M., Chen, Y. Q., Cao, Y., Ren, W. & McKee, M. (2008). Band-reconfigurable multi-UAV-based cooperative remote sensing for real-time water management and distributed irrigation control, *in Proc. IFAC World Congress*, Seoul, Korea, pp. 11744–11749.

Chao, H., Cao, Y. & Chen, Y. Q. (2009). Autopilots for small fixed wing unmanned aerial vehicles: a survey, *Int. J. Control, Automation and Systems* . Accepted to appear.

CloudCap Inc. (Accessed 2008).
    URL `http://www.cloudcaptech.com`.

Damien, D., Wageeh, B. & Rodney, W. (2007). Fixed-wing attitude estimation using computer vision based horizon detection, *in Proc. Australian Int. Aerospace Congress*, Melbourne, Australia, pp. 1–19.

Egan, G. K. (2006). The use of infrared sensors for absolute attitude determination of unmanned aerial vehicles, *Tech. Rep. MECSE-22-2006*, Monash University.

Fedro, S. Z., Allen, G. S. & Gary, A. C. (1993). Irrigation system controllers, *Series of the Agricultural and Biological Engineering Department, Florida Cooperative Extension Service, Institute of Food and Agricultural Sciences, University of Florida* .
    URL: *http://edis.ifas.ufl.edu/AE077*

Goetz, S. J. (2006). Remote sensing of riparian buffers: Past progress and future prospects, *Journal of the American Water Resources Association* **42**(11): 133–143.

gPhoto (Accessed 2008).
    URL `http://www.gphoto.org/`.

Han, Y., Dou, H. & Chen, Y. Q. (2009). Mapping river changes using low cost autonomous unmanned aerial vehicles, *in Proc. AWRA Spring Specialty Conf. Managing Water Resources Development in a Changing Climate*, Anchorage, Alaska, USA.

Han, Y., Jensen, A. M. & Dou, H. (2009). Programmable multispectral imager development as light weight payload for low cost fixed wing unmanned aerial vehicles, *in Proc. ASME Design Engineering Technical Conf. Computers and Information in Engineering*, number MESA-87741, San Diego, California, USA.

James, B. (2006). *Introduction to Remote Sensing*, 4th edn, Guilford Press.

Jensen, A. M. (2009). *gRAID: A geospatial real-time aerial image display for a low-cost autonomous multispectral remote sensing platform*, M.S. Thesis, Utah State Univeristy.

Jensen, A. M., Baumann, M. & Chen, Y. Q. (2008). Low-cost multispectral aerial imaging using autonomous runway-free small flying wing vehicles, *in Proc. IEEE Int. Conf. Geoscience and Remote Sensing Symp.*, Boston, Massachusetts, USA, pp. 506–509.

Jensen, A. M., Chen, Y. Q., Hardy, T. & McKee, M. (2009). AggieAir - a low-cost autonomous multispectral remote sensing platform: New developments and applications, *in Proc. IEEE Int. Conf. Geoscience and Remote Sensing Symp.*, Cape Town, South Africa, pp. –.

Jensen, A. M., Han, Y. & Chen, Y. Q. (2009). Using aerial images to calibrate inertial sensors of a low-cost multispectral autonomous remote sensing platform (AggieAir), *in Proc. IEEE Int. Conf. Geoscience and Remote Sensing Symp.*, Cape Town, South Africa, pp. –.

Johnson, L., Herwitz, S., Dunagan, S., Lobitz, B., Sullivan, D. & Slye, R. (2003). Collection of ultra high spatial and spectral resolution image data over California vineyards with a small UAV, *in Proc. Int. Symp. Remote Sensing of Environment*, Honolulu, Hawai, USA.

Johnson, L., Herwitz, S., Lobitz, B. & Dunagan, S. (2004). Feasibility of monitoring coffee field ripeness with airborne multispectral imagery, *Applied Engineering in Agriculture* **20**: 845–849.

Kaheil, Y. H., Gill, M. K., McKee, M., Bastidas, L. A. & Rosero, E. (2008). Downscaling and assimilation of surface soil moisture using ground truth measurements, *IEEE Trans. Geoscience and Remote Sensing* **46**(5): 1375–1384.

Kaheil, Y. H., Rosero, E., Gill, M. K., McKee, M. & Bastidas, L. A. (2008). Downscaling and forecasting of evapotranspiration using a synthetic model of wavelets and support vector machines, *IEEE Trans. Geoscience and Remote Sensing* **46**(9): 2692–2707.

Microstrain Inc. (Accessed 2008).
	URL http://www.mirostrain.com.

Paparazzi Forum (Accessed 2008).
	URL http://www.recherche.enac.fr/paparazzi/.

Picture Transfer Protocol (Accessed 2008).
	URL http://en.wikipedia.org/wiki/Picture_Transfer_Protocol.

Ren, W., Chao, H., Bourgeous, W., Sorensen, N. & Chen., Y. Q. (2008). Experimental validation of consensus algorithms for multi-vehicle cooperative control, *IEEE Trans. Control Systems Technology* **16**(4): 745–752.

Roberts, P. J., Walker, R. A. & O'Shea, P. J. (2005). Fixed wing UAV navigation and control through integrated GNSS and vision, *in Proc. AIAA Guidance, Navigation, and Control Conf. and Exhibit*, number AIAA 2005-5867, San Francisco, California, USA.

Tarbert, B., Wierzbanowski, T., Chernoff, E. & Egan, P. (2009). Comprehensive set of recommendations for sUAS regulatory development, *Tech. Rep.*, Small UAS Aviation Rulemaking Committee.

23

Possibilistic and fuzzy multi-sensor fusion for humanitarian mine action

Nada Milisavljević[1] and Isabelle Bloch[2]
[1]*Signal and Image Centre*
Royal Military Academy
Brussels, Belgium
[2]*TELECOM ParisTech (ENST)*
CNRS UMR 5141 LTCI
Paris, France

1. Introduction

Some multi-sensor data fusion applications to humanitarian mine action are presented in this chapter. Data fusion techniques may be useful to two main humanitarian mine action types: close range antipersonnel mine detection and remote sensing mined area reduction. Close range antipersonnel mine detection refers to detecting surface and subsurface anomalies that may be related to the presence of mines (e.g., detection of differences in dielectric constant, thus in the type of material, using a ground-penetrating radar) and/or detection of explosive materials. Area reduction refers to identifying the mine-free areas out of the mine-suspected areas. Data fusion for these two applications is discussed here.

In case of both humanitarian mine action types, efficient modelling and fusion of extracted features can increase the overall performance of single-sensor based processing. Yet, a wide variety of scenarios and conditions typically exists between different minefields and within a minefield itself. Therefore, a high-quality performance of humanitarian mine action tools can only be obtained using multi-sensor and data fusion approaches. In addition, as the sensors used are, actually, detectors of different anomalies that may be related to mines (and not of mines themselves), combinations of these complementary pieces of information might improve the detection and classification results. Finally, in order to consider the inter- and intra-minefield variability, uncertainty, partial knowledge and ambiguity, knowledge-based theories prove to be useful, such as belief functions in the framework of the Dempster-Shafer evidence theory (Shafer, 1976; Smets, 1994; Smets & Kennes, 1994) or possibilistic and fuzzy theory (Dubois & Prade 2006; Zadeh, 1965). We focus on the latter in this chapter. Namely, as the possibilistic and fuzzy theory allows for the flexibility in the choice of combination operators, we exploit this fact in order to account for the different characteristics of the sensors to be combined.

For close range detection, modeling and fusion of extracted features are shown, based on the possibilistic and fuzzy theory. The approaches are discussed for the case of using real data coming from three complementary sensors (ground-penetrating radar - GPR, infrared ellipse as well, A_o is the object area, and A_e is the ellipse area. Using these values, the following possibility degrees are derived (the subtraction of 5 pixels is related to the limit case - minimum 5 points needed to define an ellipse):

$$\pi_{2I}(\text{MR})=\pi_{2I}(\text{FR}) = \max\left(0, \min\left\{\frac{A_{oe}-5}{A_o}, \frac{A_{oe}-5}{A_e}\right\}\right), \tag{3}$$

$$\pi_{2I}(\text{MI})=\pi_{2I}(\text{FI}) = 1 - \pi_{2I}(\text{MR}). \tag{4}$$

Finally, the area directly provides a degree $\pi_{3I}(\text{M})$ (where M=MR$\cup$MI) of being a mine - since the range of possible anti-personnel (AP) mine sizes is approximately known, the degree of possibility of being a mine is derived as a function of the measured size:

$$\pi_{3I}(\text{M})= \frac{a_I}{a_I+0.1\cdot a_1} \cdot \exp\frac{-[a_I-0.5\cdot(a_1+a_2)]^2}{0.5\cdot(a_2-a_1)^2}, \tag{5}$$

where a_I is the actual object area on the IR image, while the approximate range of expectable mine areas is between a_1 and a_2. For AP mine detection, we set $a_1 = 15$ cm^2 and $a_2 = 225$ cm^2 (Milisavljević & Bloch, 2003; Fischer et al., 2007). Friendly objects can be of any size, meaning that the measured size is uninformative about the possibility of being a friendly object (F=FR$\cup$FI):

$$\pi_{3I}(\text{F})= 1. \tag{6}$$

MD measures. Under some conditions (Thonnard & Milisavljević, 1999; Das, 2006; Milisavljević & Bloch, 2008), it is possible to extract the object shape and area as seen by MD, as well as the burial depth. In reality, it can be difficult to adjust the sensitivity so that all the low-metal content mines are detected without causing the data saturation for high-metal content objects. In addition, in order to speed up the scanning time in large minefields, the data gathering resolution in the cross-scanning direction can be very poor. Thus, the MD information consists typically of only one measure, which is the width of the region in the scanning direction, w [cm]. As friendly objects can contain metal of any size, we define:

$$\pi_{MD}(\text{F})= 1. \tag{7}$$

For most of AP mines, the range of the expected sizes of metal in mines is between 5 cm and 15 cm), so we can assign possibilities to mines as:

$$\pi_{MD}(\text{M})= \frac{w}{20} \cdot [1 - \exp(-0.2 \cdot w)] \cdot \exp\left(\frac{w}{20}\right). \tag{8}$$

GPR measures. All three GPR measures provide information about mines (Milisavljević et al., 2003).

Regarding burial depth information (D), friendly objects can be found at any depth. On the contrary, mainly due to their activation principles, there exists a maximum depth (D_{max}, typically 25 cm) up to which AP mines can be found. Therefore, possibility distributions for mines, $\pi_{1G}(M)$, and friendly objects, $\pi_{1G}(F)$, for this measure can be modeled as follows:

$$\pi_{1G}(M)=\frac{1}{\cosh\left(\frac{D}{D_{max}}\right)^2}, \tag{9}$$

$$\pi_{1G}(F)=1. \tag{10}$$

In case of the ratio d/k between object size seen in the scanning direction, d, and its scattering function, k (which is directly related to the object shape (Capineri et al., 1998)), friendly objects can have any value of this measure. For mines, there is a range of values that mines can have, and outside that range, the object is quite certainly not a mine:

$$\pi_{2G}(M)=\exp\left(-\frac{[(d/k)-m_d]^2}{2\cdot p^2}\right), \tag{11}$$

$$\pi_{2G}(F)=1, \tag{12}$$

where m_d is the d/k value at which the possibility distribution reaches its maximum value (here, $m_d = 700$, chosen based on (Capineri et al., 1998)), and p is the width of the exponential function (here, $p = 400$).

The third GPR measure, propagation velocity, v, can provide information about object identity. For this measure, we extract depth information on a different way than in the case of the burial depth measure 0 and we preserve the sign of the extracted depth. This information indicates whether a potential object is above the surface, in which case the extracted propagation velocity should be close to $c = 3\cdot10^8$ m/s, the propagation velocity in vacuum. If the sign indicates that the object is below the soil surface, the value of v should be around the values for the corresponding medium, e.g., from $5.5\cdot10^7$ m/s to $1.73\cdot10^8$ m/s (Capineri et al., 1998) in case of sand:

$$\pi_{3G}(M)=\exp\left(-\frac{(v-v_t)^2}{2\cdot h^2}\right), \tag{13}$$

with v_t being the most typical velocity for the medium (for sand, it is $0.5\cdot (5.5\cdot10^7 +1.73\cdot10^8) = 1.14\cdot10^8$ m/s, and for air, it is equal to c), and h is the width of the exponential function (here, $h = 6\cdot10^7$ m/s). Again, friendly objects can have any value of the velocity:

$$\pi_{2G}(F)=1. \tag{14}$$

Note that all these functions have formal expressions that have been chosen in order to achieve suitable shapes and behaviours. However, other functions sharing the same properties could be used as well and we experienced that a fine tuning is not necessary.

3.2 Combination of possibility degrees

The combination is performed in two steps: the first one applies to all measures derived from one sensor, while the second one combines results obtained in the first step for all three sensors. Here, only the combination rules related to mines are considered. The issue of combination rules for friendly objects is discussed in detail in (Milisavljević & Bloch, 2008).

Let us first discuss the first step for each sensor. For IR, since mines can be regular or irregular, the information about regularity on the level of each shape measure is combined using a disjunctive operator:

$$\pi_{1IM}=\max(\pi_{1I}(MR),\pi_{1I}(MI)),\tag{15}$$

$$\pi_{2IM}=\max(\pi_{2I}(MR),\pi_{2I}(MI)).\tag{16}$$

The choice of the maximum (smallest disjunction and idempotent operator) as a t-conorm is related to the fact that the measures cannot be considered as completely independent from each other, so there is no reason to reinforce the measures by using a larger t-conorm, and the idempotent one is preferable in such situations. These two shape constraints should be both satisfied to have a high degree of possibility of being a mine (Milisavljević & Bloch, 2008). Thus, they are combined in a conjunctive way (here using a product). Finally, the object is possibly a mine if it has a size in the expected range, or if it is not in the expected range, but satisfies the shape constraint, hence the final combination for IR is:

$$\pi_I(M)=\pi_{3I}(M)+[1-\pi_{3I}(M)]\cdot\pi_{1IM}\cdot\pi_{2IM} \ .\tag{17}$$

In case of MD, as there is just one measure used, there is no first combination step and the possibility degrees obtained using (7) and (8) are directly used.

Regarding GPR, it is possible to have a mine if the object is at shallow depths and its dimensions resemble a mine and the extracted propagation velocity is appropriate for the medium. Therefore, the combination of the obtained possibilities for mines is performed using a t-norm, expressing the conjunction of all criteria. Here the product t-norm is used:

$$\pi_G(M)=\pi_{1G}(M)\cdot\pi_{2G}(M)\cdot\pi_{3G}(M) \ .\tag{18}$$

Finally, since the final possibility should be high if at least one sensor provides a high possibility and since it is better to assign a friendly object to the mine class than to miss a mine, the second combination step is performed using the algebraic sum, which is chosen for its strong disjunctive behaviour:

$$\pi(M)=\pi_I(M)+\pi_{MD}(M)+\pi_G(M)-\pi_I(M)\cdot\pi_{MD}(M)-\pi_I(M)\cdot\pi_G(M)-\pi_G(M)\cdot\pi_{MD}(M)+$$
$$+\pi_I(M)\cdot\pi_{MD}(M)\cdot\pi_G(M) \ .\tag{19}$$

3.3 Decision

The final decision can be simply obtained by thresholding the fusion result for M and providing the corresponding possibility degree as the confidence degree. As presented in (Milisavljević & Bloch, 2008), an alternative for the final decision making is to derive the combination rule for F as well, compare the final values for M and F and derive an adequate decision rule.

4. Fuzzy approach to close-range detection

In this section, we propose to introduce the fuzziness in the proposed approach. Fuzziness may be indeed attached to the measurements. An example where classes are prone to fuzziness is presented in the next section.

4.1 Fuzzy modelling of the measures

In our previous work (Section 3), all measures derived from the sensor data were considered as crisp numbers. However, they may suffer from imprecision, due to the procedures used to extract these measures. As a typical example, a threshold is applied on the IR images and measured are derived from the resulting binary regions. Although the threshold values are chosen with care, varying them by a few units may lead to variations in the derived measures.

As an original feature of the present work, we propose to model this imprecision in the fuzzy set framework and consider all measures as fuzzy numbers and not crisp ones. We detail the modeling step in this section, while introducing the fuzzy numbers in the fusion methods will be described in the next one.

The imprecision can be introduced at several levels, directly on the measures, or on the derived possibility values.

Modeling the measures as fuzzy numbers. Let us consider for instance the elongation measures. The location of the bordering pixels and the center of gravity may be subject to imprecision (for example, depending on the threshold). Let a be the minimum distance and b the maximum distance. We could estimate the typical imprecision (e.g. 2 pixels), e.g., by varying the threshold and measuring the induced variation on a and b. This could then define the 0.5-cut of a fuzzy number representing a (respectively b), as illustrated in Figure 1. We can choose simple shapes for membership functions, such as triangles, which are sufficient to represent the imprecision and only require two values to be defined: the actual measured value (a_0) in this example, and the typical imprecision ε. In this example, this leads to the following definition of the fuzzy number representing a:

$$\forall \alpha \in \mathbb{R}^+, \mu_a(\alpha) = \begin{cases} 0 & \text{if } \alpha \leq a_0 - \varepsilon, \\ \frac{\alpha - a_0 + \varepsilon}{\varepsilon} & \text{if } a_0 - \varepsilon \leq \alpha \leq a_0, \\ \frac{a_0 + \varepsilon - \alpha}{\varepsilon} & \text{if } a_0 \leq \alpha \leq a_0 + \varepsilon, \\ 0 & \text{if } \alpha \geq a_0 + \varepsilon. \end{cases} \tag{20}$$

Note that we should restrict the support to admissible values for the considered measure. For instance if the measure has to be in [0, 1], then the support of the fuzzy number should be included in [0, 1]. Otherwise we could have values with non-zero membership values (i.e. considered as possible), while they are actually not admissible.

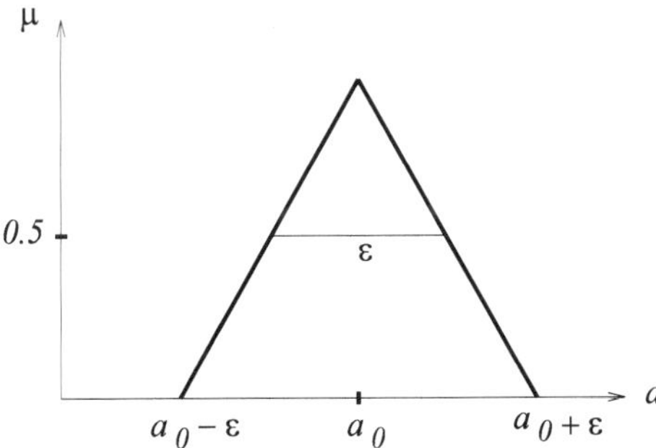

Fig. 1. Example of fuzzy number representing a

A similar model can be estimated for b, leading to a fuzzy number μ_b.

Then r_1 can be derived using the extension principle (Zadeh, 1975).

Let us recall the extension principle in the case of an operation applying on fuzzy numbers. Let μ_1 and μ_2 be the membership functions defining two fuzzy numbers (i.e., functions from the real line into [0, 1], such that all α-cuts are closed intervals, and having a unique modal value). Let $\star$ be an operation between two numbers to be extended to fuzzy numbers. The resulting fuzzy number μ is defined as:

$$\forall x \in \mathbb{R}, \mu(x) = \sup_{x_1 \in \mathbb{R}, x_2 \in \mathbb{R}, x = x_1 \star x_2} \min(\mu_1(x_1), \mu_2(x_2)). \tag{21}$$

Other extensions principles can be defined, using for instance the product instead of the minimum (Dubois & Prade, 1980).

In our particular case, we can derive a fuzzy model for r_1 from μ_a and μ_b as follows:

$$\forall \rho \in \mathbb{R}^+, \mu_{r_1}(\rho) = \sup_{\alpha \in \mathbb{R}^+, \beta \in \mathbb{R}^{+*}, \frac{\alpha}{\beta} = \rho} \min(\mu_a(\alpha), \mu_b(\beta)). \tag{22}$$

Alternatively, we can directly model r_1 as a fuzzy number. Indeed, the variations of a and b depending on the thresholds are probably correlated (for instance if a increases because the region has been expanded by the thresholding, then b increases too), and this is not taken into account in the previous model.

We reason in a similar way for all measured quantities, in order to obtain fuzzy numbers for all of them.

Summarizing, a general procedure for modeling a fuzzy number corresponding to a measure x derived from sensor data is as follows:

1. Let x_0 the actually measured value.
2. Let ε_x the typical imprecision on this measure.
3. Define a fuzzy number μ_x as a triangular fuzzy set on the real line with support $[x_0 - \varepsilon_x, x_0 + \varepsilon_x]$ and modal value x_0.
4. If needed, reduce the support to guarantee its inclusion in the set of admissible values for x.

This model is simple but robust enough. The most important feature is the ability to account for imprecision. The shape of the membership function does not need to be precisely estimated.

Introducing imprecision directly in the possibility values. An alternative way for introducing the imprecision is to model directly the possibility values as fuzzy numbers. This approach avoids frequent calls to the extension principle and leads to simpler modeling and faster computation.

Based on the imprecision estimated for each measure, it is easy to estimate the derived imprecision on the possibility values, even in a rough way (again this will be sufficient to achieve a good robustness). For instance the imprecision on $\pi_{1l}(\mathrm{MR})$ will be of the same order of magnitude as the imprecision on r_1 and r_2. So this possibility degree can be directly modeled as a fuzzy number of triangular shape with $\min(r_1, r_2)$ as modal value and a support of length 2ε, where ε is derived from the variability on r_1 and r_2 when the threshold value is varied, as explained above. For other possibility values, the imprecision should be computed by applying the same functions as for the measures (e.g. exponential, cosh, etc.).

Then the modeling steps are as follows:
1. Compute the modal values for each possibility exactly as in Subsection 3.1.
2. Introduce imprecision by defining triangular shaped membership functions, using these modal values and the typical imprecision attached to each measure derived from the data.

4.2 Fusion based on fuzzy input

For the fusion step, we propose to rely on the same approach as in Subsection 3.2, but extending it to account for fuzzy input values. Then all π and m values become fuzzy values, by using the extension principle, and their combination as well.

For instance, if r_1 and r_2 are fuzzy numbers defined by membership functions μ_{r_1} and μ_{r_2}, then Eq. 1 becomes:

$$\forall \rho \in \mathbb{R}^+, \pi_{1I}(\mathrm{MR})(\rho) = \sup_{\rho_1,\rho_2, \min(\rho_1,\rho_2)=\rho} \min(\mu_{r_1}(\rho_1), \mu_{r_2}(\rho_2)). \tag{23}$$

Note that with this definition, the minimum of two fuzzy numbers is not necessarily one of the two initial fuzzy numbers, as illustrated in Figure 2.

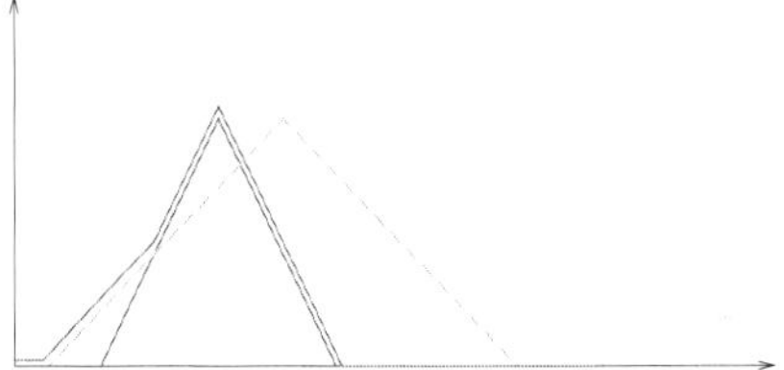

Fig. 2. Minimum of two fuzzy numbers, the result is in red.

However, a nice and useful property of this operation is that the support of the result is always included in the union of the supports of the two fuzzy numbers. In this particular example, where a possibility degree is derived from the minimum of two values in [0, 1], we get a fuzzy possibility degree, having a support in [0, 1], which is consistent.

Let us now consider Eq. 2. For computing 1 minus a fuzzy value, we will consider a membership function defining the crisp value 1 as follows: $\mu_1(1) = 1$, $\mu_1(\lambda) = 0 \ \forall \lambda \neq 1$. Then Eq. 2 can be extended using the extension principle applied to μ_1 and $\pi_{1I}(\mathrm{MR})$. Here it takes the following simple form:

$$\pi_{1I}(\mathrm{MI})(\rho) = \sup_{\rho', \rho = 1 - \rho'} \pi_{1I}(\mathrm{MR})(\rho'). \tag{24}$$

All other combinations can be defined in a similar way.

4.3 Decision

The same decision rules can be applied as in the original model (summarized in Subsection 3.3). Decision consists of both an answer on whether the considered object is a mine or a friendly object, and a confidence value. Here this confidence value can either be kept fuzzy or a crisp value can be obtained by defuzzifying the resulting possibility value (which are fuzzy numbers), for instance by using the center of gravity of these fuzzy quantities. Let μ the membership function of a resulting value, having a support $[\lambda_1, \lambda_2]$. Its defuzzification using the center of gravity is expressed as:

$$c = \frac{\int_{\lambda_1}^{\lambda_2} \lambda \mu(\lambda) d\lambda}{\int_{\lambda_1}^{\lambda_2} \mu(\lambda) d\lambda} . \tag{25}$$

Alternative methods are possible as well (e.g. taking the value having the highest membership value, etc.).

Note that a defuzzification step could be performed at any stage of the process. Considering fuzziness all through the process avoids taking crisp decisions on the possibility values at too early stages, but also leads to heavier computation.

5. Fuzzy approach to remote sensing mined area reduction

5.1 SMART system

The goal of area reduction is to determine which mine-suspected areas do not contain mines. This task is recognized as a mine action activity that should result in reduction in time and resources. Several well-known methods are in use to perform area reduction, especially using mechanical means. Most of the time, these expensive methods change and damage the environment and the ecosystem. In order to avoid this, some approaches have been developed that acquire the necessary information remotely, from air or space, using appropriate sensors associated with context information collected from the field and integrated in a geographical information system (GIS). One of these projects is the SMART project (Yvinec, 2005), funded by the European Commission/DG/INFSO and applied to Croatia. The aim of this project is to provide the human analyst with the SMART system. This system is GIS-based, augmented with dedicated tools and methods designed to use multispectral and radar data in order to assist in human analyst's interpretation of the possibly mined scene during the area reduction process. The usefulness of such image processing tools in helping photo-interpretation is in the possibility to process automatically a large amount of data and help a visual analysis (SMART consortium, 2004). The use of SMART includes a field survey and an archive analysis in order to collect knowledge about the site, a satellite data collection, a flight campaign to record the data and the exploitation of the SMART tools by an operator to detect indicators of presence or absence of mine-suspected areas. With the help of a data fusion module based on belief functions and fuzzy sets, the operator prepares thematic maps synthesizing all the knowledge gathered with these indicators. These maps of indicators can be transformed into risk maps showing how dangerous an area may be according to the location of known indicators and into priority maps indicating which areas designed to help the mined area reduction process.

Figure 3 illustrates the global SMART approach. We focus here on the fusion step, which provides an intermediary result in SMART, consisting in improved land-cover classification maps, along with confidence values. This result is exploited by the deminers together with the final result.

5.2. Data and their specificities in SMART

The available images include SAR, multispectral, high resolution optical and satellite data. SAR data were collected with the E-SAR system of the German Aerospace Centre (DLR) in fully polarimetric P- and L-band and in vv-polarization (waves are vertically transmitted and received) X- and C-band. Multispectral Daedalus data were collected with a spatial resolution of 1 m and in 12 channels, ranging from visible blue to thermal infrared. SAR and Daedalus data were geocoded. DLR also provided a complete set of RMK photographic aerial views recorded with a colored infrared film at a resolution of 3 cm. This non-geocoded data set is used as evidence to control the processing tools and for qualitative interpretation by photo-interpreters. Finally, geocoded KVR-1000 black-and-white satellite images with a resolution of 2 m, recorded before the war in Croatia, were purchased in order to assess the changes in the landscape due to the war.

The legend (expected classes in the images), derived based on the existing and gathered knowledge about the mined areas, is given in Table 1. Ground truth was provided as a set of regions (training regions and validation regions). In the fusion module, training regions are used for estimating the parameters of some of the proposed methods; validation regions are used for the evaluation of the results.

Table 2 summarizes the inputs of the fusion module. More information about these inputs can be found in (Bloch, I. et al., 2007).

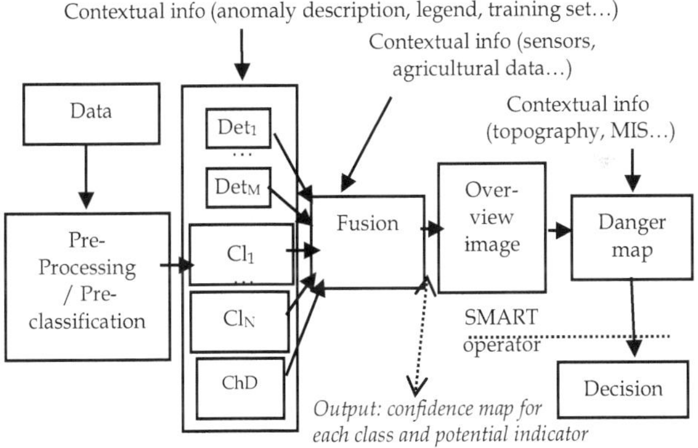

Fig 3. The global SMART approach; Det – detector, Cl - classifier, M – number of detectors, N – number of classifiers, ChD – change detection, MIS - mine information system.

Class no.	Legend
1	Abandoned agricultural land
2	Agricultural land in use
3	Asphalted roads
4	Rangeland
5	Residential areas
6	Trees and shrubs
7	Shadow
8	Water

Table 1. Expected classes in the images

Data type	Type of result
SAR	Classification with confidence images per class
SAR & Daedalus	Detection of hedges, trees, shadows, rivers, with confidence degrees, sometimes discounted
Daedalus	Supervised classification, result as a decision image
Daedalus	Region-based classification with confidence images per class
Daedalus	Belief function classification with confidence images per class
SAR & Daedalus	Binary detection of roads
SAR	River detection (binary)
Daedalus & KVR	Change detection (binary image)

Table 2. Summary of the input of the fusion module

5.3. Fuzzy fusion approach in SMART

The computations here are performed at pixel level. A final regularization step is then applied. Different fusion strategies have been developed (Bloch et al., 2007) and we present here the most promising one, a fuzzy approach.

In this approach, we choose for each class the best classifiers, based on the confusion matrix for each classifier or detector, by comparing the diagonal elements in all matrices for each class. Then, for each class, we combine the best classifiers with a maximum operator (possibly with some weights). Finally, the decision is made according to a maximum rule. This approach is very fast. It uses only a part of the information, which could also be a drawback if this part is not chosen appropriately. Some weights have to be tuned, which may need some user interaction in some cases. Although it may sound somewhat *ad hoc*, it is interesting to show what we can get by using the best parts of all classifiers.

After this first combination step, the next step which follows is knowledge inclusion. It is one of the main powers of our algorithms with respect to the commercial ones. This aspect has led to a lot of work in SMART, at different levels. Note that knowledge on the classifiers, their behaviors, etc. is already included in the previous steps. At this step, we use only the pieces of knowledge that directly provide information on the landcover classification. Other pieces of knowledge such as mine reports, etc. are not directly related to classes of interest, but rather to the dangerous areas, and are thus included in the danger map construction, which follows the fusion. Several pieces of knowledge proved to be very useful at this step. For example, some detectors are available for roads and rivers, which provide areas or lines that surely belong to these classes. There is almost no confusion, but some parts can be missing. These detections can be imposed on the classification results. As for roads, additional knowledge is used, namely on the width of the roads (based on observations from the field missions). Since the detectors provide only lines, they are dilated by the appropriate size, taking into account both the actual road width and the resolution of the images. Another type of knowledge is also very useful: the detection of changes between images taken during the project and KVR images obtained earlier. The results of the change detection processing provide mainly information about class 1, since they exhibit the fields which were previously cultivated, and which are now abandoned. These results do not show all regions belonging to class 1, but the detected areas surely belong to that class, so these results can also be imposed to the classification results.

The final step is regularization, as it is very unlikely that isolated pixels of one class can appear in another class. Several local filters were tested, such as a majority filter, a median filter, or morphological filters, applied on the decision image. A Markovian regularization approach on local neighborhoods was tested too.

The results of this fusion in its basic version are already very good (Bloch et al., 2007), due to the fact that not all information provided by the classifiers is used, but only the best part of them. Further improvements are obtained by knowledge inclusion and regularization.

6. Conclusion

Two main humanitarian action types, close-range antipersonnel mine detection and remote sensing mined area reduction, are very sensitive tasks as they both deal with human lives, which puts high demands to the quality of the sensors involved, their efficiency, robustness, detection rate... Due to the variety of situations in which mines and minefields can be found, referring to different types of terrain, of mines, of mine laying, of weather conditions etc., it is hardly possible that one sensor can cover all these situations while reaching the high detection performance needed. Thus, a solution is being sought in multi-sensor fusion systems. Taking into account that typically there is a good knowledge about these two problems, while it is difficult or even impossible to come up with some statistically relevant training data, due to a high variability of the conditions in which mines and minefields can be found, knowledge-based theories are useful. Among these theories, possibilistic and fuzzy theory provides additional advantages, discussed in this chapter, such as different fusion operators depending on the information and its characteristics. This flexibility turns out to be very helpful for the two humanitarian action types due to different characteristics of the measures and sensors that are combined.

The application of fuzzy and possibility theory to humanitarian mine action presented in this chapter also covers a wide range of possible cases, starting from crisp classes (mines/friendly objects), where we have some uncertainty/imprecision in deriving a class from the measures, modelled as possibility discriptions (Section 3), via the case when measures themselves are prone to imprecision so fuzzy measures are introduced in the possibilistic model (Section 4), up to the case where classes are fuzzy so we apply a fuzzy model (Section 5).

Although the presented modeling is related to sensors used in some of our projects (which are also sensors that are typically used for these problems worldwide), it is flexible enough to be easily adapted to new pieces of information about the types of objects and their characteristics, as well as to new sensors.

7. References

Bloch, I. (1996). Information combination operators for data fusion: a comparative review with classification. *IEEE Trans. Systens, Man and Cybernetics*, Vol. 26, pp. 52-67.

Bloch, I., Milisavljević, N. & Acheroy, M. (2007). Multisensor Data Fusion for Spaceborne and Airborne Reduction of Mine Suspected Areas. *XI International Journal of Advanced Robotic Systems*, pp. 173-186, Vol. 4, No. 2.

Capineri, L., Grande, P. & Temple, J. A. G. (1998). Advanced image-processing technique for real-time interpretation of ground penetrating radar images. *Int. J. Imaging Syst. Technol.*, Vol. 9, pp. 51-59.

Das, Y. (2006). Effects of soil electromagnetic properties on metal detectors. *IEEE Trans. Geosci. Remote Sens.*, Vol. 44, No. 6, pp. 1444-1453.

Denoeux, T. (2008). Conjunctive and Disjunctive Combination of Belief Functions Induced by Non Distinct Bodies of Evidence. *Artificial Intelligence*, Vol. 172, pp. 234–264.

Dubois, D. & Prade, H. (1980). *Fuzzy Sets and Systems : Theory and Applications.* Academic Press, New York.

Dubois, D. & Prade, H. (1985). A review of fuzzy set aggregation connectives. *Information Sciences*, Vol. 36, pp. 85-121.

Dubois, D. & Prade, H. (2004). Possibilistic logic: a retrospective and prospective view. *Fuzzy Sets and Systems*, Vol. 144, pp. 3-23.

Dubois, D. & Prade, H. (2006). Possibility theory and its applications: a retrospective and prospective view. In: *Decision theory and multi-agent planning (CISM International Centre for Mechanical Sciences)*, G. Della Riccia et al. (Eds.), Springer, pp. 89-100.

Fischer, C. et al. (2007). Detection of antipersonnel mines by using the factorization method on multistatic ground-penetrating radar measurements. *IEEE Trans. Geosci. Remote Sens.*, Vol. 45, No. 1, pp. 85-91.

Haralick, R.M. & Shapiro, L.G. (1992). *Computer and Robot Vision.* Boston, MA: Addison-Wesley Longman Publishing Co., Inc.

Milisavljević, N. & Bloch, I. (2003). Sensor fusion in anti-personnel mine detection using a two-level belief function model. *IEEE Trans. Systems, Man and Cybernetics*, Part C, Vol. 33, No. 2, pp. 269-283.

Milisavljević, N.; Bloch, I.; van den Broek, S.P. & Acheroy, M. (2003). Improving mine recognition through processing and Dempster-Shafer fusion of ground-penetrating radar data. *Pattern Recognition*, Vol. 36, No. 5, pp. 1233-1250.

Milisavljević, N. & Bloch, I. (2008). Possibilistic vs. Belief Function Fusion for Anti-Personnel Mine Detection. *IEEE Trans. Geosci. Remote Sens.*, Vol. 46, No. 5, pp. 1488-1498.

Shafer, G. (1976). *A Mathematical Theory of Evidence.* Princeton, NJ: Princeton Univ. Press.

SMART consortium (2004). SMART - final report.

Smets, P. (1994). What is Dempster-Shafer's model?. In: *Advances in the Dempster-Shafer Theory of Evidence*, R. R. Yager, M. Fedrizzi and J. Kacprzyk (Eds). New York: Wiley, pp. 5-34.

Smets, P. & Kennes, R. (1994). The transferable belief model. *Artificial Intelligence*, Vol. 66, pp. 191-234.

Thonnard, O. & Milisavljević, N. (1999). Metallic shape detection and recognition with a metal detector. in *Proc. European Workshop on Photonics applied to Mechanics and Environmental Testing Engineering (PHOTOMEC'99 - ETE'99)*, Liège, Belgium, pp. 99-103.

Yvinec, Y. (2005). A Validated Method to Help Area Reduction in Mine Action with Remote Sensing Data, *Proc. of the 4th Int'l Symposium on Image and Signal Processing and Analysis (ISPA 2005)*, Zagreb, Croatia.

Zadeh, L. (1965). Fuzzy sets. *Information and Control*, Vol. 8, No. 3, pp. 338-353.

Zadeh, L.A. (1975). The Concept of a Linguistic Variable and its Application to Approximate Reasoning. *Information Sciences*, Vol. 8, pp. 199–249.

24

Non specular reflection and depolarisation due to walls under oblique incidence

Iñigo Cuiñas, Manuel G. Sánchez and Ana V. Alejos
Universidade de Vigo
Spain

1. Introduction

Bi static radar remote sensing has became a large source of data intended to analyse the composition and the behaviour of targets placed very far away from the sensor. The possibility of determining the nature of the surface in which the electromagnetic waves have impacted, and have been reflected by, is directly correlated to the knowledge of high precision scattering models. In most of the situations, the Fresnel coefficients for reflection have been the only procedure used to obtain the electromagnetic characteristics of the target material from the reflection this surface induced in the propagating wave. That electromagnetic characterisation is quite important in remote sensing techniques, as it depends on the frequency and, so that, it could help in defining the spectral signature of each target material.

Historically, several models have been proposed in order to improve the performance of the simple Fresnel coefficients to obtain the dielectric constant of an obstacle. The aim of such models is to define the outcomes of the incidence of a propagating wave on a constructive surface (Balanis, 1989; Burnside & Burgener, 1983; Correia & Francês, 1995; Sato et al., 1997). When the reflective phenomenon is limited to the specular direction, it is commonly known as specular reflection; whereas when several directions are taken into account, the reflection concept (or, in general terms, scattering) is extended to all directions in the incidence region. Any obstacle generates its own reflection pattern, which depends on the electromagnetic characteristics of the material, the surface roughness, the frequency, and the angle of incidence. Published measured results show that there may be several scattering directions as significant as the main specular reflection one. This result leads to conclude that the received power in the specular direction could not represent the main contribution among the reflections on the surface under study. Thus, computing the dielectric constant of the surface material from just the data reflected towards the specular direction could lead to large mistakes.

The specular reflection on flat or almost flat obstacles can be considered as a well studied propagation mechanism, and several models have been enunciated to describe its performance. The interest of specular direction resides in the need of exact models and fast algorithms that are used in the radio electric planning tools. The non-specular or general reflection, although less taken into account in practice, is also a topic that focuses several works (Beckmann & Spizzichino, 1963-1987; Cuiñas et al., 2007). These results have capital importance when the application is remote sensing and the objective is to detect and even to identify possible targets by means of their scattering patterns.

A phenomenon associated to reflection, the depolarisation that could be generated when a wave beats a flat obstacle, appears to be not so fine defined and modelled. The reason of this lack of interest is probably because the typical application of reflection models has been the radio planning tools. These tools were designed for frequencies bellow millimetric bands, assigned to cellular phone or television broadcasting, and the typical obstacles (walls) are electrically flat enough to provide strong specular reflections at these frequencies. At higher frequencies, the electrical size of a given obstacle becomes larger, and the specular reflections could not be so dominant among the complete scattering arc angles.

Although there are different published research works on depolarisation, they are commonly centred at lower frequencies, and they often treat on complete radio channels, instead of the analysis of isolated obstacles. Several of these works reports the polarisation diversity gain measured at indoor radio channels: up to 15 dB at UHF band (Sánchez & Sánchez, 2000), and from 4 to 10.5 dB at 2.05 GHz (Dietrich et al., 2001); or even in outdoor environments: between 11 to 5.2 dB at 1800 MHz (Turkmani et al., 1995), depending on the type of environment: urban, suburban or rural. The contents of this chapter are not comparable to the previous work, as they are oriented to analyse the radio channel, including multipath propagation, whereas the aim of the presented work is the study of isolated flat obstacles.

This chapter summarises the theoretical approaches to the study of scattering and depolarisation generated by flat isolated obstacles, and it also outlines measurement results obtained by the authors at 5.8 GHz, involving three different constructive materials and taking into account two orthogonal polarisations for the incident wave, which comes from several angles of incidence to the obstacle.

The chapter is organised as follows. The contents of the section 2 are the theoretical basis of this work: an overview of specular reflection models, which are used to extract the electromagnetic parameters of each considered material, and the exposition of the Physical Optics model used to compute the reflection patterns.

The depolarisation indexes computation procedure is introduced in section 3. This section contains the discussion of a methodology to determine the amount of cross-polarised wave that could be generated by reflection on the surface of an obstacle in the radio channel. The proposed matrix formulation, which is inspired in the polarimetric matrix (Mott, 1986), allows the separation between the depolarisation due to the antennas and that produced by the obstacle. The depolarisation indexes indicate the fraction of the signal power that is depolarised after beating the obstacle.

The section 4 shows the results, comparing the behaviour of the Physical Optics model application when it simulates reflection patterns in similar geometrical conditions that previously measured. Finally, the section 5 contains the conclusions.

2. Scattering coefficients

The specular reflection is just a simplification of the complete, and complex, effect of walls over propagating waves. When a wave beats a wall, a scattering phenomenon is generated towards all space directions, defining a reflection pattern. However, specular reflection models are very useful to obtain an initial approach to the characteristic parameters, as permittivity and conductivity, which drive the electromagnetic behaviour of the material. These parameters could be used as input data to those more complex scattering models.

The scattering due to surfaces of any kind is a classical area of study that has its biggest impetus with the advent of radar, in the middle of 20th century. There are several models to predict the behaviour of a surface when an incident wave beats it (Ruck et al., 1970). The strategy to model their effects on the propagating wave depends on the type of roughness. Besides, the scattering by perfectly conductive surfaces is a problem analysed by several reflector antenna methods. These problems are tried to be solved by employing high frequency methods, which are valid when the reflector size (in our case, the obstacle size) is large in terms of wavelength (Scott, 1990). A combination of both radar ideas and antenna analysis methods has been applied in the work summarised at this chapter to model the reflection due to constructive walls. The formulation, based on Physical Optics, is intended to compute scattering patterns due to flat and rough surfaces, both dielectric and conductive.

2.1 Specular reflection coefficients

The behaviour of a wave when it beats an obstacle mainly depends on its permittivity. This behaviour has been modelled by different methods (Landron et al., 1993, 1997; Lähteenmäki &Karttavi, 1996; Cuiñas et al., 2001), most of them based on classical Fresnel formulation. This Fresnel theory works consistently when the obstacle thickness is electrically large and/or the material losses are huge. Typical walls are not infinite thickness, and this is the reason because several models have been enunciated to take into account this situation. Among the models based on Fresnel method, the internal successive reflections (ISR) model must be mentioned: it tries to explain transmission and reflection phenomena as a result of the coherent sum of several multipath components, generated in both boundaries between the obstacle and the free space (Burnside & Burgener, 1983). Although this model assumes that the slab surface is infinite, the results are adequate if the material sample is wide enough to contain the first Fresnel ellipsoid of the radio link.

2.2 Physical Optics model

A good characterisation of scattering due to rough or slightly rough surfaces must include all-direction effects. The Physical Optics formulation fits this condition, and it is a classical method to characterise conductive surfaces (Beckmann & Spizzichino, 1963-1987).

The electric field strength, as well as its derivative respect to surface normal, could be estimated by using Kirchhoff approximations. Thus, it can be assumed that the total field in a point on the surface is the same as in a tangent plane to the surface at the same point. The larger the roughness curvature radius is, the better the approximation is.

Assuming an electrically large obstacle, the Physical Optics model leads to a single equation to define the reflection coefficient, the general formulation of the scattering coefficient.

$$\rho = \frac{1}{4L\cos\theta_i} \int_{-L}^{L} (a\xi'-b)\cdot e^{jv_x x + jv_z \xi}\,dx \tag{1}$$

There is no general, exact and explicit solution for finite conductivity surfaces, but some approximations can be applied, which performance is better when the rough surface presents soft slopes, or large radii of curvature: in other words, when we manage rough, but smooth enough surfaces.

As a general conclusion, it can be said that scattering due to non-conductive surfaces is affected by finite conductivity only when local reflection coefficients are more influenced by its local incident angle, than by the electromagnetic properties of the scatterer.

2.3 Application of the model

The application of this formulation is suggested for three different strategies. Successive proposals have growing computational cost and complexity, but they give better concordance to actual situations (Cuiñas et al., 2007). Figure 1 depicts the three strategies.

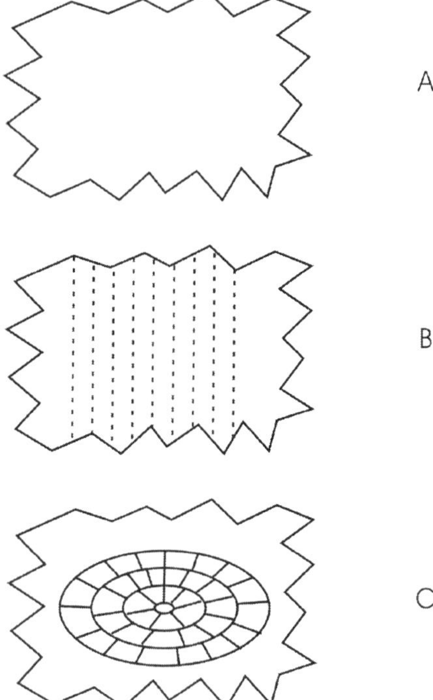

Fig. 1. Strategies to apply the Physical Optics formulation: A, direct; B, by segments; C, 3D

The first option focuses on the direct application of the Physical Optics formulation considering an incidence field on a flat surface, with an incidence angle determined by direct propagation path between the transmitter antenna and the centre of illumination. This case does not consider the radiation pattern of the antenna.

As the field strength over the illuminated area is not uniform, but varies with the angle from the centre of the beam according to the gain pattern of the antenna, a second option is proposed. This strategy consists in dividing the surface into several parallel segments of the same width and assumed infinite length. The local angle of incidence at each segment is considered when applying Physical Optics method. Thus, part of the effect of the antenna radiation pattern can be considered, as local situations are taken into account. This method may be a good choice for flat or for one-dimensional rough or periodic surfaces.

The third option consists of determining the illuminated surface on the obstacle, taking into account the radiation pattern of the transmitter antenna. That radiation pattern determines a footprint on the obstacle surface, with elliptic shape. This footprint on the surface is divided into several patches. The Physical Optics formulation is applied on each patch, considering the local incidence angle at each of them. In this situation, the application of the algorithm needs a reformulation, as the local scattered electric field has to be computed at each patch, and then all these contributions have to be coherently combined to compute the complete scattered field at any scattering direction. This idea is based on antenna analysis techniques, as (Arias et al., 1996).

3. Depolarisation indexes

The depolarisation index, for any material, at any angle of incidence and any polarisation of the transmitted waves could be defined as the fraction of the power of this wave that is received in the orthogonal polarisation. From this definition, depolarisation indexes may be computed by means of a matrix procedure. The proposed method provides a set of depolarisation indexes that characterise the reflection mechanism generated when a wave reach an obstacle with oblique incidence (Cuiñas et al., 2009).

The definition of the matrix model begins in the characterisation of the radio channel following the Friis formula, but taking into account two orthogonal polarisations, considering the presence of an obstacle in the radio path, and assuming that both transmitting and receiving antennas are pointed to the same spot on the reflecting surface. Then, the transmitting antenna is illuminating the reflection point with its maximum gain, and the receiving antenna is getting the waves from the obstacle by its maximum gain direction. The free space attenuation coefficient could be computed by substitution of the obstacle by a perfectly conductive surface.

With the model equation, the definition of the obstacle behaviour is completely independent from the effect of the free space propagation in the open links between the transmitter and the obstacle, and between the obstacle and the receiver. The matrix A^{OBS} contains the data corresponding to the obstacle effect on the complete radio channel. Figure 2 depicts the physical significance of each element in the matrix.

$$A^{OBS}\left(\theta_{inc},\theta_{obs}\right)=\begin{bmatrix} a_{VV}^{OBS}\left(\theta_{inc},\theta_{obs}\right) & a_{VH}^{OBS}\left(\theta_{inc},\theta_{obs}\right) \\ a_{HV}^{OBS}\left(\theta_{inc},\theta_{obs}\right) & a_{HH}^{OBS}\left(\theta_{inc},\theta_{obs}\right) \end{bmatrix} \tag{2}$$

The computation of depolarisation indexes is difficult, in general terms, because of the separation between the depolarisation due to the obstacle and the depolarisation due to the antennas is needed. Nevertheless, this problem could avoided with the matrix formulation that has been commented, as the elements of the obstacle matrix (A^{OBS}) are only related to the obstacle, and they do not include other effects. The depolarisation indexes can be defined from the obstacle matrix elements as:

$$DI_H\left(\theta_{inc},\theta_{obs}\right)=\frac{a_{HV}^{OBS}\left(\theta_{inc},\theta_{obs}\right)}{a_{HV}^{OBS}\left(\theta_{inc},\theta_{obs}\right)+a_{HH}^{OBS}\left(\theta_{inc},\theta_{obs}\right)}$$
$$DI_V\left(\theta_{inc},\theta_{obs}\right)=\frac{a_{VH}^{OBS}\left(\theta_{inc},\theta_{obs}\right)}{a_{VH}^{OBS}\left(\theta_{inc},\theta_{obs}\right)+a_{VV}^{OBS}\left(\theta_{inc},\theta_{obs}\right)}$$

(3)

where the sub index indicates vertical or horizontal incidence (transmission). As these indexes are computed from A^{OBS} elements, each pair of θ_{inc} and θ_{obs} would define their associated pair of depolarisation indexes.

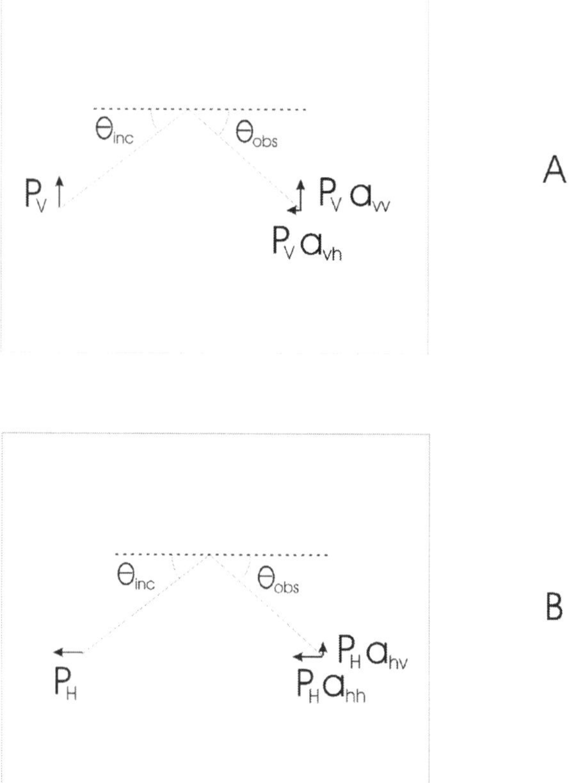

Fig. 2. Definition of matrix elements: A, vertical incidence; B: horizontal incidence

4. Results

The wellness of the proposed methods has been checked by comparing to measurement outcomes performed at 5.8 GHz (Cuiñas et al., 2007). The exposition of the results begins with the Physical Optics formulation: measured scattering patterns are compared with those computed by applying Physical Optics formulation, following the three described strategies. Then, the depolarisation indexes are computed.

4.1. Reflection pattern

The Physical Optics formulation has been applied following the three strategies previously mentioned. Direct application of the model provides reflection patterns with a very narrow main lobe and many smaller side lobes, clearly different from the measured patterns.

When the surface is perfectly conductive, the application of the Physical Optics model by segments leads to a simulation error (compared to measurement outcomes) similar to the direct use of Beckmann formulation, whereas the use of the elliptic patching (3D) model provides an improvement of around 57% in the simulated results.

The results for dielectric obstacles seem to indicate better fitting when the application takes into account the geometry of the problem; i.e. when the illuminated surface is patched in concentric rings and local contributions at each patch are coherently added at each scattering direction. This appreciation is supported by the reduction percentage in the relative simulation error obtained by applying the Physical Optics theory by segments or by elliptic patching compared to that obtained by the direct application of the Physical Optics formulation. The 3D application of the integral equation led to an improvement of the relative error of 1.5% for a brick wall obstacle, and 67% for a highly reflective smooth chip wood panel, whereas the segment application improvements were 0.2% and 62%, respectively. This result confirms the visual appreciation: the more specularly reflective the material is, the better the improvement of Physical Optics 3D method is.

4.2. Depolarisation indexes

The difference between co polar and cross polar relative powers presents strong variations in measurement results: in the vicinity of the specular direction it is commonly larger than in other directions of observation, in which the cross polar component could be even stronger than the co polar. This appreciation confirms that the non-specular directions could not be ignored when planning the network.

The difference between co polar and cross polar received power, observed towards the specular direction could be up to 58 dB, for brick walls, 28 dB for chip wood panels, and 31 dB for stone walls, at 5.8 GHz (Cuiñas et al., 2009).

Once computed the depolarisation indexes, towards the specular direction, they resulted to be up to 0.75% for the brick wall, up to 9.8% for the chip wood panel, and up to 9.27% for the stone wall. This indicates that brick wall provides reduced depolarised waves compared to the co polar reflected waves in the specular direction.

But in a general case, all scattering directions have to be considered, and not just specular one, as the reflector could be randomly located and oriented. With this aim, median depolarisation indexes for each material could be useful, being 23% and 30% for brick wall, 18% and 18.5% for chip wood panel, and 4.5% and 4% for stone wall, in horizontally and vertically polarised incident wave, respectively, at 5.8 GHz. Once several angles of observation are introduced, not just the specular ones, the depolarisation indexes grow, and differences between incident polarisations appear in the brick wall case. The brick wall is the more non isotropic material among the considered, as it presents a clearly oriented structure, whereas the chip wood panel and the stone wall are the result of the solidification of a mass, which is expected to present a more isotropic behaviour. The large median values of depolarisation indexes indicate that high depolarised waves could be generated when several scatterers are present in an environment, which is the case of indoor scenarios.

5. Conclusions

Two methods to characterise an obstacle in terms of reflection and depolarisation, and the different strategies to apply these formulations, have been summarised along this chapter, providing the references to in depth study both formulation. The behaviour of such methods have been also compared to measurement results at 5.8 GHz, using three different flat obstacles, with more or less rough surfaces, and two orthogonal linear polarisations in the incident wave.

The modelling of scattering patterns generated by flat conductive and dielectric obstacles is proposed, based on the Physical Optics formulation. Among three possible strategies of implementation, that based on elliptic patching the illuminated area on the obstacle surface appears to provide more accurate results, since it is better adapted to the real problem: it defines the finite surface in a more precise way and it takes into account the radiation pattern of the antennas employed in the experimental work. The comparison between simulation results and actual situation measurements shows the good behaviour of the algorithm. The flatter and more conductive the material is, the better the algorithm works.

When observing the depolarisation, the measured results reflect differences between co polar and cross polar received powers from 17 to 58 dB in specular directions of observation, depending on the obstacle and the angle of incidence. But the most interesting observations are in the non-specular direction, in which the cross polar received power could be even stronger than the co polar, and with amplitudes that could not be ignored. These situations indicated that up to more than 50% of the incident wave could be scattered as a cross polarised wave.

The results of this experimental work were processed by means of a matrix model to compute the depolarisation indexes. Indexes up to 10 % have been obtained in the specular direction, and median values up to 30% are reached when considering the entire scattering region.

6. References

A.M. Arias, M.E. Lorenzo, A.García-Pino, "A novel fast algorithm for Physical Optics analysis of single and dual reflector antennas". *IEEE Transactions on Magnetics*, vol.32, pp. 910-913, May 1996.

C.A. Balanis, *"Advanced engineering electromagnetics"*, John Willey & sons, 1989.

P. Beckmann, A. Spizzichino, *The scattering of electromagnetic waves from rough surfaces*, Artech House, 1963-1987.

W.D. Burnside, K.W. Burgener, "High frequency scattering by a thin lossless dielectric slab", *IEEE Transactions on Antennas and Propagation,* vol. Ap-31, no. 1, pp. 104-110, January 1983.

L.M. Correia, P.P. Francês, "Transmission and isolation of signals in buildings at 60 GHz", In: *International symposium on personal, indoor and mobile radio communications,* Toronto (Canada), September 1995.

I. Cuiñas, J.-P. Pugliese, A. Hammoudeh, M. García sánchez, "Frequency dependence of dielectric constant of construction materials at microwave and millimeter wave bands", *Microwave and Optical Technology Letters,* vol. 30, no. 2, pp. 123-124, July 2001.

I. Cuiñas, D. Martínez, M. García Sánchez, A. Vázquez Alejos, "Modelling and measuring reflection due to flat dielectric surfaces at 5.8 GHz", *IEEE Transactions on Antennas and Propagation,* vol. 55, no. 4, pp. 1139-1147, April 2007.

I. Cuiñas, M. García Sánchez, A. Vázquez Alejos, "Depolarization due to scattering on walls in the 5 GHz band", *IEEE Transactions on Antennas and Propagation,* June 2009.

C.B. Dietrich, K. Dietze, J. Randall Nealy, W.L. Stutzman, "Spatial, polarisation, and pattern diversity for wireless handheld antennas", *IEEE Transactions on Antennas and Propagation,* vol. 49, no. 9, pp. 1271-1281, September 2001.

A.K.Fung, "Scattering and depolarisation of EM waves from a rough surface", *Proceedings of the IEEE,* pp. 395-396, March 1966.

T. Kurner, D.J. Cichon, W. Wiesbeck, "Concepts and results for 3D digital terrain-based wave propagation models: an overview", *IEEE Journal on Selected Areas in Communications,* vol. 11, no. 7, September 1993.

J. Lähteenmäki, T. Karttavi, "Measurements of dielectric parameters of wall materials at 60 GHz band", *Electronics Letters,* vol. 32, no. 16, pp. 1442-1444, August 1996.

O. Landron, M.J. Feuerstein, T.S. Rappaport, "In situ microwave reflection measurements for smooth and rough exterior wall surfaces", In: *IEEE 43rd Vehicular Technology Conference,* pp. 77-80, May 1993.

O. Landron, M.J. Feuerstein, T.S. Rappaport, "A comparison of theoretical and empirical reflection coefficients for typical exterior wall surfaces in a mobile radio environment", *IEEE Transactions on Antennas and Propagation,* vol. 44, no. 3, pp. 341-351, March 1996.

M. Lebherz, W. Wiesbeck, W. Krank, "A versatile wave propagation model for the VHF/UHF range considering three-dimensional terrain", *IEEE Transactions on Antennas and Propagation,* vol. 40, no. 10, pp. 1121-1131, October 1992.

H. Mott, *Polarisation in Antennas and Radar,* John Wiley & Sons, New York, 1986.

G.T. Ruck, D.E. Barrick, W.D. Stuart, C.K. Krichbaum, Chapter 9: "Rough Surfaces", *Radar cross section handbook,* vol. 2 Plenum Press, New York, 1970.

Sato, Katsuyoshi, et al, "Measurements of reflection and transmission characteristics of interior structures of office building in the 60-GHz band", *IEEE Transactions on Antennas and Propagation,* vol. 45, no. 12, pp. 1783-1791, December 1997.

M. Sánchez, M.G. Sánchez, "Analysis of polarisation diversity at digital TV indoor receivers", *IEEE Transactions on Broadcasting,* vol. 46, no. 4, pp. 233-239. December 2000.

C. Scott, *Modern methods of reflector antenna analysis and design,* Artech House, 1990.

A.M.D. Turkmani, A. Arowjolu, P.A. Jefford, C.J. Kellett "An experimental evaluation of the performance of two-branch space and polarisation diversity schemes at 1800 MHz", *IEEE Transactions on Vehicular Technology*, vol. 44, no. 2, pp. 318-326, May 1995.

25

Dynamical Enhancement Technique for Geophysical Analysis of Remote Sensing Imagery

Iván E. Villalón-Turrubiates
University of Guadalajara Campus Los Valles
Guadalajara–Ameca Highway Km. 45.5, 46600 Ameca Jalisco, Mexico
Telephone (+52 37) 5758-0148 ext. 7267, E-Mail: villalon@ieee.org

1. Introduction

Considerable progress has been made generally in the application of remote sensing techniques to both research and operational problems for urban planning and natural resource management. Modern applied theory of image processing is now a mature and well developed research field, presented and detailed in many works.

Although the existing theory offers a manifold of statistical techniques to tackle with the particular environmental monitoring problems, in many applications areas there still remain some crucial theoretical and data processing problems. One of them is particularly related to the extraction and dynamical analysis of physical characteristics (e.g., water, land cover, vegetation, soil, humid content, and dry content) for implementation in natural resources management (modeling and planning).

The extraction of environmental physical characteristics from a particular geographical region through remote sensing data processing allows the generation of electronic signature maps, which are the basis to create a high-resolution collection atlas processed in time for a particular geographical zone. This can be achieved using a systematical tool for supervised segmentation and classification of the environmental remote sensing signatures that employs multispectral remote sensing imagery based on pixel statistics for the class description. Moreover, the analysis of a dynamical model of environmental characteristics extracted from a geographic region generates useful information for natural resource management; using the signatures map extracted from the remote sensing imagery for a particular geographic zone in discrete time the evolution study of the environmental characteristics is performed to obtain the dynamical model of the physical variables. This provides a background for understanding the future trends of the multispectral image. This chapter explores the implementation possibilities of the multispectral image classification technique with the dynamic analysis for natural resources management applications.

1.1 Remote Sensing

The goal of science is to discover universal truths that are the same yesterday, today and tomorrow. Hopefully, the knowledge obtained can be used to protect the environment and

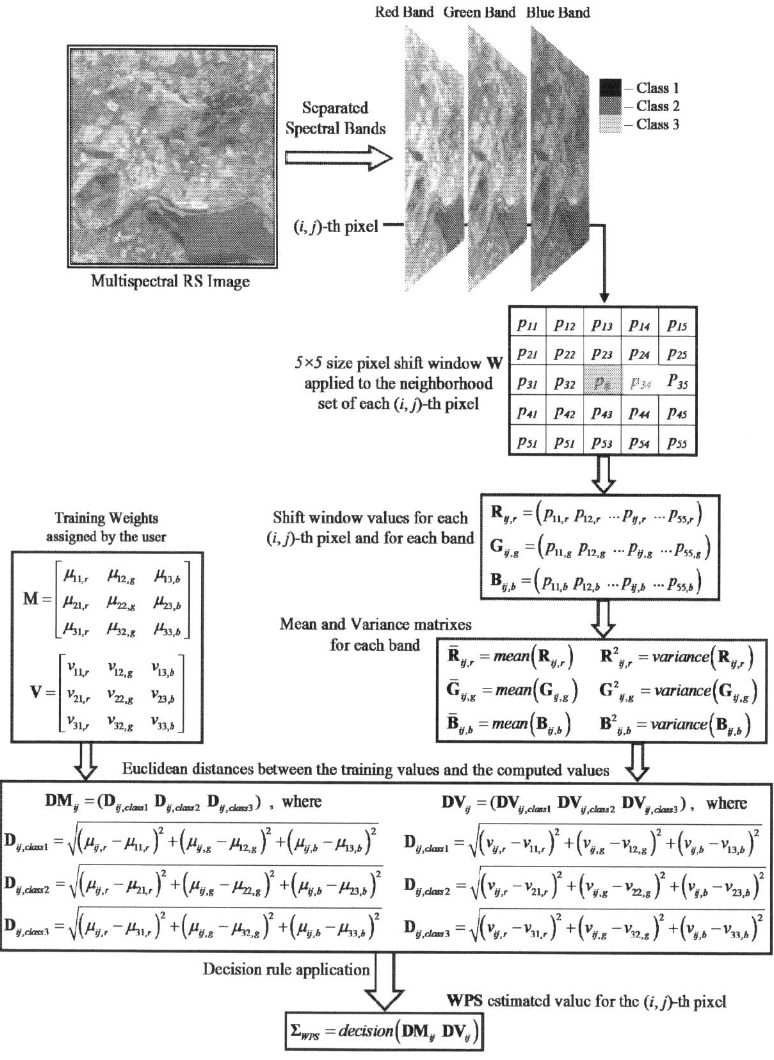

Fig. 3. Computational structure of the WPS method

The model of the equation of observation (EO) in CET is represented (Falkovich et al., 1989) as $\Sigma(\tau) = S(\Lambda(\tau)) + v(\tau)$, where $v(\tau)$ is the white observation Gaussian noise and $\tau \in \Im$, starting at τ_0 (initial moment of continuous evolution time), and the linear amplitude-modulated (Falkovich et al., 1989) model is $S(\Lambda(\tau)) = \Lambda(\tau)S_0(\tau)$, where $S_0(\tau)$ represents the deterministic "carrier" RS image frame of a given model, and $\Lambda(\tau)$ is the unknown stochastic information process to be estimated via processing of the MRS image observation data frame $\Sigma(\tau)$.

a. Continuous evolution time (CET) domain: τ-continuous time argument

b. Discrete evolution time (DET) domain: κ-discrete time argument

Fig. 4. Representation of linear dynamic systems

Is assumed that $\Lambda(\tau)$ satisfies the dynamical model specified by the following N-th order linear differential equation (Villalon & Shkvarko, 2008)

$$\frac{d^N\Lambda(\tau)}{d\tau^N} + \alpha_{N-1}\frac{d^{N-1}\Lambda(\tau)}{d\tau^{N-1}} + ... + \alpha_0\Lambda(\tau) = \beta_{N-1}\frac{d^{N-1}\xi(\tau)}{d\tau^{N-1}} + ... + \beta_0\xi(\tau) \ . \tag{3}$$

where α and β are the constant coefficients of the dynamical system model for evolution of the RSS $\Lambda(\tau)$. This stochastic model can be redefined as follows: the differential equation (3) may be transformed into a system of linear differential equations of the first order via performing the following replacement of variables

$$\begin{cases} z_1(\tau) = \Lambda(\tau), \\ z_2(\tau) = \dfrac{dz_1(\tau)}{d\tau} + \alpha_{N-1}z_1(\tau) - \beta_{N-1}\xi(\tau), \\ \dotfill \\ z_N(\tau) = \dfrac{dz_{N-1}(\tau)}{d\tau} + \alpha_1 z_1(\tau) - \beta_1\xi(\tau), \\ \dfrac{dz_N(\tau)}{d\tau} = -\alpha_0 z_1(\tau) + \beta_0\xi(\tau) \end{cases} \tag{4}$$

where

$$\mathbf{z}(\tau) = \begin{pmatrix} z_1(\tau) & z_2(\tau) & ... & z_N(\tau) \end{pmatrix}^T . \tag{5}$$

Based on the replacement of variables specified by (4), the dynamic differential equation model (3) can be now represented in a canonical vector-matrix form as follows

$$\frac{d\mathbf{z}(\tau)}{d\tau} = \mathbf{F}\mathbf{z}(\tau) + \mathbf{G}\xi(\tau) \ , \qquad \Lambda(\tau) = \mathbf{C}\mathbf{z}(\tau) \ . \tag{6}$$

where

$$\mathbf{F} = \begin{bmatrix} -\alpha_{N-1} & 1 & 0 & ... & 0 \\ -\alpha_{N-2} & 0 & 1 & ... & 0 \\ ... & ... & ... & ... & ... \\ -\alpha_1 & 0 & 0 & ... & 1 \\ \hline -\alpha_0 & 0 & 0 & ... & 0 \end{bmatrix} , \qquad \mathbf{G} = \begin{bmatrix} \beta_{N-1} \\ \beta_{N-2} \\ ... \\ \beta_0 \end{bmatrix} , \qquad \mathbf{C} = \begin{bmatrix} 1 & 0 & ... & 0 \end{bmatrix} . \tag{7}$$

The representation in the form of (6) is referred to as a canonical equation of linear dynamic system in state variables in continuous time. Here, $\mathbf{z}(\tau)$ is the state vector, the vector $\mathbf{C}$ defines a linear operator that introduces the relation between the RSS map in continuous time and the state vector $\mathbf{z}(\tau)$, $\mathbf{F}$ is a transition matrix, $\mathbf{G}$ is a transition vector, and $\xi(\tau)$ represents the white model generation noise vector characterized by the statistics, $\langle \xi(\tau) \rangle = \mathbf{0}$ and $\langle \xi(\tau)\xi^{T}(\tau') \rangle = \mathbf{P}_{\xi}(\tau)\sigma(\tau - \tau')$, respectively. Here, $\mathbf{P}_{\xi}(\tau)$ is referred to as state model disperse matrix that characterizes the dynamics of the state variances developed in continuous time $\tau\left(\tau_{0} \to \tau\right)$ starting from the initial instant τ_{0}. The dynamic model equation in the continuous time states the relation between the RSS map $\Sigma(\tau)$ extracted from the MRS scene, thus the desired dynamical RSS map $\Lambda(\tau)$ can be represented as follows (Villalon & Shkvarko, 2008)

$$\Sigma(\tau) = \mathbf{S}_0(\tau)\mathbf{C}\left(\tau\right)\mathbf{z}(\tau) + \mathbf{v}(\tau) = \mathbf{H}(\tau)\mathbf{z}(\tau) + \mathbf{v}(\tau) , \text{ where } \mathbf{H}(\tau) = \mathbf{S}_0(\tau)\mathbf{C}\left(\tau\right) . \tag{8}$$

The stochastic differential model of equations (6) and (8) allows applying the theory of dynamical filtration to reconstruct the desired RSS map in continuous time incorporating the a priori model of dynamical information about the RSS. The aim of the dynamic filtration is to find an optimal estimate of the desired RSS, $\hat{\Lambda}(\tau) = \mathbf{C}\hat{\mathbf{z}}(\tau)$, developed in continuous time $\tau\left(\tau_{0} \to \tau\right)$ via processing the RSS maps $\Sigma(\tau)$ extracted from the MRS scenes taking into considerations the a priori dynamic model of the desired RSS map specified through the state equation (6). The optimal dynamic filter when applied to the RSS maps $\Sigma(\tau)$ specified by the dynamic image model (8) must provide the optimal estimation of the desired RSS map $\hat{\Lambda}(\tau) = \mathbf{C}\hat{\mathbf{z}}(\tau)$, in which the state vector estimate $\mathbf{z}(\tau)$ satisfies the a priori dynamic behavior modeled by the stochastic dynamic state equation (6).

3.2 Mathematical model of RSS in discrete time

The canonical discrete-time solution to equation (6) in state variables for discrete time κ is expressed as follows (Falkovich et al., 1989)

$$\mathbf{z}(\kappa + 1) = \mathbf{\Phi}(\kappa)\mathbf{z}(\kappa) + \mathbf{\Gamma}(\kappa)\xi(\kappa) , \qquad \Lambda(\kappa) = \mathbf{C}(\kappa)\mathbf{z}(\kappa) , \tag{9}$$

where $\mathbf{\Phi}(\kappa) = \mathbf{F}(\tau_{\kappa})\Delta\tau + \mathbf{I}$; $\mathbf{\Gamma}(\kappa) = \mathbf{G}(\tau_{\kappa})\Delta\tau$, and $\Delta\tau$ represents the continuous time sampling interval for dynamical modeling of the RSS map in discrete time. The statistical characteristics of the a priori information are as follows

1. Generating model noise $\{\xi(\kappa)\}$:

$$\langle \xi(\kappa) \rangle = \mathbf{0} ; \quad \langle \xi(\kappa)\xi^{T}(\kappa') \rangle = \mathbf{P}_{\xi}(\kappa, \kappa') . \tag{10}$$

2. Observation (RSS map) noise $\{\mathbf{v}(\kappa)\}$:

$$\langle \mathbf{v}(\kappa) \rangle = \mathbf{0} \; ; \quad \langle \mathbf{v}(\kappa)\mathbf{v}^T(\kappa') \rangle = \mathbf{P}_v(\kappa,\kappa') \; . \tag{11}$$

3. State vector $\{\mathbf{z}(\kappa)\}$

$$\langle \mathbf{z}(0) \rangle = \mathbf{m}_z(0) \; ; \quad \langle \mathbf{z}(0)\mathbf{z}^T(0) \rangle = \mathbf{P}_z(0) \; . \tag{12}$$

where 0 argument implies the initial state for initial time instant ($\kappa = 0$). The disperse matrix $\mathbf{P}_z(0)$ satisfies the following disperse dynamic equation (Villalon & Shkvarko, 2008)

$$\mathbf{P}_z(\kappa+1) = \langle \mathbf{z}(\kappa+1)\mathbf{z}^T(\kappa+1) \rangle = \mathbf{\Phi}(\kappa)\mathbf{P}_z(\kappa)\mathbf{\Phi}^T(\kappa) + \mathbf{\Gamma}(\kappa)\mathbf{P}_\xi(\kappa)\mathbf{\Gamma}^T(\kappa) \; . \tag{13}$$

3.3 Optimal dynamic RSS filtering technique

The strategy is to design an optimal decision procedure that, when applied to all RSS observations will provide an optimal solution to the state vector $\mathbf{z}(\kappa)$ subjected to its prior defined dynamic model given by the stochastic dynamic equation (9). The estimate of the state vector optimally defined in the sense of the Bayesian minimum risk strategy (Shkvarko, 2004) in discrete time κ can be represented in the conditioned form

$$\hat{\mathbf{z}}(\kappa) = \langle \mathbf{z}(\kappa) \big| \Sigma(0), \Sigma(1), ..., \Sigma(\kappa) \rangle, \tag{14}$$

were <·> represents an ensemble averaging operator. For discrete time, the design procedure is based on the concept of mathematical induction (Falkovich et al., 1989). This is a supposition that after κ observations $\{\Sigma(0), \Sigma(1), ..., \Sigma(\kappa)\}$ the desired optimal estimate is produced, defined at the ultimate step as

$$\hat{\mathbf{z}}(\kappa) = \hat{\mathbf{z}}(\kappa) \; . \tag{15}$$

In order to use the estimate $\hat{\mathbf{z}}(\kappa)$ it is necessary to design an algorithm that produces the optimal estimate $\mathbf{z}(\kappa+1)$ incorporating new measurements $\Sigma(\kappa+1)$ according to the state dynamic equation (9). This is, we have to design an optimal decision procedure (optimal filter) that, when applied to all reconstructed RSS maps $\{\mathbf{\Sigma}(\kappa)\}$ ordered in discrete time κ $(\kappa_0 \rightarrow \kappa)$, provides an optimal reconstruction of the desired RSS map represented via the estimate of the state vector $\mathbf{z}(\kappa)$ subject to the numerical dynamic model (9). To proceed with derivation of such a filter, we first represent the state dynamic equation (9) in discrete time κ as follows

$$\mathbf{z}(\kappa+1) = \mathbf{\Phi}(\kappa)\mathbf{z}(\kappa) + \mathbf{\Gamma}(\kappa)\boldsymbol{\xi}(\kappa) \; . \tag{16}$$

3.4 Dynamic RSS map reconstruction

According to the dynamical model of equation (16), the anticipated mean value for the state vector can be expressed as (Villalon & Shkvarko, 2008)

$$\mathbf{m}_z(\kappa+1) = \langle \mathbf{z}(\kappa+1) \rangle = \langle \mathbf{z}(\kappa+1) | \hat{\mathbf{z}}(\kappa) \rangle \,. \tag{17}$$

The $\mathbf{m}_z(\kappa+1)$ is considered as the a priori conditional mean value of the state vector for the $(\kappa+1)$-*st* estimation step, thus, from equations (16) and (17) we obtain

$$\mathbf{m}_z(\kappa+1) = \mathbf{\Phi}\langle \mathbf{z}(\kappa) | \mathbf{\Sigma}(0), \mathbf{\Sigma}(1),\dots,\mathbf{\Sigma}(\kappa) \rangle + \mathbf{\Gamma}\langle \boldsymbol{\xi}(\kappa) \rangle = \mathbf{\Phi}\hat{\mathbf{z}}(\kappa) \,, \tag{18}$$

hence, the prognosis of the mean value becomes

$$\mathbf{m}_z(\kappa+1) = \mathbf{\Phi}\hat{\mathbf{z}}(\kappa) \,. \tag{19}$$

From the analysis of equations (16) thru (19), it is possible to deduce that given the fact that the particular RSS map $\mathbf{\Sigma}(\kappa)$ is treated at discrete time κ, it makes the previous reconstructions $\{\mathbf{\Sigma}(0), \mathbf{\Sigma}(1),\dots,\mathbf{\Sigma}(\kappa)\}$ irrelevant; hence, the optimal filtering strategy is reduced to the dynamical one step predictor described by the equation (16). Using these derivations, the dynamical estimation strategy can be modified to the one step optimal prediction procedure (Villalon & Shkvarko, 2008)

$$\hat{\mathbf{z}}(\kappa+1) = \langle \mathbf{z}(\kappa+1) | \mathbf{\Sigma}(0), \mathbf{\Sigma}(1),\dots,\mathbf{\Sigma}(\kappa), \mathbf{\Sigma}(\kappa+1) \rangle = \langle \mathbf{z}(\kappa+1) | \hat{\mathbf{z}}(\kappa); \mathbf{\Sigma}(\kappa+1) \rangle \,;$$
$$\hat{\mathbf{z}}(\kappa+1) = \langle \mathbf{z}(\kappa+1) | \mathbf{\Sigma}(\kappa+1); \mathbf{m}_z(\kappa+1) \rangle \,. \tag{20}$$

For the current $(\kappa+1)$-*st* discrete time estimation/prediction step, the dynamical RSS map estimate of the equation (8) in discrete time becomes

$$\mathbf{\Sigma}(\kappa+1) = \mathbf{H}(\kappa+1)\mathbf{z}(\kappa+1) + \mathbf{v}(\kappa+1) \,, \tag{21}$$

with the a priori predicted mean calculated by the equation (17) for the desired state vector given by (16). Applying the Wiener minimum risk strategy (Shkvarko, 2004) to solve the equation (21) with respect to the state vector $\mathbf{z}(\kappa)$ and taking into account the a priori information summarized by the equations (10) thru (12), we obtain the dynamic solution for the RSS map state vector

$$\hat{\mathbf{z}}(\kappa+1) = \mathbf{m}_z(\kappa+1) + \mathbf{\Theta}(\kappa+1)\left[\mathbf{\Sigma}(\kappa+1) - \mathbf{H}(\kappa+1)\mathbf{m}_z(\kappa+1)\right] \,, \tag{22}$$

where the desired dynamic filter operator $\mathbf{\Theta}(\kappa+1)$ is defined as (Villalon & Shkvarko, 2008)

$$\mathbf{\Theta}(\kappa+1) = \mathbf{K}_{\Theta}(\kappa+1)\mathbf{H}^T(\kappa+1)\mathbf{P}_v^{-1}(\kappa+1) \,, \tag{23}$$

and

$$\mathbf{K}_{\Theta}(\kappa+1) = \left[\mathbf{\Psi}_{\Theta}(\kappa+1) + \mathbf{P}_{z}^{-1}(\kappa+1)\right]^{-1} , \tag{24}$$

$$\mathbf{\Psi}_{\Theta}(\kappa+1) = \mathbf{H}^{T}(\kappa+1)\mathbf{P}_{v}^{-1}(\kappa+1)\mathbf{H}(\kappa+1) . \tag{25}$$

Finally, using the derived filter equations (22) thru (25) and the initial RSS map state model of equation (9), the optimal filtering procedure for the dynamic reconstruction of the desired RSS map can be represented in discrete time κ as

$$\hat{\mathbf{\Lambda}}(\kappa+1) = \mathbf{\Phi}(\kappa)\hat{\mathbf{z}}(\kappa) + \mathbf{\Theta}(\kappa+1)\left[\mathbf{\Sigma}(\kappa+1) - \mathbf{H}(\kappa+1)\mathbf{\Phi}(\kappa)\hat{\mathbf{z}}(\kappa)\right] ; \quad \kappa=0,1,\dots , \tag{26}$$

4. Geophysical Dynamic Laboratory

The described technique provides the dynamical RSS map based on the atlas of RSS maps extracted from MRS scenes.

The GDL method is defined in the form of equation (26) based on the atlas of RSS maps in discrete time as follows

$$\hat{\mathbf{\Lambda}}_{GDL} = \hat{\mathbf{\Lambda}}(\kappa+1) = \mathbf{\Phi}(\kappa)\hat{\mathbf{z}}(\kappa) + \mathbf{\Theta}(\kappa+1)\left[\mathbf{\Sigma}(\kappa+1) - \mathbf{H}(\kappa+1)\mathbf{\Phi}(\kappa)\hat{\mathbf{z}}(\kappa)\right] . \tag{27}$$

Here, the observation vector $\mathbf{\Sigma}$ is formed by the threshold values of the same (i, j)-th pixel through the different RSS maps of the atlas in the discrete time κ.

The estimate vector $\mathbf{z}$ is formed by the estimation values $\mathbf{\Lambda}$ one step prior in the same current discrete time. Fig. 5 shows the detailed computational structure of the GDL method for the environmental RSS extraction from MRS imagery.

5. Simulation Experiments for Hydrological Signatures

In the reported here simulation results, hydrological RSS electronic maps are extracted from high-resolution MRS images. Three level RSS are selected for this particular simulation processes, moreover, unclassified zones are also considered (2-bit classification) and are described as follows.

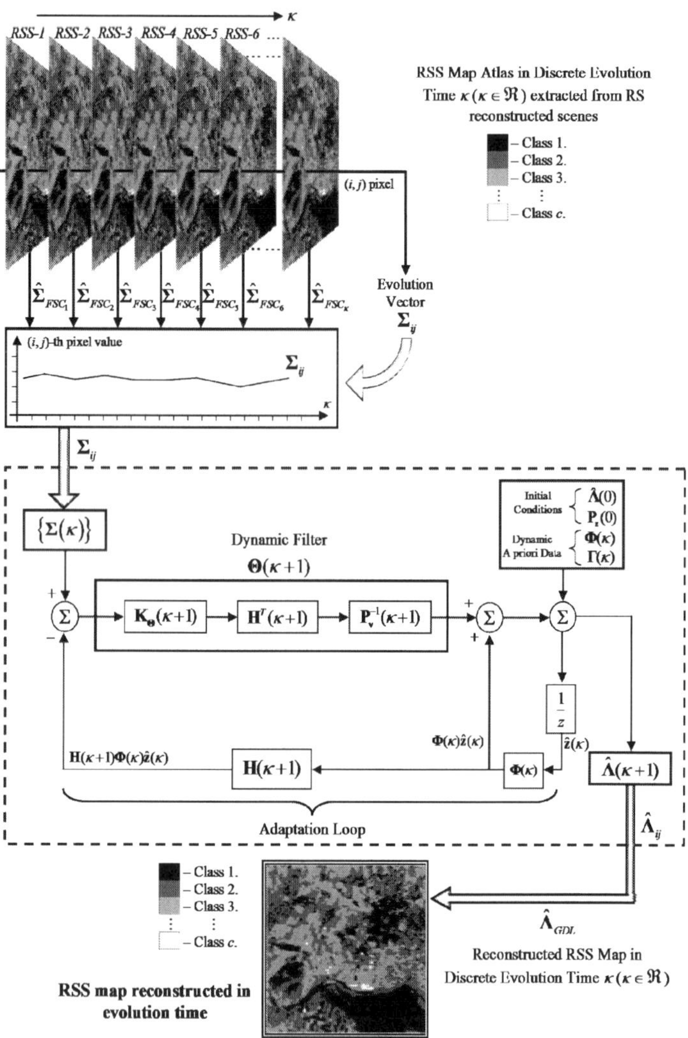

Fig. 5. Computational structure of the GDL method

■ – Black regions represents the RSS relative to the wet zones of the MRS image.

■ – Heavy-gray regions represents the RSS relative to the humid zones of the MRS.

■ – Light-gray regions represents the RSS relative to the dry zones of the MRS.

□ – White regions represent the unclassified zones of the RSS map.

5.1 Multispectral Image Classification

To analyze the overall performance of the WPS technique, a set of four high-resolution (*1024x1024*-pixels) MRS scenes in TIFF format are used, borrowed from diverse zones in Mexico.

A comparison with the results obtained with the classical WOS and MDM methods is provided. Figs. 6(a), 7(a), 8(a) and 9(a) show the four different MRS scenes, respectively.

To perform the qualitative study, Figs. 6(b), 7(b), 8(b) and 9(b) show the results obtained with the WOS method for each scene, respectively.

Figs. 6(c), 7(c), 8(c) and 9(c) show the results obtained with the MDM method for each scene, respectively.

Figs. 6(d), 7(d), 8(d) and 9(d) show the results obtained with the WPS method for each scene, respectively.

The quantitative study is performed calculating the classified percentage obtained with the WOS, MDM and WPS methods, respectively, and compared with the original class quantities from the original MRS scenes. Tables 1, 2, 3 and 4 show the quantitative results.

The theory of the WOS method defines that the classification is performed only using one band (Jensen, 2005), for this simulation the G band was used. The resulting RSS map shows a large unclassified zone, this is due to the color gradient present on the original MRS image and the lack of supervised data.

The MDM method uses the three RGB bands (Jensen, 2005). The WPS method also uses the three RGB bands to analyze the pixel-level means and variances to perform a more accurate segmentation and classification; therefore, using the statistical pixel-based information the RSS map obtained shows a high-accurate classification without unclassified zones. From the details shown in Figures 6 thru 9, the WPS method performs a more accurate and less smoothed identification of the classes.

Tables 1 to 4 show the quantitative performances. From this analysis, the WPS classified image provides a lower percentage difference from the original MRS scenes than the WOS or MDM classified images. Moreover, the WOS and MDM reveal some unclassified zones due to their respective decision rules (Johannsen et. al, 2003); the WPS method classifies all the pixels due to the use of pixel-based statistical training data.

These qualitative and quantitative results probe the overall performance of the developed WPS technique.

a. Original high-resolution MRS scene

b. RSS map extracted with the WOS method

c. RSS map extracted with MDM method

d. RSS map extracted with the WPS method

Fig. 6. Simulation results for hydrological RSS map extraction from the first MRS scene

Method →	Original	WOS method		MDM method		WPS method	
	Base [%]	%	Diff.	%	Diff.	%	Diff.
Wet	34.34	35.37	-1.03	31.93	+2.41	50.09	-15.74
Humid	32.60	26.75	+5.85	17.00	+15.61	18.37	+14.3
Dry	33.06	9.45	+23.61	47.18	-14.12	31.54	+1.51
Unclass.	-----	28.43	+28.43	3.90	+3.90	0.00	+0.00
Percentage Points Difference →		58.92%		36.04%		**31.5%**	

Table 1. Comparative table of the hydrological RSS percentages obtained with the classification methods from the first MRS scene

a. Original high-resolution MRS scene

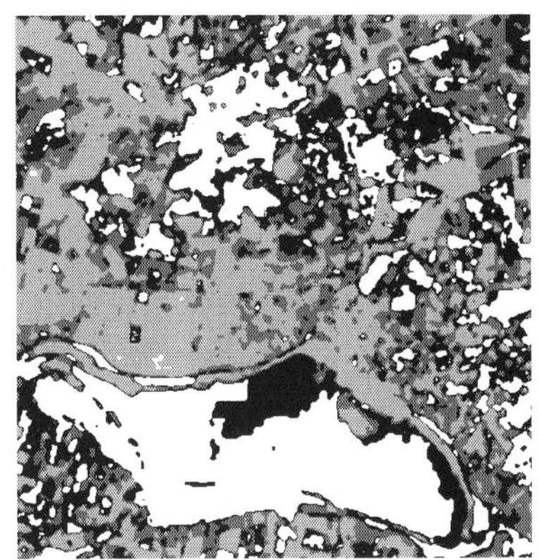
b. RSS map extracted with the WOS method

c. RSS map extracted with MDM method

d. RSS map extracted with the WPS method

Fig. 7. Simulation results for hydrological RSS map extraction from the second MRS scene.

Method →	Original	WOS method		MDM method		WPS method	
	Base [%]	%	Diff.	%	Diff.	%	Diff.
Wet	33.11	23.66	+9.45	29.53	+5.57	51.22	-18.12
Humid	33.11	21.78	+11.32	21.24	+13.87	19.79	+13.3
Dry	33.79	27.03	+6.76	50.06	-16.28	28.99	+4.83
Unclass.	-----	27.53	+27.53	1.17	+1.17	0.00	+0.00
Percentage Points Difference →		55.05%		36.88%		**36.2%**	

Table 2. Comparative table of the hydrological RSS percentages obtained with the classification methods from the second MRS scene.

a. Original high-resolution MRS scene

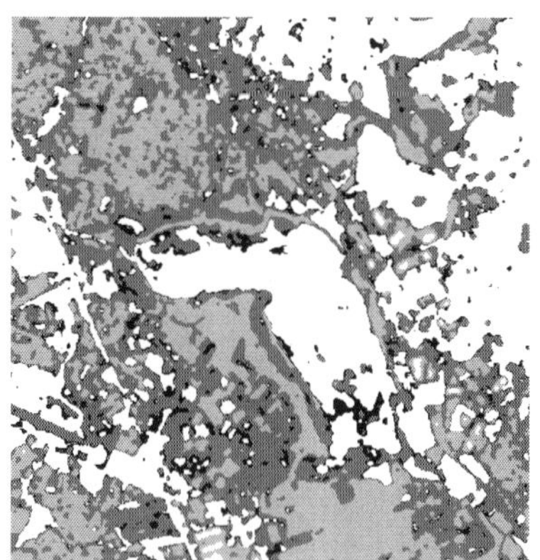

b. RSS map extracted with the WOS method

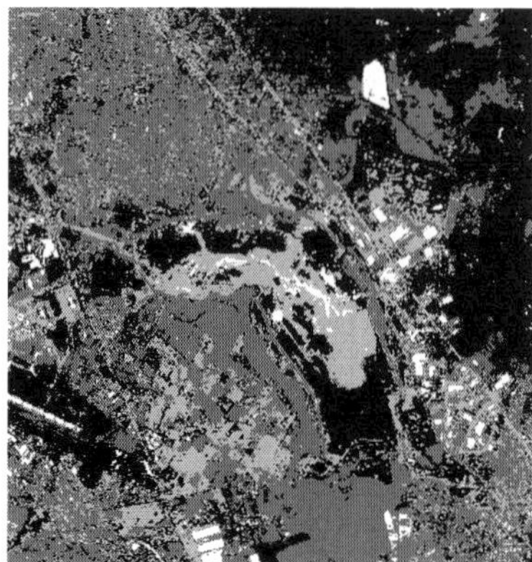

c. RSS map extracted with MDM method

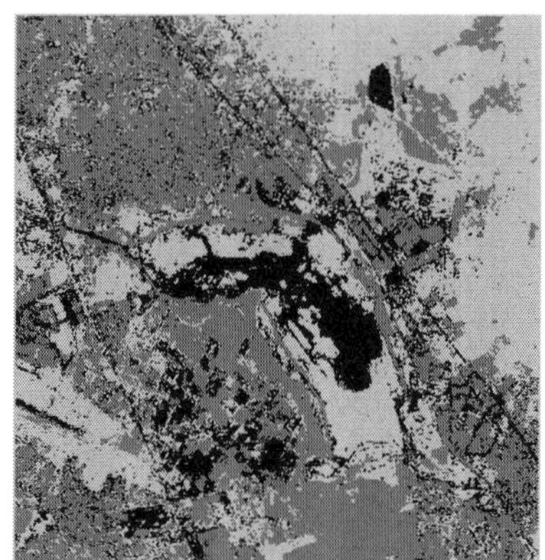

d. RSS map extracted with the WPS method

Fig. 8. Simulation results for hydrological RSS map extraction from the third MRS scene.

Method →	Original	WOS method		MDM method		WPS method	
	Base [%]	%	Diff.	%	Diff.	%	Diff.
Wet	33.02	6.87	+26.15	40.98	-7.96	21.31	+11.7
Humid	33.09	33.30	-0.22	35.38	-2.29	36.38	-3.31
Dry	33.89	23.07	+10.82	20.66	+13.23	42.31	-8.43
Unclass.	-----	36.75	+36.75	2.98	+2.98	0.00	+0.00
Percentage Points Difference →		73.94%		26.45%		**23.4%**	

Table 3. Comparative table of the hydrological RSS percentages obtained with the classification methods from the third MRS scene.

a. Original high-resolution MRS scene

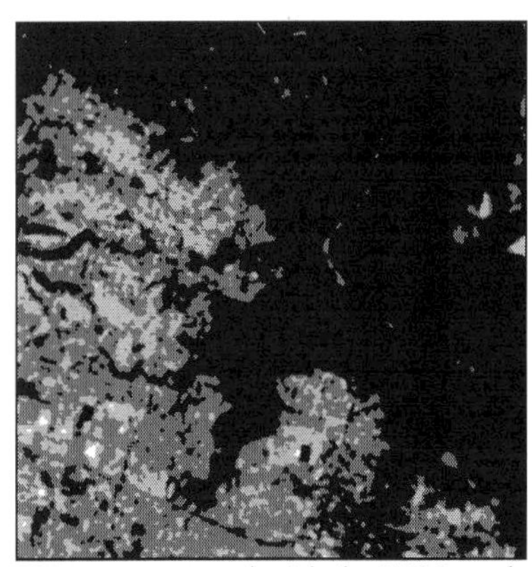

b. RSS map extracted with the WOS method

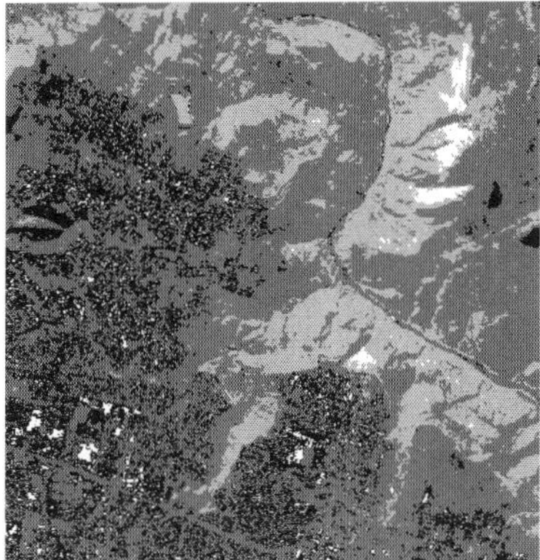

c. RSS map extracted with MDM method

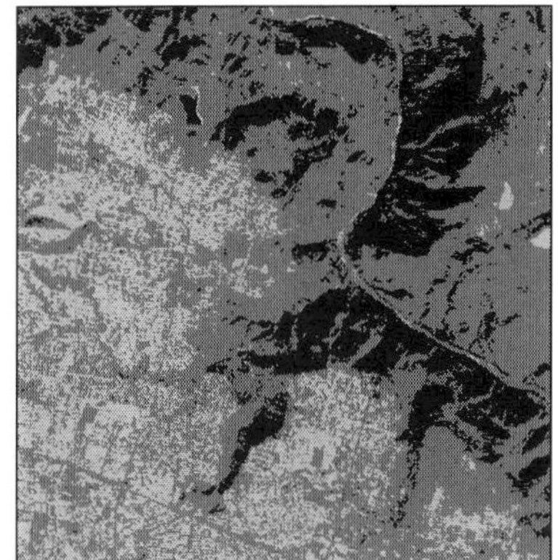

d. RSS map extracted with the WPS method

Fig. 9. Simulation results for hydrological RSS map extraction from the fourth MRS scene.

Method →	Original	WOS method		MDM method		WPS method	
	Base [%]	%	Diff.	%	Diff.	%	Diff.
Wet	32.41	65.35	-32.94	19.06	+13.35	20.62	+11.8
Humid	34.13	25.52	+8.61	60.14	-26.00	58.95	-24.82
Dry	33.46	9.03	+24.42	18.22	+15.23	20.43	+13.1
Unclass.	-----	0.10	+0.10	2.58	+2.58	0.00	+0.00
Percentage Points Difference →		66.07%		57.16%		49.6%	

Table 4. Comparative table of the hydrological RSS percentages obtained with the classification methods from the fourth MRS scene.

5.2 Dynamical Analysis

A set of hydrological RSS electronic maps were extracted from 40 MRS high-resolution images of a particular scene obtained with the same time interval (discrete time).

The GDL dynamic post-processing method is applied to the high-resolution collection of RSS map (Shkvarko & Villalon, 2007) based on the computational structure described in Fig. 5. First, the collection of 40 RSS maps (Marple,1987) collected in different time of the same scene is set for the simulation. Therefore, the discrete time $\kappa = 40$. Second, the pixel evolution vector Σ_{ij} is defined for this simulation as

$$\Sigma_{ij} = \begin{pmatrix} \hat{\Sigma}_{ij,1} & \hat{\Sigma}_{ij,2} & \cdots & \hat{\Sigma}_{ij,32} \end{pmatrix}, \tag{28}$$

where $\hat{\Sigma}$ represents the threshold values of the same (i, j)-th pixel from the 40 RSS maps. This is the observation signal to be post-processed with the dynamic post-processing method.

Third, the measurement matrix $\mathbf{H}$ and the state transition matrix $\mathbf{\Phi}$ are simplified to $\mathbf{I}$ because the equation of observation (9) and the stochastic dynamic state equation (21) are supposed to be ideal (noiseless, because the observation vector is directly extracted from the RSS maps). The dynamic filter operator (gain matrix) Θ determines the variance evolution of the observation values (28) of the dynamically reconstructed RSS. The initial conditions are the initial observation value $\Sigma(0)$ and its initial estimation $\hat{\Lambda}(0) = \Lambda\{\Sigma(0)\}$.

The GDL method specified by equation (25) is applied to estimate the ultimate value $\hat{\Lambda}$ that is the next $(\kappa + 1)$-*st* evolution time step of the observation vector Σ_{ij}. This represents the dynamic filtration of the desired RSS from the reconstructed observation data and can be expressed as

$$\hat{\Lambda}_{ij} = \hat{\Lambda}(\kappa + 1). \tag{29}$$

This process is performed through all the $\{(i, j)\}$ pixels of the 40 RSS maps to obtain a single aggregated RSS map $\hat{\Lambda}_{GDL}$. The simulation results of application of the developed GDL method are presented in Figs. 10 and 11.

Figs. 10(a) thru 10(e) show the first five high-resolution (*1024x1024*-pixel) hydrologic RSS maps extracted from the first five MRS scenes (corresponding to the Banderas Bay of Puerto Vallarta in Mexico) in different evolution time ($\kappa = 1, 2, 3, 4$ and 5), respectively.

Fig. 10(f) shows the dynamic RSS map reconstructed with the application of the GDL method for the $\kappa + 1$ time step ($\kappa = 41$) specified by the computational structure described in Fig. 5.

Fig. 11(a) shows the first original high-resolution (*1024x1024*-pixel) MRS scene ($\kappa = 1$). Fig. 11(b) shows the dynamic MRS map reconstructed with the application of the GDL method for the $\kappa = 41$ time step.

The RSS maps were reconstructed in discrete time κ, therefore, the GDL method produces the desired dynamic RSS prediction of the RSS map for the next time step ($\kappa + 1$); where $\kappa = 0, 1, \ldots$.

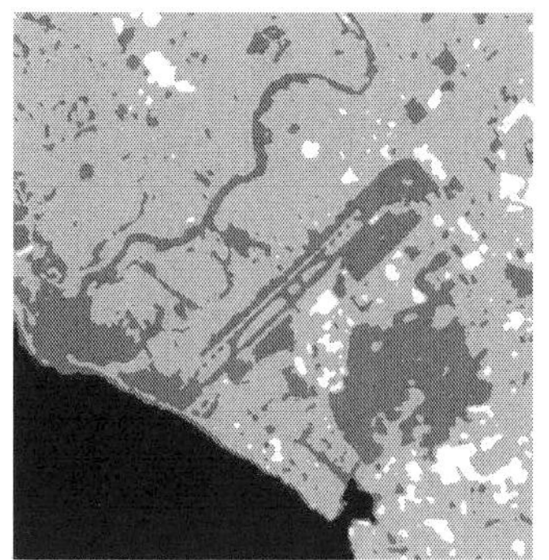

a. RSS map extracted from the MRS scene for a $\kappa = 1$ discrete time step

b. RSS map extracted from the MRS scene for a $\kappa = 2$ discrete time step

c. RSS map extracted from the MRS scene for a $\kappa = 3$ discrete time step

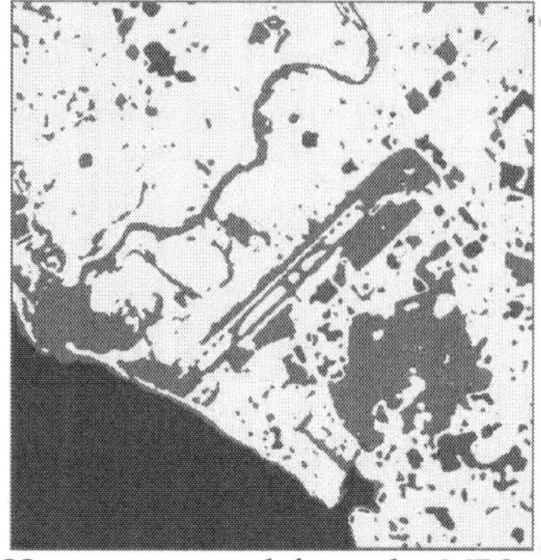

d. RSS map extracted from the MRS scene for a $\kappa = 4$ discrete time step

e. RSS map extracted from the MRS scene for a $\kappa = 5$ discrete time step

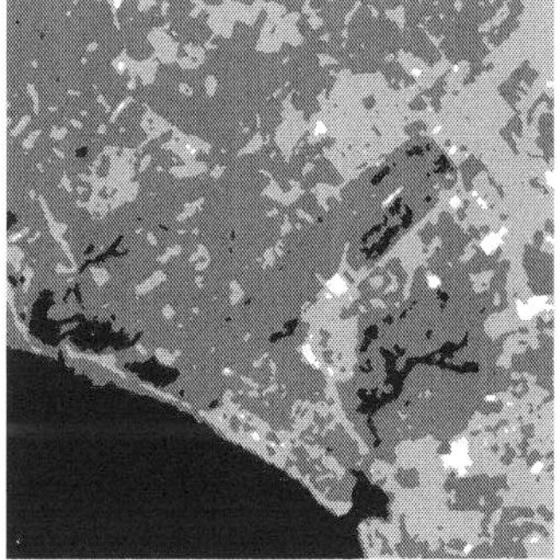

f. RSS dynamic prediction obtained with GDL for the $\kappa = 41$ discrete time step

Fig. 10. Simulation results for dynamic RSS map analysis.

a. Original high-resolution MRS scene for the $\kappa = 1$ discrete time step

b. High-resolution MRS scene predicted for the $\kappa = 41$ discrete time step

Fig. 11. Simulation results for dynamic MRS map analysis.

6. Summary of Computational Algorithms

The detailed stages of the computational algorithm for the WPS and GDL methodologies are summarized as follows.

6.1 Weighted Pixel Statistics method
1. Set the number of RSS to classify.
2. Select one point on the MRS image for each class to be classified.
3. Separate the spectral RGB band from the true-color MRS image.
4. The selected points determine the training weights that consist of the means matrix **M** and the variances matrix **V**. These matrixes contain the mean and variance of each point in the R, G and B bands, respectively.
5. For each (i, j)-th pixel in the R, G and B bands, respectively, perform the following process
   - Set a $5x5$ pixel neighbourhood shift window **W**.
   - Determine the mean of the shift window **W**.
   - Determine the variance of the shift window **W**.
   - Calculate the Euclidean distances between the means and the training means for each band and for each class (Fig. 3).
   - Calculate the Euclidean distances between the variance and the training variances for each band and for each class (Fig. 3).
   - Select the minimum class distance for the means.
   - Select the minimum class distance for the variances.
   - Perform a comparison between the class distance for the mean and the class distance for the variance, and classify the pixel according to the minimum value and the class from which is obtained.

6.2 Geophysical Dynamic Laboratory method

1. Set a collection of discrete time (κ) RSS maps from the electronic atlas extracted from the MRS imagery for a particular scene.
2. For each (i, j)-th pixel on the RSS maps perform the following process:
   - Set the pixel-based evolution vector Σ_{ij}, which contains the threshold RSS values for the pixel in discrete evolution time κ.
   - Apply the **GDL** method to the vector Σ_{ij} (Fig. 5) to obtain the dynamic prediction $\hat{\Lambda}(\kappa+1)$, which conform the matrix $\hat{\Lambda}_{ij}$.

3. The reconstructed $\hat{\Lambda}_{ij}$ matrix conform the $\hat{\Lambda}_{GDL}$ dynamic RSS image in discrete time κ.

7. Conclusion

The extraction of remote sensing signatures from a particular geographical region allows the generation of electronic signature maps, which are the basis to create a high-resolution signatures atlas processed in discrete time, and moreover, perform its dynamical analysis. This chapter analyzed the implementation possibilities of the WPS and GDL methods for hydrological resources management based geophysical applications. The extraction of hydrological RSS from high-resolution MRS imagery was reported to probe the efficiency of the developed techniques.

7.1 Multispectral Image Classification

From the simulation results one may deduce that the WOS classifier generates several unclassified zones; while the MDM classifier is more accurate because it uses more robust information in the processing (several image spectral bands), nevertheless, despite the fact that few zones are unclassified, the results have considerable density of unity pixels (sufficient for decision making based on these extracted RSS). The developed WPS method provides the high resolution environmental RSS electronic map with a high-accurate classification and without unclassified zones. This is achieved because the WPS method uses the three RGB bands to analyze the pixel-level means and variances to perform a more accurate segmentation and classification; therefore, using the statistical pixel-based information the RSS map obtained shows a high-accurate classification without unclassified zones. The resulting RSS map ensures better results in the classification achieved with the developed WPS method. This is probed by the RSS percentages obtained with the WPS method, which manifest the lowest percentage difference to those obtained with the WOS and MDM classification techniques. Also, the WPS method for RSS extraction can be applied to several MRS images of a particular geographical region obtained in different moments of time (discrete time), to generate a RSS atlas of environmental electronic maps. This process is a powerful tool for geophysical resource management.

7.2 Dynamical Analysis

The GDL method provides a possibility to perform the high-resolution intelligent analysis of the dynamic behavior or the desired environmental RSS map model with a high-accurate classification of the particular RSS map evolution.

This is achieved because the GDL algorithm aggregates the RSS map atlas information for a particular MRS scene in discrete evolution time and employs more detailed robust a priori information from the original MRS scene. The resulting dynamic RSS prediction map ensures high-accurate estimation results in the classification achieved with the developed GDL method. The reported here simulation results shows the qualitative and quantitative analysis of the overall performance of the WPS and GDL methods for remote sensing signatures analysis. The application as an auxiliary tool in Geophysical information retrieval and data interpretation for land use management and analysis are a matter of further studies.

8. References

Falkovich S.; Ponomaryov V. & Shkvarko Y. (1989). *Optimal Spatial-Temporal Signal Processing for Spread Radio Channels*, Radio and Communication Press, U.S.S.R.

Fussell J.; Rundquist D. & Harrington J. (1986). On defining remote sensing. *Journal of Photogrammetric Engineering & Remote Sensing*, Vol. 52, 1507-1511

Grewal M. & Andrews A. (2001). *Kalman Filtering: Theory and Practice using Matlab*, John Wiley & Sons, U.S.A.

Jensen J. (2005). *Introductory Digital Image Processing: A Remote Sensing Perspective*, Prentice-Hall, U.S.A.

Johannsen C.; Petersen G.; Carter P. & Morgan M. (2003). Remote sensing: changing natural resource management. *Journal of Soil & Water Conservation*, Vol. 58, 42-45

Luenberger D. (1979). *Introduction to Dynamic Systems: Theory, Models and Applications*, John Wiley & Sons, U.S.A.

Marple S. (1987). *Digital Spectral Analysis*, Prentice-Hall, U.S.A.

Mather P. (2004). *Computer Processing of Remotely-Sensed Images*, John Wiley & Sons, U.S.A.

Perry S.; Wong H. & Guan L. (2002). *Adaptive Image Processing: A Computational Intelligence Perspective*, CRC Press, U.S.A.

Shkvarko Y. (2004). Unifying regularization and bayesian estimation methods for enhanced imaging with remotely sensed data. Part I - theory. IEEE Transactions on Geoscience and Remote Sensing, Vol. 42, 923-931

Shkvarko Y. & Villalon I. (2007). Remote sensing imagery and signature field reconstruction via aggregation of robust regularization with neural computing, In: *Lecture Notes in Computer Sciences*, Blanc-Talon J.; Philips W.; Popescu D. & Scheunders P. (Ed.), pp. 235-246, Springer Verlag, Germany

Smith S. (2000). *The Scientist and Engineer's Guide to Digital Signal Processing*, Web Edition, U.S.A.

Villalon I. & Shkvarko Y. (2008). Comparative study of the descriptive experiment design and robust fused bayesian regularization techniques for high-resolution radar imaging. *Scientific and Technical Journal on Radioelectronics and Informatics*, Vol. 1, 34-48

Villalon I. (2008). Weighted pixel statistics for multispectral image classification of remote sensing signatures: performance study, *Proceedings of the 5th IEEE International Conference on Electrical Engineering, Computing Science and Automatic Control*, pp. 534-539, IEEE Press, Mexico City

Yli-Harja O.; Astola J. & Neuvo Y. (1991). Analysis of the properties of median and weighted order median filters using threshold logic and stack filter representations. *IEEE Transactions on Signal Processing*, Vol. 39, 395-410

26

Forest Inventory using Optical and Radar Remote Sensing

Yolanda Fernández-Ordóñez[1], Jesús Soria-Ruiz[2] and Brigitte Leblon[3]
[1]Postgraduate College in Agricultural Sciences (COLPOS), Mexico
[2]National Institute of Research for Forestry, Agriculture and Livestock (INIFAP), Mexico
[3]Faculty of Forestry and Environmental Management, University of New Brunswick,
Canada

1. Introduction

Forests cover approximately a third of the planet's land area, and range from undisturbed primary forests to forests managed and used for a variety of purposes. Using forests as a resource has an impact on the environment and on the economies of nations. Forests have a history of being exploited either adequately or abusively, but more efforts are now being made towards their sustainable use. The assessment of forests in terms of their extent, condition, use, and value is periodically realized at both global and country levels (Ridder, 2007). Applications relating to the monitoring of forest status and to forest management require updated and accurate inventories summarizing knowledge of land use changes related to forests, including the rates and patterns of deforestation and afforestation. This knowledge must bear more detail for use at regional and local scales. Inventories are produced in most countries at various levels and various update periodicities. The survey methods and technologies applied in each case vary according to the intended use, inherent costs, timeliness, and desired accuracy of the inventory. Sustainable Forest Management (SFM) aims to manage forest land in order to obtain products and services while simultaneously minimising any undesirable effects on the social and physical environments. Major issues concerning SFM include establishing the location, extent, volume, and status of different forest types such as permanent forest estates managed for timber, areas susceptible to reconversion, and protected areas where logging and extracting activities are to be strictly controlled (ITTO, 2006). Forest volume inventories provide valuable data for estimating above-ground biomass density and carbon stores (Zheng et al., 2007). Good estimates of forest biomass are relevant in monitoring deforestation and forest degradation, which impact the global carbon cycle (Haripriya, 2000; DeFries et al., 2006).

Traditionally aerial photographs have been used for forest inventories, but their availability and area coverage can be limited. A promising alternative is the use of images acquired by space borne sensors. Satellite images used in forest inventory were first provided by optical sensors, but their availability is still limited because their acquisition depends on solar radiation. More recently, synthetic aperture radar (SAR) imaging systems were developed. They are active sensors that produce their own incident radiation and thus are able of acquiring images whatever the weather and illumination conditions. The wavelength they used is also longer than the one of the optical sensors, allowing measuring different characteristics of the imaged area.

Thus optical and radar remote sensing (RS) provide different but complementary capabilities for estimates of general forest status such as crown height, density and shape, tree basal area, timber volume and biomass. For forest inventory some variables would be more useful *per se* than others where composition at the species level becomes relevant. Therefore, approaches that combine information obtained from multiple sources offer a promising approach towards establishing more accurate and detailed land cover inventories, particularly at regional and local levels.

This chapter is concerned with forest inventory mapping using combinations of optical and radar RS techniques and intervening field studies at cartographic scales (1:20,000) applicable to regional and local forest management. These themes are examined with respect to providing insights into the use of combined RS techniques for more comprehensive, cost-effective, and accurate forest inventories particularly in Mexico, a mega-diverse country with an increasing concern for the conservation of its biodiversity and sustainable natural resources management (Conservation International, 2008).

In the following two sections, a review of optical and radar RS as they pertain to forest inventory is presented. Then follows a section on the combination and fusion of optical and radar RS data. Ongoing and future work which aim at applying these techniques using medium and high resolution optical imagery and radar imagery in Central Mexico are described next. A paragraph on conclusions and foreseeable developments closes this chapter. The list of references is included.

2. Optical RS

Global and regional forest mapping are increasingly relying on optical RS because of the lower cost of gathering synoptic forest data. Traditional sampling methods can be costly or inapplicable due to terrain topography or field work resources and costs (Shiver and Borders, 1996). Optical sensors use wavelengths from 0.4 to 2.5 µm to measure reflected solar radiation. The range of the visible radiation, from 0.4 to 0.7 µm, also termed photosynthetically active radiation (PAR), is useful for the detection of plant photosynthesis. Visible, near-red, and infra-red wavelengths used for data imaging are of the same order of magnitude as the cell components and prevailing pigments of plants, so a variety of techniques have been established for determining tree canopy biochemistry and for deriving plant condition.

Space-borne optical sensors are passive or active, depending on the solar radiation. Data acquisition is affected by illumination conditions and is subject to atmospheric absorption and scattering, due to the applicable wavelength which is of the same order of magnitude as the size of atmospheric constituents (Gerstl, 1990). Over the years, research has provided knowledge on how various factors, such as tree foliage type and understorey vegetation, impact reflectance data in the visible, near-infrared, and infrared bands (Myneni et al., 1995). The passive satellite-borne sensor applications have focused mainly on mapping biophysical and biogeochemical information from multi-spectral images. More recently airborne optical active sensors called LiDAR that send their own incident energy have been developed. They are increasingly being used for forest applications mainly for estimating tree height, crown dimensions, canopy height of stands and canopy cover. LiDAR is not further discussed here because there is not yet an operational space-borne LiDAR sensor, and the reader is referred to recent review works which show the potential advantages of this type of sensor (Sun & Ranson, 2000); (Packalén and Mantamo, 2008).

Passive optical RS applications for forestry inventories are based on the sensitivity of optical radiation to the chlorophyllian content of the stands as detailed in Leblon (1997). However several factors can interfere such as the optical properties of the soil background, illumination and viewing geometries as well as from meteorological factors (wind, cloud). To reduce the effect of these factors on the forest canopy spectral response, single-band reflectances measured by the sensors are generally combined into vegetation indices. At least fifty different vegetation indices exist (Bannari et al., 1995). The most commonly used vegetation indices are ratios of single-band or linear-combined reflectances. Ratioing allows removal of the disturbances affecting, in the same way, reflectance in each band. As a spectral band, most ratio-based vegetation indices use either the red band, which is related to the chlorophyll light absorption or the near-infrared band, which is related to the green vegetation density. Indeed, both bands contain more than 90% of the spectral information of a plant canopy. Also, in red and near-infrared bands, the contrast between vegetation and bare soil is at a maximum. Other indices based on the short-wave infrared have been developed that are sensitive more to the moisture content of the canopy than to its chlorophyll content. It has been shown that these indices better estimated the density of managed conifer forests than those sensitive to chlorophyll (Aguirre-Salado et al., 2009).

When using optical RS images in forest inventory mapping, one should pay attention to characteristics of the phenological cycle which are directly related to both vegetation type and species diversity and thus indirectly to small-scale heterogeneity of climatic and topographic conditions in the corresponding study region. Therefore a key component for forest attribute determination is the time-series analysis of repetitive multispectral and even hyperspectral imagery (Vuolo et al., 2008; Schowengerdt, 1997).

In addition to this time dimension, one also needs to consider the spatial dimension. High spatial resolution images such as Ikonos and Quickbird provide better estimates for tree crown cover over medium resolution images from SPOT and Landsat sensors in the case of pine forests. A small scale experiment has shown that there is confusion with tree shadows when using medium resolution optical imagery, which is a drawback for forested areas in ravines or over steep slopes (Valdez-Lazalde et al., 2009). Approaches to overcome this and other drawbacks include multi-angle optical RS whereby separating sunlit and shaded leaves and using canopy architectural parameters is possible (Chen et al., 2003).

Species distribution modelling which is commonly based on ecological models has been shown to improve when incorporating spatio-temporal information derived from optical RS (Pearson, 2007). Recent work in this area (Cord et al., 2009) has shown that analytical techniques that exploit data beyond their spectral dimension to gather knowledge about phenological cycles, seasonal, and latitudinal variations over time can contribute to more accurate vegetation type and species diversity determination. This is important for inventories over natural forested areas at regional and local levels. Optical imagery covering the targeted inventory zones should be available over time. Cloud cover can impair this requirement, while radar sensors provide data even under this condition in addition to being potentially able to provide complementary information about forest structure and vegetation water content because of their different physics nature.

3. SAR RS

Microwave RS instruments are synthetic aperture radar (SAR) systems that image an area independently of solar illumination and are not affected by cloudy or dusty conditions as are optical instruments (Freeman, 1996). They operate in longer wavelengths in the optical regions of the electromagnetic spectrum giving different information on the tree canopies. The SAR sensors image the area by transmitting electromagnetic pulses in frequency bands between 1 and 100 GHz. The corresponding wavelength (λ) ranges between 1 mm and 1 m. It is of the same order of magnitude as the sizes of tree leaves, branches and canopies. By comparison, the wavelength of optical instruments ranges from 0.4 to 2.5 µm, and thus is of the same order of magnitude as the cell components and prevailing pigments of plants (Leblon & LaRocque 2008). Thereby, optical imagery is more suitable for determining tree canopy biochemistry and inferring vegetation condition, while radar images are better for determining canopy structural parameters

SAR sensors were first developed for military applications. The current satellite civil SAR sensors use one of the following bands: X-band: 12.5-8 GHz (λ: 2.4 to 4.8 cm) (TerraSAR-X); C-band: 8.0-5.0 GHz (λ: 4.8-7.5) (ERS1/2, ENVISAT, RADARSAT-1/2); and L-band: 2.0-1.0 GHz (λ: 15.0-30.0) (JERS and ALOS-PALSAR). A satellite SAR sensor called BIOMASS operating in the P-band (1.0-0.3 GHz or λ: 30.0-100.0) is in preparation by the European Space Agency. It is designed for forestry applications. As detailed in Leblon and LaRocque (2008), microwave radiations which encounter natural surfaces are either:

(i)   diffusely scattered towards the sensor to produce the backward scattering called radar backscatter or radar return (σ^0 *sigma nought*) which is the only scattering measured by the sensor;

(ii) diffusely transmitted to lower layers through forward scattering;

(iii) specularly reflected in the case of smooth surfaces.

The interactions between microwave radiations and natural surfaces depend on: (i) geometrical factors like the surface roughness as well as the incidence angle and the wavelength of the incident radiation and (ii) dielectric factors which are related to the nature and the moisture content of the target.

Because of the longer wavelength than the one for optical sensors, microwave radiation produces not only surface scattering, but also volume scattering. The intensities of both scatterings depend on the surface roughness, the incidence angle and the wavelength of the incident radiation. Volume scattering also depends on the penetration depth -a function of the wavelength- and occurs when the wave encounters dielectric discontinuities within the material of the target (Leblon & La Rocque 2008).

Radar backscatter is rather complex on tree canopies. Ulaby et al., (1990) identified eleven components of the radar backscatter of tree canopies. Fig. 1 shows nine major backscattering sources for tree canopies (Piwovar, 1997): 1- Diffuse scattering from the ground; 2- and 3- Direct scattering from various vegetation components; 4- Double- bounce vegetation-ground interaction; 5- Corner reflector between tree trunks and ground; 6- Direct backscatter from the forest canopy; 7- Volume scattering from within the forest canopy; 8- Diffuse scattering from the ground; 9- Shadowing by parts of the forest canopy of other parts of the canopy or ground.

The importance of each of the above components on the total scattering of a tree canopy depends mostly on the microwave wavelength (Leblon & LaRocque, 2008). Radar backscatter in shorter wavelengths, like X- and C-band, is mainly due to smaller tree elements (leaves or needles and small branches of the upper tree crown level) (Hoekman 1993); (Mougin et al., 1993); (Yatabe & Leckie, 1995); (Ahern et al., 1995). Therefore, shortwave bands (X- or C-band) can be used to discriminate between forest species (Drieman et al., 1989); (Hoekman, 1990); (Dobson et al., 1992a); (Mougin et al., 1993); (Rignot et al., 1994a); (Yatabe & Leckie, 1995); (Ahern et al., 1993, 1995). In particular, the C-band backscatter magnitude of red pine, jack pine and black spruce was found to be inversely related to needle length (Ahern et al., 1993); (Yatabe & Leckie, 1995). Radar backscatter of longer wavelengths (P- and L-bands), which are more penetrating, originates from the major branches of the crown, from the trunks, the double bounce scattering from the tree trunk or crown to ground and from the ground (Richards et al., 1987); (Le Toan et al., 1992); (Hoekman 1993); (Ahern et al,. 1995).

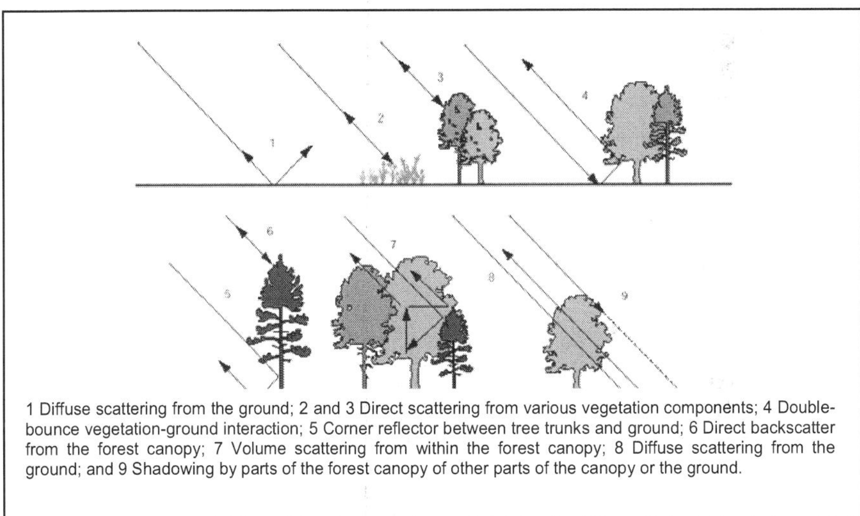

1 Diffuse scattering from the ground; 2 and 3 Direct scattering from various vegetation components; 4 Double-bounce vegetation-ground interaction; 5 Corner reflector between tree trunks and ground; 6 Direct backscatter from the forest canopy; 7 Volume scattering from within the forest canopy; 8 Diffuse scattering from the ground; and 9 Shadowing by parts of the forest canopy of other parts of the canopy or the ground.

Fig. 1. Surface and volume scattering of a SAR beam for trees (*Piwowar, 1997*).

According to Leblon et al. (2002), radar backscatter (σ^0) measurements of forested areas depend on (i) vegetation type, species, and structure (Leckie, 1990); (Dobson et al., 1992a; 1992b; 1995), (ii) vegetation biomass (Kasischke et al., 1995); (Harrell et al., 1995; 1997); (Pulliainen et al., 1996); (iii) topography and surface roughness (Shi et al., 1997) and canopy height (Riom & Le Toan, 1981); (Dobson et al., 1992b); (Harrell et al., 1997); (iv) flooding and the presence/absence of standing water (Hess et al., 1990; (Kasischke & Bourgeau-Chavez, 1997), and (iv) near-surface soil moisture or fuel moisture (French et al., 1996); (Leblon et al., 2002); (Abbott et al., 2007).

Discrimination between forest and non forest areas as well as estimation of total above-ground forest biomass or stem volume is better achieved using long wave bands (P-band) (Le Toan et al., 1992); (Dobson et al., 1992b); (Koch et al., 1992); (Rignot et al., 1994b); (Rauste et al., 1994); (Ranson & Sun, 1994); but the relationship has a logarithmic shape showing a saturation of radar backscatter beyond a certain value of biomass.

SAR sensors acquire images in one single band. By contrast, optical sensors like Landsat-TM acquire images in a multispectral mode, i.e., several images at different wavelengths at the same time. However, radar sensors can acquire images under various incidence angles which might highlight differences. The most recent radar sensors like ENVISAT-1, ALOS-PALSAR TerraSAR-X, and RADARSAT-2 also allow the acquisition of images under different polarizations.

This will also allow the production of colour composites like for multispectral optical images. The polarization of the transmitted and returned radiation is an important property, which can be used as a target separator in SAR image analyses. Polarization refers to the position of the locus of the electric field vector in the plane perpendicular to the direction of propagation of the radar microwave radiation. Current SAR systems transmit horizontally polarized (H) or vertically polarized (V) EM waves, which upon interacting with a target generate a backscattered wave with a horizontal polarization (H) or a vertical polarization (V).

The possible combinations are (transmit polarization given first): HH, VV, HV and VH. With multi-polarization sensors, the following imaging capabilities are thus possible: single polarization (HH or VV), dual polarization (HH and HV, VV and VH, or HH and VV); and quad-polarization (HH, VV, HV, and VH). Some multi-polarization sensors (RADARSAT-2, ALOS-PALSAR and TerraSAR-X) have also a polarimetric mode, that provides not only the intensity information in each polarization (HH, VV, HV, and VH), but also the phase information. As a result, the complexity of SAR images processing increases with the level of complexity of the polarization information as it is shown in Fig. 2 in the case of the RADARSAT-2 sensor.

The information provided by multi-polarized or polarimetric SAR images has been investigated, either directly with airborne and satellite radar data, or through simulation models, with respect to their ability to discriminate land cover or forest types (Freeman & Durden, 1998); (Touzi et al., 2004; (Freeman, 2007), to estimate forest growth (e.g., Izzawati et al., 2006), biomass (e.g., Le Toan et al., 1992) or forest architecture (Lucas et al., 2006). Although the anticipated potential of the polarimetric SAR images like those acquired by RADARSAT-2 for mapping forest inventory, clear cuts, fire scars and biomass has been rated as moderate, merging polarimetric SAR data with other data sources is a promising investigation approach (van der sanden et. al., 2005). This is discussed in the following sections of this chapter.

4. Forest Inventory Research in Mexico Using Optical and SAR Data

National forest inventories are labour-intensive and costly endeavours in need of advanced technologies to reduce costs and time schedules and to produce information at the levels required by all involved actors. The National Forest Inventory (NFI) in Mexico pertains to the federal government and its general terms are regulated by law. The cartographic scale used in the last inventory, NFI 2000, is 1:250000.

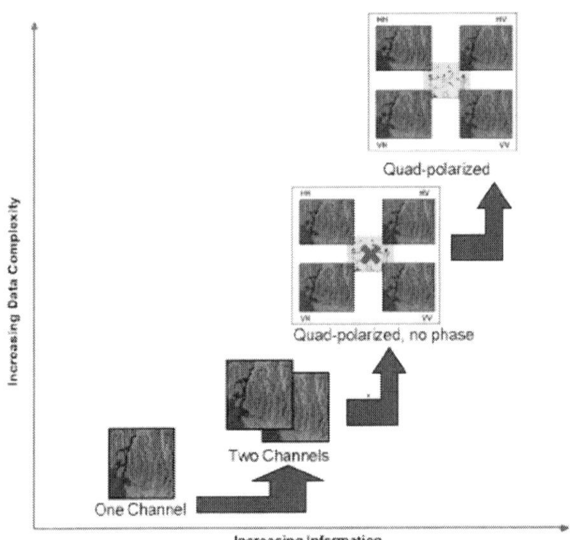

Fig. 2. Information sequence contained in a Radarsat-2 image (*courtesy of G. Staples, MDA*)

There is, however, a need for establishing more specific forest inventories at the state level, and is the responsibility of state and municipal governments. For the State of Mexico the most recent forest inventory was completed in 1994. The state is located in central Mexico and it contains 75% of the country's population, and although the physical environment has been modified in recent years there is no up-to-date, reliable, official assessment of forest resources. Information about land cover change has been obtained recently but it is not oriented toward forest inventory (Escalona-Maurice, 2006). Many regions have been transformed by deforestation resulting from various activities including illegal logging, the opening of new agricultural areas, and the development of upper-class residential areas within or near forested regions.

Since 2004, the responsibility for producing an NFI has lain with two government agencies, the National Forestry Commission (CONAFOR) and the Ministry of Environment and National Resources (SEMARNAT). The 2009 inventory will provide valuable forest and soils information but with insufficient detail to adequately support regional or local protection, conservation, and afforestation activities. To our knowledge, there are currently no operational applications of radar RS within Mexican government agencies oriented towards forest inventory or forest management. We consider that SAR and optical RS technologies used together will contribute to satisfying the time requirements, accuracy, and finer levels of detail that are required for forest inventory.

4.1 Forest Inventory

In Mexico, as in most countries, forest inventories at different levels of scale are a current concern. There is a worldwide public awareness of the need for sustainable development. However, in developing countries this need is not so easy to meet when faced with other immediate priorities and with a scarcity of economic resources. Even so, periodic national information requested by FAO for global forest resource assessments must be provided. At a global level the requirement is for a picture of existing forests, statistics, and derived trends. At the national level the request includes both the country as a whole and the major administrative subdivisions of the territory. Information items include land use and the changes in forested areas, estimates of existence and growth, and diagnosis of the health status of forests. Information about the conservation of natural ecosystems and of biodiversity is also sought (Toledo & Ordoñez, 1993).

Mexico, with an area of nearly two million km², is one of three mega-diverse countries (along with the United States and Colombia) with coastlines on the Atlantic and the Pacific. However a growing population and ensuing changes in land use pose many threats to the permanence of this biodiversity, particularly in forested regions. In spite of inadequate funding and technical difficulties due to the extent and varied topography of the country, there has been considerable progress in the construction of the NFI. The last NFI dates back to the year 2000 (Palacio et al., 2000). The cartography was based on a classification scheme based on visual interpretation of Landsat ETM+ colour composite images using the national digital base map. The accuracy assessment was supported with high resolution digital aerial photography. The analysis of deforestation was based on a classification which included four forest classes: (i) primary and secondary tropical and temperate; (ii) scrubland; (iii) man-made covers (grazing land, agricultural cropland, and human settlements); and (iv) other covers (Mas et al., 2002). The NFI 2000 provides the current frame of reference for lower-scale inventories which will be used for the 2009 inventory.

The 2009 inventory, which is 80% complete in the stratified sample stage and as yet unpublished, will use optical images (AVHRR, MODIS, LANDSAT and SPOT (Sandoval Uribe, 2007). The national mapping scale is being maintained at 1:250000 but a new scale of 1:50000 for reporting at the state level is being adopted. Information will also be reported at the ecological region level, using the official ecological zonification scheme. The results are to be included in a web information system (CONAFOR, 2009). The study area for the investigation reported in this chapter is located approximately within the bounding rectangle of Region 13 Temperate Mountain Areas, which includes the ecosystems of (i) Forest and (ii) High and Medium Jungle, with reported national volume/ha values for these ecosystems of 41.86 and 39.81, respectively (Fig. 3). The preceding discussion motivates and justifies the research project reported in section 4.2.

4.2 Research Project

The general research objective of our current research project is to produce a methodology to improve forest inventory assessment and results through the combined use of different sources of data. The methodology should be amenable to translation for use at regional and state levels elsewhere in the country. Particular objectives include the following: (i) to assess the benefits of using polarimetric SAR data in this region, which includes forested areas in the mountain chain "Eje Neovolcánico Transversal"; and (ii) to evaluate techniques for the fusion of optical and SAR data. Part of this project is included in the RADARSAT SOAR #2755 project. The main objective of the Canadian Space Agency SOAR program is to assess and explore operationally viable solutions to current problems and issues, including those related to the environment. Initiatives include fostering research and development of new applications using RADARSAT-2 data, particularly those of fully polarimetric modes (MDA et al., 2009).

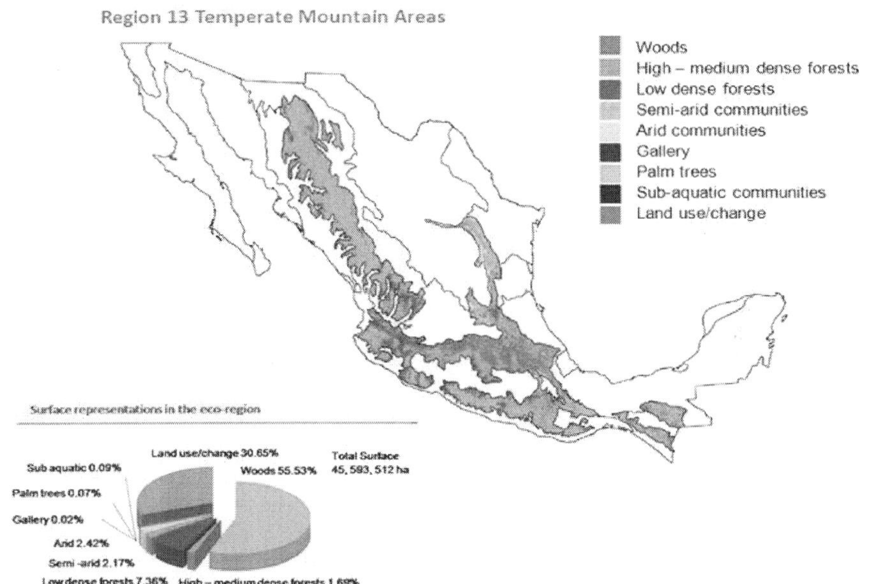

Fig. 3. National Forest Inventory Region 13 Temperate Mountain. (*CONAFOR, Mexico*)

Although the concept of data fusion is intuitively understood, its exact meaning varies among researchers and among scientific and technological domains. Therefore the term should be formally defined and its various aspects discussed. Fusion has been taken as a synonym of merge, combine, integrate, classify, and synergize. The range of application objectives, methods, and techniques in RS is very diverse, so it is unlikely that a single axis (whether wavelength, acquisition means, image, spatial resolution, mathematical tools, or application domains) can be favoured upon which a fusion definition can be based.

Increased quality of the information is often the goal when using data and information from different sources, but quality in RS depends on the particular objectives to be attained and on whom the end-consumer of the results is. Quality can refer to the enhancement of previous results, of method efficiency, or of results accuracy, all of which are clearly tied to the application and its intended users (Fasbender et al., 2009). Hall and Llinas (1997) (cited by Wald, 1999) consider information quality in terms of "data fusion techniques (that...) combine data from multiple sensors, and related information from associated databases, to achieve improved accuracy and more specific inferences that could be achieved by the use of a single sensor alone."

The notion of the pixel has frequently been used to address data fusion "at the pixel level", but it is often required to consider the merging of information at levels other than the pixel. The information content of a pixel is not comparable in optical and SAR images. Wald (1999) proposed a definition which suits the objectives of this chapter and of our research, since he contends that fusion is a framework, not to be confined to the sensors' signals, nor restricted to the peculiarities of the sensor systems: "data fusion is a formal framework for the alliance of data originating from different sources. It aims at obtaining information of greater quality; the exact definition of 'greater quality' will depend upon the application." The processing tools and methods thus comprise no part of this definition of fusion as a framework, even though they are relevant to a particular implementation or methodology of a data fusion effort. The terms "merge" and "combine" are intentionally considered to be looser concepts, allowing for the description of processes and methods in a general way without getting into details. The reader is referred to the ample literature for the ongoing discussion on the definition of the fusion concept and its many and fruitful uses (Pohl & van Genderen, 1998); (Nikolakopoulos, 2004); (Wang et al., 2005).

Our current work is part of a broader on-going action to map land cover and to assess land cover changes in the State of Mexico (Soria-Ruiz et al., 2009). Forest inventory is an emerging research line within this broader action. Previous related work combined Landsat-ETM and SPOT producing a map atlas (Soria-Ruiz et al., 2005). The selected study area for this stage of the project is the Zinacantepec Municipality in the State of Mexico. This area is well known from extensive field surveys containing over 60 control points, making it suitable for analyzing natural and deciduous stands (mixed and homogeneous species) and other types of vegetation considered by the national forest inventory. The terrain includes flatland and part of the "*Nevado de Toluca*" volcano.

Previous tests using Radarsat-1 data and Landsat-ETM in Central Mexico for land cover in Central Mexico indicated a global map accuracy with optical imagery alone of 78%, whilst fusing optical and SAR resulted in an accuracy of 89%, with a kappa coefficient improvement from 0.72 to 0.86%. So the fusion of Landsat ETM+ and RADARSAT-1 encouraged more tests of fusion for forest classification and for other land covers as well (Soria-Ruiz & Fernandez-Ordonez, 2004).

Partial classification results, for four classes, using Landsat ETM images from 2000-2003 are shown in Fig. 4.

The main goals set out for the period 2009-10 using other optical and SAR data are:

(i)   Continued update of the land cover map base with high form SPOT, Ikonos, and Quickbird;

(ii) Integration of multi-polarization Radarsat-2 C-band images over the summer and winter seasons to investigate the applicability in this area of HH, VV, and HV polarizations. At least six scenes will be used, including polarimetric, fine quad-pol HH and VV, and HV and VH;

(iii) Derivation of polarization signatures and relating them to forest parameters for the genera *Abies, Pinus,* and *Quercus,* which are largely represented in the area and are of importance for forestry;

(iv) Accuracy assessment of the classifications; and

(v) Assessment of other polarimetric techniques (like interferometry) for different types of stands, including commercial/natural, homogeneous/mixed, and older/younger.

Recent work by Viergever et al., (2007) has investigated biomass estimation for tropical forests in Central America using backscatter and fully polarimetric data from AIRSAR. C-, L-, and P-bands were used; vegetation heights were underestimated due to the heterogeneity of the vegetation cover. In another case, confusion was observed in backscatter values for P and L bands for the same type of vegetation (Freeman and Durden

1998, Freeman 2007). SAR and polarimetric SAR allow using interferometry techniques that are suitable for tree height and density (Papathanassiou & Cloude 2001). For local forest inventory a high accuracy of classification (on the order of 80 % or better) is desirable. Our work will explore these technical possibilities, and assess their cost-effectiveness, although the long repeat cycle for the satellite overpass leads to de-correlation that restricts the use of such techniques. Since different wavelengths have different penetration properties in forest targets data from other radar sensors such as TerraSAR-X and ALOS-PALSAR (L-band) will also be used.

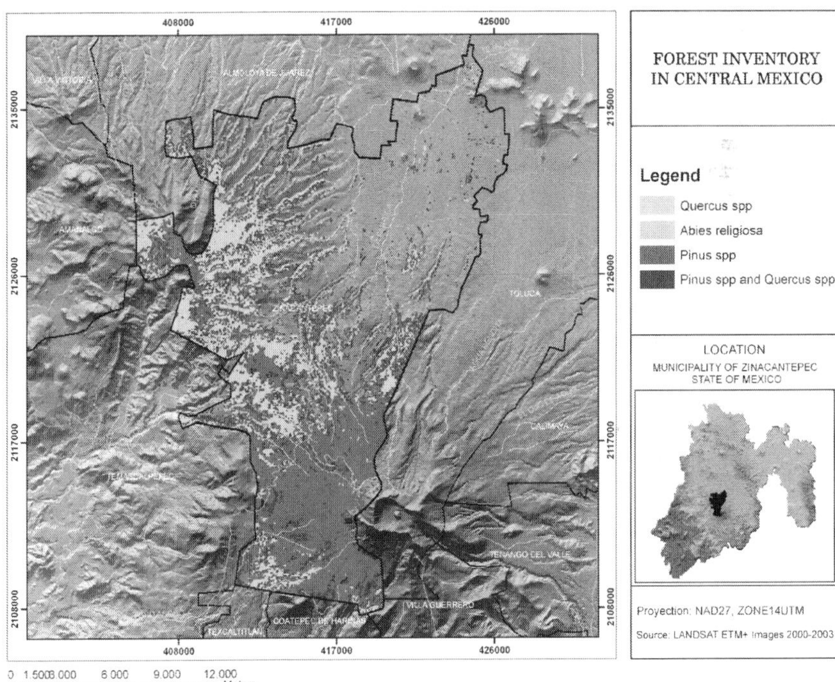

Fig. 4. Forest inventory in the Zinacantepec Municipality, State of Mexico, 2008.

5. Conclusion

State and local governments in Mexico have expressed a need for assessment of forest resources at state and municipal level. Bio-conservation and ecological research have data requirements at even smaller cartographic scales, less than 1:10000. Currently, techniques for gathering information at this level are performed on a per-project basis under various limitations and at great cost especially in mountain regions.

The avenue of investigation undertaken in our work is how best to combine optical and SAR RS techniques profiting from the different but complementary information they provide as it was shown in the current chapter. Our long term objective is to determine whether polarimetric radar RS in complementary use with optical RS offer conveniences for forest inventory and ecosystem studies in Central Mexico. The study will be based on SPOT, Quickbird and Ikonos images and RADARSAT-2 C-band polarimetric SAR images. These images offer new capabilities in forest inventories, because they are acquired in four polarizations and have the phase information. SAR RF for forests is a challenging research field with many unsolved issues pertaining to forest inventories in mega-diverse areas. Our long term objective is to determine what contribution SAR RS is able to make with respect to forest inventory and ecosystem studies in Central Mexico. However, it has been shown that an even better SAR sensor for forestry applications should operate in the P-band, like the BIOMASS sensor that is currently designed by the European Space Agency. Further projects will be undertaken once the images from this new satellite will become available. Also, interferometry and polarimetric interferometry have been shown to be well suitable for tree height and canopy density measurements from SAR images. Such techniques will be more efficient once images from the new German satellite system called TANDEM-X will be available. The system allows acquiring two images on the same area in almost the same time allowing computing interferograms without temporal decorrelations

6. References

Abbott, K.N.; Leblon, B; Staples, G.C.; Maclean, D.A. & Alexander, M.E. (2007). Fire danger monitoring using RADARSAT-1 over northern boreal forests. *International Journal of Remote Sensing*. Vol. 28 , Issue 6 (March 2007). 1317-1338. ISSN 0143-1161.

Ahern, F.J.; D.G.; Leckie & J.A. Drieman, J.A. (1993). Seasonal changes in relative C-band backscatter of Northern forest cover types, *IEEE Transactions on Geoscience and Remote Sensing*, Vol. 31, Issue. 3, (May 1993). 668-680, ISSN 0196-2892.

Ahern, F.J.; Landry, R.; Paterson, J.S.; Boucher, D. & McKirdy, I. (1995). Forest land cover information content of multi-frequency multi-polarized SAR data of a boreal forest, *Proc. 17th Canadian Symposium on Remote Sensing*, June 1995, Saskatoon, Saskatchewan, Canada, 537-549. ISSN 1712-7971.

Aguirre-Salado, C.A.; Valdez-Lazalde, J.R.; Angeles-Perez, G.; De Los Santos-Posadas, H.M.; Haapanen, R. & Aguirre-Salado, A.I. (2009). Mapping aboveground tree carbon in managed *patula* pine forests in Hidalgo, Mexico. *Agrociencia*, Vol 43, (February-March 2009), 209-220, ISSN 1405-3195.

Bannari, A.; Morin, D.R.; Huete A.R. & Bonn, F. (1995). A review of vegetation indices, *Remote sensing reviews*, No. 13, 95-120. ISSN 0275-7257.

Chen, J.M.; Liu, J.; Sylvain G.; Leblanc, S.G.; Lacaze, R. & Roujean, J.L. (2003). Multi-angular optical remote sensing for assessing vegetation structure and carbon absorption. *Remote Sensing of Environment*, Vol. 84, Issue 4 (April 2003) 516–525, ISSN 0034-4257.

CONAFOR, Mexican National Forestry Commission. (2009). National Forest Inventory System SNIF. Accesed 19.02.09
http://148.223.105.188:2222/gif/snif_portal/index.php?option=com_content&task=view&id=4&Itemid=59

Conservation International (Content Partner). Mark McGinley (Topic Editor). (2008). *Biological diversity in the Madrean pine-oak woodlands.0020* In: Encyclopedia of Earth. Eds. Cutler J. Cleveland (Washington, D.C.: Environmental Information Coalition,

National Council for Science and the Environment). First published in the Encyclopedia of Earth May 3, 2007; Last revised August 22, 2008; Accessed 02.01.09 http://www.eoearth.org/article/Biological_diversity_in_the_Madrean_pine-oak_woodlands

Cord, A.; Schmidt, M. & Dech, S. (2009). Potential and limitations of multi-temporal Earth observation data to improve modeled results of tree species distribution in Mexico. *Proceedings of* 33rd *International Symposium on Remote Sensing of the Environment*, May 4-8, 2009, Stresa, Lago Maggiore, Italy.

DeFries, R.; Achard, F.; Brown, S.; Herold, M.; Murdiyarso, D.; Schlamadinger, B. & De Souza, C. Jr. (2006). *Reducing Greenhouse Gas Emissions from Deforestation in Developing Countries: Considerations for Monitoring and Measuring.* Global Observation of Forest and Land Cover Dynamics. Report Series GOFC-GOLD-34, (August 2006). Vol. 29, No. 5 (October 2003) 633–649, ISSN 1712-7971.

Dobson, M.C.; Pierce, L.; Sarabandi, K.; Ulaby, F.T. & Sharik, T. (1992a). Preliminary analysis of ERS-1 SAR for forest ecosystem studies, *IEEE Transactions on Geoscience and Remote Sensing*, Vol. 30 Issue 2, (March 1992) 203-210. ISN 0196-2892.

Dobson, M.C.; Ulaby, F.T.; Le Toan, T.; Beaudoin, A. & Kasischke, E.S. (1992b). Dependence of radar backscatter on coniferous forest biomass, *IEEE Transactions on Geoscience and Remote Sensing*, Vol. 30, Issue 2 (March 1992) 412-416. ISSN 0196-2892.

Dobson, M. C.; F. T. Ulaby, & L. E. Pierce, (1995). Land-Cover Classification and Estimation of Terrain Attributes Using Synthetic Aperture Radar. *Remote Sensing of Environment,* Vol. 51, No. 1, (January 1995) 199-214, ISSN 0034-4257.

Drieman, J.A.; Leckie, D.G. & Ahern, F.J. (1989). Multitemporal C-sar for forest typing in Eastern Ontario, *Proceedings IEEE International Geoscience and Remote Sensing Symposium IGARSS'89 & 12th Canadian Symposium on Remote Sensing*, Vol. 3, 1376 – 1378, Vancouver, British Columbia, Canada, July 1989.

Drieman, J.A.; Ahern, F.J. & Corns, I.G.W. (1989). Visual interpretation of multipolarization C-sar imagery of Alberta Boreal Forest, *Proceedings IEEE International Geoscience and Remote Sensing Symposium IGARSS'89 & 12th Canadian Symposium on Remote Sensing*, Vol. 3, 1401-1405, Vancouver, British Columbia, Canada, July 1989.

Escalona-Maurice, M.J. (2006). Landscape of the Texcoco Municipality: morphological and functional analysis in a cartographic model framework. State of Mexico. Mexican United States. (In Spanish: El Paisaje del Municipio de Texcoco: análisis morfológico y funcional en el marco de un modelo cartográfico. Estado de México. Estados Unidos Mexicanos). *Doctoral thesis.* Departamento de Geografía, Universidad de Alcalá, España. 529 p.

Freeman, A. (2007) Fitting a two-component scattering model to polarimetric SAR data from forests. *IEEE Transactions on Geoscience and Remote Sensing,* Vol. 45, Issue 8, (August) 2583 – 2592. ISSN 0196-2892.

Harrell, P.A.; Bourgeau-Chavez, L.L.; Kasischke, E.S.; French, N. H. F. & Christensen, N.L. (1995). Sensitivity of ERS-1 and JERS-1 radar data to biomass and stand structure in Alaskan boreal forest, *Remote Sensing of Environment,* Vol. 54, Issue 3 (December 1995) 247-260, ISSN 0034-4257.

Harrell, P.A.; Kasischke, E.S.; Bourgeau-Chavez, L.L.; Haney, E. & Christensen, N.L. (1997). Comparison of approaches to estimate of aboveground biomass in southern pine forests using SIR-C data, *Remote Sensing Environment. Vol.* 59, Issue 3 (February 1997) 223-233. ISSN 0034-4257.

Hess, L.; Melack, J.M. & Simonett, D.S. (1990). Radar detection of flooding beneath the forest canopy, *International Journal of Remote Sensing* Vol. 11, No. 7, 1313-1325. ISSN 0143-1161.

Hoekman, D.H. (1990). Radar remote sensing data for applications in forestry, *Ph.D. thesis*, Wageningen Agricultural University, The Netherlands, 277 p.

Hoekman, D.H. (1993) Radar signature and forest vegetation, In: *Land Observation by Remote Sensing: Theory and Applications*, J.G.P.W. Clevers and H.J. Buiten and (Eds), Current Topics in Remote Sensing, vol.3, Gordon and Breach Science Pub. Ch. 11, 219-235. ISBN 2-88124-939-6, New York, NY, USA.

Kasischke, E. & Bourgeau-Chavez, L. (1997). Monitoring South Florida wetlands using ERS-1 SAR imagery. Photogrammetric Engineering and Remote Sensing, Vol. 63 No. 3, (MONTH) 281-291. ISBN 0099-1112

Kasischke, E.S.; Christensen, N.L. Jr. & Bourgeau-Chavez, L.L. (1995). Correlating radar backscatter with components of biomass in loblolly pine forests, *IEEE Transactions on Geoscience and Remote Sensing*, Vol. 33, Issue 3 (May 1995) 643-659, 1995. ISSN 0196-2892.

Koch, B.; Foerster, B. & Ammer, U. (1992). The use of polarimetric radar remote sensing for application in forestry, Proc. European Int. Space Year Conf. "Remote Sensing for Environmental Monitoring and Resource Management", Munich (Germany), April 1992, ESA SP-341, 789-794.

Leblon, B. 1997. Soil and plant optical properties, The Remote Sensing Core Curriculum, vol. 4, module 9, http://www.r-s-c-c.org/rscc/Volume4/Leblon/leblon.htm

Leblon, B.; Kasischke, E. S.; Alexander, M. E.; Doyle, M. & Abbott, M. (2002) Fire danger monitoring using ERS-1 SAR images in the case of northern boreal forests.. Natural Hazards, 27, pp. 231-255.

Leblon, B. & LaRocque, A. (2008). *Radar Polarimetry*, online course, 11 chapters Accessed 26.09.08 http://extend.unb.ca/oalp/courses/for4304.php

Leckie, D.G. (1990). Advanced in remote sensing technologies for forest surveys and management. *Canadian Journal of Forestry Research.* 20: 464-483. ISSN 0045-5067.

Le Toan, T.; Beaudoin, A.; Riom, J. & Guyon, D. (1992). Relating Forest Biomass to SAR Data. *IEEE Transactions on Geoscience and Remote Sensing,* Vol. 30, Issue 2, (March 1992) 403-411, ISSN 0196-2892.

Lucas, R.M.; Lee, A.C & Williams, M.L. (2006). Enhanced Simulation of Radar Backscatter From Forests Using LiDAR and Optical Data. *IEEE Transactions on Geoscience and Remote Sensing*, Vol. 44, Issue 10, (October 2006) 2736-2754. ISSN 0196-2892.

Mas, J.F.; Velázquez, A.; Palacio-Prieto, J.L.; Bocco, G.; Peralta, A. & Prado, J. (2002). Assessing forest resources in Mexico: Wall-to-wall land use/cover mapping. Photogrammetric Engineering and Remote Sensing, Vol. 68, No 10, (October 2002) 966-968, Accessed 07.04.09
http://www.asprs.org/publications/pers/2002journal/october/

MDA, CSA & CCRS. (2009). The Science and Operational Applications Research for RADARSAT-2 Program (SOAR). Canada.

Accessed 22.03.09 http://www.radarsat2.info/outreach/soar/index.asp

Mougin, E.; Lopes, A.; Karam, M.A. & Fung, A.K. (1993). Effect of tree structure on X-band microwave signature of conifers. *IEEE Transactions on Geoscience and Remote Sensing*, Vol. 31, Issue 3, (May 1993) 655-667. ISSN 0196-2892.

Myneni, R.B.; Hall, F.G.; Sellers, P.J.; Marshak, A.L. (1995). The interpretation of spectral vegetation indices. *IEEE Transactions on Geoscience and Remote Sensing*, Vol. 33, Issue 2, (March 1995) 481 – 486, ISSN 0196-2892.

Nikolakopoulos, K.G. (2004). Comparison of four different fusion techniques for IKONOS data. (2004) 1401-1405. *Proceedings of IEEE International Geoscience and Remote Sensing Symposium, 2004. IGARSS'04,* Vol. 4, 2534 – 2537, ISBN 0-7803-8742-2, Anchorage, Alaska, USA. September 2004.

Packalén, P. & Maltamo, M. (2008). Estimation of species-specific diameter distributions using airborne laser scanning and aerial photographs. *Can. J. For. Res.* Vol. 38. (May 2008). 1750–1760, ISSN 1712-7971.

Palacio P. J.L.; Bocco, G; Velázquez, A.; Mas, J.F.; Takaki, F.; Victoria, A.; Luna, L.; Gómez, G.; López, J.; Palma, M.; Trejo, I.; Peralta, A.; Prado, J.; Rodríguez, A.; Mayorga, R. & González, F. (2000). The Current Condition of Forest Resources in Mexico: Results From the National Forest Inventory (In Spanish). La Condición Actual de los Recursos Forestales en México: Resultados del Inventario Forestal Nacional 2000. *Investigaciones Geográficas*, Boletín del Instituto de Geografía, UNAM. No. 43: 183-203. ISSN 0188-4611.

Papathanassiou, K.P. & Cloude, S.R. (2001). Single-baseline polarimetric SAR interferometry, *IEEE Transactions on Geoscience and Remote Sensing*, Vol. 39, Issue 11 (November 2001) 2352–2363. ISSN 0196-2892.

Pearson, R.G. (2007). Species' Distribution Modeling for Conservation Educators and Practitioners. Synthesis. American Museum of Natural History. Accessed 07.01.08 http://ncep.amnh.org.

Piwowar, J.M.. (1997). Radar Image Characteristics 22. Complex Scattering Examples. *Geomatics International, RADARSAT Distance Learning Program. uregina.ca/piwowarj/geog309/21-RadarImagery.pps*

Pohl, C. & van Genderen, J. L. (1998). Multisensor image fusion in remote sensing: concepts, methods and applications, International *Journal of Remote Sensing,* Vol. 19, No. 5, (March 1998), 823-854. ISSN 0143-1161.

Pulliainen, J.T.; Mikkelä, P.J.; Hallikainen, M. T. & Ikonen, J.P. (1996). Seasonal dynamics of C-band backscatter with applications to biomasss and soil moisture estimation. *IEEE Transactions on Geoscience and Remote Sensing*, Vol. 34, Issue 3, (May 1996) 758-770. ISSN 0196-2892.

Ranson, K.J. & Sun, G. (1994). Mapping biomass of a Northern forest using multifrequency SAR data, *IEEE Transactions on Geoscience and Remote Sensing*, Vol. 32, Issue 2, (March 1994) 388-396. ISSN 0196-2892.

Rauste, Y.; Häme, T.; Pulliainen, J.; Heiska, K. & Hallikainen, M. (1994). Radar-based forest biomass estimation, *International Journal of Remote Sensing,* Vol. 15, No. 14, (September 1994) 2797-2808. ISSN 0143-1161.

Richards, J.A.; Sun, G.Q. & Simonett, D.S. (1987). L-band radar backscatter modeling of forest stands, *IEEE Transactions on Geoscience and Remote Sensing*, Vol. 25, Issue 4, (July 1987) 487-498. ISSN 0196-2892.

27

Application of Multi-Frequency Synthetic Aperture Radar (SAR) in Crop Classification

Jiali Shang, Heather McNairn, Catherine Champagne and Xianfeng Jiao
Research Branch, Agriculture and Agri-Food Canada
Canada

1. Introduction

The application of remote sensing to agriculture has traditionally focused on the use of data from optical sensors such as Landsat Thematic Mapper (TM) and SPOT. Due to cloud and haze interference, however, optical images are not always available at phenological stages important for crop discrimination. When gaps in data acquisition occur during critical growth periods, classification accuracies using optical data are often inadequate (Jewell, 1989; McNairn et al., 2002; Blaes et al., 2005). Mid to late season optical images are essential to achieve accurate crop classification, and this dependency on late-season data reduces the ability to deliver early-season crop acreage estimates (McNairn et al., 2008a, b; Shang et al., 2006, 2008). These constraints seriously impede the use of optical data for operational annual crop mapping. Unlike visible and infrared wavelengths which are sensitive primarily to plant biochemical properties, longer-wavelength microwave energy responds to the large-scale structural attributes of vegetation, including the size, shape and orientation of the leaves, stems, and fruits. The dielectric properties of the vegetation canopy also influence the magnitude of the radar backscatter. These diverse sensitivities suggest that the integration of data from optical and radar sensors will generate a synergistic effect. Recent research has found that this complementarity, in most cases, provides enough information to separate a wide variety of crop types when an integrated optical-radar dataset is used (McNairn et al., 2008a; Shang et al., 2006, 2008).

1.1 Airborne multi-frequency SAR applications to agriculture

Although an integrated optical-radar approach can consistently discriminate crops, the continued dependency on optical data, particularly in cloud-prone regions, is less than ideal for operational delivery of crop information. A radar-only approach to crop discrimination and acreage estimation would provide an operational advantage. Until recently, however, the successful development of a radar-only approach to crop classification has been hindered because of the availability of radar data with only limited dimensionality. Single frequencies, and for some sensors single polarizations, do not provide enough information for accurate discrimination even when multi-temporal acquisitions are exploited (Shang et al, 2006, 2008). Research campaigns based on airborne SAR acquisitions have explored the benefits of multi-frequency SAR for crop discrimination. The Jet Propulsion Laboratory's

AirSAR is a fully polarimetric SAR sensor operating at P- (0.45 GHz), L- (1.25 GHz), and C- (5.31 GHz) bands (Lee and Pottier, 2009). The value of multi-frequency fully polarimetric data has been demonstrated for many land applications. For example, Rao et al. (1993) studied multi-frequency (P-, L-, C-band) polarimetric AirSAR data over corn fields. It found that the mean polarization phase difference increases with increasing wavelength. Lemonie and associates (1994) used the AirSAR data to study the contribution of multi frequency radar to increased agricultural class separabilities. The study by Baronti and associates (1995) carried out a three-frequency (P-, L-, and C-band) AirSAR data analysis. It found that P-band data are effective only in discriminating broad classes of agricultural landscapes. The integration of L- and C-band helps reveal finer class details.

Much research on the advantages of multi-frequency SAR has also been conducted with radar scatterometers. For example, Snoeij et al. (1990) used C- and X-band airborne SAR data to study the general behaviour of the radar signature of different European test sites as a function of frequency. The study conducted by van Leeuwen (1992) used six-frequency (L-band at 1.2 GHz; S-band at 3.2 GHz; C-band at 5.3 GHz; Ku1-band at 13.7 GHz; and Ku2-band at 17.3 GHz) radar scatterometer data over beet and wheat fields to examine the physical meaning of radar model (CLOUD-model: Attema & Ulaby, 1978) parameters in relation to crops. More recently Inoue et al. (2002) studied multi-frequency (Ka-, Ku-, X-, C- & L-band) radar backscattering signatures over paddy rice fields and their relationship with rice canopy growth variables. This research demonstrated that the backscatter coefficients of higher frequency bands (Ka and Ku) are highly correlated with the weight of heads. Lower frequency bands such as L-band, are better correlated with fresh biomass while C-band is better correlated with leaf area index.

1.2 Space-borne multi-frequency SAR applications to agriculture

Airborne SAR systems enabled radar scientists to develop methods to derive information of interest from multi-frequency and multi/full-polarization SAR, often under controlled experimental conditions. Airborne sensors provide a far greater signal to noise ratio and much higher spatial resolution compared with spaceborne sensors. However, these airborne platforms are not suited for large scale operational campaigns. With their wide swaths and repeat orbits, spaceborne sensors provide a cost effective solution for operational activities.

Since the launch of the first spaceborne radar system in 1965, Radar Evaluation Pod (REP), many spaceborne radar sensors have been launched (Lacomme et al., 2001). Earth Resources Satellite (ERS-1 launched in 1987 and ERS-2 launched in 1995) data have been used to develop many applications in agriculture, wetlands and forestry (Ban & Howarth, 1999; Bouman et al., 1992; Engdahl et al., 2001; Kohl et al., 1994; Michelson et al., 2000; Paudyal et al., 1995; Wang et al., 1998). The frequency-polarization (C-HH) of RADARDAT-1 (launched in 1995) was selected to maximize information for marine applications including sea ice and ocean features. Nevertheless, scientists developed the use of these data for a wide range of land applications including agriculture (McNairn et al., 1998a, b, c; Phoompanich et al., 2005; Ribbes & Toan, 1999; Shang et al., 2006).

The next generation European C-Band sensor, ASAR, has further advanced the use of SAR for agricultural mapping. The availability of dual like-polarizations (HH-VV) or dual like-cross polarizations (HH-HV or VV-VH) with ASAR, assists in providing more information on vegetation type and condition. Radar backscatter varies from one polarization to another since interaction is dictated by the transmitting polarization relative to the horizontal and vertical structure of the canopy. Consequently sensors which have polarization diversity provide more information on both crop structure and crop condition. The launch of ALOS PALSAR (L-band: 1.27 GHz) in 2006, followed by TerraSAR-X (X-band: 9.6 GHz) and RADARSAT-2 (C-band: 5.405 GHz) in 2007, marked the beginning of the multi-frequency spaceborne SAR era. Availability of data from this suite of satellites has accelerated the use of SAR for land applications. When used together, a multi-frequency dataset from multiple SAR platforms holds the promise to provide exceptional information for agriculture. Using a Canadian example, this chapter demonstrates the application of integrated L-, C-, and X-band SAR for crop mapping.

2. Study sites and data collection

2.1 Study sites

In 2006 Agriculture and Agri-Food Canada (AAFC) established two research sites close to their Ottawa (Ontario, Canada) research station. The Canadian Food Inspection Agency (CFIA) site is a controlled experimental site within the city of Ottawa (centred at 45°13′N, 75°46′W). The second site, Casselman, is a region of private land ownership (centred at 45°37′N, 75°01′W) approximately 50 km east of Ottawa. Both sites support non-irrigated dry land farming with one crop grown during the relatively short May to September growing season. The size of the fields in this part of Canada tends to be relatively small, 20 ha on average. These sites are typical of the crop mix found in this part of Canada, with production acreages primarily consisting of corn, soybean, cereal and pasture-forage.

2.2 Satellite data collection

(a) CFIA site

Satellite data were acquired from optical sensors (Landsat-5) as well as SAR sensors (RADARSAT-1, Envisat-ASAR, and ALOS PALSAR) throughout the 2006 growing season. Acquisitions were targeted to capture crop growth stages of importance for crop discrimination using optical and SAR sensors (Table 1).

Date	Resolution (m)	Mode	Polarization	Incidence Angle
Landsat-5 TM				
June 5	30			
July 7	30			
August 22	30			
Envisat ASAR				
May 27	30	IS3	VV, VH	25.8° - 31.2°
June 9	30	IS1	VV, VH	14.5° - 22.1°
July 1	30	IS3	VV, VH	25.8° - 31.2°
July 14	30	IS1	VV, VH	14.5° - 22.1°
August 5	30	IS3	VV, VH	25.8° - 31.2°
September 18	30	IS4	VV, VH	30.8° - 36.1°

RADARSAT-1				
May 18	30	S1	HH	24° - 31°
July 5	30	S1	HH	24° - 31°
August 22	30	S1	HH	24° - 31°
ALOS PALSAR				
May 19	10	PLR	Polarimetric	21.5°
July 4	10	PLR	Polarimetric	21.5°
August 19	10	PLR	Polarimetric	21.5°

Table 1. Satellite data acquired over CFIA during the 2006 growing season

(b) Casselman site

Optical and radar satellite data were collected over the Casselman site during the 2008 growing season. No cloud-free (less than 20% cloud cover) Landsat data were available due to poor weather conditions throughout the summer of 2008. The SPOT sensor was programmed in two week windows through the entire 2008 season. This acquisition strategy yielded four SPOT-4 images. Six TerraSAR-X scenes were also acquired. Due to the late start of the TerraSAR-X project, X-band data acquisition did not commence until mid July. Table 2 gives the details of each satellite acquisition.

Date	Resolution (m)	Mode	Polarization	Incidence Angle
SPOT-4				
June 5	20			
July 7	20			
August 22	20			
RADARSAT-2				
May 27	10	FQ19	quad-pol	38.3° - 39.8°
June 9	10	FQ19	quad-pol	38.3° - 39.8°
July 1	10	FQ19	quad-pol	38.3° - 39.8°
July 14	10	FQ19	quad-pol	38.3° - 39.8°
TerraSAR-X				
July 19	6	Stripmap	VV, VH	43.6° - 44.6°
July 30	6	Stripmap	VV, VH	43.6° - 44.6°
August 10	6	Stripmap	VV, VH	43.6° - 44.6°
August 21	6	Stripmap	VV, VH	43.6° - 44.6°
August 26	6	Spotlight	HH, VV	53.9°
September 1	6	Stripmap	VV, VH	43.6° - 44.6°

Table 2. Satellite data acquired over Casselman during the 2008 growing season

2.3 Ground data collection

For both sites, ground truth observations were collected twice over the growing season, once in early July and once in mid August. The second visit provided an opportunity to check for errors which might have occurred during the first field visit. During the second visit, variations in crop growth condition, harvesting, and emergence of under seeded crops were also noted. Underseeding is a cropping system where a primary species is seeded with a successive species that emerges at the end of the primary species growth cycle, for instance, where annual cereal crops are underseeded with a perennial forage crop such as alfalfa. Wheat in this region is usually harvested between late July and mid August depending on the planting date. Underseeded wheat fields are thus characterized by rapid growth of forage after wheat harvest. Harvesting of corn and soybean typically occurs near the end of October.

In 2006, a total of 240 fields were visited. Table 3 gives details on the number of fields surveyed per crop.

Crop Type	Number of Training Fields	Number of Testing Fields
Cereal	16	17
Corn	35	35
Soybean	30	30
Forage/Pasture	38	39

Table 3. Ground truth data used for CFIA 2006 crop classification

For the Casselman site, a total of 247 fields were surveyed during the 2008 growing season. The distribution of field surveyed is documented in Table 4.

Crop Type	Number of Training Fields	Number of Testing Fields
Cereal	17	18
Corn	45	45
Soybean	41	41
Forage/Pasture	33	34

Table 4. Ground truth data used for Casselman 2008 crop classification

3. Data pre-processing

3.1 Atmospheric correction of optical data

Atmospheric correction was applied to all optical data to retrieve the at-surface reflectance using the Atcor algorithm in PCI software (Richter, 2004). The Atcor algorithm uses the MODTRAN 4.2 radiative-transfer code for the radiance to reflectance conversion (Champagne et al., 2005).

3.2 Speckle filtering of SAR data

Speckle is an inherent phenomenon for coherent systems such as SARs. To suppress speckle, adaptive radar filters should be applied prior to classification of SAR data. All ALOS PALSAR, RADARSAT-2, and TerraSAR-X data were speckle filtered, using a 5 X 5 Gamma-MAP filter.

For the Casselman study site, two pairs of TerraSAR-X (TSX) and RADARSAT-2 (RSAT-2) imagery collected close in time were selected for comparison. Table 6 documents the classification results derived using data acquired over Casselman during the 2008 growing season.

Sensors	Frequency	Polarization Used for Comparison	Date	Pasture/ Forage	Soybean	Corn	Wheat	Overall Accuracy
TSX	X-band	VV/VH	July 16	69.1	58.2	48.0	85.2	59.9
RSAT-2	C-band	VV/VH	July 19	47.1	64.7	50.7	42.8	54.2
TSX	X-band	VV/VH	August 9	59.1	79.4	71.0	61.8	71.0
RSAT-2	C-band	VV/VH	August 10	39.6	73.8	56.2	36.8	57.4

Table 6. Comparison of single-date TerraSAR-X and RADARSAT-2 crop classification accuracy (producer's) over the Casselman site from the 2008 growing season

For both acquisition windows (mid July, early August), X-band data outperformed the C-band data. When comparing overall accuracies, the mid July X-band data produced a crop map with an accuracy 5.7% higher than for C-band data acquired only three days later. Comparing data acquired in early August, X-band provided significantly better accuracies - an overall accuracy of 71% or 13.6% higher than the C-band data. In August the X- and C-Band data were acquired only one day apart. Examining the individual class accuracies, X-band performed better in identifying all crop types later in the season. Among all crop types, X-band provided dramatically higher accuracies for wheat. A 42.4%increase for mid July and 25% increase for early August are noted for the wheat class, when X-band results are compared with those of C-band. For the same wheat class, X-band data also performed better than C-band later in the growing season. At mid season (mid July), results derived from X- and C-band are similar for corn, with a difference of less than 3%.

5.2 Multi-frequency classification comparison
To evaluate the benefits of a multi-frequency SAR approach for crop classification, four datasets from the CFIA site were analyzed. Table 7 provides the classification accuracies derived using single-frequency (L- or C-band) and two-frequency (L- and C-band) approaches.

Sensors (Date)	Frequency	Polarization Used for Comparison	Pasture/ Forage	Soybean	Corn	Wheat	Overall Accuracy
1 ALOS (May20)	L-band	VV/VH	24.1	57.5	84.1	1.4	49.7
1 ASAR (May27)	C-band	VV/VH	60.0	50.7	81.7	4.9	55.7
1 ASAR (May27) + 1 ALOS (May20)	C- & L-band	VV/VH	60.9	61.9	70.8	40.8	60.6
1 ALOS (July 5)	L-band	VV/VH	18.0	67.0	86.7	11.8	54.0
1 ASAR (July 1)	C-band	VV/VH	70.6	65.2	88.8	33.0	68.1
1 ASAR (July 1) + 1 ALOS (July 5)	C- & L-band	VV/VH	61.1	77.6	92.4	48.4	73.9

2 ALOS (May20, Jul5)	L-band	HH	55.8	49.3	83.8	9.2	54.8
2 RS1 (May18, Jul5)	C-band	HH	79.8	54.0	61.0	8.1	52.8
2 ALOS (May20, Jul5) + 2 RS1 (May18, Jul5)	C- & L-band	HH	87.5	73.2	82.1	24.8	69.8
2 ALOS (May20, Jul5)	L-band	VV/VH	54.2	67.4	83.6	37.5	64.7
2 ASAR (May27, Jul1)	C-band	VV/VH	74.1	61.2	91.6	52.6	72.2
2 ASAR (May 27, Jul 1) + 2 ALOS (May20, Jul 5)	C- & L-band	VV/VH	72.4	84.5	95.7	69.6	82.9

Table 7. Producer's accuracies using multi-frequency SAR classifications from the 2006 data acquired over CFIA

Results in Table 7 clearly confirm the benefits of a multi-frequency solution for crop identification. For a single-date dual-polarization (VV/VH) comparison, there is an increase of 4.9% in overall accuracy in late May compared to result derived using ASAR alone. When compared with result derived from May ALOS, there is an increase of 10.9%. When early July L- and C-band data are integrated together in the classifier, a similar improvement in accuracy (5.8%) was observed. When multiple dates (one in May and one in July) of L- and C-band were used a 15% gain in accuracy was observed, even though only a single polarization (HH) was used. Two dates of dual-frequency and dual-polarization SAR from PALSAR (L-band) and ASAR (C-band) produced a map with an overall accuracy of 82.9%. For the Casselman site, comparisons were made between classifications using a single frequency (C- or X-band) and results achieved by integrating these two frequencies (Table 8). The multi-temporal X-band data on its own was capable of identifying crops with an overall accuracy of 84.9%. Consequently, adding C-band SAR to the classification brought only modest improvements in overall accuracy. Nevertheless C-band did assist in boosting accuracies for most individual crop classes.

Sensors (Date)	Frequency	Polarization Used for Comparison	Pasture/ Forage	Soybean	Corn	Wheat	Overall Accuracy
4 RSAT-2	C-band	VV/VH	66.2	82.9	76.1	64.0	75.4
5 TerraSAR-X	X-band	VV/VH	83.2	83.6	87.3	84.1	84.9
4 RSAT-2 + 5 TerraSAR-X	C- & X-band	VV/VH	84.1	86.8	89.9	85.6	87.3

Table 8. Producer's accuracies of multi-frequency SAR classification from 2008 growing season over Casselman

6. Conclusions and future research

A multi-year and multi-site study by Agriculture and Agri-Food Canada demonstrated the improvements brought by integrating multiple frequencies (L-, C-, and X-band) of SAR data for crop classification. Penetration into the crop canopy is dependent upon SAR frequency and results indicate that the differences in this depth between frequencies are advantageous for crop identification. The case study presented here concludes that when multi-temporal multi-frequency SAR data are used, satisfactory crop classification (above 85% accuracy) can be achieved using a SAR-only dataset.

Even with these promising results, further improvements in accuracy would be desirable prior to implementing a radar-alone solution for crop classifications. The acquisition planning associated with the datasets used in the research was limited by several factors. TerraSAR-X data collection did not begin until mid season due to a late start in the project. In future growing seasons, a more complete data set will be collected. This study also did not permit comparisons among all three frequencies as TerraSAR-X, ASAR, RADARSAT and PALSAR data were not all collected over either site. The programming of PALSAR in concert with TerraSAR-X and RADARSAT-2 was not successful. In future growing seasons, all three sensors have been programmed over the Casselman site. These methods will also be evaluated in future growing seasons over a third site in the Canadian prairies, which will represent a more complex cropping system with a greater variety of crops. Lastly, acquisitions of data in RADARSAT-2's polarimetric mode will permit assessment of polarimetric parameters derived from multi-frequency SAR for improved crop classification.

7. Acknowledgments

Funding of this research project was provided by the Canadian Space Agency's Government Related Initiatives Project (GRIP) and AAFC Research Branch's A-base project. The TerraSAR-X data were provided under the TerraSAR-X AO project LAN0337. The authors wish to acknowledge the assistance of Dr. Ridha Touzi (Canada Centre for Remote Sensing) in sharing his expertise on processing of the PALSAR data. We would like to thank JAXA for having provided the ALOS data under the PI project 228.

8. References

Attema, E.P.W. & Ulaby, F.T. (1978). Vegetation model as a water cloud. Radio Science, Vol. 13, pp. 357-364.

Ban, Y. & Howarth, P. J. (1999). Multitemporal ERS-1 SAR data for crop classification: a sequential-masking approach. *Canadian Journal of Remote Sensing*, Vol. 25, pp. 438-447.

Berger, Z.; Irving, R. & Clark, J. (2009). Exploration applications of RADARSAT imagery in the Foothills and Western Canadian Basin, Intermap whitepapers: <http://www.intermap.com/uploads/1170701249.pdf>.

Blaes, X., Vanhalle, L, and Defourny, P. (2005). Efficiency of crop identification based on optical and SAR image time series, Remote Sensing of Environment 96: 352-365.

Bouman, B.A.M. & Uenk, D. (1992). Crop classification possibilities with radar in ERS-1 and JERS-1 configuration. *Remote Sensing of Environment*, Vol. 40, pp. 1-13.

Champagne, C., Shang, J., and McNairn, H. (2005). Exploiting spectral variation from crop phenology for agricultural land-use classification, Proceedings of SPIE Optics and Photonics, Vol. 5884, San Diego, CA, USA, July–August, 2005 (CD ROM).

Engdahl, M.E.; Borgeaud, M. & Rast, M. (2001). The use of ERS-1/2 Tandem interferometric coherence in the estimation of agricultural crop heights. IEEE Transactions on Geoscience and Remote Sensing, Vol. 39, No. 8, pp. 1799-1806.

Freeman, A. & Durden, S.L. (1998). A three-component scattering model for polarimetric SAR data. IEEE Transactions on Geoscience and Remote Sensing, Vol. 36, pp. 963-973.

Hill, M.J.; Ticehurst, C.J.; Lee, J.S.; Grunes, M.R.; Donald, G.E. & Henry, D. (2005). Integration of optical and radar classifications for mapping pasture type in western Australia. IEEE Transactions on Geoscience and Remote Sensing, Vol. 43, pp.1665-1681.

Inoue, Y.; Kurosu, T.; Maeno, H.; Uratsuka, S.; Kozu, T.; Dabrowsk-Zielinska, K. & Qi, J. (2002). Season-long daily measurements of multifrequency (Ka, Ku, X, C, and L) and full-polarization backscatter signatures over paddy rice field and their relationship with biological variables. Remote Sensing of Environment, Vol. 81, pp. 192-204.

Kohl, H.G.; Nezry, E.N.; Mróz, M. & De Groof, H. (1994). Towards the integration of ERS SAR data in an operational system for rapid estimates of acreage at the level of the European Union, Proceedings of the 1st Workshop on ERS-1 Pilot Projects, pp.433-441, Toledo, Span, June1994.

Lacomme, P.; Marchais, J-C. & Normant, E. (2001). Air and Spaceborne Radar systems: an introduction. William Andrew Publishing, Norwich, NY, USA, 504p.

Lee, J & Pottier, E. (2009). Polarimetric Radar Imaging: From Basics to Applications, 433p., 9781420054972, CRC/Taylor & Francis Group, New York, NY, USA.

Lemonie, G.G.; de Grandi, G.F. & Sieber, A.J. (1994). Polarimetric contrast classification of agricultureal fields using MASTRO1 AirSAR data. International Journal of Remote Sensing, Vol. 15, pp. 2851-2869.

McNairn, H.; Wood, D.; Gwyn, Q.H.J.; Brown, R.J. & Charbonneau, F. (1998a). Mapping tillage and crop residue management practices with RADARSAT, Canadian Journal of Remote Sensing, Vol. 24, No. 1, pp. 28-35.

McNairn, H.; Brown, R.J. & Wood, D. (1998b). Incidence angle considerations for crop mapping using multi-temporal RADARSAT data, Proceedings of the 20th Canadian Symposium on Remote Sensing, pp. 211-214, Calgary, AB, Canada, May 1998.

McNairn, H.; Wood, D. & Brown, R.J. (1998c). Mapping crop characteristics using multitemporal RADARSAT images, Proceedings of 1st International Conference: Geospatial Information in Agriculture and Forestry, Vol. 2, pp. 501-507, Orlando, USA, June 1998.

McNairn, H.; van der Sanden, J.J.; Brown, R. & Ellis, J. (2000). The potential of RADARSAT-2 for crop mapping and assessing crop condition, Proceedings of 2nd International conference on Geospatial Information in Agriculture and Forestry, Vol. 2, pp. 81-88, Lake Buena Vista, Florida, USA.

McNairn, H, Ellis, J., van der Sanden, J J, Hirose, T. and Brown, R J. 2002. Providing crop information using RADARSAT-1 and satellite optical imagery. International Journal of Remote Sensing 23:851-870.

McNairn, H.; Champagne, C., Shang, J., Holmstrom, D. and Reichert, G. (2008a). Integration of optical and Synthetic Aperture Radar (SAR) imagery for delivering operational annual crop inventories. Photogrammetry and Remote Sensing (in press: doi:10.1016/j.isprsjprs.2008.07.006)

McNairn, H., Shang, J., Jiao, X., and Champagne, C. (2008b). The Contribution of ALOS PALSAR Multi polarization and Polarimetric Data to Crop Classification, IEEE Transactions on Geoscience and Remote Sensing, accepted with modifications.

Michelson, D.B.; Liljeberg, b.M. & Pilesjo, P. (2000). Comparison of algorithms for classifying Swedish landcover using Landsat TM and ERS-1 SAR data. Remote Sensing of Environment, Vol. 71, No. 1, pp. 1-15.

Pal, M. & Mather, P.M. (2003). An assessment of the effectiveness of decision tree methods for land cover classification. Remote Sensing of Environment, Vol. 86, pp. 554-565.

Paudyal, D.R.; Eiumnoh, A. & Aschbacher, J. (1995). Multitemporal analysis of SAR images over the tropics for agriculture applications, *Proceedings of the International Geoscience and Remote Sensing Symposium*, pp. 1219-1221, 0-7803-2567-2, Firenze, Italy, July 1995, IEEE.

Phoompanich, S.; Anan, T. & Phongam, S. (2005). Multi-temporal RADARSAT for crop monitoring, Proceedings of 26th Asian Conference on Remote Sensing, Hanoi, Vietnam, November 2005
< http://www.aars-acrs.org/acrs/proceeding/ACRS2005/Papers/AGC2-3.pdf>.

Rao, K.S.; Rao, Y.S. & Gurusamy, R. (1993). Frequency dependence of polarization phase difference and polarization index for vegetation covered fields using polarimetric AirSAR data, *Proceedings of the International Geoscience and Remote Sensing Symposium*, pp. 37-39, Tokyo, Japan, August 1993, IEEE, Piscataway, N.J., U.S.A.

Ribbes, F. & Toan, T.L. (1999). Rice field mapping and monitoring with RADARSAT data. International Journal of Remote Sensing, Vol. 20, No. 4, pp. 745-765.

Rulequest Research (2008). Data Mining Tools See5 and C5.0 [Online]. Available by Rulequest Research http://www.rulequest.com/see5-info.html.

Shang, J., McNairn, H., Champagne, C. and Jiao, X. (2008). Contribution of multi-frequency, multi-sensor, and multi-temporal radar data to operational annual crop mapping. International Geoscience and Remote Sensing Symposium, Boston (Massachusetts), 7-11 July 2008, 4 pp. Invited paper.

Shang, J., Champagne, C., and McNairn, H. (2006). Agriculture land use using multi-sensor and multi-temporal Earth Observation data. In Proceedings of the MAPPS/ASPRS 2006 Fall Specialty conference, San Antonio, Texas, USA.

Snoeij, P.; Swart, P.J.F. & Attema, E.P.W. (1990). The general behavior of the radar signature of different European test sites as a function of frequency and polarization, *Proceedings of the 10th International Geoscience and Remote Sensing Symposium*, pp. 2315-2318, May 1990, College Park, Maryland, USA.

Skriver, H.; Svendsen, M.T. & Thomsen, A.G. (1999). Multitemporal C- and L-band polarimetric signatures of crops. IEEE Transactions on Geoscience and Remote Sensing, Vol. 24, No. 5, pp. 2413-2429.

Van Leeuwen, H.J.C. (1991). Multifrequency and multitemporal analysis of scatterrometer radar data with respect to agricultural crops using the Cloud model, *Proceedings of the 11th International Geoscience and Remote Sensing Symposium*, Vol. 4, pp. 1893-1897, Helsinki, Finland.

Wang, J.; Shang, J.; Brisco, B. & Brown, J. (1998). Evaluation of multidata ERS-1 and multispectral Landsat imagery for wetland detection in southern Ontario. *Canadian Journal of Remote Sensing*, Vol. 24, No. 1, pp. 60-68.

28

Multispectal Image Classification Using Rough Set Theory and Particle Swam Optimization

Chih-Cheng Hung, Hendri Purnawan, Bor-Chen Kuo* and Scott Letkeman
Southern Polytechnic State University, Marietta, GA 30060 USA
*National Taichung University, Taichung, Taiwan

Abstract

This chapter provides an exploration of the rough set theory and particle swarm optimization for multispectral image classification. The rough set theory, particle swarm optimization and K-means clustering algorithm are briefly described. Two multispectral image classification algorithms based on the rough set theory and particle swarm optimization are proposed: Algorithm 1 is a multispectral image classification approach based on the rough set theory which uses upper and lower bounds for the class description, and Algorithm 2 is a hybrid rough K-means algorithm for image classification. The rough set theory is used to extract classification rules and establish the lower and upper bounds for data clustering with the K-means algorithm. This algorithm is able to deal with vagueness in image date, but since its ability to determine some key parameters is limited, partial swarm optimization must subsequently be used to locate optimal values for those parameters. Experimental results show that the proposed algorithms perform well and improve the classification in the blurred and vague areas of the image. A comparison of Algorithm 1 with the parallelepiped classifier, where the former uses the concept of cuts and the later uses the maximum and minimum values, is performed. Preliminary experimental results show that the proposed classifiers are effective for multispectral image classification.

1. Introduction

In the real world, data representation is most often imperfect, in the sense that the data may be either incomplete or redundant. Philosophers, logicians and mathematicians have dealt with this problem for a long time. In recent years, propelled by the advent of the computer, the problem of imperfect knowledge has been becoming an important topic for computer scientists engaged in artificial intelligence research, especially those involved with knowledge discovery from databases, expert systems, and pattern recognition.

Our research is focused on rough set as a tool for image processing, or more precisely, for image segmentation. Many techniques for image segmentation have been developed over time. There are clustering, edge detection, region growing and even more advanced techniques that use neural networks. In general, image segmentation techniques can be categorized as supervised or unsupervised. Supervised techniques require previously known truth data for training purposes, while unsupervised techniques have no such requirement

In this chapter, the classical rough set theory is reviewed in section 2. Particle swarm optimization is then introduced in section 3. The Davies-Bouldin measure for cluster validity is also described in this section. The K-means algorithm is briefly sketched in section 4. Multispectral image classification using rough set theory is discussed in section 5. A hybrid algorithm which combines the K-means algorithm, rough set and particle swarm optimization is given in section 6. Experimental results are shown in section 7. The conclusion and future work then follow.

2. Rough Set Theory

Rough set theory [5] is a mathematical tool that deals with the uncertainty of the data. The theory consists of finite sets, equivalence relations and cardinality concepts. As the theory matures and more applications reap the benefits of the concept, an abundance of related theorems and algorithms are being incorporated to extend rough sets theory.

It was introduced by Pawlak in the early 1980's and has been argued to overlap with other theories, such as statistics, evidence theory and fuzzy set. Furthermore, rough set is said to complement fuzzy set, a theory introduced by Zadeh in the early period. Rough set and fuzzy set were both introduced to deal with imprecise information however; fuzzy set deals with vagueness, while rough set deals with coarseness. Rough set does not need as much preliminary knowledge about the data where as fuzzy set requires knowledge of the possible values in advance. Basically, when using rough set, the data itself is used to come up with the approximation in order to deal with the imprecision within. It can therefore be considered a self-sufficient discipline.

Rough set mainly deals with data analysis in table format. The approach is generally to pre-process the data in the table and then to analyze them. Reducts are extracted with an algorithm and finally rules are generated based on the reducts. Rough set does not support analog values in the table attributes; therefore discretization must be performed in advance in order to evaluate the table. The following subsections will use a simple example to illustrate the concept of rough set theory.

2.1 Information Systems

In essence, an information system is a set of objects represented in a data table (attribute - value system). Each row contains an object and each column represents a measurable attribute for each object. Formally, an information system is a pair $A = (U, A)$ where U is a non- empty finite set of objects representing the universe and A is a non-empty finite set of attributes such that a: $U \rightarrow V_a$ for every a $\in$ A. The set V_a is the set of values for a.

Object Index	Salary	Age
u_1	80	30
u_2	30	23
u_3	80	40
u_4	50	45
u_5	80	55
u_6	50	45
u_7	30	60
u_8	100	35

Table 1. shows an information system which is a collection of salary and age attributes.

2.2 Decision Systems

If an information system has an additional attribute, namely a decision attribute, then it becomes a decision system. The decision attribute is associated with the object classification outcome, and it may depend on several other attributes. Formally, a decision system is a piece of information whose form is $A = (U, A \cup \{d\})$, where d $\in A$ is the decision attribute. A decision attribute called "Class" has been added as shown in Table 2, where M and E denote Manager and Employee, respectively. The table was modified from the original [12].

Object Index	Salary	Age	Class
u_1	80	30	M
u_2	30	23	M
u_3	80	40	E
u_4	50	45	M
u_5	80	55	E
u_6	50	45	E
u_7	30	60	M
u_8	100	35	E

Table 2. A decision system where each row is classified into a class. Salary and Age are the condition attributes, while class is the decision attribute.

2.3 Indiscernibility

Objects in information and decision systems may be indistinguishable from one another based on a set of attributes B that belongs to A ($B \subseteq A$). A set of objects is indiscernible or equivalent when their attributes are related by an equivalence relation. An equivalence relation is a relation on a set B when it is:

1. Reflexive (if a R a, then R is reflexive).
2. Symmetric (if a R b then b R a, then R is symmetric).
3. Transitive (if a R b and b R c, then a R c, thus R is transitive).

For an information system $A = (U, A)$, there is an equivalence relation for any of the sets B $\subseteq$ A. The equivalence relation can be formalized as

$$IND_A(B) = \{(x, x') \in U^2 \mid \forall a \in B, \ a(x) = a(x')\}$$

Referring to table 4, the *IND* relations for *Salary* can be written as shown.

$$IND(Salary) = \{\{u_1, u_3, u_5, u_8\}, \{u_2, u_4, u_6, u_7\}\}$$

It is impossible to write the *IND* relations for *Salary* until discretization is completed.

2.4 Discretization

Discretization is not directly related to rough set theory. It is simply a preprocessing technique. Discretization is associated with information loss. In general, when it is too coarse (i.e. longer interval), there is too much information loss or noise in the data. However, it is better for the classification capability of unseen objects. When the discretization is more fine (i.e. shorter interval), less noise exists in data, but classification capability of unseen objects may be impaired.

In our decision system table, both Salary and Age need to be discretized. The set of possible Salary and Age values, respectively referred to as *s* and *a* from here on, is given by

$$V_s = [15, 120)$$
$$V_a = [18, 65)$$

The lower and upper bounds of the attribute's interval are extended to cover possible values. For example, the Age attribute is extended to include likely working ages from age 18 through 65.

The set of values of *s* and *a* in U is

$$s(U) = \{30, 50, 80, 100\}$$
$$a(U) = \{23, 30, 35, 40, 45, 55, 60\}$$

The intervals obtained for *s* are

$$[30, 50); [50, 80); [80, 100)$$

The intervals obtained for *a* are

$$[23, 30); [30, 35); [35, 40);$$
$$[40, 45); [45, 55); [55, 60).$$

Boundary intervals such as [15, 30) and [100, 120) should not be used since one can not discern anything for this data set.

The intervals introduce a set of cuts, which are defined as (s, c) where $c \in V_s$ and (a, c) where $c \in V_a$. If the cut is taken based on the mid-point of each interval, the set of cuts **P** obtained for *s* and *a* are respectively

$$(s, 40); (s, 65); (s, 90);$$
$$(a, 26.5); (a, 32.5); (a, 37.5);$$
$$(a, 42.5); (a, 50); (a, 57.5)$$

The next step is to find the set of minimal cuts that can discern all of the objects that are needed. It turns out that the problem of finding the irreducible set of cuts **P** in the decision system is NP-complete while the effort to find the optimal set of cuts **P** in a decision system is NP-hard [5].

However, there are heuristics that can be used to find the optimal set of cuts **P** in practical time. One of them is the Maximal Discernability heuristic [1], [5], which is demonstrated here. The algorithm to construct table A^* from A is listed in the following steps:

1. Each column in table A^* is a Boolean variable of the corresponding column in A. If each pair of objects can be discerned by the Boolean variable, then assign value 1, else assign 0.
2. Choose a column from A^* that has a maximal number of 1's and delete all the rows which contain a 1 in the selected column.
3. Repeat step 2 and continue until all columns and rows are consumed.

$\mathcal{A}^*$	p_1^s	p_2^s	p_3^s	p_1^a	p_2^a	p_3^a	p_4^a	p_5^a	p_6^a	d
u1, u3	0	0	0	0	1	1	0	0	0	1
u1, u5	0	0	0	0	1	1	1	1	0	1
u1, u6	0	1	0	0	1	1	1	0	0	1
u1, u8	0	0	1	0	1	0	0	0	0	1
u2, u3	1	1	0	1	1	1	0	0	0	1
u2, u5	1	1	0	1	1	1	1	1	0	1
u2, u6	1	0	0	1	1	1	1	0	0	1
u2, u8	1	1	1	1	1	0	0	0	0	1
u3, u4	0	1	0	0	0	0	1	0	0	1
u3, u7	1	1	0	0	0	0	1	1	1	1
u4, u5	0	1	0	0	0	0	0	1	0	1
u4, u6	0	0	0	0	0	0	0	0	0	1
u4, u8	0	1	1	0	0	1	1	0	0	1
u5, u7	1	1	0	0	0	0	0	0	1	1
u6, u7	1	0	0	0	0	0	0	1	1	1
u7, u8	1	1	1	0	0	1	1	1	1	1
new	0	0	0	0	0	0	0	0	0	0

Table 3. Decision System A*.

The following example clarifies the process of constructing table A^* from A mentioned in step 1 of the algorithm. Each cut previously obtained is assigned a Boolean variable, which in turn is used as a condition attribute in table A^*.

For example, (s, 40) is assigned Boolean variable P_1^s. Each object pair in A^*, is derived from table A by looking at the decision attribute. Objects that do not have the same decision attribute should be paired up. For example, u_1 and u_3 are paired up since they have different decision values (i.e. M and E).

The resulting table A^* created from A is shown in Table 3.

The optimal cut chosen is (s, 65), (a, 32.5) and (a, 50). These cuts are then used to discretize the decision table 2. The rank can be assigned using the following rules:

1   If s < 65, value 0 is assigned to s, else assign value of 1.
2   If a < 32.5, value 0 is assigned to a.
3   If ($32.5 \leq a < 50$), value 1 is assigned to a.
4   If $a \geq 50$, value 2 is assigned to a.

A discretized table can be produced by applying the condition to each analog value in the table.

Index	Salary	Age	Class
u_1	1	0	M
u_2	0	0	M
u_3	1	1	E
u_4	0	1	M
u_5	1	2	E
u_6	0	1	E
u_7	0	2	M
u_8	1	1	E

Table 4. Discretized Decision System.

2.5 Lower and Upper Approximations in Rough Set and Accuracy

Let U be the non-empty finite set and R be an equivalence relation. The pair $A = (U, R)$ is an approximation space. The equivalence relation R on U leads to a partition of the objects in the universe U. The idea here is to partition the objects that have the same outcome, or in other words, to partition objects that have the same decision attribute. However, this may not always be as easy as stated. There will be objects with the same condition attributes (in the same equivalence class), but different decision attributes. Therefore one can not define every set precisely.

In cases where the set can not be defined precisely, it can be approximated. This is where rough set emerges. Let us assume that there is an information system $A = (U, A)$, a set of attributes $B \subseteq A$, and a set of objects $X \subseteq U$. Using the set of attributes B, one can approximate the objects X into:

1.   Lower Approximation: the set of objects that can be classified as member X with certainty. Formally stated as

$$\underline{B}(X) = \bigcup \{E_i \in U^B : E_i \subseteq X\}$$

(2.1)

2.   Upper Approximation: the set of objects that can possibly be classified as a member X. Formally stated as

$$\overline{B}(X) = \bigcup \{E_i \in U^B : E_i \cap X \neq \{\}\}$$

(2.2)

Between the lower and upper approximation, one can define the set of objects that cannot be classified into X decisively. This set is also known as the *B-boundary region* of X.

There is a coefficient that reflects the accuracy of approximation,

$$\alpha_B(X) = \frac{|\underline{B}(X)|}{|\overline{B}(X)|}$$

(2.3)

where $|X|$ denotes the cardinality of $X \neq \varnothing$. When $\alpha_B = 1$, the X is crisp with respect to B, otherwise, X is rough with respect to B.

For our example, the boundary region would be for object u_4 and u_6 since they can not be discerned. The lower and upper approximations can be written as

$$\underline{B}(M) = \{u_1, u_2, u_4, u_7\}$$
$$\overline{B}(M) = \{u_1, u_2, u_4, u_6, u_7\}$$

$$\alpha_B(M) = \frac{|\underline{B}(M)|}{|\overline{B}(M)|} = \frac{4}{5} = 0.8$$

In general, the value of α reflects the accuracy of decision rules obtained.

2.6 Reducts

One way to increase computation efficiency is to reduce the size of data by reducing attributes that need to be taken into account. Only attributes that do not contribute to the classification result can be omitted such that the indiscernibility relation remains intact. The set of remaining attributes is the minimal set and is called a reduct.

Although finding the equivalence class is a relatively straightforward computation process, finding reducts with minimal attributes is known to be NP-hard. Fortunately, there are heuristics that allow minimal reducts to be computed in reasonable time.

2.7 Discernibility Matrix

Computing the reducts of an information system $A = (U, A)$ can be started by creating the indiscernibility matrix. This matrix is a symmetric $n \times n$ matrix where each entry c_{ij} is defined as

$$c_{ij} = \{a \in A \mid a(x_i) \neq a(x_j)\} \quad for \quad i, j = 1, ..., n$$

(2.4)

Each cell in the matrix holds the set of attributes where objects x_i and x_j are discernable. The cell would have empty set when.

1    $x_i = x_j$, that is case for diagonal cells.

2   For decision systems, when the decision attribute of objects x_i and x_j are equal, or formally, $d(x_i) = d(x_j)$.

2.8 Discernibility Functions

Based on the discernibility matrix, a discernibility function can be immediately obtained. It is constructed using Boolean expressions from the discernibility matrix, defined as

$$f_A(a_1^*,...,a_m^*) = \wedge \{\vee \ c_{ij}^* \mid 1 \leq j \leq i \leq n, \ c_{ij} \neq \{\}\}$$

(2.5)

where $a_1^*,...,a_m^*$ are related to attributes $a_1,...,a_m$. The attributes may be transformed during discretization process. $c_{ij}^* = \{a^* \mid a \in c_{ij}\}$ is the set of Boolean variables.

Once the discernibility function f_A is formed, it can be further developed using Boolean algebra simplification.

2.9 Decision Rules

When applying rough set for supervised learning, we need to construct a set of rules from the training data, such that new or unseen objects can be separated into known classes.

A basic method for forming the decision rules is begun by finding the reducts of the decision table. Then for each reducts $R = \{r_1,...,r_n\}$, we generate the decision rule by taking the conjunction $(r_1 = r_1(u)) \wedge .... \wedge (r_2 = r_2(u))$ as the predecessor. Next we take the decision attribute d with value $d(u)$ as the successor and format the predecessor and successor values as

$$(r_1 = r_1(u)) \wedge .... \wedge (r_2 = r_2(u)) \Rightarrow d = d(u)$$

(2.6)

Rule induction is about deciding which attributes should be included in the predecessor of the rule. Rules obtained can always be minimized, but it will introduce noise and may poorly classify the unseen objects.

Once the rules are obtained, they can be used to classify the objects that were unseen before. The basic steps involved can be outlined as follows [1].

1.   Apply the existing rules to the new objects so that it can determine which rules actually are a fit to the new objects.
2.   If none of the rules are matched, then fallback a must be chosen, or the objects would be classified as undefined.
3.   If more than one rule is applicable, then a negotiation among the rules must be performed to decide which one to be used.

For the discernibility function extracted from the decision table 2, we obtain the following sets of decision rules by:

1. If (a < 32.5), then d = M
2. If (s > 65) and (32.5 ≤ a < 50), then d = E
3. If (s < 65) and (32.5 ≤ a < 50), then d = M
4. If (s > 65) and (a ≥ 50), then d = E
5. If (s < 65) and (32.5 ≤ a < 50), then d = E
6. If (s < 65) and (a ≥ 50), then d = M

3. Particle Swarm Optimization

PSO was originally introduced by Kennedy and Eberhart [21]. The algorithm was inspired by a sociological observation of a flock of birds behavior while searching for food. Each member of the flock moves with a direction and speed influence by its own previous state and that of the as a whole flock.

PSO consists of a swarm (collection) of particles searching through the solution space. Each particle holds information that can potentially become the solution. Each particle has a position and velocity that are mutually affecting those of other particles. Each particle will adjust its parameter according to the swarm's best outcome, while still considering its own experience. Therefore, at any instance, the following information is maintained by each particle.

- x_i, the current position of the particle;
- v_i, the current velocity of the particle; and
- y_i, the personal best position of the particle (*pbest*); the best position visited so far by the particle.
- $\hat{y}$, the global best position of the swarm (*gbest*); the best position visited so far by the entire swarm.

The search performed by the swarm is either to maximize or minimize the objective function *f(x)*. The personal best position (*pbest*) is obtained by evaluating the following.

$$y_i(t+1) = \begin{cases} y_i(t) & \text{if } f(x_i(t+1)) \geq f(y_i(t)) \\ x_i(t+1) & \text{if } f(x_i(t+1)) < f(y_i(t)) \end{cases} \tag{3.1}$$

The global best position (*gbest*) is obtained by using

$$\hat{y}(t) \in \{y_0, y_1, ..., y_s\} = \min\{f(y_0(t)), f(y_1(t)), ..., f(y_s(t))\} \tag{3.2}$$

After each iteration, the current position (x_i) and velocity (v_i) are recalculated using

$$v_i(t+1) = \omega v_i(t) + c_1 r_1(t)(y_i(t) - x_i) + c_2 r_2(t)(\hat{y}(t) - x_i(t)) \tag{3.3}$$

$$x_i(t+1) = x_i(t) + v_i(t+1) \tag{3.4}$$

where ω is the inertia weight which reflects the memory of previous velocities. $y_i(t) - x_i$ (cognitive component) represents the particle's own experience as to where the best solution is. $\hat{y}(t) - x_i$ (social component) represents the direction of the entire swarm towards the best solution. The c_1 and c_2 are acceleration constants. $r_1(t)$, $r_2(t)$ are in the distribution of $U(0,1)$ which will be a random number between 0 and 1.

In image classification, the PSO algorithm is used to optimize the objective functions that are mainly to:

- Minimize the distance between pixels and cluster means for each cluster.
- Maximize the distance between clusters.

In unsupervised training, there is no prior knowledge of the number of clusters. Therefore the cluster validity is determined by the objective functions. In the algorithm, the Davies-Bouldin index is used as the means to evaluate the result of each iteration.

3.1 Cluster Validity – Davies-Bouldin Index.

The accuracy or validity of the classification results need to be measured using certain criteria. As a prerequisite, a set of objects needs to possess a natural group structure. In our image classification algorithm outlined in section 5, the Davies-Bouldin (DB) index is used as the aid in parameter tuning. Our objective function is to minimize the DB index, since a smaller index value indicates compact and well-separated clusters. The similarity index between two clusters C_i and C_j can be expressed as [17]

$$R_{ij} = \frac{s_i + s_j}{d_{ij}} \tag{3.5}$$

where s_i and s_j are a measure of distance within a cluster, and d_{ij} is the distance between cluster i and j. The s_i is defined as [17]

$$s_i = \left(\frac{1}{n_i} \sum_{x \in C_i} \left\| \mathbf{x} - \mathbf{m}_i \right\|^r \right)^{1/r} \tag{3.6}$$

where n_i is the number of pixels in the cluster C_i. The distance between two clusters d_{ij} is defined as [17]

$$d_{ij} = \left(\sum_{k=1}^{l} \left| m_{ik} - m_{jk} \right|^q \right)^{1/q} \tag{3.7}$$

where l is the number of clusters and m represents the mean distance.
Let R_i be defined as [17]

$$R_i = \max_{j=1,\dots,m, j \neq i} R_{ij}, \qquad i = 1,\dots,m \tag{3.8}$$

Then the DB index is defined as [17]

$$DB_m = \frac{1}{m}\sum_{i=1}^{m} R_i \tag{3.9}$$

4. The K-means algorithm for multispectral image classification

The K-means algorithm is one of the simplest and most efficient unsupervised learning algorithms to solve clustering problems in image segmentation. In this algorithm, random cluster means are assigned and repeatedly modified throughout the process in order to minimize the squared error function. Suppose there are N pixels in an image to be classified into m clusters. Each pixel v_i , where $1 \le i \le N$, is assigned to one of the clusters c_j , where $1 \le j \le m$, based on squared Euclidean distance of each pixel to each cluster mean.

$$d(v,c) = \sum_{i=1}^{N}\sum_{j=1}^{m}(v_i - c_j)^2 \tag{3.10}$$

Upon the completion of the assignment, each new cluster mean is calculated using

$$c_j = \frac{\sum_{v \in c} v_i}{n} \tag{3.11}$$

where $1 \le i \le N$, and n is the number of pixels in cluster c_j . The process ends when c_j stabilizes.

The weakness of K-means is that it is dependent on the initial selection of the cluster means and it may be trapped into locally optimal results. However, running the algorithm repeatedly and randomly selecting different sets of cluster means may offset the problem. In a paper by Hung and Germany [19] it is shown that the local optimal results may also be avoided by assigning the cluster means based on distribution of patterns in histogram of an image.

5. Multispectral Image Classification using Rough Set Theory

Multi-spectral images can be analyzed using rough set theory. However, since all the attribute values are analog, the discretization process is required. Multispectral images contain multiple bands, for example the RGB color band.

Object Index	R	G	B	Class
u_1	149	148	143	1
u_2	154	155	150	1
u_3	159	160	155	1
u_4	174	171	164	2
u_5	164	161	154	2
u_6	179	183	186	3
u_7	159	165	163	3
u_8	178	184	182	3

Table 5. Decision System Table for multi-spectral image.

The values of the condition attributes are obtained from the image data shown in figure 1, while the values of the decision attributes are obtained from 'ground truth' data..
Each object has three condition attributes, Red (R), Green (G) and Blue (B) which are associated with a decision attribute. The decision attributes signify the following:

1 Class 1 represents land
2 Class 2 represents village
3 Class 3 represents water.

The value of each attribute ranges from 0 to 255, hence the training data from Table 5 can be expressed as:

V_R={0, 149, 154, 159, 164, 174, 178, 179, 255}
V_G={0, 148, 155, 160, 161, 165, 171, 183, 184, 255}
V_B={0, 143, 150, 154, 155, 163, 164, 182, 186, 255}

Based on the above intervals, the following set of cuts are obtained.

For the R attribute:
(r, 151.5); (r, 156.5); (r, 161.5); (r, 169); (r,176); (r,178.5)
For the G attribute:
(g, 151.5); (g, 157.5); (g, 160.5); (g, 163); (g, 168); (g, 177); (g, 183.5)
For the B attribute:
(b, 146.5); (b, 152); (b, 154.4); (b, 159); (b,163.5); (b, 173); (b, 184)

The optimal set of cuts needs to be selected now. There are many ways to perform the selection. For decision table $A = (U, A \cup \{d\})$, a local method can be used as [1]:

Input: The consistent decision table A.
Output: The semi-minimal set of cuts D consistent with A.
Method: Initialize the binary tree variable T with the empty tree. Label the root by the set of all objects U and fix the status of the root to be unready.

while there is a leaf marked by unready **do**

    begin for any unready leave N of the tree **T**

        begin

            if objects labeling N have the same decision value **then**

            begin

                Replace the object set at N by its common decision

                Change the status of N to ready.

            end

            else

            begin

                Compute the value W^N (a,c) for all cuts from $\mathbf{C}_A$ and search for cut $(a^*,\ c^*)$ maximizing the function $W^N(.)$ i.e.

$$(a^*, c^*) = \arg \max_{(a,c)} W^N (a,c)$$

                Replace the label of N by $(a^*,\ c^*)$ and mark it as ready;

                Create two new nodes N1 and N2 with status unready as the left and right subtrees of N, where:

$$N_1 = \{u \in N : a^*(u) < c^*\} \text{ and}$$

$$N_2 = \{u \in N : a^*(u) \geq c^*\}$$

            end

        end

    end

return T

By applying the algorithm above to the image data as shown in Table 6, the following details are derived. For each cut of the R, G and B attributes, we find the cut that yields the maximum number of pairs. The search gives us (g, 160.5) as the optimal solution which yields 15 pairs.

The cut (g, 160.5) divides the set into two, $X_1 = \{u_1, u_2, u_3\}$ and $X_2 = \{u_4, u_5, u_6, u_7, u_8\}$. Notice that X_1 actually consists of objects of the same class, so the search ends. The search continues for X_2. Three sets of cuts are found from the R, G and B attributes for X_2. All of the cuts, (r, 176), (g, 177) and (b, 173) yield the same number of objects (4 pairs). We only need to select one, and the one chosen is (r, 176). Again, this cut divides the set into two, $Y_1 = \{u_4, u_5, u_7\}$ and $Y_2 = \{u_6, u_8\}$. Y_2 consists of objects of the same class, so the search ends. The search continues for Y_1. The cut that can discern the most from Y_1 is (r, 161.5).

The cut (r, 161.5) divides Y_1 into two sets, $Z_1 = \{u_4, u_5\}$ and $Z_2 = \{u_7\}$. The search ends since both sets contain objects of the same class.

The set of cuts selected are:

$$(r, 161.5);\ (r, 176.0)$$
$$(g, 160.5)$$

It appears that our data set only requires two attributes to be fully discerned. Note that different discretization methods will obtain different results. For example, if a naïve algorithm was used, the B attribute will be considered in generating the cuts.

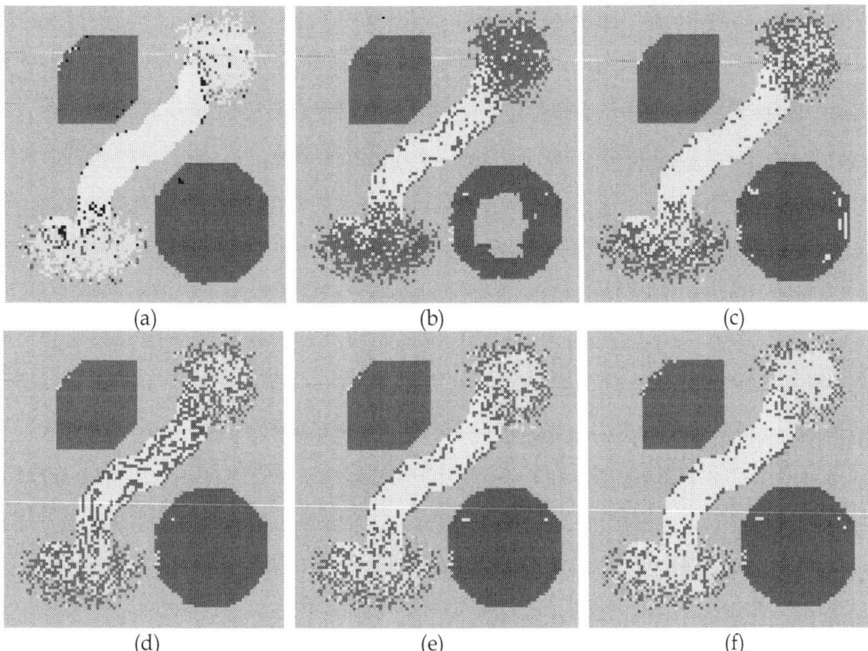

(a)   (b)   (c)

(d)   (e)   (f)

Fig. 7. Rough K-Means for artificial image. (a) Ground Truth, (b) w_{lower} = 0.55 and Threshold = 0.45 Accuracy = 68.17 % (c) w_{lower} = 0.65 and Threshold = 0.35 Accuracy = 82.83 %, (d) w_{lower} = 0.75 and Threshold = 0.25 Accuracy = 78.92 % (e) w_{lower} = 0.85 and Threshold = 0.15, Accuracy = 85.11 %, and (f) w_{lower} = 0.95 and Threshold = 0.05, Accuracy = 89.54 %.

Experiments using the K-means and rough K-means PSO algorithms are performed. For the comparison, the number of iteration is limited to 50 and the tolerance is set to 0.001. The result shown in Figure 8 is selected from the best outcome of 20 runs of the K-means and rough K-means algorithms. For the rough K-means PSO, 10 particles are used to explore the search space. Comparing the results of the K-means, rough K-means and rough K-means PSO algorithms, it reveals that although the improvement can be made, it is in the order of more or less 5 %. It is not very significant, but we should note that the rough K-means PSO achieve the optimal results independent of initial mean selections.

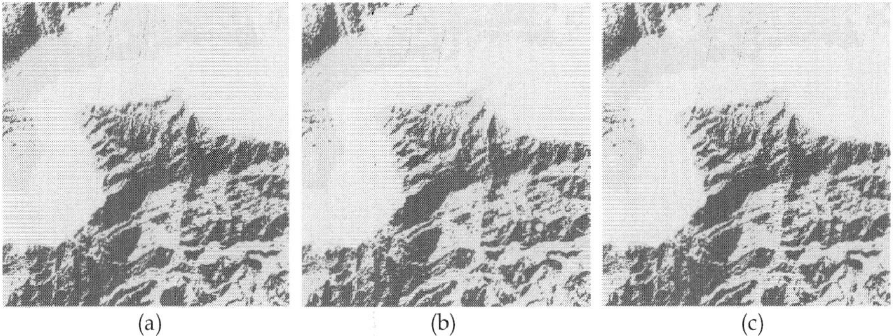

<table>
<tr><td align="center">(a)</td><td align="center">(b)</td><td align="center">(c)</td></tr>
</table>

Fig. 8. Comparison of the results. (a) K-Means Accuracy = 56.26 %, (b) Rough K-Means w_{lower} = 0.95 and Threshold = 0.05, Accuracy = 51.42 %, and (c) Rough K-Means PSO, Accuracy = 59.96 %.

While running, the algorithm is tuned by keeping track of the DB index and adjusting the PSO particle accordingly to calculate the new mean. Figure 9 shows the DB index tracking for Figure 8(a) and 8(c).

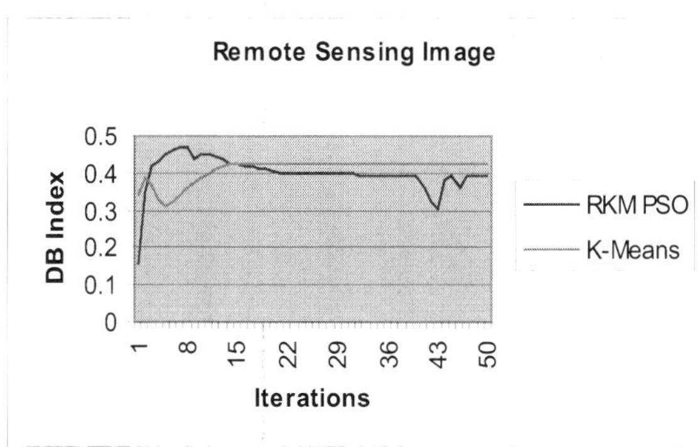

Fig. 9. K-means and rough K-means PSO DB index tracking for Figure 8(a) and 8(c).

Referring to Figure 9, it is apparent that the K-means algorithm eventually converges and locks into a certain mean value. The rough K-means PSO shows a better capability to search for solutions, because there are about 10 particles to keep track of global best and adjust the velocity towards the best solution in every iteration. Similar results are obtained from the remaining tests performed. The resulting improvement, however, is not as obvious as those shown in the artificial image (Figure 10). Part of the reason is because the artificial image does not have enough roughness. Hence, it is not difficult for K-means to perform well in this case. Figure 11 shows the tracking of the DB index.

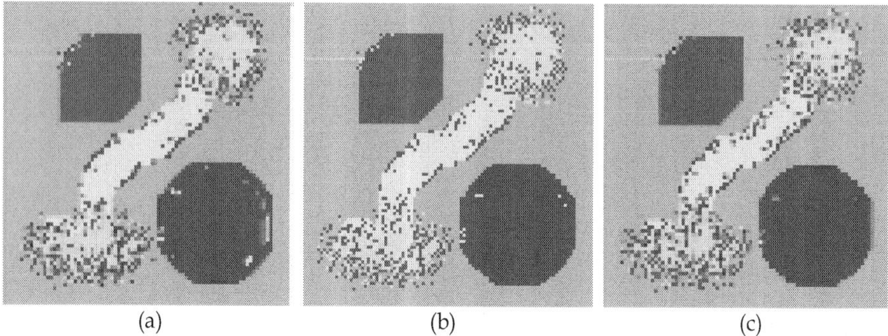

(a)   (b)   (c)

Fig. 10. Comparison of the results. (a) K-means Accuracy 90.12 %, (b) Rough K-means w_{lower} = 0.95 and Threshold = 0.05, Accuracy = 89.55 %, and (c) Rough K-means PSO, Accuracy = 90.65 %

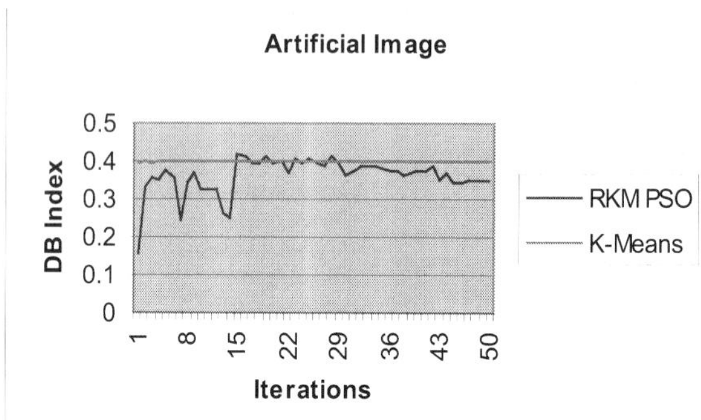

Fig. 11. The K-means and rough K-means PSO DB index tracking for the artificial image shown in Figure 10.

For the planet image, there is no ground truth data available. However, the visual inspection reveals improvement. Based on the results shown in Figure 12, we can see that the K-means algorithm actually has some difficulty in the segmentation of the blurred or rough boundaries. The rough K-means PSO however, appears to be able to discern the rough boundaries, and therefore comes up with a much more rounded shape for the planet. The outer shape of the planet appears sharper, more rounded and less distorted. Figure 13 shows the DB index tracking of the planet image.

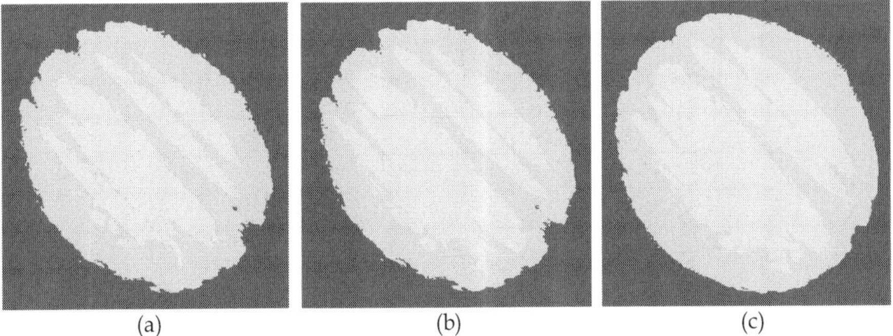

(a)   (b)   (c)

Fig. 12. Experimental results for planet image, (a) K-means, (b) rough K-means w_{lower} = 0.95 Threshold = 0.05, and (c) rough K-means PSO.

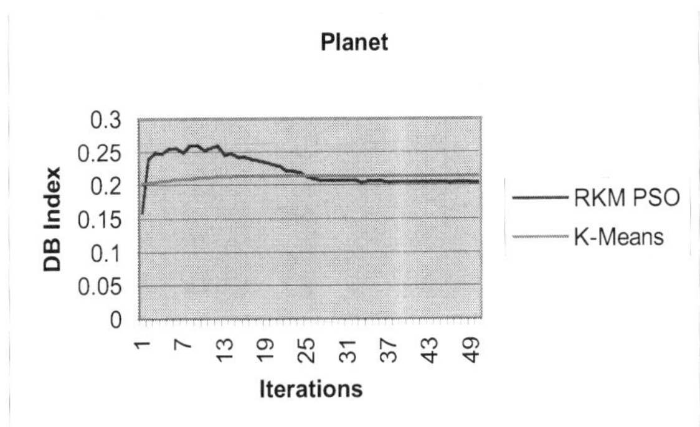

Fig. 13. The K-means and rough K-means PSO DB index tracking for the planet image shown in Figure 12.

8. Conclusions and Future Work

Image classification and segmentation by applying rough set theory may be approached from two different perspectives: unsupervised or supervised methods. From the experiment, it is generally shown that a supervised classification achieves better results as compared with the unsupervised methods. However, it should be noted that unsupervised classification may be preferred because it requires less prior knowledge. The K-means algorithm can be enhanced by using the rough set theory for image classification, however it has a practical limitation by itself, since the parameters (w_{lower} and the threshold value) are difficult to tune manually. To solve this problem, the PSO is used for tuning these parameters. The algorithm is tested on several and in general its improved noise immunity can be seen in the results especially when the images have rough boundaries or noisy details. For future work, the ant colony optimization and differential evolution algorithms will be explored for tuning the parameters in the rough K-Means algorithm.

9. References

[1] Bazan, J., H.S. Nguyen, S.H. Nguyen, P. Synak, and J. Wroblweski, "Rough Set Algorithms in Classification Proble," In L. Polkowski, S. Tsumoto, and T. Lin (Editors), Rough Set Methods and Applications. New York, Physica-Verlag Heidelberg New York (2000): pp 49-88.

[2] Bazan, J. and M. Szczuka, "The Rough Set Exploration System," In J. Peter and A. Skowron (Editors), Transactions on Rough Sets III. Springer (2005): pp 37-55.

[3] Bazan, J.and M. Szczuka, "RSES and RSESlib - A collection of tools for Rough Set Computations," Lecture Notes in Artificial Intelligence 3066, Heidelberg, Springer (2000): pp. 592-601.

[4] Kim, D., "Data classification based on tolerant rough set." Pattern Recognition, 34 (2001): pp. 1613-1624.

[5] Komorowski, J., Z. Pawlak, L. Polowski, and A. Skowron, "Rough Sets: A Tutorial," In S. K. Pal and A. Skowron (Editors), Rough Fuzzy Hybridization. Springer (1999): pp 3-98.

[6] Lillesand, T. M. and R. W. Kiffer, Remote Sensing and Image Interpretation (3rd ed), John Wiley & Sons, Inc., 1994.

[7] Lingras, P., "Unsupervised Rough Set Classification Using GAs," Journal of Intelligent Information Systems, 16. (2001): pp 215-228.

[8] Nguyen, H. S., Data Regularity analysis and applications in data mining, Ph.D thesis, Supervisor B. Chlebus, Warsaw University, 1999.

[9] Pawlak, Z., "Rough Set," International Journal of Information and Computer Science, 11. (1982): pp 341-356.

[10]Pawlak, Z., J. Grzymala-Busse, R. Slowinski and W. Ziarko, "Rough Sets," Communication of ACM, 38 (1995): pp 89-95.

[11] Schowengerdt, R. A., Remote Sensing: Models and Methods for Image Processing (2nd ed), Academic Press, 1997.

[12] Qin, Z., G. Wang, Y. Wu, and X. R. Xue, „A Scalable Rough Set Knowledge Reduction Algorithm," Heidelberg , Springer - Verlag, (2004): pp 445-454.

[13] Petrosino, A. and G. Salvi, "Rough Fuzzy set based scale space transforms and their use in image analysis." International Journal of Approximate Reasoning, 41 (2006): pp 212-228.

[14] Seul M., L. O'Gorman, and M. J. Sammon, Practical Algorithms for Image Analysis, Cambridge University Press, 2000.

[15] Sonka M., V. Hlavac, and R. Boyle, Image Processing, Analysis, and Machine Vision (2nd ed), Brooks/Cole Publishing Company, 1999.

[16] Das S., A. Abraham and S. K. Sakar, "A Hybrid Rough Set – Particle Swarm Algorithm for Image Classification." Hybrid Intelligent System 6th Conference, 6 (2006): pp 26.

[17] Theodoris S. and K. Koutroumbas, Pattern Recognition (3rd Ed), Academic Press, 2006.

[18] Jang Q. and S. S. R. Abidi, "A hybrid of conceptual clusters, rough sets and attribute oriented induction for inducing symbolic rules," Machine Learning and Cybernetics Conference, 9 (2005): pp 5573 – 5578.

[19] Hung, C. C. and G. Germany, "K-means and Iterative Selection Algorithms in Image Segmentation," in proceedings of IEEE Southeast Conference, Session 1: Software Development, Jamaica, West Indies, April 4 - 6, 2003.

[20] Lingras, P and C. West, "Interval Set Clustering of Web Users with Rough K-Means," Journal of Intelligent Information Systems, 23.1 (2004): pp 5-16.

[21] Kennedy, J and R. C. Eberhar, "Particle Swarm Optimization," Proceedings of IEEE International Conference on Neural Networks, Perth, 1995.

[22] Pal, N. R., K. Pal, J. M. Keller and J. C. Bezdek, "A possibilistic fuzzy c-Means clustering algorithm," IEEE Transaction on Fuzzy Systems, 13. 4 (2005): pp 517-530.

[23] Dariusz, M. and T. W. Sławomir, "Standard and Genetic k-means Clustering Techniques in Image Segmentation," The 6th International Conference on Computer Information Systems and Industrial Management Applications, (2007): pp 299-304.

[24] Bazan, J., R. Latkowski, S. H. Nguyen, P. Synak, A. Wojna, M. Wojnarski and J. Wróblewski, Rough Set Exploration System Software. Version 2.2. http://logic.mimuw.edu.pl/~rses

[25] Hung, C. C., H. Purnawan, and B.-C. Kuo, "Multispectral Image Classification Using Rough Set Theory and the Comparison with Parallelepiped Classifier," The IEEE International Conference on Geosciences and Remote Sensing (IGARSS), Barcelona, Spain, July 23-27, 2007 (CD).

[26] Hung, C. C. and H. Purnawan, "A ybrid Rough K-means Algorithm and Particle Swarm Optimization for Image Classification," Springer-Verlag, Lecture Notes in Computer Science (LNAI 5317 pp. 585-593), MICAI 2008: Advances in Artificial Intelligence.

29

Derivative analysis of hyperspectral oceanographic data

Elena Torrecilla and Jaume Piera
Marine Technology Unit, Spanish National Research Council (UTM - CSIC)
Spain
Meritxell Vilaseca
Centre for Sensors, Instruments and Systems Development,
Technical University of Catalonia (CD6 - UPC)
Spain

1. Introduction

It has long been recognized that transmission of light in sea water is essential to the productivity of the oceans. It provides the energy necessary for ocean currents, and the majority of marine life is supported by the thin layer of warm water near the ocean's surface. Light plays a decisive role in the primary formation of biomass by oceanic chlorophyll-bearing marine plants, through the process of photosynthesis, which is the basis of the entire marine food chain. Light transmission is therefore a key factor in the ecology of the upper ocean and biogeochemical cycling, since it has a strong influence on the dynamics of the chemical compounds. In addition, variability of light in the sea is strongly influenced by the distribution of the components in the water column, which varies across horizontal and vertical space and time scales.

The subdiscipline of ocean optics, which concerns the assessment of the propagation of light through the oceanic water column and surface, has become fundamental for understanding water dynamics and composition. Over the last decades, there has been a noticeable increase in the number of coastal and open-ocean studies using in situ and remote sensing optical measurements. Recent advances in monitoring surface and underwater optical properties have been specially related to the progressive shift from using multispectral to high spectral resolution (hyperspectral) acquisition systems. Hyperspectral technology has opened the possibility for optical oceanographers to more accurately characterize complex oceanic environments (Chang et al., 2004). A broad range of hyperspectral sensors, covering from several hundreds to thousands of contiguous spectral bands, have recently been developed for different monitoring applications. The use of high sample rate hyperspectral systems may provide the potential for mapping phytoplankton functional types, including the detection of harmful algal blooms (Nair et al., 2008; Craig et al., 2006).

The higher spectral resolution provides the opportunity to better perform spectral shape analysis of oceanic optical measurements. Techniques such as derivative spectroscopy, which serves to enhance subtle features in spectra, have been extensively used to assess information regarding optically significant water constituents. Derivative of hyperspectral data can yield more information than traditional analysis based on ratios of discrete spectral bands (multispectral approaches). However, when hyperspectral measurements are to be used for further analysis, the uncertainties of the measurement system must be taken into account. Appropriate calibration strategies must be followed, since the accuracy of the measurements relies on the associated uncertainties of the measurement system. A well-calibrated spectroradiometer includes a characterization process, describing the instrument behavior in terms of the spectral sensitivity, the signal-to-noise ratio, the dark current, the wavelength calibration, the nonlinearity, the temperature dependence of the measurements and the spectral scattering, or what is called the spectral stray-light (Brown et al., 2006). The stray-light of a spectroradiometer is described as the unwanted background radiation, scattered due to imperfections in the dispersive device and other internal optical elements.

In ocean optics applications, special attention must be paid to corrections for the inherent distortions of the hyperspectral sensors (e.g. noise, spectral stray-light), because the errors in the measured radiance distributions may be potentially significant and lead to inaccurate retrievals of water properties. This issue becomes even more important when commonly derivative spectroscopy is used to explore subtle features in shape of hyperspectral data (Torrecilla et al., 2008a). Derivative analysis is notoriously sensitive to noise and smoothing techniques must be used to overcome this problem. Therefore, in order to make an optimal application of the derivative analysis, it is worth noting that a suitable selection of the smoothing and derivative parameters (filter size and band separation) must be done according to the resolution of each type of hyperspectral data (Torrecilla et al., 2007). An effort must be made to determine the best trade-off between denoising and the ability to resolve spectral features of interest.

There are two major goals in this chapter. The first one is to offer a brief review of relevant advances that have been achieved using hyperspectral technology in oceanography and a description of some important issues that must be considered when derivative spectroscopy is applied to data acquired by hyperspectral sensors. The second goal is to provide some results demonstrating the feasibility of applying derivative spectroscopy to hyperspectral measurements of remote-sensing reflectance in open-ocean environments, in particular to identify phytoplankton pigment assemblages. In this part, a simulation-based framework is proposed to address this issue and special attention is given to the role of involved parameters when derivative analysis of hyperspectral data is performed. For instance, the effect of spectral sensitivity of each high spectral resolution instrument considered on the results of the derivative-based approach is also discussed.

2. Ocean Optics: an overview

In order to understand light propagation in the ocean, which is an optically complex environment, the radiative transfer theory was defined as the theoretical framework (Kirk, 1994; Mobley, 1994). Optical properties of sea water are classified into inherent and apparent optical properties (IOPs, AOPs) and are related via the radiative transfer equation (RTE). Most of the quantities are shown in Figure 1 and have wavelength dependence.

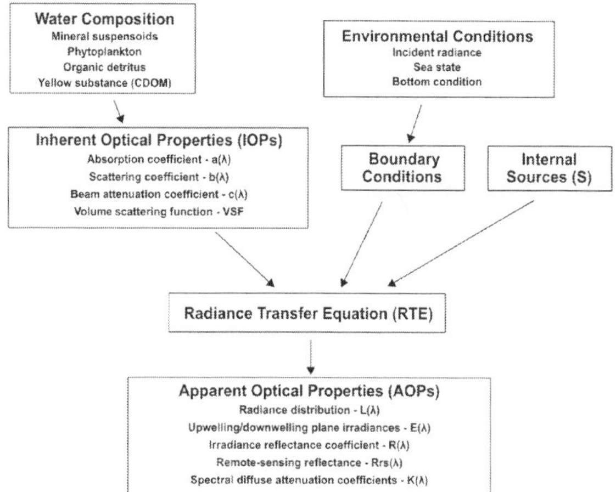

Fig. 1. Schematic diagram illustrating the relationships among the ocean optics quantities, based on Figure 3.27 of Mobley, 1994.

Spectral absorption and scattering coefficients are inherent optical properties (IOPs), i.e. they are only affected by the constituents of the aquatic medium and are independent of the ambient illumination conditions. IOPs are additive and their variation is generally attributed to variability in four constituents of the aquatic medium in natural oceanic waters: pure sea water, phytoplankton, detritus and gelbstoff or colored dissolved organic matter (CDOM). In contrast, the apparent optical properties (AOPs), such as spectral radiance and irradiance, are not additive and depend both on the medium and on the directional structure of the ambient light field. Several types of models have been reported to predict AOPs given field measurements of IOPs and environmental conditions ("direct models") or to estimate IOPs given experimental measurements of AOPs ("inverse models") (IOCCG, 2000).

In situ and remote sensed optical observations of IOPs and AOPs are two complementary oceanographic applications that offer the potential of being direct proxies for important biological and biogeochemical variables in the ocean (Chang et al., 2006). Remote sensing provides global-scale optical data, whereas in situ measurements are useful for calibration of remote sensing measurements and for obtaining higher temporal resolution data. Both types of observations of the ocean are important for providing relevant information regarding the relative concentrations of optically significant constituents in the water column. Several studies have traditionally focused on the development of bio-optical algorithms linking measurable optical properties to the primary pigment in phytoplankton, chlorophyll-*a*, a proxy for the phytoplankton biomass (Bricaud et al., 1998; Reynolds et al., 2001; O'Reilly et al., 2000), and on the detection of phenomena such as marine harmful algal blooms (HABs) (Cullen et al., 1997). However, other approaches have been more recently applied to estimate the surface concentration of particulate organic carbon (POC) from optical measurements (Stramski et al., 2008), to derive the vertical distribution of phytoplankton communities in the open-ocean based on the near-surface chlorophyll-*a* content (Uitz et al., 2006), and to map phytoplankton functional types (PFTs) from ocean-color data (Nair et al., 2008), which serve as a contribution to the mapping of biodiversity in marine phytoplankton on the global scale.

3. Hyperspectral Data Acquisition Systems in Oceanography

Technological advances and innovations in instrumentation have given rise during the last few decades to a broad range of optical sensors for different monitoring applications in oceanography, including active and passive systems. Passive radiometric systems have been widely used to measure properties such as radiance and irradiance distributions, and solar-induced fluorescence. They can operate only with sunlight and are restricted to measurements within the euphotic depth (i.e. to a depth where light intensity falls to 1% of that at the surface). On the other hand, active radiometric systems use an internal light source to measure other optical properties such as absorption, scattering and fluorescence. These systems are able to provide optical measurements at deeper depths and at wavelengths that have reduced penetration in the ocean (e.g. the UV portion of the electromagnetic spectrum). However, their use involves a higher power consumption.

The rapid maturing of optical instrumentation has led to the most significant recent advance which is the development of hyperspectral sensors. The ability to measure surface and underwater light fields at hundreds of narrow and closely spaced wavelength bands, with a resolution better than 10 nm, has become one of the most powerful and fastest growing technologies in the field of ocean optics. Relative to multispectral measurements, hyperspectral sensors potentially enable the extraction of more detailed spectral information than is possible with any other type of in situ and remote sensing optical data. Figure 2 shows examples of phytoplankton absorption (IOP) and remote-sensing reflectance spectra (AOP), gathered by multi- and hyperspectral systems. Hyperspectral patterns of absorption or reflectance across wavelengths provide more information about spectral singularities.

The advantage of a hyperspectral over a multispectral inversion for phytoplankton species identification stems from the fact that more accurate spectral information, such as spectral features related to characteristic phytoplankton pigment absorption peaks, is resolved. In the past, analyses of single band-ratios obtained in discrete multispectral bands were employed to resolve a variety of water column properties (e.g. chlorophyll-*a* concentration (O'Reilly et al., 2000), the presence of specific species and other water constituents (Sathyendranath et al., 2004), etc.). However, the advent of hyperspectral sensors provided ample spectral information and facilitated the identification of water column constituents, such as phytoplankton species, which are spectrally unique. For instance, Craig et al. (2006) assessed the feasibility of remote detection and monitoring of the toxic dinoflagellate, *Karenia brevis*, applying two numerical methods to in situ hyperspectral measurements of remote-sensing reflectance. Bracher et al. (2008) recently adapted the differential optical absorption spectroscopy (DOAS) technique for retrieving the absorption and biomass of two major phytoplankton groups of cyanobacteria and diatoms from high spectrally resolved data of the satellite sensor SCIAMACHY (Scanning Imaging Absorption Spectrometer for Atmospheric Chartography). Lohrenz et al. (2008) directly explored the hyperspectral patterns of remote-sensing reflectance to better characterize water mass properties in coastal areas, and Gagnon et al. (2008) performed sea-bed mapping of benthic assemblages in optically complex shallow waters.

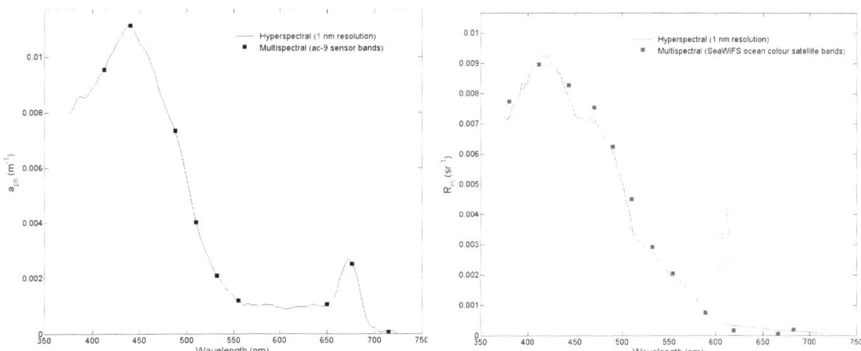

Fig. 2. (Left) Hyperspectral (solid continuous line, 1 nm resolution) and multispectral (closed circles) phytoplankton absorption spectra. (Right) Hyperspectral (solid line, 1 nm resolution) and multispectral (closed circles) remote-sensing reflectance spectra. (Absorption and multispectral remote-sensing reflectance data were provided by Dariusz Stramski and Rick Reynolds, Scripps Institution of Oceanography, from field measurements during the ANT-XXIII/1 expedition on R/V Polarstern in the eastern Atlantic in 2005 (Stramski et al., 2008)).

3.1 Hyperspectral Sensors

Hyperspectral sensors generally use diffraction gratings or linear variable optical filters as a dispersive element to separate light into specific wave bands centered on desired wavelengths, and different scanning mechanisms to generate one- or two-dimensional high-resolution spectra or images. Mechanical-scanning spectrometers have been traditionally employed to obtain a spectrum by using a single photo-detector and rotating the dispersive element over a given specific spectral region after passing through the sample. However, more recent spectrometers use a fixed grating and multi-element array detectors that allow an entire spectrum to be acquired over some finite spectral region simultaneously.

Array spectrometers are being widely used as a tool for rapid measurements of spectral distributions in oceanographic applications (Figure 3) in which acquisition speed is an important issue. Spectra can be measured using three types of arrays: charge-coupled device (CCD), photodiode or complementary metal-oxide semiconductor (CMOS). High performance is available in all technologies today when they are designed properly, and each have their own strengths and weaknesses. For instance, CCD arrays show a slightly higher dynamic range (sensitivity) and a lower system noise than CMOS ones, whereas CMOS ones offer more integration, smaller system sizes and higher speeds. The proper selection of the type of array depends on many parameters (e.g. pixel dimensions, sensitivity, spectral range coverage, dynamic range, saturation exposure, integration time) and on the specific application. Other important advantages of array spectrometers are the non moving parts, the robustness and the low production costs. A key challenge has also been to construct small size array spectrometers without compromising performance. However, they involve several drawbacks compared with mechanical-scanning spectrometers, such as the fixed wavelength resolution, the lower sensitivity and the higher stray-light radiation (described in more detail below), due to the lack of an output slit and the integral illumination over the full wavelength range.

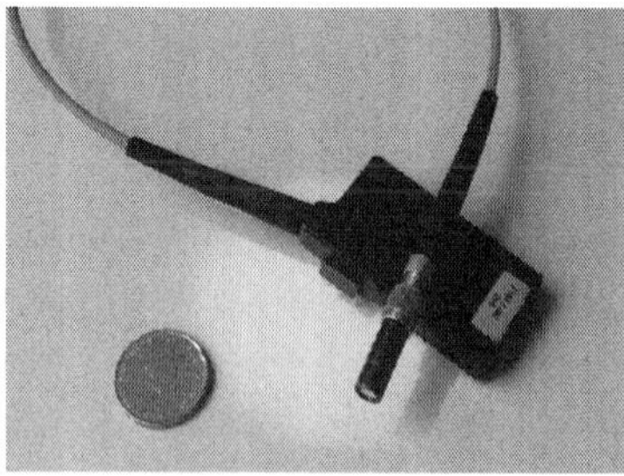
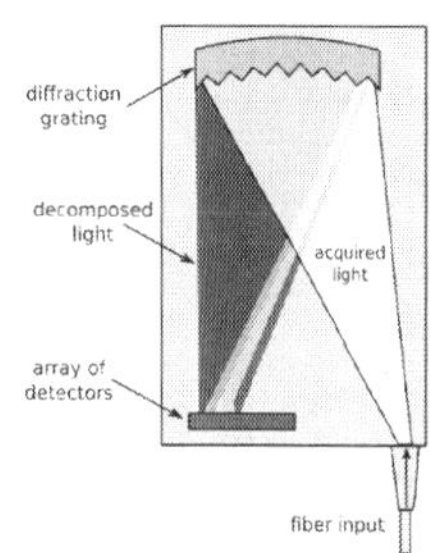

Fig. 3. Example of a miniature hyperspectral CMOS-array spectrometer designed for oceanographic monitoring purposes and its interior view (Pons et al., 2007). The concave grating splits the light collected through the input fiber and images each component in one detector in the array.

New technologies and the miniaturization of electro-optical components have permitted the development of accurate, low-cost and energy-efficient hyperspectral sensors designed to measure oceanic high spectral resolution IOPs and AOPs (absorption, total upwelling or downwelling radiance and irradiances, etc.). In the design and development of an instrument for hyperspectral measurements in the ocean, it is necessary to consider several optimal characteristics required of its detection system. The wavelength coverage should be broad, ideally from 350 to 750 nm. A low-light-level detector should be selected with a large dynamic range. Therefore, its responsivity should be high over the whole spectral region covered, especially in the blue and red spectral portions, and the dark signal should be low. Furthermore, in order to minimize a variety of external perturbations during measurement (e.g. changing sky conditions, surface wave noise producing high-frequency fluctuations in irradiance, ship shadow effect), the scan time of the system should be fast. The fastest devices are multi-element array spectrometers with a fixed grating system.

3.2 Oceanographic Platforms for Hyperspectral Sensors

Currently available miniature spectrometers based on linear array detectors represent best possible compromises between size, cost, resolution, range and performance. As a result, their use is no longer limited to bench-top applications but they are being deployed on fully fledged sea-going observational platforms. New interdisciplinary ocean observing platforms incorporating hyperspectral sensors are being deployed (Dickey et al., 2006). Various satellite-based hyperspectral imagers are currently available: NASA's Hyperion sensor on the EO-1 satellite, ESA's CHRIS sensor on PROBA satellite or the U.S. Air Force Research Lab's FTHSI sensor on the MightySat II satellite. In addition, examples of existing hyperspectral airborne sensors providing high spatial resolution images of specific areas are the Ocean Portable Hyperspectral Imager for Low-Light Spectroscopy (PHILLS, Davis et al., 2002), the Compact Airborne Spectrographic Imager (CASI), the Airborne Visible Infrared Imaging Spectrometer (AVIRIS), etc. Other in situ ocean observing platforms for which hyperspectral sensors may be suitable are ships and moored buoys (Kuwahara et al., 2007). These platforms offer continuous measurements at a high spatial and temporal resolution delivering data even under cloudy conditions, thereby complementing discrete observations by satellites and aircrafts and providing the data essential for calibration and validation purposes. Furthermore, Lagrangian platforms that follow a particular water mass (e.g. floats and drifters) can provide hyperspectral measurements in oceanic regions not normally accessible by satellites or oceanographic research vessels. The goal of these research initiatives is to determine the material composition of the water mass under investigation through the accurate analysis of simultaneous optical and hydrographic measurements.

The number of newly emerging ocean observing platforms capable of incorporating hyperspectral instrumentation is continuously growing. For instance, new autonomous underwater vehicles (AUVs) have been developed recently with integrated equipment for adaptive sampling and the ability to perform a wide-range of preprogrammed monitoring surveys over extended periods of time (Perry & Rudnick, 2003). Another member of this family of instruments are the special AUVs called gliders that propel themselves through the water by changing their buoyancy with a minimal expenditure of energy. The glider provides a wide spatial coverage measuring – among others - the optical properties during periods of three or four weeks (Woods, 2009). Another approach to monitor the variability of optical properties in the ocean is the use of profiling systems which can, for instance, be part of some of the emerging cabled observatories. Instrument platforms such as the Vertical Profiler System (VPS), part of Neptune Canada (Neptune, 2009), the world's first regional-scale underwater ocean observatory plugged directly into the Internet, offer great power and bandwidth for gathering large quantities of hyperspectral measurements.

The amount of hyperspectral data sets available collected using all these types of platforms, covering a wide range of temporal and spatial scales, will increase in the near future. New interdisciplinary research initiatives are being successfully carried out such as the Hyperspectral Coastal Ocean Dynamics Experiment (HyCODE, 2009) conceived to exploit the new capabilities of hyperspectral ocean color sensors. Within this framework, several collaborative short-term and long-term field experiments, involving the simultaneous use of hyperspectral devices deployed in situ and remotely, are being conducted for calibrating, groundtruthing and relating subsurface optical properties to remote sensing ocean color measurements. These investigations are essential to develop and validate optical models and to further our understanding of the processes that contribute to the temporal and spatial variability of IOPs and AOPs in the ocean. One of the most important challenges of HyCODE experiment is the standardization of the quality control protocols, including the definition of procedures to perform detailed analysis of the reliability of measurements and an evaluation of the impact of instrumentations inaccuracies.

4. Hyperspectral Data Pre- and Post-Processing Methods in Oceanography

4.1 Data Quality Assurance

When hyperspectral measurements can be performed and are to be used for further analysis, the uncertainties of the measurement system must always be taken into account. Appropriate calibration and pre-processing strategies must be undertaken, since the accuracy of the measurements relies on the associated uncertainties of the measurement system. The clearest example in optical oceanography is perhaps the in situ radiometric measurements carried out during several field campaigns, which are still currently essential to provide a proper calibration and validation of remote sensing measurements collected by ocean color satellites and airborne platforms (Zibordi et al., 2001; McClain et al., 2004; Clark et al., 2003).

A well-calibrated high spectral resolution array-based spectrometer, for both in situ and remote sensing applications, should include likewise its characterization and an evaluation of all meaningful sources of uncertainty (Lewis, 2008; Voss et al., 2008). It is therefore always necessary to describe the instrument's behavior in terms of responsivity, signal-to-noise ratio (SNR), dark current, nonlinearity, temperature dependence of measurements and spectral scattering, or what is called the spectral stray-light of an instrument.

Changes in the thermal response of these systems are due to the silicon used for the arrays of detectors. For instance, within the framework of the SORTIE (Spectral Ocean Radiance Transfer Investigation Experiment) project, corrections to in situ hyperspectral radiometric measurements were performed and demonstrated to be very repeatable (McLean, 2008). The thermal characterization was carried out using shutters and installing a thermistor on the diode array. The spectral scattering or stray-light radiation of an array-based spectrometer is described as the unwanted background radiation that has been scattered due to imperfections in the fixed dispersive element and other optical elements, such as higher-order diffraction gratings, surfaces and internal baffles. The characterization of an instrument's response for measurement errors arising from its spectral stray-light can be performed using a set of monochromatic spectral line sources covering the entire instrument's operational spectral range. This method is based on computing the ratio of the spectral stray-light signal to the total signal within the bandpass of an array-based spectrometer (Brown et al., 2006). Figure 4 shows an example of the effectiveness of this spectral stray-light correction method applied to the hyperspectral CMOS-array spectrometer from Figure 3 (Torrecilla et al., 2008a). The correction method has been proved to be effective using a calibration broadband light source and a green absorption bandpass filter. The stray-light signals outside the filter's bandpass region are clearly reduced by more than two orders of magnitude, to a level of 10^{-4}.

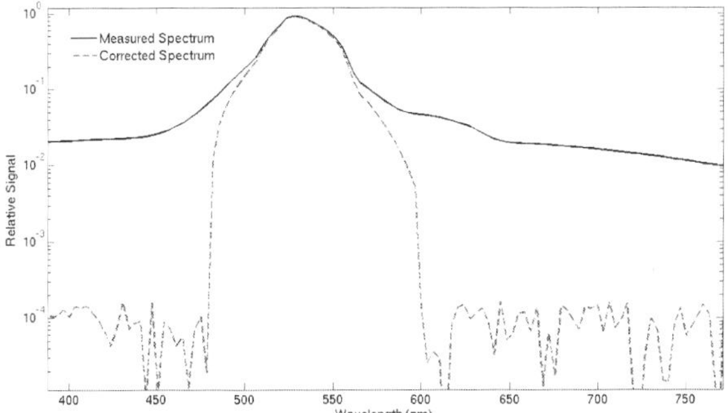

Fig. 4. Spectral stray-light correction applied to the miniature hyperspectral CMOS-array spectrometer in Figure 3, using a broadband source with a bandpass filter. Normalized measured and corrected signals. The y-axis is a logarithmic scale.

A data pre-processing and corrections for the spectral stray-light must be given special attention for underwater optical measurement purposes, in which very weak signals are collected and the errors in the measured radiometric distributions may be potentially significant, leading to inaccurate retrievals of water properties. For instance, Figure 5 (left panel) depicts measured and spectral stray-light corrected signals corresponding to field underwater light measurements gathered by a miniature hyperspectral CCD-array spectrometer at two different depths at a test site in the Alfacs Bay (Ebre Delta, NW Mediterranean). It is worth noting that the difference in percentage (Figure 5, right panel) between measured and spectral stray-light-corrected data at each depth has a spectral dependency (i.e. it is not the same for each spectral band). Therefore, the amplitude and shape of signals is modified when the spectral stray-light correction is applied. This is an issue that becomes very important when spectral shape analysis techniques such as the central topic of this chapter derivative spectroscopy, which is described below are used to explore subtle features in hyperspectral data.

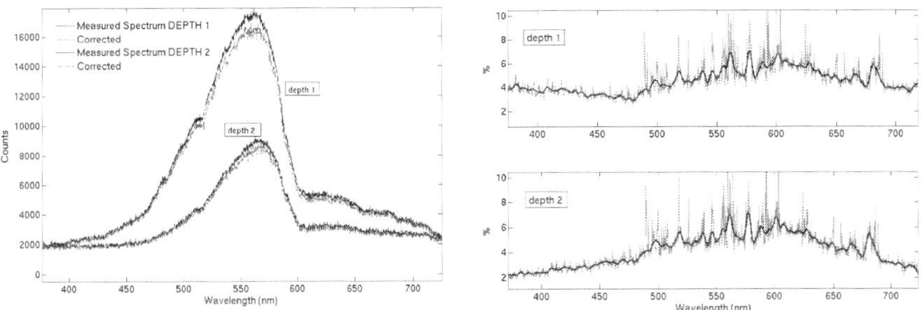

Fig. 5. (Left) Spectral stray-light correction applied to field underwater light measurements at two depths, acquired by a miniature hyperspectral CCD-array spectrometer. (Right) Difference in percentage between measured and corrected signals at each depth.

In addition to the above-mentioned corrections, new spectrometric devices and deployment techniques are now being developed, in order to optimize spectrometer's throughput and accuracy. Some spectrometers are designed with a two-dimensional area detector. They offer a significant improvement in the signal-to-noise ratio by averaging a vertical row of pixels to obtain each value of the whole spectrum at each wave band. Another solution is a new deployment technique called the multicast technique, which consists in performing and processing several casts in order to achieve the highest accuracy when underwater radiometric measurements are collected by hyperspectral devices (McLean, 2008). These are critical issues that should be tackled in addition to the radiometric and wavelength calibrations that must be carried out periodically using standard and highly characterized lamps (e.g. at the National Institute of Standards and Technology [NIST]).

4.2 Derivative Analysis

The emerging hyperspectral technology in remote sensing and in situ optical oceanography leads to a need for ongoing evaluation and improvement of hyperspectral processing methods. The higher spectral resolution provides the opportunity to develop and evaluate advanced methods of spectral shape analysis, such as derivative spectroscopy, that better distinguish subtle features in spectra and may be critical for discriminating optically significant water constituents.

Derivative spectroscopy has been commonly used in the analysis of hyperspectral data using different computation algorithms (Tsai & Philpot, 1998; Ruffin et al., 2008). The process of estimating derivative spectra can be addressed using a finite divided difference algorithm, named "finite approximation", which consists in computing the changes in curvature of a given spectrum over a sampling interval ($\Delta\lambda$) or band separation (BS), defined as $\Delta\lambda = \lambda_j - \lambda_i$, where $\lambda_j > \lambda_i$. The first and the nth derivative are obtained using Eqs. 1 and 2, respectively:

$$\left.\frac{ds}{d\lambda}\right|_i \approx \frac{s(\lambda_i) - s(\lambda_j)}{\Delta\lambda} \tag{1}$$

$$\left.\frac{d^n s}{d\lambda^n}\right|_j = \frac{d}{d\lambda}\left(\frac{d^{(n-1)}s}{d\lambda^{(n-1)}}\right) \tag{2}$$

Derivative analysis can be applied to hyperspectral measurements of both inherent and apparent oceanographic optical properties (e.g. absorption, irradiance, remote-sensing reflectance). It is a useful tool for enhancing spectral features, which are often relevant, for instance, because they can be related to absorption bands of pigments present in the considered water samples (Figure 6, left panels). Derivatives of hyperspectral data are less affected by possible spectral fluctuations of skylight conditions. However, successful extraction of spectral details of interest through derivative spectroscopy depends on the band separation chosen. The importance of selecting a suitable band separation ($\Delta\lambda = BS$) in each case stems from the fact that spectral data features of interest with a smaller scale than the band separation will not be preserved in the derivative results. Figure 6 (right panel) shows different derivatives of the spectrum shown in the left panel, computed according to different finite band resolutions or band separations (BS).

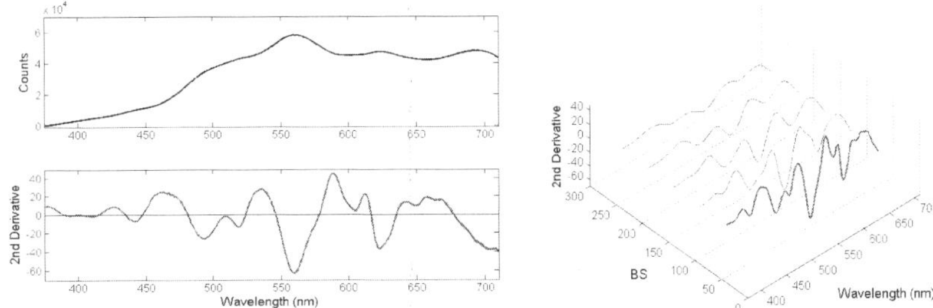

Fig. 6. (Left) Original spectrum and its second derivative of a water sample containing a unique phytoplankton algal culture (*Alexandrium minutum*), acquired by a miniature hyperspectral CCD-array spectrometer. (Right) Second derivatives computed for several values of band separation (BS), each of them leading to spectral features at different scales.

Noise level in hyperspectral data can be considerable, as the small amount of energy gathered by the narrow bandwidth may be exceeded by the intrinsic sensor's noise. In order to make an optimal application of derivative analysis, which is a technique clearly sensitive to noise, smoothing techniques must be applied previously to derivative computation of hyperspectral data (Vaiphasa, 2006). A number of smoothing algorithms have been developed within the last few decades (e.g. Savitzky-Golay, Kawata-Minami or mean-filter smoothing). In all approaches the smoothing level applied depends on the size of the filter window (WS). It is therefore worth noting that an appropriate selection of the smoothing and derivative parameters (filter size and band separation) must be done according to the resolution of each type of hyperspectral data (Torrecilla et al., 2007). An important effort must be made to determine the best compromise between denoising and the ability to resolve fine spectral details of interest.

The advantages offered by hyperspectral measurements of remote-sensing reflectance ($R_{rs}(\lambda)$) in combination with derivative spectroscopy have been recently exploited for various purposes in optical oceanography. For example, Louchard et al. (2002) assessed qualitative and quantitative information regarding major sediment pigments of benthic substrates from derivatives of shallow marine hyperspectral $R_{rs}(\lambda)$. In addition, a bathymetric algorithm was suggested through the analysis of the magnitude of some derivative peaks. Craig et al. (2006) detected a toxic algal bloom from the analysis of the fourth derivative of phytoplankton absorption spectra, estimated from in situ hyperspectral measurements of $R_{rs}(\lambda)$ using a quasi-analytical inversion algorithm. The monitoring of these harmful algal blooms is possible because some accessory pigments are unique to individual phytoplankton taxa and can be better differentiated in hyperspectral absorption spectra than in multispectral spectra with a limited number of wavelengths. The advantage of a hyperespectral over a multispectral inversion for identification of other algal blooms was also confirmed by Lubac et al. (2008). In this case, the inversion and quantitative assessment are directly based on the analysis of the position of the maxima and minima of the second derivative of $R_{rs}(\lambda)$.

4.3 A model-based approach to evaluating the efficacy of processing methods

Most of the studies dedicated to derivative analysis of hyperspectral remote-sensing reflectance have traditionally focused on the use of ratios of discrete values of derivative spectra to characterize or classify oceanic environments. However, more advanced and contemporary methods are beginning to take advantage of the information contained in the whole derivative spectrum. For example, Filippi (2007) proposes a combination of derivative spectroscopy and artificial neural network (ANN) algorithms. The derivative-neural approach has been proven to be effective for providing bathymetry, bottom type and constituent concentration estimations from remote-sensing reflectance $(R_{rs}(\lambda))$ measurements.

In order to test the effectiveness of such a methodology, or the one described in this chapter, it would be desirable to have a large number of observed oceanographic hyperspectral data sets, covering a wide range of environmental conditions. However, because valid hyperspectral data sets are still difficult to obtain and often unavailable, oceanic radiative transfer (RT) models are used. Hydrolight is an example of a radiative transfer numerical model (Mobley, 1994; Mobley & Sundman, 2008), which computes radiance distributions and derived quantities (e.g. $R_{rs}(\lambda)$) given water column IOPs and other oceanographic environmental conditions. The Hydrolight code employs mathematically sophisticated invariant imbedding techniques to solve the radiative transfer equation (RTE, Figure 1) and offers the possibility of performing numerical simulations in controlled environments (Albert & Mobley, 2003; Kempeneers et al., 2005). For instance, Figure 7 depicts several Hydrolight-generated hyperspectral $R_{rs}(\lambda)$ spectra, corresponding to different open-ocean scenarios, each dominated by a single phytoplankton group.

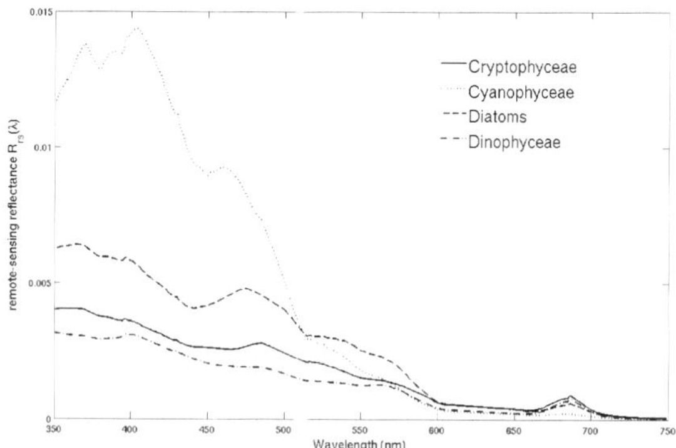

Fig. 7. Results of Hydrolight radiative transfer simulations, showing four remote-sensing reflectances $R_{rs}(\lambda)$ (1 nm resolution) obtained under different composition conditions. Labels indicate each dominant phytoplankton group considered in each case.

One of the logical steps for improving the Hydrolight-based approaches, when simulated hyperspectral $R_{rs}(\lambda)$ are to be used as a basis for further validation of some processing techniques, is to conduct sensor experiments. With the goal of exploring the potential of any processing technique, the response of the sensor with which the $R_{rs}(\lambda)$ acquired would be acquired must be taken into account. More accurate and realistic retrievals will be carried out from hyperspectral oceanographic data if, as has been stated as a priority in Section 4.1, the effect of the sensor in terms of noise, sensitivity, spectral resolution, stray-light, etc. is included in the simulation-based approach. This is one of the main issues that will be discussed in the next section.

5. Experimental Results

5.1 Experimental Design

In the following study, we investigate the potential offered by hyperspectral sensors and derivative spectroscopy to identify phytoplankton assemblages from hyperspectral measurements of remote-sensing reflectance ($R_{rs}(\lambda)$) in open-ocean waters. A simulation-based framework was used to achieve this goal, which includes the use of the Hydrolight-Ecolight version 5 radiative transfer model (Mobley & Sundman, 2008). The analysis methodology of this research is presented schematically in Figure 8.

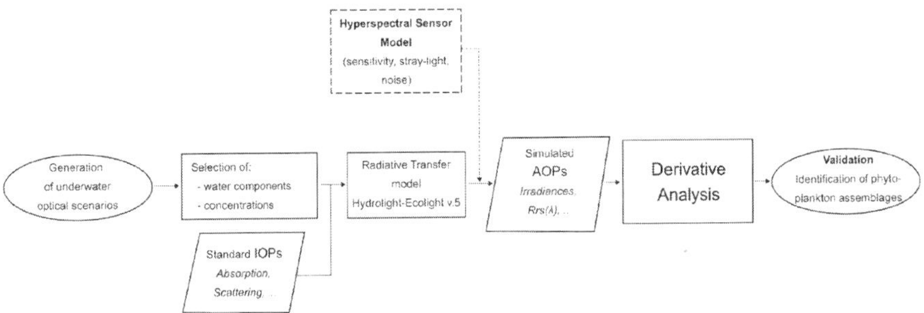

Fig. 8. Flowchart showing the methodology followed for testing the automatic identification of phytoplankton assemblages from hyperspectral $R_{rs}(\lambda)$ and derivative spectroscopy.

The generation of a set of Hydrolight-simulated hyperspectral $R_{rs}(\lambda)$ spectra (1 nm resolution) takes into account the constituents present in the water column and their inherent optical properties (absorption, scattering, etc.). The sensor's behavior can also be included in the $R_{rs}(\lambda)$ computation (see dashed box in Figure 8). The next step defined in the diagram is the computation of the second derivative of normalized $R_{rs}(\lambda)$ spectra (i.e. $R_{rs}(\lambda)$ normalized by the $R_{rs}(\lambda)$ obtained at $\lambda=555nm$). A normalized $R_{rs}(\lambda)$ is used as an input in order to emphasize the shape singularities of each spectrum before derivative spectroscopy. Previous to derivative analysis, a mean filter smoothing type is also applied to normalized $R_{rs}(\lambda)$ data. It consists of a simple average of points within the chosen filter window. As stated in Section 4.2, the efficient way to accomplish the smoothing and derivative is to carefully adapt the filter size and the sampling interval (BS) to better match the scale of the spectral features of interest in each case.

Finally, the goal of validating the potential of hyperspectral $R_{rs}(\lambda)$ measurements for identifying phytoplankton assemblages is addressed through the comparison of second

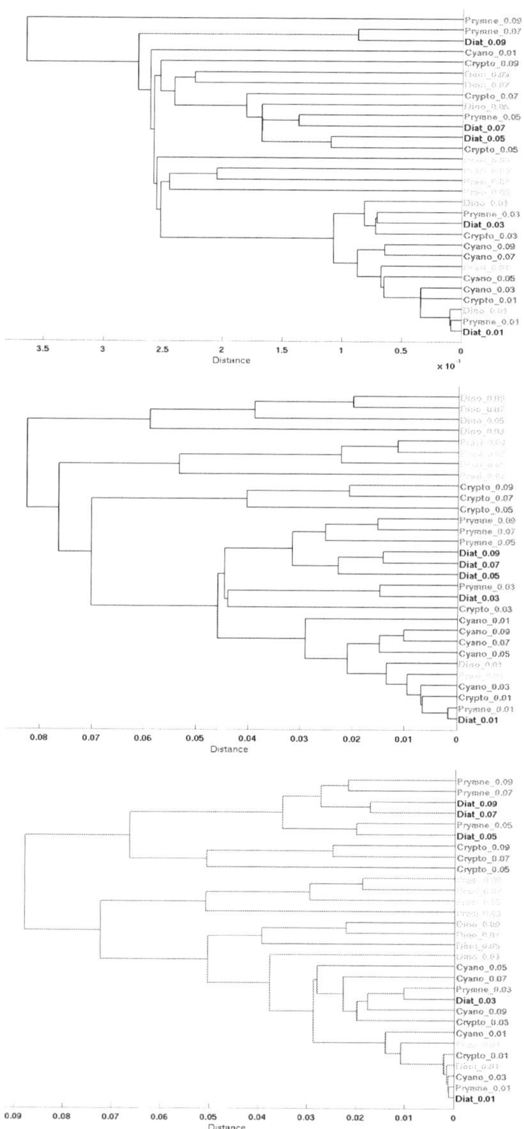

Fig. 10. Cluster analyses based on: (top) raw hyperspectral normalized $R_{rs}(\lambda)$, (centre) second derivative of hyperspectral normalized $R_{rs}(\lambda)$ spectra and (bottom) second derivative of hyperspectral normalized $R_{rs}(\lambda)$ spectra using analyzing parameters that are too coarse (i.e. derivative band separation and smoothing filter size).

5.3 Impact of the sensor's response on the proposed automatic identification

A sensor sensitivity analysis was also conducted to test the effect of the sensor's response, with which the radiometric measurements would hypothetically be acquired in the proposed methodology (Figure 8) to identify different phytoplankton assemblages from derivatives of hyperspectral $R_{rs}(\lambda)$ spectra.

The response of a miniature hyperspectral CMOS-array spectrometer (Figure 3) designed for oceanographic monitoring purposes (Pons et al., 2007; Torrecilla et al., 2008a) was characterized in terms of sensitivity. The spectral sensitivity curve obtained was applied to both simulated hyperspectral water-leaving radiance ($L_w(\lambda)$) and downwelling irradiance ($E_d(\lambda)$), the ratio of which was used to estimate the new data set of $R_{rs}(\lambda)$ spectra. Each new $R_{rs}(\lambda)$ spectrum (Figure 11, red curve in the left panel) now has a spectral resolution of 3 nm approximately, in contrast to the 1 nm resolution of Hydrolight-modeled $R_{rs}(\lambda)$, since it has been adapted to the sensor's spectral resolution. The magnitude of $R_{rs}(\lambda)$ spectra has not been essentially modified after the consideration of the sensor's spectral sensitivity. This is because remote-sensing reflectance is an apparent optical property obtained as the ratio of two radiometric properties. This rationing serves as an effective way to remove the effects of the magnitudes of $E_d(\lambda)$, $L_w(\lambda)$, and other possible external light fluctuations. However, the small differences between $R_{rs}(\lambda)$ spectra with 1 and 3 nm spectral resolution become more evident on the results of derivative analysis (Figure 11, right panel) and will play an important role on the further clustering analysis.

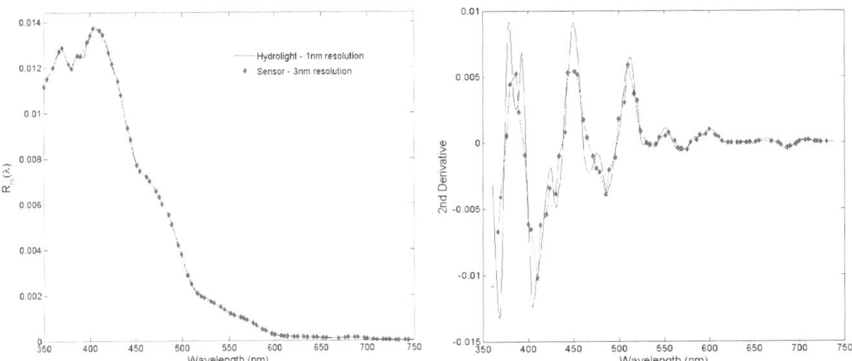

Fig. 11. (Left) An example of a simulated hyperspectral remote-sensing reflectance, $R_{rs}(\lambda)$, before and after applying the modeled sensor's response (black and red curves, respectively). (Right) Second derivatives of spectra in the left panel.

Analogously to the previous subsection, derivative and clustering analysis was carried out with this new data set of hyperspectral-sensor $R_{rs}(\lambda)$. As can be seen in Figure 12, a proper identification and clustering of similar phytoplankton assemblages was only possible when derivative spectra were considered (right panel). Again, a cluster was also created by grouping all those cases corresponding to low concentrations (0.01 mg/m³).

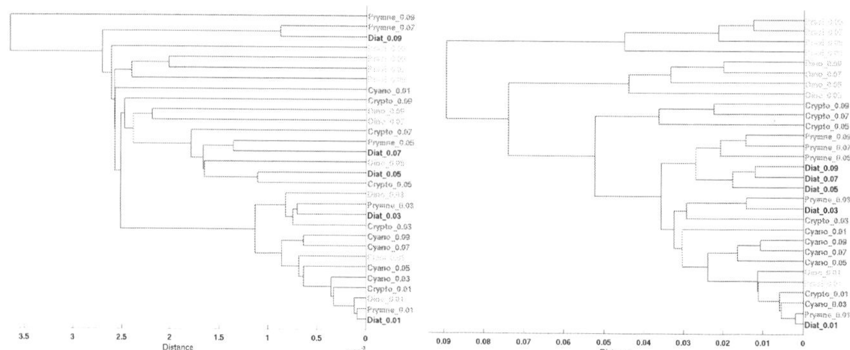

Fig. 12. Results of analog derivative and cluster analysis when the hyperspectral sensor's response in terms of sensitivity is considered based on: (left) raw hyperspectral-sensor normalized $R_{rs}(\lambda)$ and (right) second derivative of hyperspectral-sensor normalized $R_{rs}(\lambda)$ spectra.

Though similar results were obtained, even including the effect of the sensor's response in terms of sensitivity, it should be pointed out that the applied derivative spectroscopy needed to be adapted. Suitable values of smoothing filter size and derivative band separation were selected because of the different spectral resolution of the new $R_{rs}(\lambda)$ data set. Furthermore, based on the results from this experiment, the selected low-cost and miniature hyperspectral CMOS-array spectrometer (Pons et al., 2007) has been confirmed as a potential tool for water component detection and monitoring.

6. Conclusion and Future Work

Derivative spectroscopy has been satisfactorily applied to numerical simulations of hyperspectral remote-sensing reflectance ($R_{rs}(\lambda)$) corresponding to different open-ocean environments. Its feasibility for identifying phytoplankton pigment assemblages in comparison with the use of raw $R_{rs}(\lambda)$ spectra or traditional ratios of discrete spectral bands (Torrecilla et al., 2008b) has been confirmed using a validation approach based on hierarchical cluster analysis (HCA). Simulation-based experiments have also been performed, including a model of the response in terms of sensitivity of a hyperspectral sensor, with which radiometric measurements would hypothetically be acquired. The aim was to analyze the sensor's effect on the proposed methodology of analysis. For each simulated hyperspectral data set, a proper adaptation of parameters involved in derivative analysis (i.e. smoothing filter size and derivative band separation) was necessary according to the characteristics of the signals and particularly their spectral resolution.

The experiments yield promising results when all the information contained in the second derivative of hyperspectral $R_{rs}(\lambda)$ spectra is considered. This method can therefore provide a means for optical oceanographers to better characterize complex oceanic waters, detect harmful algal blooms or map phytoplankton functional types from hyperspectral oceanographic information. The recent advances in hyperspectral technology (e.g.

miniaturization and power supply reduction) have given rise to a great number of sensor configurations that are suitable for incorporating in a large number of in situ and remote sensing platforms of oceanographic observing systems (e.g. satellites, gliders), which will allow these challenges to be overcome once hyperspectral data become more available.

Future research will focus on experiments based on larger field hyperspectral data sets, since the potential of integrating hyperspectral $R_{rs}(\lambda)$ measurements and derivative spectroscopy has been emphasized in this work. However, the design of a more realistic approach when simulated-based experiments are employed would be useful for better validating the efficacy of the proposed method. An effort of detailed consideration of several the above-mentioned factors involved in the analysis process is suggested. For instance, the possibility of having some variability of the inherent optical properties (IOPs) used as an input in the radiative transfer modeling should be explored. Furthermore, the accuracy of the derivative-based method for identifying different phytoplankton compositions would be improved if distortion experiments, caused by the sensor (noise, spectral stray-light, thermal effects, etc.) were included in the simulation-based approach.

7. Acknowledgements

This study was supported by the projects HIDRA (PIE06-301102) and ANERIS (PIF08-015) funded by the Spanish Ministry of Science and Innovation.

8. References

Albert, A. & Mobley, C. D. (2003). An analytical model for subsurface irradiance and remote sensing reflectance in deep and shallow case-2 waters. *Opt. Express.*, 11(22), 2873–2890.

Bracher, A.; Vountas, M.; Dinter, T.; Burrows, J.P.; Röttgers, R. & Peeken, I. (2008). Quantitative observation of cyanobacteria and diatoms from space using PhytoDOAS on SCIAMACHY data. *Biogeosciences Discussions*, 5, 4559-4590.

Bricaud, A.; Morel, A.; Babin, M.; Allali, K. & Claustre, H. (1998). Variations of light absorption by suspended particles with chlorophyll a concentration in oceanic (case 1) waters: Analysis and implications for bio-optical models. *J. Geophys. Res.*, 103, 31,033–31,044.

Brown, S. W.; Eppeldauer, G. P. & Lykke, K. R. (2006). Facility for spectral irradiance and radiance responsivity calibrations using uniform sources. *Applied Optics*, Vol. 45, No. 32, pp. 8218-8237.

Chang, G. C.; Mahoney, K.; Briggs-Whitmire, A.; Kohler, D.; Mobley, C.; Moline, M.; Lewis, M.; Boss, E.; Kim, M.; Philpot, W. & Dickey, T. (2004). The New Age of Hyperspectral Oceanography. *Oceanography*, 17(2), 22-29.

Chang, G. C.; Dickey, T. & Lewis, M. (2006). Toward a global ocean system for measurements of optical properties using remote sensing and in situ observations. In: *Remote Sensing of the Marine Environment: Manual of Remote Sensing, Volume 6, Chapter 9*, edited by J. Gower, 285-326.

Clark, D. K.; Yarbrough, M. A.; Feinholz, M. E.; Flora, S.; Broenkow, W.; Kim, Y. S.; Johnson, B. C.; Brown, S. W.; Yuen, M. & Mueller, J. L. (2003). MOBY, a radiometric buoy for performance monitoring and vicarious calibration of satellite ocean color sensors: measurement and data analysis protocols. *Ocean Optics Protocols for Satellite Ocean Color Sensor Validation, NASA Technical Report Series, Revision 4, Volume 6*, Mueller, J. L., Fargion, G. S. and McClain, C. R., Greenbelt, MD, NASA Goddard Space Flight Center, NASA/TM-2003-211621, 3-34.

Craig, S. E.; Lohrenz, S. E.; Lee, Z.; Mahoney, K. L.; Kirkpatrick, G. J.; Schofield, O. M. & Steward, R. G. (2006). Use of hyperspectral remote sensing reflectance for detection and assessment of the harmful alga, Karenia brevis. *Applied Optics*, Vol. 45, No. 21, pp. 5414-5425.

Cullen, J.J.; Ciotti, A.M.; Davis, R.F. & Lewis, M.R. (1997). Optical detection and assessment of algal blooms. *Limnol. Oceanogr.*, 42 (5), 1223–1239.

Davis, C. O.; Bowles, J.; Leathers, R.A.; Korwan, D.; Downes, T.V.; Snyder, W.A.; Rhea, W.J.; Chen, W.; Fisher, J.; Bissett W.P. & Reisse, R.A. (2002). Ocean PHILLS hyperspectral imager: Design, characterization, and calibration. Opt. Express, 10(4), 210–221.

Dickey, T.; Lewis, M. & Chang, G. (2006). Optical oceanography: recent advances and future directions using global remote sensing and in situ observations. *Reviews of Geophysics*, 44, RG1001, 1-39.

Filippi, A. M. (2007) Derivative-Neural Spectroscopy for Hyperspectral Bathymetry Inversion. *The Professional Geographer*, 59 (2), 236-255.

Gagnon, P.; Scheibling, R.E.; Jones, W. & Tully, D. (2008). The role of digital bathymetry in mapping shallow marine vegetation from hyperspectral image data. *Int. Journal of Remote Sensing*, 29, 3, 879–904.

HyCODE (2009). Retrieved on 1 July 2009. <http://www.opl.ucsb.edu/hycode.html>

IOCCG, International Ocean-Colour Coordinating Group (2000). Remote sensing of ocean colour in coastal, and other optically-complex waters, edited by S. Sathyendranath, IOCCG Rep. 3, pp 140, Dartmouth, N. S., Canada.

Jain, A.K.; Murty, M. N. & Flynn, P. J. (1999). Data clustering: A review. *ACM Comput. Surveys*, 31, 3, 264-323.

Kempeneers, P.; Sterckx, S.; Debruyn, W.; De Backer, S. ; Scheunders, P.; Park, Y. & Ruddick, K. (2005). Retrieval of oceanic constituents from ocean color using simulated annealing, *Proceedings of the 25th International Geoscience and Remote Sensing Symposium (IGARSS)*, Seoul.

Kim, M. & Philpot, W. (2006) Ocean Optical Phytoplankton Simulator. Retrieved on 12 June 2009. <http://ceeserver.cee.cornell.edu/wdp2/OOPS/oops_manual.html>

Kirk, J. T. O. (1994). *Light and Photosynthesis in Aquatic Ecosystems*, 2nd ed., pp. 509, Cambridge Univ. Press, New York.

Kuwahara, V.S.; Chang, G. & Dickey, T.D. (2007). Innovations in ocean optics for coastal and open ocean mooring applications, *Proceedings of IEEE/OEE Oceans Conference and Exhibition, OCEANS'07 Europe*, Aberdeen.

Lewis, M. (2008) Spectral Ocean Radiance Transfer Investigation Experiment (SORTIE), *Pan Ocean Remote Sensing Conference*, Guangzhou, China.

Lohrenz, S.E.; Cai, W., Chen, X. & Tuel, M. (2008). Characterizing water mass properties in river dominated coastal waters using underway hyperspectral remote sensing reflectance, *Proceedings of Ocean Optics XIX Conference*, Barga, Italy.

Louchard, E. M.; Reid, R. P.; Stephens, C. F.; Davis, C. O.; Leathers, R. A.; Downes, T. V. & Maffione, R. (2002). Derivative Analysis of Absorption Features in Hyperspectral Remote Sensing Data of Carbonate Sediments. *Optics Express*, 10(26),1573–1584.

Lubac, B.; Loisel, H.; Guiselin, N.; Astoreca, R.; Felipe Artigas, L. & Mériaux, X. (2008) Hyperspectral and multispectral ocean color inversions to detect Phaeocystis globosa blooms in coastal waters. *J. Geophys. Res.*, 113, C06026.

McClain, C. R.; Feldman, G. C. & Hooker, S. B. (2004). An overview of the SeaWiFS project and strategies for producing a climate research quality global ocean bio-optical time series. *Deep Sea Research II*, 51, 5-42.

McLean, S. (2008). Radiometric Data Processing Course, *Ocean Optics XIX Conference*, Barga, Italy.

Mobley, C. D. (1994) *Light and water: radiative transfer in natural waters*, pp. 592, Academic Press, San Diego, CA.

Mobley, C. D. & Sundman, L. (2008). *Hydrolight 5.0 Users' guide*. Sequoia Scientific, Inc., Bellevue, WA.

Nair, A.; Sathyendranath, S.; Platt, T.; Morales, J.; Stuart, V.; Forget, M.; Devred, E. & Bouman, H. (2008). Remote sensing of phytoplankton functional types. *Remote Sensing of Environment*, 112, 3366–3375.

Neptune (2009). Retrieved on 26 June 2009. <http://neptunecanada.ca>

O'Reilly, J. E.; Maritorena, S.; Siegel, D. A. et al (2000). Ocean Color Chlorophyll a Algorithms for SeaWiFS, OC2, and OC4: Version 4. *SeaWiFS Postlaunch Technical Report Series, Volume. 11, SeaWiFS Postlaunch Calibration and Validation Analyses, Part 3*, edited by Hooker, S. B. and Firestone, E. R., Greenbelt, Maryland, NASA/TM-2000-206892, 9–27.

Perry, M. J. & Rudnick, D.L. (2003). Observing the oceans with autonomous and Lagrangian platforms and sensors: The role of ALPS in sustained ocean observing systems. *Oceanography*, 16(4), 31–36.

Pons, S.; Aymerich, I.F.; Torrecilla, E. and Piera, J. (2007). Monolithic spectrometer for environmental monitoring applications, *Proceedings of IEEE/OEE Oceans Conference and Exhibition, OCEANS'07 Europe*, Aberdeen.

Reynolds, R. A.; Stramski D. & Mitchell, B. G. (2001). A chlorophyll-dependent semianalytical model derived from field measurements of absorption and backscattering coefficients within the Southern Ocean. *J. Geophys. Res.*, 106, 7125–7138.

Ruffin, C.; King, R.L. & Younan, N.H. (2008). A combined derivative spectroscopy and Savitzky-Golay filtering method for the analysis of hyperspectral data. *GIScience and Remote Sensing*, 45 (1), 1-15.

Sathyendranath, S.; Watts, L.; Devred, E.; Platt, T.; Caverhill, C. & Maass, H. (2004). Discrimination of diatoms from other phytoplankton using ocean colour data. *Mar. Ecol. Prog. Ser.*, 272, 59– 68.

Stramski, D.; Reynolds, R.A.; Babin, M.; Kaczmarek, S.; Lewis, M.R.; Röttgers, R.; Sciandra, A.; Stramska, M.; Twardowski, M.S.; Franz, B.A. & Claustre, H. (2008). Relationships between the surface concentration of particulate organic carbon and optical properties in the eastern South Pacific and eastern Atlantic Oceans. Biogeosciences 5: 171-201.

Torrecilla, E.; Aymerich, I.F.; Pons, S. & Piera, J. (2007). Effect of spectral resolution in hyperspectral data analysis. *Proceedings of International Geoscience and Remote Sensing Symposium*, pp. 910-913, Barcelona.

Torrecilla, E.; Pons, S.; Vilaseca, M.; Piera, J. & Pujol, J. (2008a). Stray-light correction of in-water array spectroradiometers. Effects on underwater optical measurements. *Proceedings of IEEE/OEE Oceans Conference and Exhibition*, Quebec.

Torrecilla, E.; Stramski, D.; Reynolds, R. A.; Piera, J. & Millan Nunez, E. (2008b). Identification of phytoplankton pigment assemblages using spectral shape analysis of hyperspectral remote-sensing reflectances, *Proceedings of Ocean Optics XIX Conference*, Barga, Italy.

Tsai, F. & Philpot, W.D. (1998). Derivative analysis of hyperspectral data. *Remote Sensing of Environment*, 66, 1, 41-51.

Uitz, J.; Claustre, H.; Morel, A. & Hooker, S. (2006). Vertical distribution of phytoplankton communities in open-ocean: An assessment based on surface chlorophyll. *Journal of Geophysical Research*, 111, CO8005, doi:10.1029/2005JC003207.

Vaiphasa, C. (2006). Consideration of smoothing techniques for hyperspectral remote sensing. *Journal of Photogrammetry and Remote Sensing*, 60, 2, 91-99.

Voss, K.; Gordon, H.; Lewis, M.; Johnson, C.; Yarbrough, M.; Flora, S.; Feinholz, M., & Trees, C. (2008). Radiometry and Uncertainties from SORTIE (Spectral Ocean Radiance Transfer Investigation and Experiment), *NASA Carbon Cycle and Ecosystems Joint Science Workshop*.

Wood, S. (2009). Autonomous underwater gliders. In: *Underwater Vehicles, Chapter 26*, edited by A. V. Inzartsev, 499-524, In-Tech, Vienna, Austria.

Zibordi, G.; D'Alimonte, D.; van der Linde, D.; Berthon, J.-F.; Hooker, S.B.; Mueller, J.L.; Lazin, G. & McLean, S. (2002). *The Eighth SeaWiFS Intercalibration Round-Robin Experiment (SIRREX-8), September–December 2001. NASA Tech. Memo. 2002-206892, Volume 21.*, edited by Hooker, S.B. & Firestone, E.R., pp. 39, NASA Goddard Space Flight Center, Greenbelt, MD.

30

Electromagnetic Scattering Analysis of Simple Targets Embedded in Planar Multilayer Structures: Remote Sensing Applications

Sidnei J. S. Sant'Anna[1,2], J. C. da S. Lacava[2] and David Fernandes[2]

[1]*Instituto Nacional de Pesquisas Espaciais*
[2]*Instituto Tecnológico de Aeronáutica*
Brazil

1. Introduction

The potential and usefulness of SAR (Synthetic Aperture Radar) data for land uses and land cover change applications have been demonstrated over the past few decades. More recently, polarimetric SAR data have been widely used in electromagnetic modeling and inverse problems. Moreover, the combined use of numerical modeling and inversion methods has enabled computerized quantitative information extraction of the polarimetric data. In particular, the electromagnetic modeling can be used to increase the knowledge of non-metallic target reflectivity. The problem of detecting the many types of non-metallic land mines in the presence of widely varying soil background is an example of one of the most difficult subsurface sensing problems.

The target electromagnetic scattering characterization (van Zyl et al., 1987; Cloude & Pottier, 1996; Freeman & Durden, 1998; Sant'Anna et al., 2007a) might influence some remote sensing practical applications, as SAR image classification, sensor calibration among others. Consequently, understanding the target scattering mechanism plays a valuable role in the analysis and interpretation of polarimetric SAR data. In order to improve the insight into scattering mechanism subject, several electromagnetic models have been proposed in literature (Lin & Sarabandi, 1999; Israelsson et al., 2000; Zurk et al., 2001). Usually, in these models complex targets are modeled by a collection of targets having simple and canonical geometries. For instance, a three can have its elementary parts as trunk, branches and leaves, built through a set of dielectric cylinders and discs. Besides, depending on the modeling the simple targets have different sizes, shapes, orientations and dielectric properties. Therefore, the study of electromagnetic scattering of simple targets is fundamental to enhance the comprehension of the electromagnetic characteristics of more complex ones.

Keeping this issue in mind, the purpose of this work is to show the potentiality and applicability of an electromagnetic model within remote sensing subject. This modeling is based on scattering elements embedded in a multilayer planar structure, which is illuminated by an elliptically polarized plane wave at oblique incidence. Therefore, the chapter is organized as follow. In Section 2, the multilayer structure under analysis is described jointly with an explanation of the electromagnetic modeling employed to solve the scattering problem. The electromagnetic model usefulness is illustrated in Section 3, through some remote sensing applications. The Section 4 is devoted to draw the conclusions concerning the obtained results.

2. Electromagnetic Model

The electromagnetic model is based on the determination of the electromagnetic fields scattered by a multilayer planar structure that is excited by plane waves. The structure under analysis is composed of $N+2$ isotropic, linear and homogenous layers stacked up in z direction. The layers are assumed to be unbounded along the x and y directions. The lower layer, having complex permittivity ε_g and complex permeability μ_g, is denoted as ground layer and occupies the negative-z region. The next N layers are characterized by thickness ℓ_n, complex permittivity ε_n and complex permeability μ_n, where $1 \leq n \leq N$. The planar interface $z = d_N$ separates the N-th layer from free space (the upper layer). Elements supporting electric (**J**) or magnetic (**M**) surface current densities are printed at arbitrary positions on each one of the $N+1$ interfaces of the structure. These elements will behave as scattering elements. The development is based on a global right-handed rectangular coordinate system located on the top of the ground layer (interface $z = 0$) and lying on the xy-plane. The geometry of the planar multilayer structure is depicted in Figure 1.

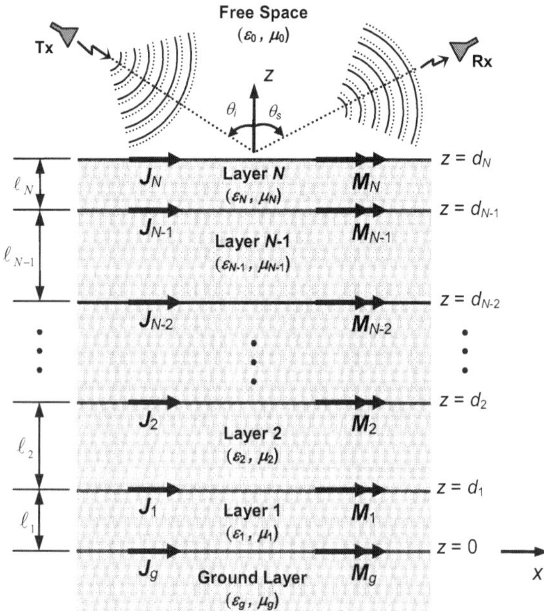

Fig. 1 – Geometry of the planar structure with $N+2$ layers (lateral view).

This electromagnetic modeling for multilayer structures is completely described in (Sant'Anna, 2009) and it can be synthetically summarized in the flowchart illustrated on Figure 2. In this figure it is shown step by step the entire approach to obtain the far electromagnetic fields scattered by the multilayer structure. Starting from the wave equation in each layer, the methodological flowchart highlights the Green's functions acquirement and the integral equations formulation, which are solved applying the method of moments.

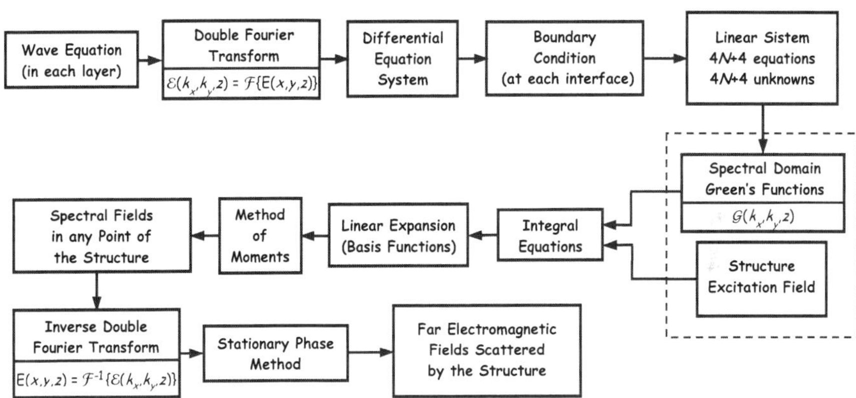

Fig. 2 – Methodological flowchart.

2.1 Electromagnetic Fields in the Structure

The electromagnetic fields in a multilayer structure can be determined through the methodological approach described in (Lacava et al., 2002). According to this methodology, which employs the spectral domain full-wave technique, the structure is treated as a boundary value problem, where the surface current densities induced on the printed elements (electric or magnetic) are the virtual sources of the scattered fields. Since the layers of the structure are free of sources, assuming time dependence of the form $e^{i\omega t}$, the wave equations for the n-th layer is written as

$$\nabla^2 E_n(x,y,z) + k_n^2\, E_n(x,y,z) = 0 \,, \tag{1}$$

$$\nabla^2 H_n(x,y,z) + k_n^2\, H_n(x,y,z) = 0 \,, \tag{2}$$

where for free space $n = 0$ and for ground layer $n = g$, $k_n^2 = \omega^2 \mu_n \varepsilon_n$ gives the wave number in the n-th layer, ω is the angular frequency, and the vectors $E_n(x, y, z)$ and $H_n(x, y, z)$ denote the complex electric and magnetic fields, respectively (bold face letters represent vectors). The wave equations can be solved in the spectral domain using the double Fourier transform. In this work, the Fourier transform pair is defined as

$$\mathcal{F}(k_x, k_y, z) = \int\limits_{-\infty}^{+\infty}\!\!\int F(x,y,z)\, e^{i(k_x x + k_y y)}\, dx\, dy \,, \tag{3}$$

$$F(x,y,z) = \frac{1}{4\pi^2} \int\int_{-\infty}^{+\infty} \mathcal{F}(k_x, k_y, z) e^{-i(k_x x + k_y y)} \, dk_x \, dk_y \, , \tag{4}$$

where the function $F(x,y,z)$ represents the fields $E_n(x,y,z)$ and $H_n(x,y,z)$. Applying the double Fourier transform to (1) and (2) yields a differential equation system whose solution, in terms of the field components, is given by

$$\mathcal{E}_{n\tau\vartheta}(k_x, k_y, z) = \mathbb{e}_{n\tau\vartheta}(k_x, k_y) e^{i\gamma_{n\tau} z} \, , \tag{5}$$

$$\mathcal{H}_{n\tau\vartheta}(k_x, k_y, z) = \mathbb{h}_{n\tau\vartheta}(k_x, k_y) e^{i\gamma_{n\tau} z} \, , \tag{6}$$

with

$$\gamma_{n\tau} = (-1)^\tau \sqrt{k_n^2 - (k_x^2 + k_y^2)} \, , \quad \mathrm{Im}(\gamma_{n\tau}) \leq 0 \tag{7}$$

where $\mathbb{e}_{n\tau\vartheta}(k_x, k_y)$ and $\mathbb{h}_{n\tau\vartheta}(k_x, k_y)$ are the amplitudes of the transformed field components, k_x and k_y are the spectral variables, $\gamma_{n\tau}$ is the τ-th propagation constant in the n-th layer, $\vartheta = x, y$ or z, and $\mathrm{Im}(\ )$ means the imaginary-part of a complex function. The variable τ, which defines the wave propagation direction, can assume values 1 or 2. Only the former value, representing propagation in the positive-z direction, occurs in the upper layer (free space). In this case the wave propagation constant will be denoted by γ_0. For the ground layer, on the other hand, τ is equal to 2, i.e., a wave propagating in the negative-z direction and the propagation constant is represented by γ_g. For the inner layers, however, both values of τ will occur, resulting in $\gamma_{n1} = -\gamma_{n2}$ and the wave propagation constant is denoted by γ_n.

Interesting relations among the amplitudes of the transformed fields are derived by introducing the inverse Fourier transform of (5) and (6) in the Maxwell's curl equations. Such that the amplitudes of the transversal components (x and y directions) are written as functions of the amplitude of the longitudinal ones (z direction). By enforcing the boundary conditions for the electromagnetic fields at each interface a set of 4N+4 equations with equal number of unknowns is obtained. The analytical solution of this system leads to the spectral domain Green's functions. These functions, jointly with the transformed surface current densities, allow the determination of the transformed fields at any point of the multilayer structure. The transformed electromagnetic field components are expressed by

$$\mathcal{E}_{n\vartheta}(k_x, k_y, z) = \sum_{\upsilon} G_{\vartheta\upsilon x}^{(n)}(k_x, k_y, z) j_{\upsilon x}(k_x, k_y) + \sum_{\upsilon} G_{\vartheta\upsilon y}^{(n)}(k_x, k_y, z) j_{\upsilon y}(k_x, k_y) +$$

$$\sum_{\upsilon} Q_{\vartheta\upsilon x}^{(n)}(k_x, k_y, z) m_{\upsilon x}(k_x, k_y) + \sum_{\upsilon} Q_{\vartheta\upsilon y}^{(n)}(k_x, k_y, z) m_{\upsilon y}(k_x, k_y), \tag{8}$$

$$\mathcal{H}_{n\vartheta}(k_x, k_y, z) = \sum_{\upsilon} P_{\vartheta\upsilon x}^{(n)}(k_x, k_y, z) j_{\upsilon x}(k_x, k_y) + \sum_{\upsilon} P_{\vartheta\upsilon y}^{(n)}(k_x, k_y, z) j_{\upsilon y}(k_x, k_y) +$$

$$\sum_{\upsilon} R_{\vartheta\upsilon x}^{(n)}(k_x, k_y, z) m_{\upsilon x}(k_x, k_y) + \sum_{\upsilon} R_{\vartheta\upsilon y}^{(n)}(k_x, k_y, z) m_{\upsilon y}(k_x, k_y), \tag{9}$$

where $G_{\vartheta\upsilon\varsigma}^{(n)}(k_x,k_y,z)$ and $Q_{\vartheta\upsilon\varsigma}^{(n)}(k_x,k_y,z)$ represent, respectively, the spectral domain Green's functions of electric type in the n-th layer, which relate the ς ($\varsigma = x$ or y) components of the electric field to the transformed surface current densities $j_{\upsilon\varsigma}(k_x,k_y)$ and $m_{\upsilon\varsigma}(k_x,k_y)$ located at the interface d_υ, with $\upsilon \in \{g, 1, 2, \ldots, N\}$, and $P_{\vartheta\upsilon\varsigma}^{(n)}(k_x,k_y,z)$ and $R_{\vartheta\upsilon\varsigma}^{(n)}(k_x,k_y,z)$ are the respective spectral domain Green's functions of magnetic type. Note that $d_g = 0$ for the structure defined in Figure 1.

In order to demonstrate the mathematical aspect of the spectral domain Green's functions, it is shown in equations (10) and (11) examples of electric and magnetic function types, respectively, for free-space region. These equations are obtained when the structure is composed by four layers ($N = 2$).

$$G_{x2x}^{(0)}(k_x,k_y,z) = -\frac{4\omega^2}{(k_x^2+k_y^2)}\left\{\frac{k_x^2\gamma_0\Theta_2}{\Delta_1}+\frac{k_y^2\omega\mu_0\Xi_2}{\Delta_2}\right\}e^{-i\gamma_0 z}, \tag{10}$$

$$R_{x2x}^{(0)}(k_x,k_y,z) = -\frac{4\omega^2}{(k_x^2+k_y^2)}\left\{\frac{k_y^2\omega\varepsilon_0\Theta_5}{\Delta_1}+\frac{k_x^2\gamma_0\Xi_5}{\Delta_2}+\right\}e^{-i\gamma_0 z}, \tag{11}$$

where

$$\Delta_1 = -4\omega^3 e^{-i\alpha_0}\{\varepsilon_1\gamma_2(\varepsilon_2\gamma_0\cos\alpha_2+i\varepsilon_0\gamma_2\sin\alpha_2)(\varepsilon_g\gamma_1\cos\alpha_1+i\varepsilon_1\gamma_g\sin\alpha_1)+$$
$$\varepsilon_2\gamma_1(\varepsilon_0\gamma_2\cos\alpha_2+i\varepsilon_2\gamma_0\sin\alpha_2)(\varepsilon_1\gamma_g\cos\alpha_1+i\varepsilon_g\gamma_1\sin\alpha_1)\}, \tag{12}$$

$$\Delta_2 = 4\omega^3 e^{-i\alpha_0}\{\mu_1\gamma_2(\mu_2\gamma_0\cos\alpha_2+i\mu_0\gamma_2\sin\alpha_2)(\mu_g\gamma_1\cos\alpha_1+i\mu_1\gamma_g\sin\alpha_1)+$$
$$\mu_2\gamma_1(\mu_0\gamma_2\cos\alpha_2+i\mu_2\gamma_0\sin\alpha_2)(\mu_1\gamma_g\cos\alpha_1+i\mu_g\gamma_1\sin\alpha_1)\}, \tag{13}$$

$$\Theta_2 = \gamma_2\{\varepsilon_1\gamma_2(\varepsilon_1\gamma_g\sin\alpha_1-i\varepsilon_g\gamma_1\cos\alpha_1)\sin\alpha_2-$$
$$\varepsilon_2\gamma_1(\varepsilon_1\gamma_g\cos\alpha_1+i\varepsilon_g\gamma_1\sin\alpha_1)\cos\alpha_2\}, \tag{14}$$

$$\Theta_5 = -\omega\varepsilon_2\{\varepsilon_1\gamma_2(\varepsilon_g\gamma_1\cos\alpha_1+i\varepsilon_1\gamma_g\sin\alpha_1)\cos\alpha_2-$$
$$\varepsilon_2\gamma_1(\varepsilon_g\gamma_1\sin\alpha_1-i\varepsilon_1\gamma_g\cos\alpha_1+)\sin\alpha_2\}, \tag{15}$$

$$\Xi_2 = \omega\mu_2\{\mu_1\gamma_2(\mu_g\gamma_1\cos\alpha_1+i\mu_1\gamma_g\sin\alpha_1)\cos\alpha_2-$$
$$\mu_2\gamma_1(\mu_g\gamma_1\sin\alpha_1-i\mu_1\gamma_g\cos\alpha_1)\sin\alpha_2\}, \tag{16}$$

$$\Xi_5 = \gamma_2\{\mu_2\gamma_1(\mu_1\gamma_g\cos\alpha_1+i\mu_g\gamma_1\sin\alpha_1)\cos\alpha_2-$$
$$\mu_1\gamma_2(\mu_1\gamma_g\sin\alpha_1-i\mu_g\gamma_1\cos\alpha_1)\sin\alpha_2\}, \tag{17}$$

with $\alpha_0 = \gamma_0 d_2$, $\alpha_1 = \gamma_1 d_1$, $\alpha_2 = \gamma_2 (d_2-d_1)$, $d_1 = \ell_1$ and $d_2 = \ell_1 + \ell_2$. It can be noted that the Green's functions carry information of electromagnetic parameters from all structure layers.

2.2 Method of Moments (MoM)

Once the Green's functions are derived, the next step is to set up integral equations constrained to the required boundary conditions. The integral equations arise as a combination of the Green's functions and the structure excitation fields, as can be seen from the dashed box in Figure 2. The boundary conditions are related to the total electric and magnetic tangential fields to each respective scattering element. The subsequent development will be described only when electric elements are present in the structure. However, it could be also applied to magnetic elements, taking into account its specific nature. Therefore, the integral equation is a statement of the boundary condition requiring that the total electric field tangential to the each of the electric element is zero (Newman & Forrai, 1987). That is,

$$z \times [E^i(x,y,d_\upsilon) + E^r(x,y,d_\upsilon)] = -z \times E^s(x,y,d_\upsilon) \text{, on } S_\upsilon \tag{18}$$

where S_υ is the support region of the electric element, $E^s(x,y,d_\upsilon)$ denotes the scattered electric field excited by the current on S_υ, $E^i(x,y,d_\upsilon)$ stands for the incident electric field and $E^r(x,y,d_\upsilon)$ identifies the field that are reflected by the multilayer structure in the absence of electric elements. The incident and reflected fields define the excitation mechanism of the structure, which in this analysis is due to an elliptically polarized plane wave at an arbitrary incidence angle.

The currents induced on the electric elements by these fields are unknown. To solve the integral equations (18), with the unknown surface currents, the method of moments (MoM) is applied. This method is one the most popular numerical techniques used to analyze the radiation and scattering from complex structures. In the MoM, first the surface current is linearly expanded in a set of basis functions with unknown coefficients

$$j_\upsilon(k_x,k_y) = x \sum_{m=1}^{M_{L\upsilon}} \sum_{n=1}^{N_{L\upsilon}} I^x_{m\upsilon n\upsilon} \, j_{m\upsilon}(k_x,k_y) + y \sum_{m=1}^{M_{L\upsilon}} \sum_{n=1}^{N_{L\upsilon}} I^y_{m\upsilon n\upsilon} \, j_{n\upsilon}(k_x,k_y), \tag{19}$$

where ML and NL control the expansion modes in x and y directions on each interface layer, respectively, $I^\varsigma_{m\upsilon n\upsilon}$ are the complex coefficients in the ς direction ($\varsigma = x$ or y) that need to be determined, and $j_{m\upsilon}(k_x,k_y)$ and $j_{n\upsilon}(k_x,k_y)$ are the Fourier transform of the surface current density components, which are defined only over the electric element. Applying the Galerkin technique (whereby the test functions are chosen to be identical to the basis ones) the integral equation is reduced to a system of simultaneous linear equations, which can be compactly written in matrix form as $[V] = [Z][I]$. In this notation $[V]$, $[Z]$ and $[I]$ denote, respectively, the excitation matrix, the impedance matrix and the coefficient matrix. For example, the integral equation referring to the scattered fields at the interface $z = d_\upsilon$ can be written as

$$4\pi^2 \iint_{S_\upsilon} E^s_{0x}(x,y,d_\upsilon) J^*_{pn}(x,y)\,dx\,dy = \sum_{\upsilon=0}^{N} \sum_{m=1}^{M_{L\upsilon}} \sum_{n=1}^{N_{L\upsilon}} I^x_{m\upsilon n\upsilon} \int_{-\infty}^{+\infty} \int_{-\infty}^{+\infty} G^{(0)}_{x\upsilon x} \, j_{m\upsilon} \, j^*_{pn} \, dk_x \, dk_y +$$

$$\sum_{\upsilon=0}^{N} \sum_{m=1}^{M_{L\upsilon}} \sum_{n=1}^{N_{L\upsilon}} I^y_{m\upsilon n\upsilon} \int_{-\infty}^{+\infty} \int_{-\infty}^{+\infty} G^{(0)}_{x\upsilon y} \, j_{n\upsilon} \, j^*_{pn} \, dk_x \, dk_y \,, \tag{20}$$

$$4\pi^2 \iint_{S_\upsilon} E^s_{0y}(x,y,d_\upsilon) J^*_{qn}(x,y)\,dx\,dy = \sum_{\upsilon=0}^{N} \sum_{m=1}^{M_{L\upsilon}} \sum_{n=1}^{N_{L\upsilon}} I^x_{m\upsilon n\upsilon} \int_{-\infty}^{+\infty} \int_{-\infty}^{+\infty} G^{(0)}_{y\upsilon x} \, j_{m\upsilon} \, j^*_{qn} \, dk_x \, dk_y +$$

$$\sum_{\upsilon=0}^{N}\sum_{m=1}^{M_{L\upsilon}}\sum_{n=1}^{N_{L\upsilon}} I_{m\upsilon n\upsilon}^{y} \int_{-\infty}^{+\infty}\int_{-\infty}^{+\infty} \mathcal{G}_{y\upsilon y}^{(0)}\; j_{n\upsilon}\; j_{qn}^{*}\; dk_{x}\, dk_{y}\;, \tag{21}$$

where the left side of (20) and (21) defines the $[V]$ matrix, the double integrals are related to the $[Z]$ matrix and the variables p and q range equally to the variables m and n, respectively. It is important to point out that for notation simplicity the functions $\mathcal{G}_{\vartheta\upsilon\varsigma}^{(0)}(k_{x},k_{y},d_{n})$, $j_{m\upsilon}(k_{x},k_{y})$ and $j_{pn}^{*}(k_{x},k_{y})$ were written without their respective variables in equations (20) and (21).

As abovementioned, following the same procedure it is possible to get similar expressions when magnetic elements are used. Nevertheless, the scattered electric field, the transformed surface electric current density and the Green´s function $\mathcal{G}_{\vartheta\upsilon\varsigma}^{(0)}$ should be changed by the scattered magnetic field, the transformed surface magnetic current density and the Green´s function $\mathcal{R}_{\vartheta\upsilon\varsigma}^{(0)}$, respectively.

The double integrations in equation (20) and (21) must be performed numerically, usually in a very inefficient and time-consuming way. In order to improve the computation efficiency some mathematical simplifications are employed. These simplifications include the evaluation of the even and the odd properties of the Green's functions, the change of the coordinate system (rectangular to polar: $k_{x} = \beta \cos\alpha$ and $k_{y} = \beta \sin\alpha$) and the asymptotic extraction technique.

2.3 Asymptotic Expressions

In microwave remote sensing applications the major interest consist on the target radiated power and it's carried information. Therefore, is necessary the determination of the far electromagnetic fields scattered by the target. These fields scattered by the multilayer structure are computed based on asymptotic expressions, which are derived from the stationary phase method (Collin & Zucker, 1969). The electric far-field, using this method, is given by

$$\boldsymbol{E}_{0}(r,\theta,\phi) \cong -\frac{i\,k_{0}}{2\pi}\frac{e^{-ik_{0}r}}{r}\cot\theta\left\{\hat{\theta}\,\mathbf{e}_{0z}(k_{xe},k_{ye}) - \hat{\phi}\,\eta_{0}\mathbb{h}_{0z}(k_{xe},k_{ye})\right\}, \tag{22}$$

in a spherical coordinate system, where the intrinsic impedance of free space is represented by η_{0}, $k_{xe} = k_{0}\sin\theta\cos\phi$ and $k_{ye} = k_{0}\sin\theta\sin\phi$ are the stationary phase points, k_{0} is the wave number of the excitation wave, $\mathbf{e}_{0z}(k_{xe},k_{ye})$ and $\mathbb{h}_{0z}(k_{xe},k_{ye})$ are the amplitudes of the transformed far-field components (in the z direction) in free space and r characterizes the distance between the receiving antenna and the target. Note that the far magnetic field $H_{0}(r,\theta,\phi)$ might be obtained from $E_{0}(r,\theta,\phi)$. For instance, considering a particular structure consisting of four layers ($N = 2$) and having electric and magnetic elements printed on each interface, the $\mathbf{e}_{0z}(k_{xe},k_{ye})$ and $\mathbb{h}_{0z}(k_{xe},k_{ye})$ are expressed by

$$e_{0z}(k_{xe},k_{ye}) = \frac{4\omega^2}{\Delta_1}\{(k_{xe}\Theta_0)j_{gx} + (k_{ye}\Theta_0)j_{gy} + (k_{xe}\Theta_1)j_{1x} + (k_{ye}\Theta_1)j_{1y} +$$

$$(k_{xe}\Theta_2)j_{2x} + (k_{ye}\Theta_2)j_{2y} + (k_{ye}\Theta_3)m_{gx} + (-k_{xe}\Theta_3)m_{gy} +$$

$$(k_{ye}\Theta_4)m_{1x} + (-k_{xe}\Theta_4)m_{1y} + (-k_{ye}\Theta_5)m_{2x} + (k_{xe}\Theta_5)m_{2y}\}, \tag{23}$$

$$h_{0z}(k_{xe},k_{ye}) = \frac{4\omega^2}{\Delta_2}\{(-k_{ye}\Xi_0)j_{gx} + (k_{xe}\Xi_0)j_{gy} + (-k_{ye}\Xi_1)j_{1x} + (k_{xe}\Xi_1)j_{1y} +$$

$$(k_{ye}\Xi_2)j_{2x} + (-k_{xe}\Xi_2)j_{2y} + (k_{xe}\Xi_3)m_{gx} + (k_{ye}\Xi_3)m_{gy} +$$

$$(k_{xe}\Xi_4)m_{1x} + (k_{ye}\Xi_4)m_{1y} + (k_{xe}\Xi_5)m_{2x} + (k_{ye}\Xi_5)m_{2y}\}, \tag{24}$$

with

$$\Theta_0 = -\varepsilon_1\varepsilon_2\gamma_1\gamma_2\gamma_g, \tag{25}$$

$$\Theta_1 = -\varepsilon_2\gamma_1\gamma_2(\varepsilon_1\gamma_g\cos\alpha_1 + i\varepsilon_g\gamma_1\sin\alpha_1), \tag{26}$$

$$\Theta_3 = \omega\varepsilon_1\varepsilon_2\varepsilon_g\gamma_1\gamma_2, \tag{27}$$

$$\Theta_4 = \omega\varepsilon_1\varepsilon_2\gamma_2(\varepsilon_g\gamma_1\cos\alpha_1 + i\varepsilon_1\gamma_g\sin\alpha_1), \tag{28}$$

$$\Xi_0 = -\omega\mu_1\mu_2\mu_g\gamma_1\gamma_2, \tag{29}$$

$$\Xi_1 = -\omega\mu_1\mu_2\gamma_2(\mu_g\gamma_1\cos\alpha_1 + i\mu_1\gamma_g\sin\alpha_1), \tag{30}$$

$$\Xi_3 = \mu_1\mu_2\gamma_1\gamma_2\gamma_g, \tag{31}$$

$$\Xi_4 = \mu_2\gamma_1\gamma_2(\mu_1\gamma_g\cos\alpha_1 + i\mu_g\gamma_1\sin\alpha_1), \tag{32}$$

where the terms Δ_1, Δ_2, Θ_2, Θ_5, Ξ_2 and Ξ_5 are given, respectively, by equations (12) to (17). It is important to observe that all terms of equations (23) and (24) are obtained on stationary phase points and the transformed surface current densities were written without their respective variables.

2.4 Simple Target

In order to complete the electromagnetic modeling development, simple target embedded in a multilayer structure consisting of four layers is considered. The simple targets are represented by electric and magnetic rectangular dipoles having infinitesimal thickness, oriented along the x direction and printed on interface $z = d_2$. The dipoles size $2a$ and $2b$ ($a \gg b$) in the x and the y directions, respectively. In this particular situation, the surface current density along the y direction is neglected since the dipoles width is considered to be very thin. Thus, for either dipole only the $\left[Z_{p2m2}^{x2x2}\right]$ matrix needs to be evaluated, being represented by $[Z_{pm\zeta}]$, where the variable ζ defines the type of dipole, electric ($\zeta = e$) or magnetic ($\zeta = m$). This evaluation involves the Green's functions $G_{x2x}^{(0)}$ (electric dipole) or $R_{x2x}^{(0)}$ (magnetic dipole). After the aforementioned mathematical simplifications (subsection 2.2), the $[Z_{pm\zeta}]$ matrix becomes

$$\left[Z_{pm\zeta}\right] = -\frac{1}{4\pi^2} \int\limits_{-\infty}^{+\infty} \int\limits_{-\infty}^{+\infty} A_{x2x}^{(0)}(k_x,k_y,d_2)\, d_{m2}(k_x,k_y)\, d_{p2}^{*}(k_x,k_y)\, dk_x\, dk_y =$$

$$= \frac{1}{\pi^2}\left\{ \int\limits_{\beta=0}^{+\infty} \int\limits_{\alpha=0}^{\pi/2} [A_{1\zeta}(\cos\alpha)^2 + A_{2\zeta}(\sin\alpha)^2]\, R_{pm\,\zeta}\, d\alpha\, d\beta + \right.$$

$$\left. \int\limits_{\beta=0}^{+\infty} \int\limits_{\alpha=0}^{\pi/2} \frac{i\, A_{3\zeta}}{\omega}(\cos\alpha)^2\, R_{pm\,\zeta}\, d\alpha\, d\beta \right\}, \tag{34}$$

where $A_{x2x}^{(0)}$ represents the spectral Green's function and d_{m2} means the transformed surface current density. The spectral Green's function will be defined by $G_{x2x}^{(0)}$ or by $R_{x2x}^{(0)}$ in the electric and magnetic dipole cases, respectively. Similarly, the transformed surface current density will be described by j_{m2} or m_{m2}. In each dipole case, the factors $A_{1\zeta}$, $A_{2\zeta}$, $A_{3\zeta}$ and $R_{pm\zeta}$ are expressed by

$$A_{1e} = \frac{4\omega^2 \beta \gamma_0 \Theta_2}{\Delta_1} - \frac{i\, A_{3e}}{\omega}, \tag{35}$$

$$A_{2e} = \frac{4\omega^3 \beta \mu_0 \Xi_2}{\Delta_2}, \tag{36}$$

$$A_{3e} = \frac{\beta^2}{\varepsilon_0 + \varepsilon_2}, \tag{37}$$

$$A_{1m} = \frac{1}{\omega}\left(\frac{\beta \gamma_0 \Xi_5}{\Delta_2} - i\, A_{3m} \right), \tag{38}$$

$$A_{2m} = \frac{\beta \varepsilon_0 \Theta_5}{\Delta_1}, \tag{39}$$

$$A_{3m} = \frac{\beta^2}{\mu_0 + \mu_2}, \tag{40}$$

$$R_{pm\,e} = \pi^2\, b^2\, J_0^2(bk_y)\,\text{sinc}^4(k_x \Delta x/2)\cos[k_x(x_m - x_p)], \tag{41}$$

with $R_{pmm} = R_{pme}$.

Additionally the equation (41) is obtained from the modeling of the surface electric current density as the summation of piecewise-linear subdomain basis functions (rooftop functions) taking into account the edge condition. In this equation $\text{sinc}(\chi) = \sin(\chi)/\chi$, $J_0(.)$ stands for the zero-order Bessel function of the first kind and $\Delta x = 2a/(M+1)$, where M is the number of current expansion modes.

From (34) to (41) it is noted that the first double integral of $[Z_{pm\zeta}]$ is dependent on the operating frequency, whereas the second one is not. The former is the well-known Sommerfeld integrals, which exhibit singularities in the form of branch points and poles; as such, their computation requires careful attention. The poles (generally complex) correspond to surface and leaky waves that can be excited in the layers. According to (Marin et al., 1990) the number of surface wave poles and their locations depends on the thickness of the layers, their electromagnetic parameters and the wave number. For a multilayer structure and depending on the thickness of the layers, the Green's functions might present hundreds of poles, making their integration a formidable task. Since the number of poles and their locations are not known beforehand (Newman & Forrai, 1987), the use of a deformed path to compute the integrations seems to be an efficient way to avoid this problem. Therefore, in this work a parabolic path was chosen to compute the integrations.

Using the rooftop function as basis function, the excitation matrices $[V_{pe}]$ and $[V_{pm}]$, for this particular structure, are defined only over the S_2 region, which is limited by $-a$ and $+a$ in x-direction and by $-b$ and $+b$ in y-direction being given by

$$\left[V_{p\varsigma}\right] = \iint_{S_2} F(x,y,d_2)\,\mathcal{D}^*_{p2x}(x)\,\mathcal{D}^*_{p2y}(y)\,dx\,dy =$$

$$= 2\,(B_\varsigma \sin\varphi_i + C_\varsigma \cos\varphi_i)\,\mathrm{sinc}^2(\Psi_1 \Delta x / 2)\;e^{i\Psi_1 x_p} \int_0^{+b} \frac{\cos(\Psi_2 y)}{\sqrt{1-(y/b)^2}}\,dy\,, \quad (42)$$

where variable ς (e or m) defines the type of dipole as defined in equation (34). Therefore, the function $F(x,y,d_2)$ can represent the scattered field $E^s_{0x}(x,y,d_2)$ or $H^s_{0x}(x,y,d_2)$ and the complex conjugate of the surface current density $\mathcal{D}^*_{p2\varsigma}$ will stand for $J^*_{p2\varsigma}$ or $M^*_{p2\varsigma}$, with $\varsigma = x$ or y. The constants B_ς, C_ς, Ψ_1 and Ψ_2 are given by

$$B_e = -(E^i_h e^{i\Omega} + E^r_h e^{-i\Omega})\,, \tag{43}$$

$$C_e = -\cos\theta_i\,(E^i_v e^{i\Omega} - E^r_v e^{-i\Omega})\,, \tag{44}$$

$$B_m = -\eta_0^{-1}(E^i_v e^{i\Omega} + E^r_v e^{-i\Omega})\,, \tag{45}$$

$$C_m = -\eta_0^{-1}\cos\theta_i\,(-E^i_h e^{i\Omega} + E^r_h e^{-i\Omega})\,, \tag{46}$$

$$\Psi_1 = k_0 \sin\theta_i \cos\phi_i\,, \tag{47}$$

$$\Psi_2 = k_0 \sin\theta_i \sin\phi_i\,, \tag{48}$$

with $\Omega = k_0 d_2 \cos\theta_i$, E^i_p and E^r_p standing for incident and reflected electric fields having linear p polarization (horizontal or vertical) and the angles θ_i and ϕ_i define the direction of the incident wave. In equations (43) to (48) the subscript v and h stand for vertical and horizontal linear polarization. The solution of integral equation leads to coefficient matrix $[I]$, from which it is possible to obtain the surface current density expression over the scattering element. This density jointly with the spectral Green's function permits the evaluation of the electromagnetic fields in spectral domain.

3. Remote Sensing Applications

This section is devoted to some remote sensing applications, which can be derived from the electromagnetic modeling described in section 2. All results are referred to a simple target (electric and/or magnetic dipoles) embedded in a four layers structure ($N = 2$).

3.1 Scattering Matrix

The determination of analytical expressions for the dipoles scattering matrices is the first application considered in this section. The scattering matrix is the most important parameter in polarimetric SAR analysis, since it provides complete information about the scattering mechanism. All polarimetric features that describe the target scattering can be derived from it. The scattering matrix can be seen as a mathematical characterization of the target scattering. This matrix linearly relates the electric fields of the wave scattered (E^s) by a target to the ones of the incident wave (E^i). As consequence the scattering matrix can be seen as a linear operator between E^s and E^i, given by

$$\left[E^s\right] = \begin{bmatrix} E_v^s \\ E_h^s \end{bmatrix} = \frac{e^{-ik_0 r}}{r} \begin{bmatrix} S_{vv} & S_{vh} \\ S_{hv} & S_{hh} \end{bmatrix} \begin{bmatrix} E_v^i \\ E_h^i \end{bmatrix} = \frac{e^{-ik_0 r}}{r} [S][E^i], \tag{49}$$

where the subscript v and h stand for vertical and horizontal linear polarization. In order to determine the scattering matrix of electric and magnetic dipoles embedded in a multilayer structure, as illustrated in Figure 3, the procedure defined in (Sant'Anna et al., 2007a) will be followed. This procedure directly relates the elements of the scattering matrix to the components of the scattered electric field; a standard spherical coordinate system (r, θ, ϕ) is chosen to coincide with a coordinate system (k, v, h) defined in terms of horizontal and vertical linear polarization components, as stated in (Ulaby & Elachi, 1990).

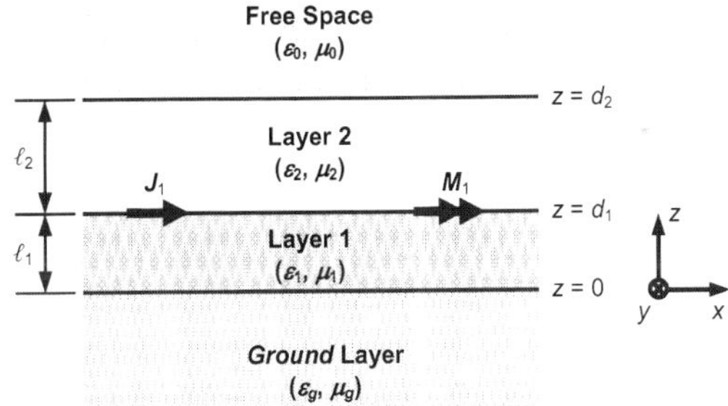

Fig. 3 – A lateral view of the multilayer structure under analysis.

The incident wave sets up currents on the scattering elements (dipoles), which in turn re-radiates a scattered wave. The currents induced on the dipoles depend on the incident wave polarization. Therefore, from the reciprocity theorem, relation (50) between the components of the transformed far electric field in free space due to each linear polarization

$$\mathbf{e}_{0z}^{(h)}(k_{xe}, k_{ye}) = -\eta_0 \mathbf{h}_{0z}^{(v)}(k_{xe}, k_{ye}), \tag{50}$$

can be obtained, where the superscripts (v, h) are associated to polarization of the incident wave. Consequently, for an electric planar dipole the induced currents due to horizontal and vertical linear polarizations are, then, related by

$$j_{1xe}^{(h)}(k_{xe}, k_{ye}) = -C_1 \left(\frac{k_{ye}}{k_{xe}}\right) j_{1xe}^{(v)}(k_{xe}, k_{ye}). \tag{51}$$

This relation leads to the following scattering matrix

$$[S_{ed}] = \frac{\mathbb{C}_2}{\cos\phi_s}\begin{bmatrix} (\cos\phi_s)^2 / \mathbb{C}_1 & \sin\phi_s \cos\phi_s \\ \sin\phi_s \cos\phi_s & \mathbb{C}_1(\sin\phi_s)^2 \end{bmatrix}, \tag{52}$$

where the constants $\mathbb{C}_1$ and $\mathbb{C}_2$ are given by

$$\mathbb{C}_1 = \frac{\Delta_1 \Xi_1}{\Delta_2 \Theta_1}\eta_0 \quad \text{and} \quad \mathbb{C}_2 = -\frac{i2\omega^2 k_0^2 \eta_0 \Xi_1}{\pi \Delta_2}\cos\theta_s \, j_{1xe}^{(v)}(k_{xe}, k_{ye}).$$

Similarly for the magnetic dipole, the relation between the induced currents and its corresponding scattering matrix are expressed, respectively, by

$$m_{1xe}^{(h)}(k_{xe}, k_{ye}) = -\mathbb{C}_3\left(\frac{k_{xe}}{k_{ye}}\right)m_{1xe}^{(v)}(k_{xe}, k_{ye}), \tag{53}$$

$$[S_{md}] = \frac{\mathbb{C}_4}{\sin\phi_s}\begin{bmatrix} (\sin\phi_s)^2 / \mathbb{C}_3 & -\sin\phi_s \cos\phi_s \\ -\sin\phi_s \cos\phi_s & \mathbb{C}_3(\cos\phi_s)^2 \end{bmatrix}, \tag{54}$$

where the constants $\mathbb{C}_3$ and $\mathbb{C}_4$ are represented by

$$\mathbb{C}_3 = \frac{\Delta_1 \Xi_4}{\Delta_2 \Theta_4}\eta_0 \quad \text{and} \quad \mathbb{C}_4 = -\frac{i2\omega^2 k_0^2 \eta_0 \Xi_4}{\pi \Delta_2}\cos\theta_s \, m_{1xe}^{(v)}(k_{xe}, k_{ye}).$$

Note that the constants $\mathbb{C}_2$ and $\mathbb{C}_4$, which are directly related to the currents induced on the dipoles, affect equally all the elements of the scattering matrices. The terms Δ_1, Δ_2, Θ_1, Θ_4, Ξ_1 and Ξ_4, given by equations (12), (13), (26), (28), (30) and (32) respectively, are evaluated at stationary phase points. From equations (52) and (54) it can be also observed that the scattering matrix expressions are derived for any direction of the scattered wave.

In (Sant'Anna et al., 2008a) it was used two polarimetric features as comparative metrics between the electric and magnetic scattering matrices. These features included the α-angle derived from the eigenvalue/eigenvector decomposition (Cloude & Pottier, 1997) and a similarity measurement based on square of correlation coefficient, which was proposed in (Yang et al., 2001). The results were obtained in (Sant'Anna et al., 2008a) for a structure composed of layers having the same electromagnetic parameters of free space. They showed that even though the dipoles present different electromagnetic nature the dominant α-angle for both are equal to 45°. Therefore, based on this feature it was not possible to differentiate the two dipoles, despite their electromagnetic characteristics. On the other hand, measuring the correlation coefficient it was shown that it can be used as a discriminating feature between the two scattering matrices. This statement is promptly verified when the four layers are assumed to be the free space. In this particular case the constants $\mathbb{C}_1$ and $\mathbb{C}_3$ are reciprocal real values yielding a correlation coefficient equal to zero, i.e. the scattering matrices are non-similar.

3.2 Simple Target Characterization

The simple target characterization is the second application considered in this work within remote sensing context. The characterization is made using several electromagnetic attributes that are measured at far-field condition. The measurements are done based on a particular configuration of the four layers structure, in which the dielectric parameters of layers 1 and 2 are made equals. That is, the structure is then composed by three dielectric layers (the free-space region, one inner layer and the ground layer) and the dipoles (electric and magnetic) are printed on interface $z = d_2$, as depicted in Figure 4.

In antenna theory the directivity function $\mathbb{D}(\theta, \phi)$ is one of several parameters used to characterize or define an antenna. According to (Balanis, 1997) this function is defined as the ratio of the radiation intensity $\mathbb{U}(\theta, \phi)$ in a given direction from the antenna to the radiated intensity average over all direction. The $\mathbb{D}(\theta, \phi)$ is expressed by

$$\mathbb{D}(\theta,\phi) = \frac{4\pi\,\mathbb{U}(\theta,\phi)}{\mathbb{P}_i}, \tag{55}$$

where $\mathbb{P}_i = \int_\Omega \mathbb{U}(\theta,\phi)\,d\Omega$ is the average radiated power within the solid angle Ω and the $\mathbb{U}(\theta, \phi)$ function is defined, for a multilayer structure, as

$$\mathbb{U}(\theta,\phi) = \frac{1}{2\eta_0}\left(\frac{k_0\cot\theta}{2\pi}\right)^2 \left\{ \left|\mathfrak{e}_{0z}(k_{xe},k_{ye})\right|^2 + \eta_0^2\left|\mathfrak{h}_{0z}(k_{xe},k_{ye})\right|^2 \right\}, \tag{56}$$

with the transformed electric and magnetic longitudinal amplitudes the $\mathfrak{e}_{0z}(k_{xe},k_{ye})$ and $\mathfrak{h}_{0z}(k_{xe},k_{ye})$ being expressed by equations (23) and (24), respectively.

Considering the structure of Figure 4 and assuming that the size of dipoles are infinitesimal the function $\mathbb{D}(\theta, \phi)$ is promptly computed. Since, in this case, the surface current density can be considered as constant. The directivity function for a structure characterized by $\varepsilon_r = 2$, $\mathrm{tg}\delta = 0$, $\varepsilon_{rg} = 1$ and $\mathrm{tg}\delta_g = 1.0\times10^{+15}$ as dielectric parameters values, it was evaluated at 2.25 GHz for printed infinitesimal electric and magnetic dipoles. The respective three-dimensional graphic of $\mathbb{D}(\theta, \phi)$ are illustrated in Figures 5 and 6 for two values of inner layer thickness, $\ell = 4$ mm $(0.03\lambda_0)$ and $\ell = 40$ mm $(0.3\lambda_0)$.

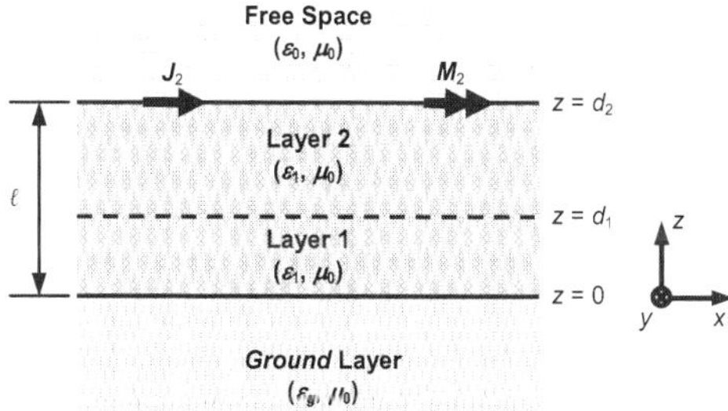

Fig. 4 – A lateral view of the multilayer structure under analysis.

From these figures it can be notice that the radiation patterns are completely different for both dipoles. This fact it was expected due to the dissimilar electromagnetic feature of the dipoles. Theoretically, the radiation pattern can be used to differentiate structures which present these kinds of dipoles. In addition, it can be observed that the shapes as well as the radiation levels are modified by the variation of the inner layer thickness. The radiation levels modification may be associated to the modes (guided and leaky) existing within the structure. Increasing the layer thickness more guided modes might be established inside the structure, consequently decreasing its scattered power.

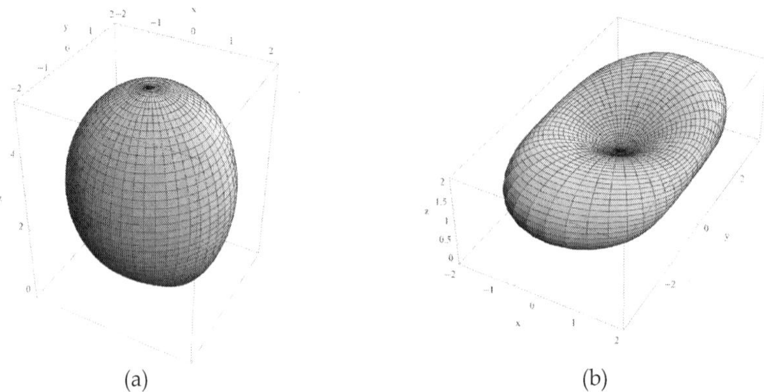

(a)   (b)

Fig. 5 – Three-dimensional graphic of $\mathbb{D}(\theta, \phi)$ for infinitesimal electric dipole: (a) $\ell = 4$ mm and (b) $\ell = 40$ mm.

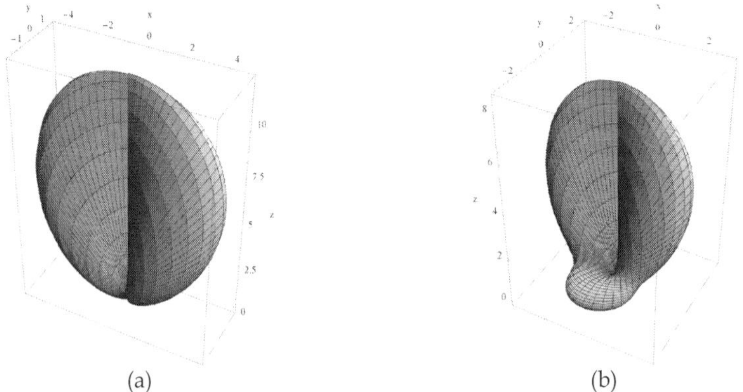

(a)   (b)

Fig. 6 – Three-dimensional graphic of $\mathbb{D}(\theta, \phi)$ for infinitesimal magnetic dipole: (a) $\ell = 4$ mm and (b) $\ell = 40$ mm.

The *SCS* is other parameter that can be used on the characterization of target scattering properties. In general, it is function of the operating frequency, wave polarization, incident and observation angles, target geometry and its dielectric properties. According to (IEEE, 1993) the target *SCS* can be expressed by

$$SCS(\boldsymbol{k}_s, \boldsymbol{k}_i) = \lim_{r \to \infty} \left[4\pi r^2 \frac{\left| \boldsymbol{E}^s(\boldsymbol{k}_s) \right|^2}{\left| \boldsymbol{E}^i(\boldsymbol{k}_i) \right|^2} \right], \tag{57}$$

where $\boldsymbol{E}^i(\boldsymbol{k}_i)$ is the electric field of incidence plane wave in $\boldsymbol{k}_i$ direction and $\boldsymbol{E}^s(\boldsymbol{k}_s)$ is the electric field of the scattered wave in $\boldsymbol{k}_s$ direction, measured at a distance r of the target.

The multilayer structure (Figure 4) containing dipoles that their length and with are equal to 50 mm and 1 mm, respectively, is used to evaluate the SCS parameter. The dipole length is oriented along the x-direction while its width is oriented along the y-direction. For SCS analysis a basis structure characterized by $\varepsilon_r = 2.33$, $tg\delta = 1.2 \times 10^{-1}$, $\ell = 50$ cm, $\varepsilon_{rg} = 5$ and $tg\delta_g = 2.0 \times 10^{-1}$ is used. The evaluation, conducted at 1.25 GHz, uses a vertically polarized incident wave at oblique direction defined by $\theta_i = 40°$ and $\phi_i = 60°$. The SCS analysis comprises the influence of the inner layer by varying its thickness (ℓ) and its relative permittivity (ε_r). The SCS resulting curves are plotted in azimuth (θ_s fixed at θ_i and ϕ_s ranging from $-180°$ to $+180°$) and in elevation (θ_s ranging from $-90°$ to $+90°$ and ϕ_s fixed at ϕ_i) directions. It is important to mention that for each analysis the others structure parameters were kept constant.

The plots of azimuth SCS for the ℓ variation are depicted in Figure 7. It can be observed that the SCS results for the structure containing the electric dipole are greater than those for the magnetic one. An inversion of the greatest and the lowest SCS curves, relative to both structures, is also observed in Figure 7. That is, for instance, the $\ell = 0.2\lambda_0$ leads to the greatest SCS values for the structure having the electric dipole, however conduct to the lowest SCS values for the structure with the magnetic dipole. From Figure 7b it can be seen that the greatest RCS level occurs when $\ell = 1.0\lambda_0$ and the ℓ values $0.1\lambda_0$, $0.2\lambda_0$, and $0.5\lambda_0$ present the lowest RCS levels, where the scattering of the dipole seems to be mandatory. The amplitude of the SCS variation (it is about 4 dB) is similar for both structures.

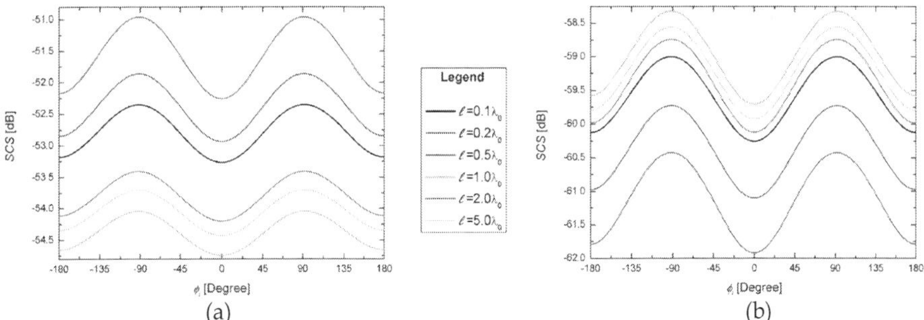

Fig. 7 – The azimuth SCS curves for different ℓ values: (a) electric dipole and (b) magnetic dipole.

The effects of the inner layer relative permittivity (ε_r) on the SCS curves are shown for structure with electric dipole in Figure 8 and for structure that has the magnetic dipole in Figure 9. In both figures are presented the SCS in azimuth and elevation directions. In general, similarly to ℓ variation, the SCS values for the structure having the electric dipole are greater than those for the magnetic dipole. For the structure with the magnetic dipole is noted that there is not alteration in the curves shape, only in their SCS levels. These SCS

levels increase when the ε_r grows. However, the *SCS* curves shape as well as their magnitudes for the structure having the electric dipole are modified by the inner layer ε_r. From the azimuth *SCS* curves (Figures 8a and 9a) it can be seen that, for ε_r values greater than 4, the curves seem to show a kind of complementary. Since, in the ϕ_i angles where occur the maximum *SCS* levels for structure containing the electric dipole are found the minimum *SCS* level for the other structure and vice-versa. The *SCS* is sensitive to electromagnetic properties of the scattering element since; in general, both dipoles bring different responses.

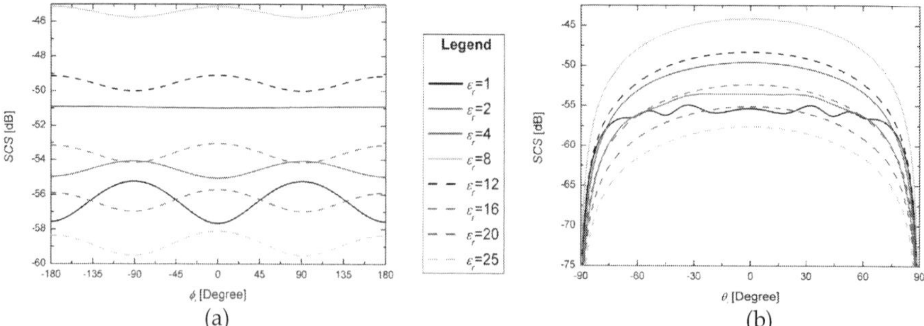

Fig. 8 – The electric dipole *SCS* curves for different of ε_r values: (a) azimuth and (b) elevation.

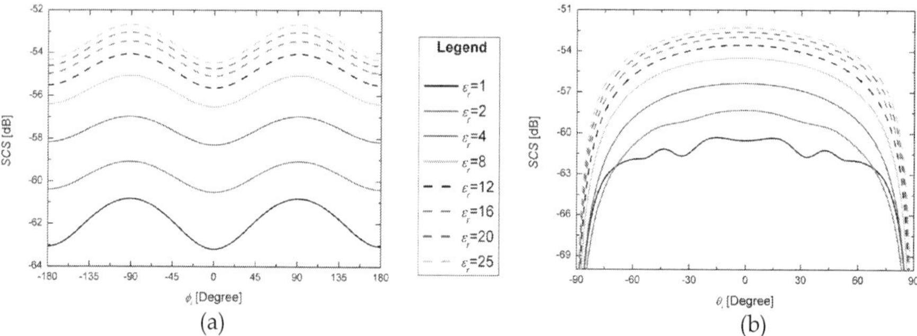

Fig. 9 – The magnetic dipole *SCS* curves for different of ε_r values: (a) azimuth and (b) elevation.

Several parameters have been proposed to assist the interpretation and the classification of polarimetric SAR data. For instance, the entropy, the anisotropy and α–angle derived from Cloude-Pottier's target decomposition theorem (Cloude & Pottier, 1997) has been widely used for this purpose. The α–angle can be also used as characterization parameter since it provides information about the type of scattering mechanism. In this way, the α–angle was evaluated at 1.25 GHz for the same multilayer structure used on the *SCS* analysis. In Figure 10 is showed the α–angle variation in terms of the inner layer relative permittivity (ε_r).

From the evaluation of α–angle it is verified that the average α–angle varies completely different from that expected with α–theoretical (around 45°). This fact it is observed for both structures. It probably occurs due to the existence of cross components in the scattering matrix of the structures. In contrast, the computed α–dominant values are almost constant and very close to 45° and to 60° for the structure containing the electric dipole and magnetic one, respectively. Therefore, according to (Cloude & Pottier, 1997), the scattering mechanism of the structure having the magnetic dipole can not be assumed as a scattering mechanism produced by an electric dipole. It leads to state that the layers introduce a certain degree of anisotropy to scattering mechanism of the magnetic dipole since its α–dominant value is characteristic of anisotropic scatter.

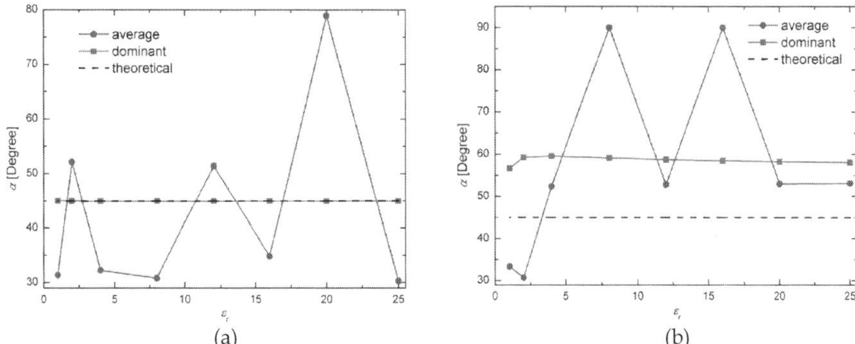

(a)   (b)

Fig. 10 – The α–angle variation in terms of ε_r for structure with: (a) electric and (b) magnetic dipoles.

According to (van Zyl et al., 1987) the polarimetric response is another way to represent the target *SCS*. In this graphical representation, the *SCS* is plotted in function of ellipticity and orientation angles of the transmitted electromagnetic wave. The polarimetric responses for the multilayer structures containing the electric and the magnetic dipoles when $\varepsilon_r = 25$ are presented, respectively, in Figures 11 and 12. The polarimetric response is sensitive to electromagnetic properties of the scattering element since both dipoles bring different responses. Besides, it can be observed that the two polarimetric responses seem to be complementary, because the point where occurs the maximum value of polarimetric response for one dipole corresponds to the point of minimum value for the other dipole. This complementary relation can be though as the duality that there exist for both dipoles. The magnitude of the polarimetric response for electric dipole is greater than for the magnetic one for this value of ε_r, as can be sustained by Figures 8 and 9.

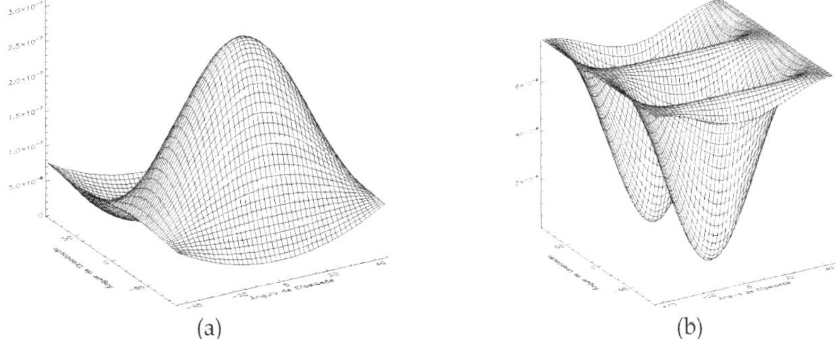

(a)   (b)

Fig. 11 – Polarimetric response of electric dipole: (a) co-polarized and (b) cross polarized.

In general, from Figure 16 can be noticed that all regions are better classified when the φ angles feature is used, this is corroborated by the values of overall accuracy and kappa coefficient. From a statistical z-test with 95% of significance level is observed that the classification result of φ image is much better than that obtained with β image, showing that discriminatory power carried by φ feature is superior to one gathered by β feature, for the simulated image used in this study. It is noted that the regions border are better defined in φ image classification, probably due to their smallest class variances.

(a)　　　　　　　　　　　　　　　　　　　　(b)

Fig. 16 – Classified images arising from: (a) β and (b) φ features.

Orig./Class	A	B	C	D
A	96.22	0.02	0.05	3.71
B	0.03	97.34	0.07	2.56
C	0.00	0.15	96.85	3.00
D	8.69	3.66	7.86	79.79

Overall Accuracy = 0.93

$\hat{\kappa} = 0.90$ and $\hat{\sigma}_{\kappa}^2 = 7.46 \times 10^{-6}$

Table 2. Confusion matrix for β image.

Orig./Class	A	B	C	D
A	97.80	0.17	0.12	1.91
B	0.05	99.76	0.02	0.17
C	0.07	0.95	98.10	0.88
D	1.54	4.39	0.98	93.09

Overall Accuracy = 0.97

$\hat{\kappa} = 0.96$ and $\hat{\sigma}_{\kappa}^2 = 2.97 \times 10^{-6}$

Table 3. Confusion matrix for φ image.

The region D, where dipole local orientation angles are randomly distributed, presents the greatest confusion in both classification results, being the worst classified region. The region B (dipole oriented preferentially at 30°) obtained the best region discrimination. The regions A (dipoles at 10°), B and C (dipoles at 20°) are very well differentiate by the simple threshold method, using β and φ features. The β image classification is characterized by the largest misclassification of the region D values. The classification results point out that these two features gather the proper information to distinguish targets whose present differences in their scatter orientation about the radar line sight.

The variation of the soil moisture content might be symbolized by variations on the relative permittivity and on the loss tangent of the structure layers. As consequence, simulated images containing regions that are distinct by these layers dielectric parameters should be used aiming for the retrieval of soil moisture from SAR data studies. Keeping this idea in mind the second image set was generated. Images having five different regions were simulated for a SAR operating at C-band (5.3 GHz) and X-band (9.6 GHz). The dielectric parameters of the multilayer structure of each image region are presented in Table 4.

Using the HH and VV images were computed two polarimetric attributes derived from the backscattering coefficient. The former was proposed in (Singh & Dubey, 2007) and is called by polarization discrimination ratio (PDR). The later was employed in (Tadono et al., 1999) and is denoted here as polarimetric description square root (PDS). These attributes are, respectively, given by

$$PDR = \frac{\sigma_{vv}^0 - \sigma_{hh}^0}{\sigma_{vv}^0 + \sigma_{hh}^0},\tag{58}$$

$$PDS = \sqrt{\sigma_{vv}^0\,\sigma_{hh}^0},\tag{59}$$

where σ_{pq}^0 represents the backscattering coefficient for pq polarization.

Region Location	ε_r	$\tan\delta_r$	ε_{rg}	$\tan\delta_g$	Dipole Orientation
Upper Left	2.33	1.2×10^{-4}	5.0	2.0×10^{-1}	$10°$
Lower Right	2.33	1.2×10^{-4}	5.0	2.0×10^{-1}	$30°$
Central	2.33	1.2×10^{-4}	5.0	2.0×10^{-1}	TR[a]
Lower Left	4.00	1.2×10^{-1}	8.0	$2.0\times10^{+1}$	TR
Upper Right	2.33	1.2×10^{-4}	8.0	$2.0\times10^{+1}$	TR

(a) TR means totally random.

Table 4. Dielectric characteristic of the structure layers of each image region.

The attribute images are depicted in Figures 17 and 18 for C- and X-bands, respectively. In these figures are also shown their respective *ICM* (Correia et al., 1998) classification results. It is important to mention that it was assumed a bivariate Gaussian distribution to model the joint density of the *PDR* and *PDS* attributes. From the classification results it can be noted that the five image regions were completely distinguished in both bands. The overall accuracy reached 99.75% for C-band images and 99.70% for X-band images, showing that the *PDR* and *PDS* attributes properly combined can be used to discriminate image regions having different dielectric characteristics.

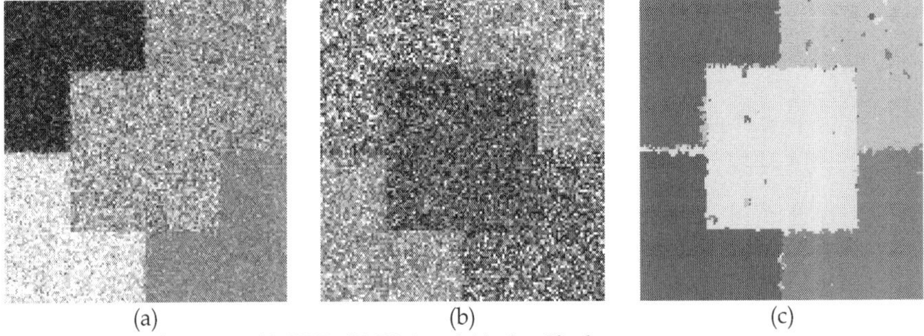

(a)   (b)   (c)

Fig. 17 – C-band images: (a) *PDR*, (b) *PDS* and (c) classified.

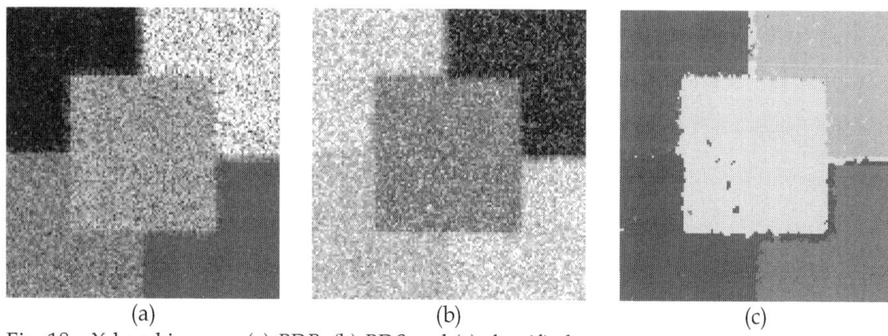

(a)   (b)   (c)

Fig. 18 – X-band images: (a) *PDR*, (b) *PDS* and (c) classified.

4. Conclusions

A scattering electromagnetic model for planar multilayer structure is introduced in this work. The methodological approach used to determine the electromagnetic fields existing in this type of structure is concisely described. The electromagnetic model potentiality is demonstrated through some remote sensing applications. These applications include the establishment of analytical expression for the scattering matrix of electric and magnetic dipoles, the simple target electromagnetic characterization and the 1-look SAR polarimetric image simulation.

The characterization of radar scattering from complex targets in terms of a linear combination of scattering from simpler ones, consists in an effective tool for the analysis of polarimetric data. From this point of view, knowledge of the scattering matrix from simpler targets is crucial to extract information gathered by polarimetric SAR data. Besides, the closer to the exact solution the determination of the scattering matrix is, the more precise the scattering analysis will be. From polarimetric SAR images and using the knowledge of scattering matrix it is possible, for instance, to develop mathematical models for natural targets and to derive several polarimetric target descriptors. It is also possible to improve the classification accuracy of land cover and use, to identify the scattering mechanisms intrinsic on an ensemble of pixels, and to built up a sensor calibration processes. In this context, the computation of scattering matrix plays a central rule in the polarimetric radar remote sensing imagery.

The scattering characterization of dipoles embedded in multilayer structure was carried out by evaluate the directivity function, the scattering cross section (*SCS*), the polarimetric response and the α–angle derived from Cloude-Pottier's target decomposition theorem. The directivity function revealed that the both infinitesimal dipoles irradiate differently, indicating their distinct electromagnetic nature. Besides, from the directivity function it could be the influence of the inner layer thickness on the radiated power. The *SCS* curves analysis demonstrated how the inner layer thickness (ℓ) and the relative permittivity (ε_r) of a multilayer structure can affect the target characterization. The ℓ parameter affects only the *SCS* levels while the ε_r influences the *SCS* shape as well as its level. In general, the *SCS* shape is dissimilar for the electric and magnetic dipoles. This statement is also verified from the graphic of polarimetric response. The estimated average α–angle for both dipoles showed great deviation from the theoretical one. On the other hand, when the dominant α–angle is used the estimative for the electric dipole is in accordance with expected one. However, for the magnetic dipole the estimated dominant α–angle values are constant and greater than 45° (typical value for an electric dipole), indicating a degree of anisotropy in the scattering mechanism.

The developed image simulation process is not restricted to monostatic airborne sensors, it can be employed to orbital sensors as well as in bistatic configuration without any additional complexity. The simulated images can be generated at several operating frequencies, containing different spatial resolution and pixel spacing among others SAR parameters. The generated images were evaluated according to several measurements commonly employed in SAR data analysis. It was checked by using a simple regression linear model whether the mean value (μ) and the standard deviation (σ) of the data exhibit the linear relationship $\sigma = 0.5227 \times \mu$ within homogeneous areas. From regression linear analysis it can be concluded that this relationship holds for all simulated images. The equivalent number of looks was estimated, proving that the simulated data have only one look as simulated. It was also assessed the discriminatory capability of the image regions by applying two classification approaches, one using a simple threshold procedure and other based on the *ICM* classifier. Consequently, from the evaluation results it can be concluded that the simulation process is working properly, since the results are in accordance with those presented in the literature.

Furthermore this simulation process can be used to improve the understanding of SAR data properties in different situation, images can be generated to do theoretical as well as practical studies in several SAR subjects. From these kind of simulated image can be developed specific tools for digital image processing and to perform studies relative to SAR data calibration. In the remote sensing context still catches a glimpse the use of the model in ground penetrating radar (GPR) studies. Finally, from the obtained results it can be concluded that the presented electromagnetic model is a useful and powerful tool to be utilized in several remote sensing issues.

5. References

Balanis, C. A. (1997). *Antenna theory: analysis and design*. New York, John Willey & Sons Inc.

Cloude, S. R. & Pottier ,E. (1996). A review of target decomposition theorems in radar polarimetry. *IEEE Transactions on Geoscience and Remote Sensing*, Vol. 34, No. 2, 498-512.

Cloude, S. R. & Pottier E. (1997). An entropy based classification scheme for land applications of polarimetric SAR, *IEEE Transactions on Geoscience and Remote Sensing*, Vol. 35, No. 1, 68-78.

Collin, R. E. & Zucker, F. J. (1969). *Antenna Theory: Part 1*, New York, McGraw-Hill Book Company.

Correia, A. H.; Freitas,C. C.; Frery, A. C.; & Sant'Anna, S. J. S. (1998). A user friendly statistical system for polarimetric SAR image classification. *Revista de Teledetección*, Vol. 10, 79–93.

Freeman, A. & Durden, S. L. (1998). A three component scattering model for polarimetric SAR data. *IEEE Transactions Geoscience and Remote Sensing*, Vol. 36, No. 3, 963-973.

Institute of Electrical and Electronics Engineers (1993). *IEEE Std 145–1993: IEEE standard definitions of terms for antennas*. New York.

Israelsson, H.; Ulander, L. M. H.; Martin, T.; Askne, J. I. H. (2000). A coherent scattering model to determine forest backscattering in the VHF. *IEEE Transactions on Geoscience and Remote Sensing*, Vol. 38, No. 1, 238–248.

Lacava, J. C. S.; Proaño De la Torre, A. V. & Cividanes, L. (2002). A dynamic model for printed apertures in anisotropic stripline structures, *IEEE Transactions on Microwave Theory*, Vol. 50, No.1, 22–26.

Lee, J. S.; Schuler, D. L. & Ainsworth, T. L. (2000). Polarimetric SAR data compensation for terrain azimuth slope variation. *IEEE Transactions on Geoscience and Remote Sensing*, Vol. 38, No. 5, 2153-2163.

Lin, C. Y. & Sarabandi, K. (1999). A Monte Carlo coherent scattering model for forest canopies using fractal-generated trees. *IEEE Transactions on Geoscience and Remote Sensing*, Vol. 37, No. 1, 440–451.

Marin, M. A.; Barkeshli, S. & Pathak, P. H. (1990). On the location of proper and improper surface wave poles for the grounded dielectric slab. *IEEE Transactions on Antennas and Propagation*, Vol. 38, No.4, 570–573.

Newman, E. H. & Forrai, D. (1987). Scattering from microstrip patch. *IEEE Transactions on Antennas and Propagation*, Vol. AP–35, No.3, 245–251.

Sant'Anna, S. J. S., Lacava, J. C. S. & Fernandes, D. (2007a). Closed form expressions for scattering matrix of simple targets in multilayer structures, in *Proceedings of International Geoscience and Remote Sensing Symposium*, Barcelona, Spain, July 2007, 714–717.

Sant'Anna, S. J. S., Lacava, J. C. S. & Fernandes, D. (2007b). A useful tool to simulate polarimetric SAR images. *Proceedings of Eleventh URSI Commission F Triennial Open Symposium on Radio Wave Propagation and Remote Sensing*. Rio de Janeiro, Rio de Janeiro.

Sant'Anna, S. J. S., Lacava, J. C. S. & Fernandes, D. (2008a). Evaluation of the scattering matrix of flat dipoles embedded in multilayer structures. *PIERS Online*, Vol. 4, No. 5, 536–540.

Sant'Anna, S. J. S., Lacava, J. C. S. & Fernandes, D. (2008b). From Maxwell's equations to polarimetric SAR images: a simulation approach. *Sensor*, Vol. 8, No. 11, 7380–7409.

Sant'Anna, S. J. S., Lacava, J. C. S. & Fernandes, D. (2008c). Analysis of simulated polarimetric SAR images generated by a multilayer electromagnetic scattering model., *Proceedings of International Geoscience and Remote Sensing Symposium*, Boston, July 2008, 1245–1248.

Sant'Anna, S. J. S. (2009). Modelagem do espalhamento eletromagnético de estruturas multicamadas com aplicações em sensoriamento remoto por micro-ondas, *PhD. Thesis* (in Portuguese), ITA, São José dos Campos.

Singh, D. & Dubey, V. (2007). Microwave bistatic polarization measurements for retrieval of soil moisture using an incident angle approach, *Journal of Geophysics and Engineering*, n. 4, 75–82.

Tadono, T.; Qong, M.; Wakabayashi, H.; Shimada, M.; Kobayashi, T. & Shi, J.   (1999). Preliminary studies for estimating surface soil moisture and roughness based on a simultaneous experiment with CRL/NASDA airbone SAR (PI-SAR), *Proceedings of Asian Conference on Remote Sensing*. Hong Kong.

Ulaby, F. T. & Elachi, C. (1990). *Radar polarimetry for geoscience applications*. Norwood, Artech House.

Yang, J.; Peng, Y. N. & Lin, S. M. (2001). Similarity between two scattering matrices, *Electronics Letters*, Vol. 37, No. 3, 193–194.

van Zyl, J. J.; Zebker, H. A. & Elachi, C. (1987). Imaging radar polarization signatures: theory and observation. *Radio Science*, Vol. 22, No. 4, 529-543.

Zurk, L. M.; Koistinen, P.; Sarvas, J. & Holmström, L. (2001). Electromagnetic scattering model for forest remote sensing. *Research Reports A38*, University of Helsinki, Helsinki.

31

On Performance of S-band FMCW Radar for Atmospheric Measurements

Turker Ince

Izmir University Of Economics
Turkey

1. Introduction

At low microwave frequencies, radars respond to spatial variations in the index of refraction of the air often characterized by the atmospheric refractive-index structure parameter, C_n^2. While remote measurement of the atmospheric boundary layer (ABL) is complicated by the low radar reflectivity of clear-air turbulence structures, frequency-modulated, continuous-wave (FMCW) radars, having tremendous sensitivity and spatial resolution compared to their pulsed counterparts, have proven to be a solution to this problem. While Doppler capability can be added to FMCW radars (Strauch et al., 1976), the unique strength of this technology lies in its ability to monitor the atmospheric refractive-index structure parameter, C_n^2, with unparalleled resolution in height and time. Since the first high-resolution atmospheric FMCW radar, developed by Richter (1969), a number of FMCW radars have been developed for high-resolution atmospheric probing (Chadwick et al., 1976; Eaton et al., 1995; Hirsch, 1996), which have provided valuable information about the fine structure and dynamics of clear-air turbulence. The advent of S-band, FMCW radars has opened a new research field, which was reviewed by Gossard (1990).

Microwave backscattering from the clear-air ABL is generally a combination of Bragg scattering from small-scale spatial variations in refractive index caused by turbulence and Rayleigh scattering from insects, birds, dust, or other airborne particles whose size is much smaller than the radar wavelength. At S-band frequencies, radars measure clear-air backscatter from both refractive index turbulence and from insects. Based on the experimental observations (Wilson et al., 1994; Eaton et al., 1995; Ince et al., 2003; Martin and Shapiro, 2007), Rayleigh scattering from insects can be the dominant source of clear-air returns in both the convective boundary layer (CBL) and the nocturnal boundary layer (NBL). Depending on an application, radar return from airborne insects may be considered either as source of clutter (such as contamination of weather radar measurements) or as desirable passive tracers of the clear air (Wilson et al., 1994). Due to its very fine spatial and temporal resolution, S-band FMCW radars are able to resolve individual insects and discriminate between discrete particle echoes and distributed clear-air echo.

Despite FMCW radars being excellent tools for turbulence studies in the convective ABL and in the capping inversion region (Eaton et al., 1995), relatively few atmospheric studies reported in the literature have used FMCW radar data extensively for both quantitative and qualitative analysis of clear-air features observed. While S-band FMCW radars have been designed with the capability to obtain height and time resolutions of 1 m and 1 s, respectively, the degree to which these resolution limits are obtained in practice depends upon the properties of the atmospheric echo itself. Range measurement errors due to non-zero Doppler velocities and finite coherence of the clear-air echo have implications on both spatial resolution and Doppler estimation of FMCW radars. Although the theoretical background for FMCW radars is well established, their measurement limitations for atmospheric targets and performance impact for atmospheric applications has received limited attention in the literature.

The detailed review of the theory of operation of S-band FMCW radar and its application to atmospheric boundary layer profiling can be found in the literature (Eaton et al., 1995; Ince et al., 2003). In this study, performance aspects of atmospheric FMCW radars including the effects of range-Doppler ambiguity, parallax typical of two-antenna radar systems, and near-field operation are discussed. The data collected during recent field experiments by the high-resolution S-band FMCW radar, described in Ince et al. (2003), is used to illustrate system performance. Analysis of FMCW radar signatures of atmospheric targets detectable at S-band frequencies has been performed by extracting both qualitative morphological information and quantitative information on the intensity of backscattered power from clear-air turbulence.

2. Principles

Frequency-modulated, continuous-wave (FMCW) radars may be thought of as a limiting case of pulse-compression radar where the duty cycle of the transmitted waveform approaches 100%. They operate by transmitting a long, coded waveform of duration T and bandwidth B. The improvement factor they gain over pulsed radars of equivalent range resolution is given by the time-bandwidth product of the waveform BT, which is often referred to as the compression gain. In FMCW systems, this gain can be very large, exceeding 60 dB. While several types of frequency coding may be used to yield the bandwidth B, linear frequency modulation is the simplest and most commonly used method in atmospheric FMCW radars.

Consider an FMCW radar that transmits constant amplitude linear FM signal of the form

$$s(t) = \exp(jwt + j(a/2)t^2), \qquad -T/2 < t < T/2 \tag{1}$$

where w is the radian frequency and a is the chirp rate in rad/s^2. The echo from an atmospheric target is essentially a delayed, attenuated, and possibly Doppler-shifted replica of the transmitted signal. The echo is demodulated by mixing it with a portion of the transmitted signal and low-pass filtering the result. The resulting beat frequency for a point target at range R_0, moving at radial velocity u_r, is given by

$$f_r(t) = -(\frac{2u_r}{\lambda} + \frac{2R_0}{c}\dot{f} + \frac{4u_r}{c}\dot{f}t), \tag{2}$$

where λ is the electromagnetic wavelength and $\dot{f} = a/2\pi = B/T$ is the chirp rate in Hz/s. Here, the first term represents the Doppler frequency shift, f_D, due to radial motion, and the second term represents the shift due to the nominal range of the target, which is exploited primarily by the FMCW radar. The final term represents a defocusing due to the dilation of the reflected signal's bandwidth due to the dependence of Doppler frequency to carrier frequency. For lower atmospheric velocities, the effect of the last term on the bandwidth of the echo may be safely ignored. Thus, for the case of stationary radar and targets, the beat frequency is linearly proportional to range.

Signal processing of the echo usually involves matched filtering, which is most commonly implemented for all ranges simultaneously through spectral analysis via an FFT algorithm. In this case, the output of the matched filter, which gives the response of the linear FMCW radar to a moving point target, can be expressed as

$$y(R) = \frac{\sin\left[\pi\left(f_D T + \left(R - R_0\right)/\Delta R\right)\right]}{\pi\left(f_D T + \left(R - R_0\right)/\Delta R\right)}, \tag{3}$$

where f_D is the Doppler frequency, and $\Delta R = c/(2B)$ is the expected range resolution for a transmitted bandwidth $B = \dot{f}T$. It is worth noting that this result is equivalent to impulse response of linear FMCW radar.

For FMCW Doppler radar, Doppler information can be retrieved on a sweep-to-sweep basis by analyzing the sequence of echoes from a particular range, as discussed by Strauch et al. (1976). In this case, the sampling frequency is the reciprocal of the sweep period T, the Nyquist Doppler frequency is $1/2T$, and resulting in an unambiguous Doppler velocity interval of $|u_r| \leq \lambda/4T$ assuming that range measurement errors due to Doppler can be ignored. Conventionally, two-dimensional FFT is performed on received signals to obtain range and Doppler frequency (velocity) spectrum.

The received mean power of a bistatic weather radar from a distributed atmospheric target of uniform reflectivity η, for circularly symmetric Gaussian shape antenna patterns, is given by Doviak & Zrnic (1993),

$$\overline{P}(R) = \frac{P_t G_t G_r \lambda^2 \Theta_1^2 \Delta R}{(512 \ln 2)\pi^2 R^2}\eta, \tag{4}$$

where P_t is transmitted power, G_t and G_r are transmit and receive antenna gains, Θ_1 is the one-way 3 dB beamwidth, ΔR is range resolution, and λ is radar wavelength. To better estimate η from the measured echo power at the output of receiver, the radar system must be calibrated. Based on Tatarskii's theory of electromagnetic wave propagation in a turbulent atmosphere (Tatarskii, 1961) and Ottersten's work in his 1969 landmark paper (Ottersten, 1969), clear-air radars can measure the refractive-index structure parameter (C_n^2)

from radar backscattered power. For Bragg scattering from refractive index fluctuations due to homogeneous, isotropic turbulence, η is commonly related to C_n^2 by Ottersten' well-known equation:

$$\eta = 0.38 C_n^2 \lambda^{-1/3}.$$ (5)

This equation assumes that the Bragg wavenumber, $k_B = 4\pi / \lambda$ lies within the inertial subrange. Note that C_n^2 in (5) is defined as an ensemble average. In practice, such averaging is performed over space or time assuming ergodicity. Therefore, for S-band radars, Bragg scatter from clear air depends much less on the wavelength of the radar than that of Rayleigh scatter (λ^{-4} dependence) by airborne particles.

3. Performance Analysis

3.1 Range and Doppler

FMCW radar performance characteristics, such as range-Doppler coupling, can be seen from analysis of (3), which is also ambiguity function for linear FMCW signal. The presence of both range and Doppler terms in the argument of the sinc function illustrates the effect of target motion on the radar's ability to locate. Figure 1 shows a typical two-dimensional time delay (range) – Doppler (velocity) ambiguity surface for a linear FMCW signal with time-bandwidth product of 1000. As seen from Figure 1, maximum unambiguous Doppler can be increased (by decreasing sweep period) at the expense of range ambiguity, or alternatively increasing maximum range results in Doppler ambiguity. By rearranging the terms in the argument in (3), the apparent range of the target can be expressed as $R_{app} = R_0 - f_D T \Delta R$, where it is easy to see that range mislocation by one resolution cell occurs when $f_D T = 1$, or $u_r = \pm \lambda / 2T$ which also corresponds to the Nyquist velocity interval for FMCW Doppler radar. Therefore, targets with unambiguously measured velocities are misregistered by no more than one half a resolution cell, or misregistration occurs when target velocities are aliased.

Additionally, the coherence of the atmospheric target during the sweep interval will limit range resolution of linear FMCW radar. For complex moving targets and for volume scattering, the coherence time (or the reciprocal of the Doppler spectral width) of the echo will limit resolution. A distribution of Doppler velocities observed over an integration time, T, will yield a distribution of apparent ranges. The resulting rms spread in range is determined by the transformation of the Doppler spectrum to the range domain using the relationship $R_{app} = R_0 - f_D T \Delta R$,

$$\sigma_R = \sigma_f (T \Delta R),$$ (6)

where σ_f is the Doppler spectral width of the echo. From this relation, it is apparent range resolution and sensitivity are optimized by matching the sweep time to the reciprocal of the Doppler bandwidth of the echo. In this case, the range spreading is equal to the range resolution, and the entire Doppler spectrum is confined to one range bin. No improvement in sensitivity or in resolution is achieved by increasing sweep time beyond this value, as the resulting echo simply spreads to adjacent range bins.

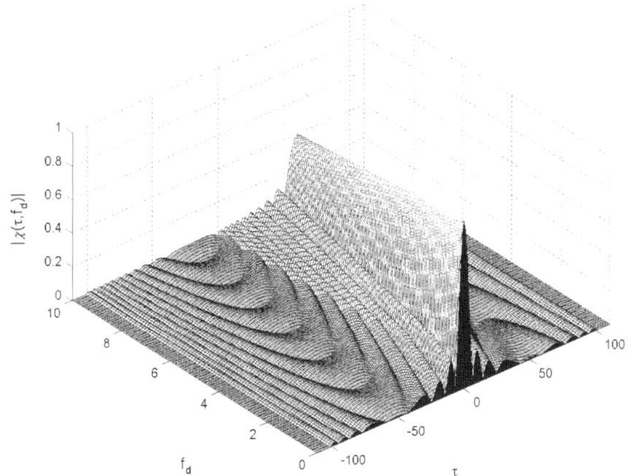

Fig. 1. The ambiguity surface for a linear FMCW signal with BT = 1000.

To maximize resolution and sensitivity, it is desirable to make both bandwidth and sweep time as large as possible. The effective value of sweep time is however constrained by the coherence time of the atmospheric echo. For example, a sampling volume with rms radial velocity of 1 m/s has a spectral width of 20 Hz at 3 GHz implying a coherence time of approximately 50 ms.

3.2 Parallax and Near-field Operation

From the weather radar equation (4), the received mean power can be correctly estimated only if radar targets are in the far field of antenna, that is $R > 2D^2 / \lambda$, where D is the diameter of the antenna. Additionally, for bistatic FMCW radar employing two spatially separated antennas, a correction of the received power is necessary at near ranges to account for the reduced beam overlap. For Gaussian shaped beams aligned with their axes parallel, the antenna parallax function, or fractional beam overlap is given by Ince et al. (2003),

$$C_p(R) = \exp(-2\ln 2 \frac{d^2}{\Theta_1^2 R^2}),$$

(7)

where d is the separation distance between transmit and receive antennas, and Θ_1 is the one-way half-power beamwidth. Note that parallax results in an apparent reflectivity reduction and parallax error of backscattered power becomes significant within the radiating near-field of the antennas such that assumption of a far-field Gaussian beam pattern is not valid. It is also worth noting that the far-field criterion, $R_f = 2D^2 / \lambda$, is based on a conservative specification of maximum phase error, significant effects on the shape of the main lobe are not evident until $R \approx R_f / 4$ (Hansen, 1985).

FMCW atmospheric radars can measure boundary layer targets at minimum distances as close as 50 m. However, this minimum range lies in the Fresnel region in which antenna gain and pattern shape vary significantly with distance, and from (4) radar reflectivity is underestimated for targets in the near field. Based on the universal near-field reflectivity correction by Sekelsky (2002), 1 dB reduction in reflectivity at $R_f / 4$ is indicated, and at closer ranges ($R < R_f / 4$) near-field gain reduction becomes much more significant. Since at closer ranges the beam shape is no longer approximately Gaussian and is strongly dependent upon the particular antenna design, it is necessary to apply appropriate reflectivity correction for parallax and near-field gain reduction depending upon the parameters of radar system. In Figure 2, the relative importance of near field gain reduction and parallax for the parameters of the S-band FMCW radar system in Ince et al. (2003) is compared, and in this case parallax has the primary influence at close ranges.

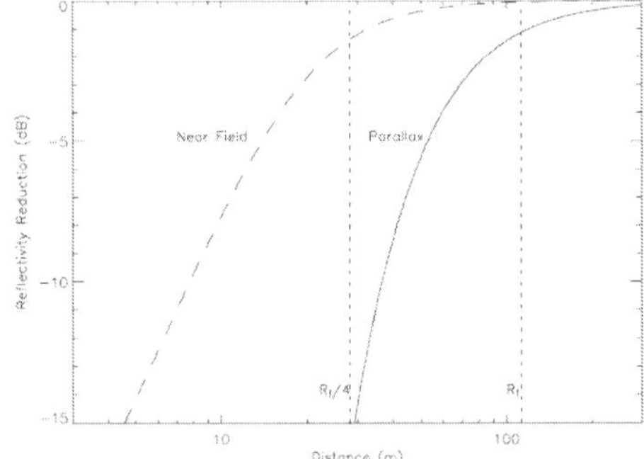

Fig. 2. Reflectivity reduction due to parallax (solid line) and near field (dashed line) effects for the parameters of the FMCW system in Ince et al. (2003).

4. Experimental Measurements

Data collected by the University of Massachusetts' high-resolution FMCW radar (Figure 3) during recent field experiments is used to illustrate system performance and analyze radar signatures of atmospheric targets detectable at S-band frequencies. Detailed description of this system and data processing can be found in Ince et al. (2003). The S-band FMCW radar system implements an internal calibration loop (by injecting into the receiver an attenuated and delayed sample of the high power amplifier output) to perform calibration of the atmospheric echo and estimate volume reflectivity.

Fig. 3. The University of Massachusetts S-Band FMCW radar system mounted on flatbed truck (electronics are contained within the cab).

Figure 4 shows one day continuous radar record of diurnal cycle of the ABL beginning at 14:27 local time (CDT, 19:27 UTC). The radar echo is expressed in terms of the logarithm of microwave C_n^2 obtained using (5), however, this representation is only meaningful for the clear-air component of the backscatter described by the Bragg scattering mechanism. In addition to distributed Bragg scatter from clear air, Figure 4 shows strong point echoes, which are in this study assumed to be entirely due to Rayleigh scatterers. It is worth noting that the distribution of Rayleigh scatter in reflectivity time-height image appears to reveal qualitatively additional boundary layer structure not otherwise detectable at S-band. However, the effect of much stronger Rayleigh echo (due to much stronger wavelength dependence, λ^{-4} as opposed to $\lambda^{-1/3}$ for Bragg scattering) on quantitative backscattered signal information is adverse. Interestingly, with high spatio-temporal resolution capability of FMCW radar, by reducing time averaging to below 1 s in Figure 4, it becomes possible to see undulations of O(10 m) in the radar echo as a consequence of Doppler-induced range measurement error due to the flapping of bird's wings. It is estimated that under clear sky conditions vertical velocities of atmospheric echoes in general will contribute misregistration of about one range bin at most.

From Figure 4, four-hour time period of convective ABL (between 14:30 to 18:30 local time) is extracted to perform quantitative analysis of FMCW radar reflectivity. In this case, the collocated radiosonde (operated by NCAR/ATD) measurements of temperature and humidity show typical characteristic of convective ABL bounded above by dry air. In the three panels of Figure 5 respectively reflectivity, Doppler (vertical) velocity, and the correlation coefficient of successive echoes for one-hour of convective ABL are shown. The latter two products are the result of the pulse-pair processing (Doviak & Zrnic, 1993) averaged over 20 pulses (~1 s averaging).

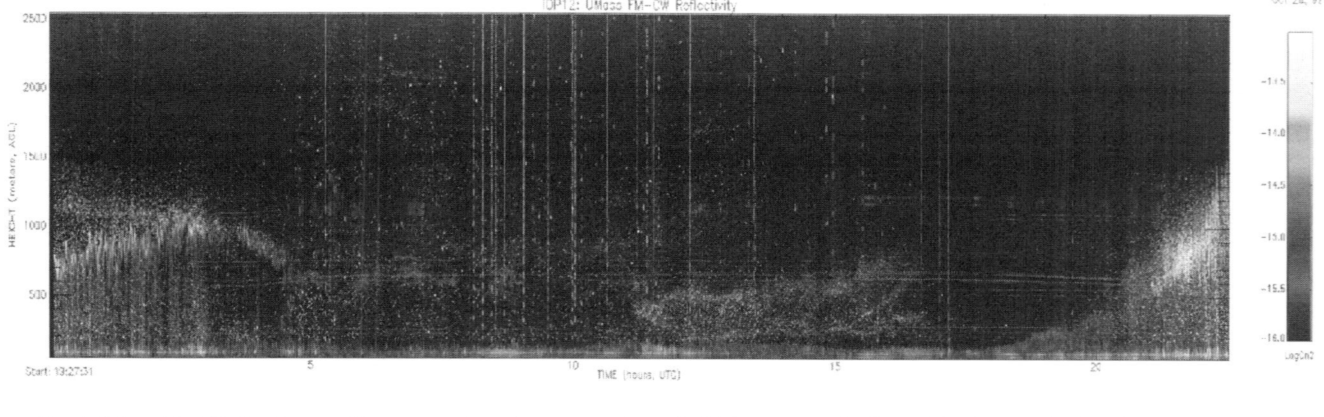

Fig. 4. S-band FMCW radar profiler image showing approximately 24-hour record of the ABL during CASES-99 experiment on 26 Oct beginning at 14:27 CDT.

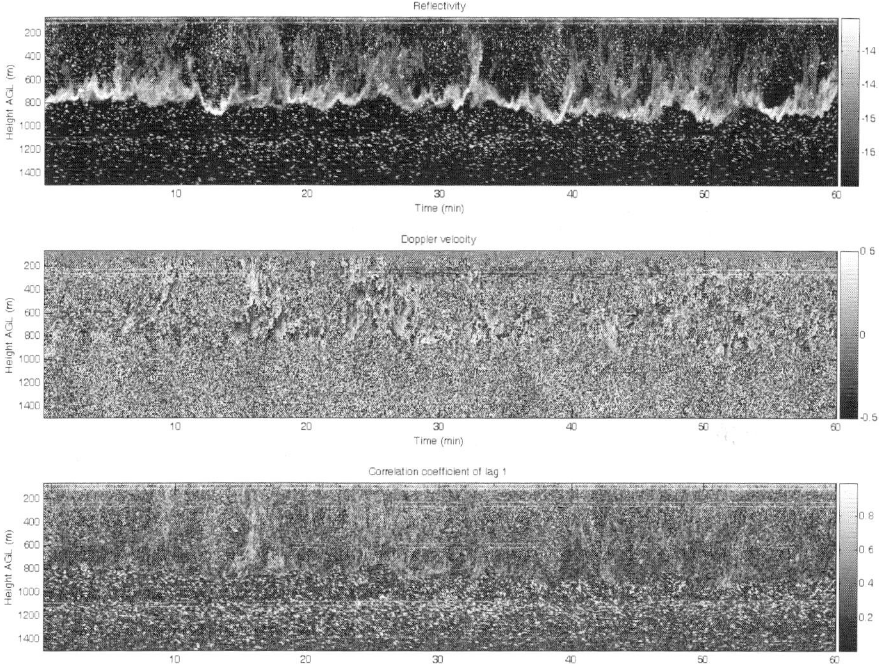

Fig. 5. First hour of convective ABL echo from Figure 4 showing reflectivity, Doppler velocity, and single-lag correlation coefficient.

The reflectivity image shows that initially significant Rayleigh backscatter is observed both above and below the capping inversion which peaks near 1000 m altitude at about 17:00 local time. After this time, both the Rayleigh scatter and the distributed Bragg scatter below the inversion decrease significantly, and the strong echo at the top of the boundary layer disappears. The velocity image which is derived from the phase of the single-lag covariance shows structure for some, but not all of the clear-air echo (due to velocity aliasing). The single-lag covariance panel shows extremely

high correlations for the Rayleigh scattering while the correlation coefficient for the Bragg scattering is relatively lower on average (due to larger spectral width of clear-air echo).

To discriminate the distributed clear-air echo from particulate scatter, a signal processing method based on the correlation coefficient of successive echoes can be applied to time-series of backscattered power. According to this method, magnitude of the sweep-to-sweep correlation coefficient is compared to an empirically determined threshold (in this case, a value of 0.8 is used) and high-correlation points (due to Rayleigh scatter) are removed and filled with estimated data by using least-squares interpolation (without affecting the statistics of the atmospheric clear-air echo). This processing allows estimation of atmospheric component of the clear air backscatter. Alternatively, median filtering can be applied to the two-dimensional radar reflectivity image to differentiate morphologically differing Bragg scatter (distributed echoes) and Rayleigh scatter (dot echoes). However, in case of large density of particle echoes, their morphology will be similar to distributed scatter, hence the filter will not be able to separate two sources of scattering.

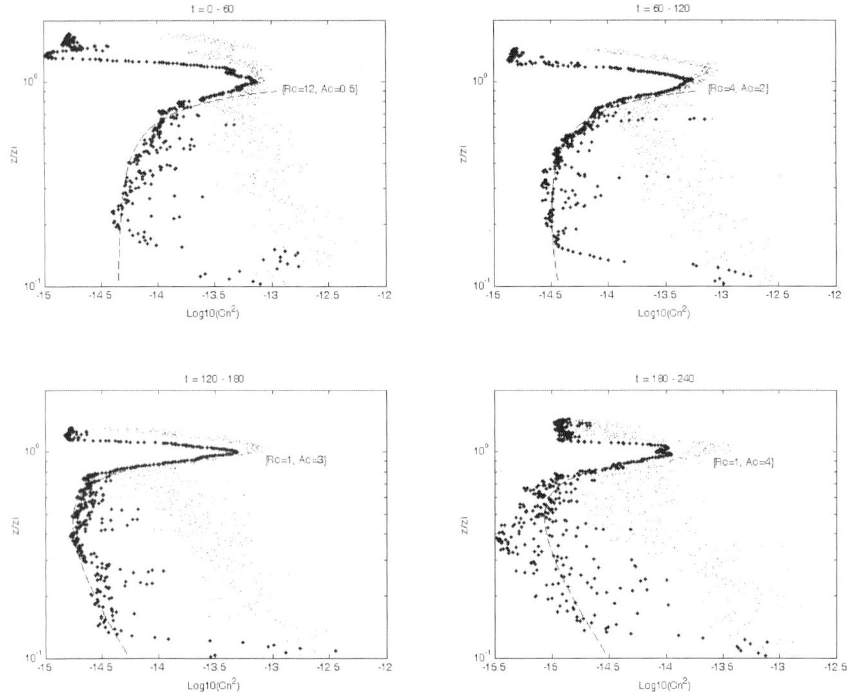

Fig. 6. Vertical profiles of estimated mean C_n^2 (large data points) obtained from post-processing over four consecutive 60-min periods. The small dotted profiles show the mean reflectivity including both Bragg and Rayleigh echo. The dashed lines are normalized profiles of C_n^2 based on the Fairall's model (1987).

In Figure 6, the computed vertical profiles of mean C_n^2 (estimated from postprocessed reflectivity) for four consecutive 60-min segments are plotted against height. The small dotted profiles show the corresponding mean reflectivity over the same intervals including both Bragg and Rayleigh echo. The star (*) profiles indicate estimated clear air component of the vertical profile over the 60-min interval. The vertical axis of each profile is scaled by the boundary layer depth, z_i, which can simply be obtained from the maximum of the reflectivity. Given the mean horizontal winds, the vertical profiles are roughly equivalent to a streamwise spatial average over approximately 10 km. Due to an increased number of insect echoes in the afternoon boundary layer and insect plumes (tracing sudden vertical motion from updrafts), Rayleigh scatterers dominate the observed profiles of mean reflectivity.

In this case, FMCW radar estimates of C_n^2 can be used to quantitatively test theoretical predictions on the ABL. In the free-convection boundary layer, C_n^2 is expected to follow a $z^{-4/3}$ power law, where z is height above ground level (Wyngaard & LeMone, 1980). From Figure 6, radar estimated C_n^2 increases with height inside convective mixed layer (between $0.2z_i$ and $0.9z_i$ the mean reflectivity follows a $z^{2/3}$ profile, a discrepancy of z^2) with an expected sharp turbulence induced peak at the top of the mixed layer ($\sim z_i$). Note in this case strong ground clutter affects C_n^2 measurements below $0.1z_i$. This observed discrepancy in C_n^2 from its mixed-layer prediction may be due to entrainment effects, which can be predicted by a model developed by Fairall (1987) based on a top-down and bottom-up diffusion approach in the entraining, convective boundary layer. In Figure 6, by choosing appropriate values for R_c and A_c parameters (respectively top-down and bottom-up components) of the model, the predicted normalized C_n^2 profiles is fitted to the vertical profiles of mean reflectivity from experimental data. Additionally, theoretical models on the shape of the reflectivity profile in the inversion region have been developed by Wyngaard & LeMone (1980). The existence of the reflectivity peak is known from in situ measurements (Wyngaard & LeMone, 1980) and from large-eddy simulation (Muschinski et al., 1999).

In Figure 7, log-variance of estimated C_n^2 over four consecutive 60-min periods is compared against the radiosonde-based potential temperature profiles. It can be seen that much higher variance occurs across the top of the convective boundary layer (corresponding to the height of the strongest potential temperature gradient) due to strong refractive index turbulence, and peak C_n^2 variance increases as the potential temperature jump across the inversion increases.

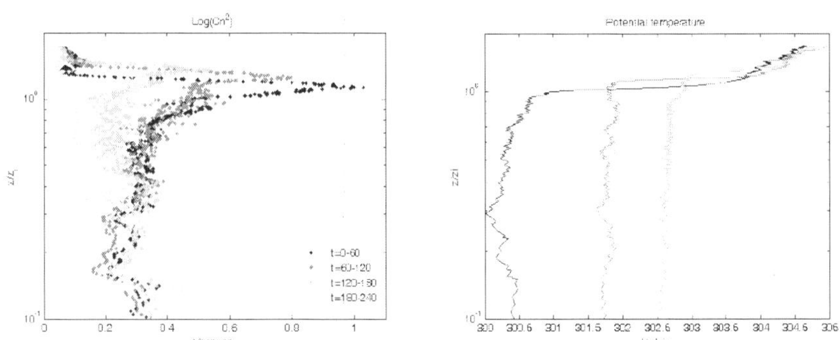

Fig. 7. The log-variance of estimated C_n^2 over four consecutive 60-min periods (left), and corresponding potential temperature profiles measured by radiosonde (right).

In Figure 8, radar reflectivity time-height image (starting at 0650 UTC in Figure 4) of nocturnal boundary layer turbulence below and insect layer within a low-level jet in the

NBL is shown. Collocated radiosonde profile of wind speed taken at 0658 UTC is shown in Figure 9. The FMCW reflectivity shows intensified turbulence and wave activity from surface extended to heights of about 150 m where sounding shows evidence of strong shear. Additionally, radiosonde wind measurements shows a low-level jet formed between heights 150 and 550 m with maximum winds of over 16 m/s at approximately 300 m height. In this case, a layer of insects detected by the radar at heights of the wind speed maximum helps reveal qualitatively this additional boundary layer structure.

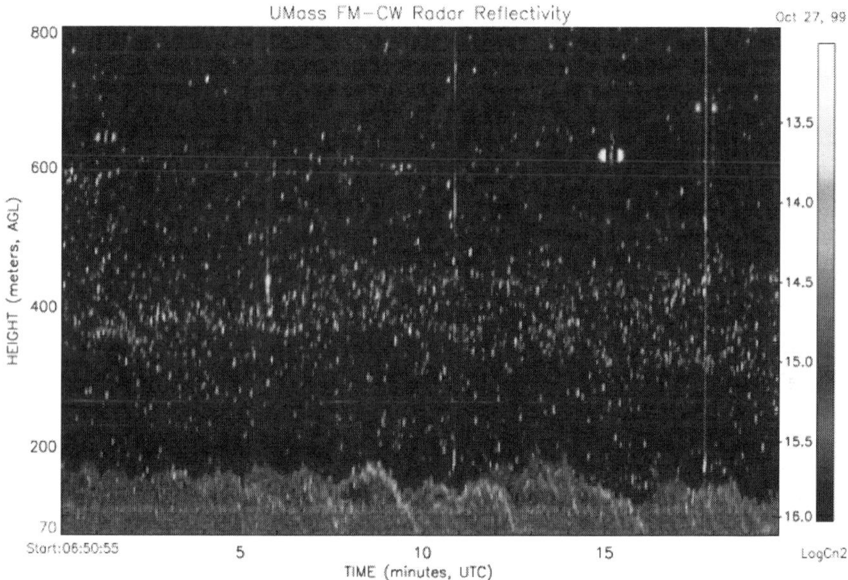

Fig. 8. Nocturnal boundary layer turbulence below and insects within a low-level jet observed at 0650 UTC.

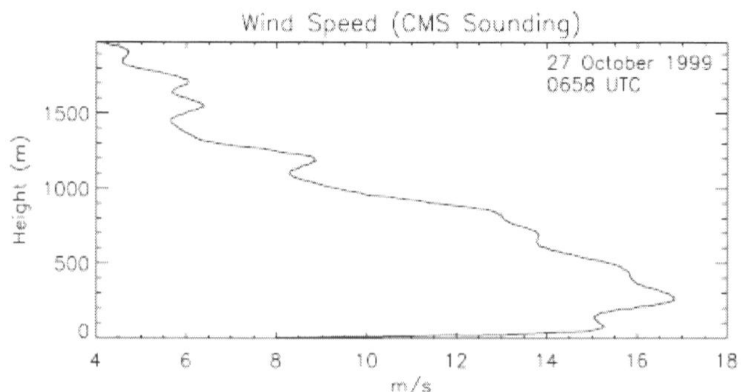

Fig. 9. Collocated radiosonde profile of wind speed at 0658 UTC.

5. Conclusion

In this study we have discussed and illustrated system performance of FMCW radar detection of atmospheric targets at S-band frequencies. In particular, the effects of non-zero Doppler velocities and finite coherence of the atmospheric echo on both spatial resolution of FMCW radars and on Doppler estimation have been presented. Additionally, we have considered parallax errors in reflectivity and near-field operation. For the clear-air atmosphere measurements, the S-band radar's sensitivity to Bragg and Rayleigh scattering and the effect of Rayleigh scatterers on vertical profiles of mean radar reflectivity have been illustrated. Experimental data collected by the high-resolution atmospheric FMCW radar has been used to perform analysis of S-band radar signatures of atmospheric targets and to quantitatively test theoretical predictions on the atmospheric boundary layer.

6. References

Chadwick, R.B.; Moran, K.P.; Strauch, R.G.; Morrison, G.E. & Campbell, W.C. (1976). Microwave radar wind measurements in the clear air. *Radio Science*, vol. 11, pp. 795–802.

Doviak, R.J. & Zrnic, D.S. (1993). *Doppler Radar and Weather Observations*. 2d ed. Academic Press, ISBN-10 : 0122214226.

Eaton, F.D.; McLaughlin, S.A. & Hines, J.R. (1995). A new frequency-modulated continuous wave radar for studying planetary boundary layer morphology. *Radio Science*, vol. 30, pp. 75–88.

Fairall, C. W. (1987). A top-down and bottom-up diffusion model of Ct2 and Cq2 in the entraining convective boundary layer. *J. Atmos. Sci.*, vol. 44, pp. 1009-1017.

Gossard, E.E. (1990). Radar research on the atmospheric boundary layer, In: *Radar in Meteorology*, edited by D. Atlas, pp. 477-527, Amer. Meteor. Soc., Boston.

Hansen, R. C. (1985). Aperture theory, In: *Microwave Scanning Antennas*, vol. 1, edited by R. C. Hansen, Penninsula, Los Altos, Calif.

Hirsch, L. (1996). Spaced-antenna-drift measurements of the horizontal wind speed using a FMCW-radar-RASS. *Contr. Atmos. Phys.*, vol. 69, pp. 113–117.

Ince, T. ; Frasier, S.J.; Muschinski, A. & Pazmany, A.L. (2003). An S-band frequency-modulated continuous-wave boundary layer profiler: Description and initial results. *Radio Science*, vol. 38, no. 4, pp. 1072–1084.

Martin, W.J. & Shapiro, A. (2007). Discrimination of bird and insect radar echoes in clear air using high-resolution radars. *J. Atmos. Oceanic Technol.*, vol. 24, pp. 1215–1230.

Muschinski, A.; Sullivan, P.P.; Wuertz, D.B.; Hill, R.J.; Cohn, S.A.; Lenschow, D.H. & Doviak, R.J. (1999). First synthesis of wind-profiler signals on the basis of large-eddy simulation data. *Radio Science*, vol. 34, pp. 1437-1459.

Ottersten, H. (1969). Atmospheric structure and radar backscattering in clear air. *Radio Science*, vol. 12, no. 4, pp. 1179–1193.

Richter, J. (1969). High-resolution tropospheric radar sounding. *Radio Science*, vol. 4, pp. 1261–1268.

Sekelsky, S. M. (2002). Near-field reflectivity and antenna boresight gain corrections for millimeter wave atmospheric radars. *J. Atmos. Oceanic Technol.*, vol. 19, pp. 468–477.

Strauch, R.G.; Campbell, W.C.; Chadwick, R.B. & Moran, K.P. (1976). Microwave FM-CW Doppler radar for boundary layer probing. *Geophys. Res. Lett.*, vol. 3, pp. 193–196.

Tatarskii, V. I. (1961). *Wave Propagation in a Turbulent Medium*. McGraw-Hill, New York.

Wilson, J.W.; Weckwerth, T.M.; Vivenkanandan, J.; Wakimoto, R.M. & Russell, R.W. (1994). Boundary layer clear-air radar echoes: Origin of echoes and accuracy of derived winds. *J. Atmos. Oceanic Technol.*, vol. 11, pp. 1184–1206.

Wyngaard, J.C. & LeMone, M.A. (1980). Behavior of the refractive index structure parameter in the entraining convective boundary layer. *J. Atmos. Sci.*, vol. 37, pp. 1573-1585.

32

Use of Lidar Data in Floodplain Risk Management Planning: The Experience of Tabasco 2007 Flood

Ramos J., Marrufo L. and González F.J.
Instituto de Ingeniería, Universidad Nacional Autónoma de México
Av. Universidad 3000, 04510 Coyoacán, México, D.F.

1. Introduction

Floodplains along rivers are vulnerable and suffer frequent socio-economical losses as consequence of large rainfall events. The risk of flooding is function of several factors such as: climate, the material that makes up the banks of the stream, the channel slope and the topography. The risk of floodplains is increased as a result of the human tendency to use the river land to cope with community demands. As the river plains are synonymous with fertile soils with high phreatic levels, their transformation into arable areas or into urban areas are guaranteed to produce an economical input improving human life. This situation makes flood less important since the benefits produced outweigh the damage; sometimes damage could be reduced or mitigated by constructing barriers. The actual problem is when the flooding generates devastation and the benefits are minimised. This is a situation that could be present at any moment through the development of the region (Maza and Franco, 1997). This is the case for many of the watersheds in the Gulf of Mexico including the Grijalva Basin, one of the largest in Mexico, which suffered severe flooding in 2007 affecting almost 70 % of the Tabasco state. As a consequence around one million residents lost their homes and the economic impact was calculated in millions of pesos.

The modification of the river landscape is determinant because it alters the natural condition of the area dredging water bodies, change drainage, building impermeable surfaces and constructing barriers. These modifications make it difficult to contain the water in the river channel when prolonged heavy rains take place as is the case in recent times, where flooding becomes more frequent, extreme, damaging and costly (Werritty, 2006).

A frequent problem is that floodplains lose their capability to preserve the hydraulic conditions to prevent floods and to function as areas to support ecosystems. In general, floodplains provide important habitats for plants and animals, and also these areas can retain water, reduce the flow velocity of a river and minimise erosion reducing the flood impact downstream (Moss and Monstadt, 2008). However, it is not enough to study the water movement in the watershed in order to determine the amount of water arriving to the river channel and to be transported during the flood event. It is also necessary to know the dynamics of the river and to determine the spatial-temporal distribution of flow as well as the water level reached in the floodplains (Stabel and Löffler, 2003). The latter is addressed through the use of flood maps (if they exist) but most of them are theoretical predictions. Other maps are gathered using remote sensing data obtained by tracking the movement of the water on the land. However, it is difficult to find a direct relation between ground elevation and the highest expected river level.

The research reported in this chapter presents a contribution to calculate water levels and flooding maps using Geographic Information Systems (GIS) to combine existing datasets, water elevation survey data and hydrological models. There is a major emphasis to use high accuracy elevation data to model and map a particular flood event (past or future), to assess the risk of flooding in a particular region (Jones, 2004; Jones et al., 1998). LiDAR (Light Detection And Ranging) technology has the ability to produce accurate elevation data throughout the flood plain, in particular flat areas with slopes of less than 50 cm per kilometre as is the case of the coastal flatplains in Tabasco. LiDAR data allows a better calibration of hydrological models, to map the flood risk and water levels helping decision makers to cope with an emergency and to implement strategies after the event (Filin, 2003; Webster & Stiff, 2008).

2. Gathering topographic data by Digital Elevation Models

As floodplains are neither static nor stable, one needs to consider at least three aspects related to the land surface characteristic to describe a flood: (a) topography, (b) geomorphology, particularly when unconsolidated fluvial deposits are the base material, and (c) hydrology and the extent of recurring floods (DRD, 1991). These three aspects require the acquisition of different sources of information that in the case of geographic data is quite difficult to achieve using traditional terrestrial surveys and photogrammetry due to time consuming and spatial constraints (Lijian et al., 2008). Thus, there is a need to know the dynamics of the river and to determine the spatial-temporal distribution of flow. The use of remote sensing techniques has demonstrated to be an efficient tool to map the flood extent (Horritt & Bates, 2002; Jensen, 2000). These maps have been used to validate 1D or 2D flood models; however it is not enough to study the water movement in the watershed. It is also necessary to know the water level reached in the floodplains in order to determine the amount of water arriving to the river channel and to be transported during the flood event. In this case, the use of 3D information is a better option since it offers: river depths, river surface and banks topography, vegetation height, hydraulic structures and identifies different objects present in the river channel (Mohammadzadeh & Valadan Zoej, 2008). One alternative to acquire altimetry measures as well as identifying land surfaces features is the time-of-light information from a LiDAR system mounted on air- or- space based platforms.

2.1 LiDAR data

LiDAR data can acquire high precision and density elevation data of terrain surface since this technology involves laser altimeters (i.e. Nd:YAG laser or pulsed gallium-arsenide diode laser), a Global Positioning System (GPS) as a position reference to track the aircraft trajectory, and an Inertial Measurement Unit (IMU) as the main component of Inertial Navigation System (INS) to calculate the position, orientation, and velocity of a moving object without external references (Webster & Dias, 2006a). Thus, knowing the angle and the absolute position of the instruments it is possible to establish the location (x,y,z) of the laser pulse destination precisely and accurately obtain a real representation of the ground.

The LiDAR aircraft flew at 500 m to 1 km above the surface, thus the accuracy of the laser altimeters is around a few centimetres in range (Weitkamp, 2005; Raber et al., 2007). The use of illumination with visible light from artificial sources has been used for active optical detection of objects considering relative distance values. These distance values can be determined measuring the transit time of radiation from the source to the object and back. This means that when a short pulse of radiation is transmitted in a medium, the radiation scattered from an object is going to be sensed as a function of the distance of the object at the surface. Of course the sensed signal is not going to have the same length as the transmitted pulse, but it will be extended in time (Weitkamp, 2005). This technique is also suitable for microwaves and sound sources and time-resolving detection systems. LiDAR has sensors for visible (λ= 400 – 700 nm), ultraviolet (λ= 225 – 400 nm) and infrared (λ= 700 – 1200 nm) wavelengths. For this reason, LiDAR not only provides elevation data but also provides information on pressure, temperature, turbulence and wind for gases in the upper, middle and low atmosphere, and can be used to determine hydrospheric and land use parameters (oil slicks, phytoplankton, water depth, turbidity, type of land use, health of vegetation, tree-stem diameters, biomass, and leaf area index, amongst others).

The extraction of topographic parameters using LiDAR gives advantages compared with other techniques such as photogrammetry since the high density data acquired avoids loss information. Also, it is not influenced by the shadow angle of the sun and it is not dependent on the ground control since it can gather a huge quantity of control points.

The high vertical accuracy of LiDAR enables the production of Digital Elevation Models (DEM), Digital Surface Models (DSM) and a Digital Orthophoto map (DOM) with high resolution and precision at a low cost compared with traditional techniques (Lijian et al., 2008) Despite the high accuracy of DEM, it is important to identify the errors associated with the LiDAR data in order to reduce them (Raber et al. 2007; Webster & Dias, 2006a). The errors can be associated with several causes, some of them are related to the system (characteristics of the platform and laser) and conditions (terrain characteristics, vegetation and atmospheric conditions) (Filin, 2001). Also, Raber (2006) noticed the influence of the land cover and slope of the surface. To solve this Cobby et al. (2001) proposed a segmentation method that was tested by Raber et al. (2002) finding that the vertical accuracy is improved. Other errors are associated with the interpolation method and the data model utilised to create the DEM (Hodgson & Bresnahan, 2004). In order to reduce the systematic error, Webster et al. (2006b) suggested to perform an independent validation. In fact LiDAR processing data includes several steps such as: (a) data collection, (b) data pre-treatment, (c) filter and classification of point cloud, (d) fusion and application, and (e) ground-object extraction and modelling.

3. River and floodplain and the 2007 flooding

The present case study is based on the analysis of the Grijalva River Basin flooding in 2007, affecting almost all of Tabasco state resulting in enormous socio-economical losses.

3.1 Grijalva River Basin: characteristics

The Grijalva River Basin (GRB) is one of the largest basins in Mexico, low-lying in the Tabasco state at the Villahermosa city, and in the river delta at the Gulf of Mexico. The side branches of the Grijalva River have high runoff percentages (20%) associated with their medium permeability with little vegetation or low permeability with dense vegetation. The Grijalva River is born in the Chiapas state changing its name to Mezclapa after the Malpaso dam, to recover its name after it crosses Villahermosa city to its end at the Gulf of Mexico.

The Grijalva River is a mature river crossing Tabasco; an alluvial plain that made it to divagate generating broad plains, wide flows and meanders. Also, it has been divided over time creating two of the most important branches of the Carrizal River in 1881, and the Samaria River in 1932. Other important side branches of the Grijalva coming from the Chiapas Mountain Chain are the Pichucalco, Puyacatengo, Teapa and the La Sierra rivers (figure 1). The Grijalva River's total length is approximately 640 km and in its main branch is the country's fourth largest hydroelectric plants in the Chiapas state (GET, 2008; INEGI, 2000).

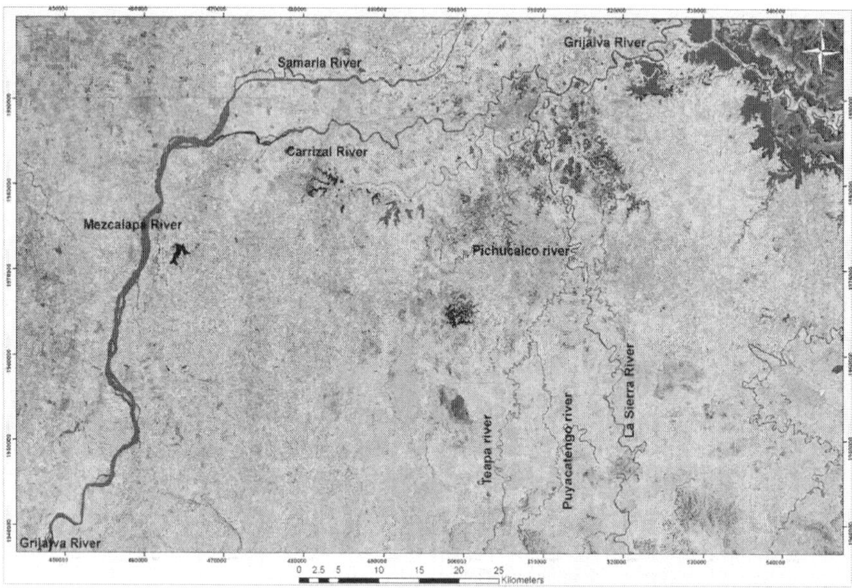

Fig. 1. The Grijalva River and tributaries around Villahermosa city, Tabasco. Landsat TM image from 28/03/2001

Tabasco is divided into two physiographical regions: Llanura Costera del Golfo Sur (Coastal plain at the south Gulf) covering almost all the state (94%) with the Sierras de Chiapas and Guatemala (Chiapas and Guatemala mountain chain) covering the remaining 6%. The main characteristics of the coastal plains are a gradual sedimentation area refilled with detritus material and eroded zones with low elevations (less than 60m), the majority related to small subtracted dominated sand and limestone highland rivers. This made the zone a source of important lacustrine, palustrine and litoral deposits linked to extended floodplains, abandon streams and coastal lagoons.

The climate is warm humid subtropical with maritime influence. The mean annual temperature is between 24 to 28°C with a maximum and minimum mean annual temperature of 28-30 °C in May and 21-25 °C in January, respectively. The rainy season is from June to October with a mean annual precipitation in the coast of 1,500 mm with a gradual increment to the intercontinental zones till 5,000 mm. Two rainfall peaks are present, one in June with 220 mm, and the other in September with 348 mm; some authors consider another peak in October with 370 mm. This generates a cloudy atmosphere with a relative humidity around 80 to 86%. These mean annual values are a result of the extensive plains that commonly produce air masses from the sea that bring the major contribution to precipitation (INEGI, 2000).

3.2 Protection along the Grijalva River

The economic activity in Tabasco (mainly petroleum and agropecuary industries) expanded a rapid growth of most cities, the most important one being Villahermosa, capital of the state. This city is bordered by the Grijalva River (Carrizal, Mezcalapa and de La Sierra tributaries) and is quite close to the Gulf of Mexico giving at it an important role. Originally, Villahermosa was located in highland rivers surrounded by undisturbed rain forest and water bodies. However, the rapid growth of population changed the physiography of the city by dredging water bodies, modifying the course of the rivers and removing the vegetation from the surface to build urban areas (figure 2) (Oropeza, 2004).

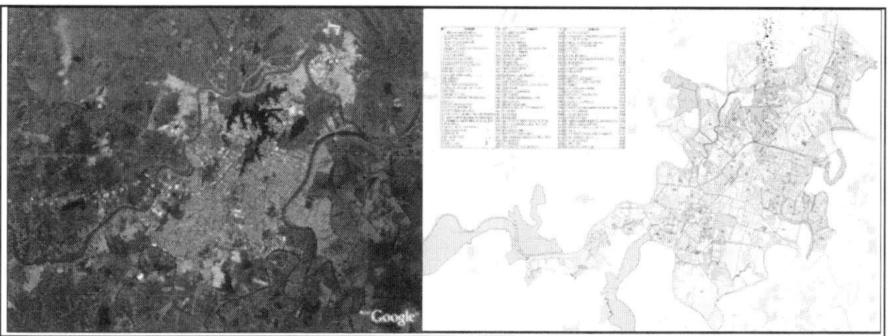

Fig. 2. (a) City of Villahermosa in 2003 (Google image 15/03/2003) and (b) proposal for the growth population for Villahermosa after mid 2007 (PEDUT, 2007)

As Villahermosa was developed along the Carrizal, Mezcalapa and de La Sierra Rivers the risk of flooding is at its greatest in events of prolonged heavy rainfall. In order to protect the city from floods the water regime was disrupted constructing technical structures such as levees, barriers, dykes, roads, etc. In this manner, after the 1955 flooding barriers were constructed along the Samaria River, and the Malpaso and La Angostura dams were built to regulate runoff and reduce the risk of flooding in Villahermosa (Maza, 1997). Years later, the development of the transport communication system in the state and, particularly, in Villahermosa made it possible to construct levees and barrages, change of the river flow and dredging of the rivers. Thus there was a dramatic change in the dynamics of the Grijalva River (GET, 2008) increasing the flood risk since the water levels overpass the channel and the levees act as retainers of water.

In 1997, Maza studied the hydraulic conditions in the Grijalva-Usumacinta Basin and attending to the socioeconomic development in the region he proposed several alternatives to protect against flooding. These actions were reviewed after the 1999 flood. This was an extreme event with approximately a 20-year return period that affected the Tabasco state with loss of human lives, houses and an impact on the economy. Some of the proposal considered the restitution of channels, the use of lagoons and swamps for regulation, and the improvement of current infrastructure (dams and barriers constructed along the rivers Carrizal and Grijalva in the Villahermosa city). Some tasks were done, but others were unfinished and others were unattended. In particular, for La Sierra River it was evident that hydraulic measures were not taken along the river maintaining its unregulated character. This was the reason that in the summer and autumn of 2007, several precipitation events registered along the Gulf of Mexico decreased the free water surface in the rivers, overtopping them in zones with instable soils or where the barriers failed.

3.3 The greatest flood in Tabasco 2007

In the summer and autumn of 2007, several precipitation events were registered along the Gulf of Mexico. The majority of this rain was related to 37 tropical depressions and 3 cold fronts but also it was rain due to the hurricane Dean (August 22nd) and Noel (October 28th). Exceptionally strong rainfalls took place on three periods: (a) October 10-11 associated with the cold front No.2 plus a low pressure event below Guatemala, (b) October 22-24 associated with the cold front No.4 located in Tabasco and Chiapas, and (c) October 28-30 associated with a cold air mass in the southeast of the Gulf of Mexico plus the cold front No.4 and the Tropical storm Noel (GET, 2008). After the strong rainfalls in the upper and middle GRB (specifically in the Ocotepect town in the Chiapas Mountain Chain) with 4000 mm of accumulated rainfall in 24 hours, the increase of the water levels in the river system and dams generated by the end of October and the beginning of November, caused severe flooding, which impacted 70% of the Tabasco state in Mexico, covering all urban and plain areas. In particular, the Villahermosa capital city of the Tabasco state was covered by water in around 80% of its territory due to the flooding of the Carrizal and Grijalva rivers that go through the city, and also by rivers coming from mountain areas, in the border between Tabasco and Chiapas, such as La Sierra and Pichucalco. As a result, the economic effect was huge as thousands of residential and non-residential properties were destroyed, and jobs, crops, animals, and infrastructure were lost. The total volume of overflow was around 1,800 million of m^3 for a return period of 100 years (Domínguez et al., 2009).

Different causes have been given for this disaster for example the failure of the protection system (levees), thousands of inhabitants allocated in the river divagation areas as a result of the rapid population growth, and modifications in the river flow, amongst others. On one side, as Tabasco was an alluvial plain with highland rivers, the risk of inundation was lower. However, most of the low-lying alluvial plains have been filled to allow urban developments. In consequence, the floodplain area has been increased over a long time. As floodplains are areas subject to inundation by lateral overflow water from rivers with which they are associated (Tockner and Stanford, 2002), the effect of its increment is more significant, particularly the greatest flooding in 2007 with a 100 year return period event (figure 3).

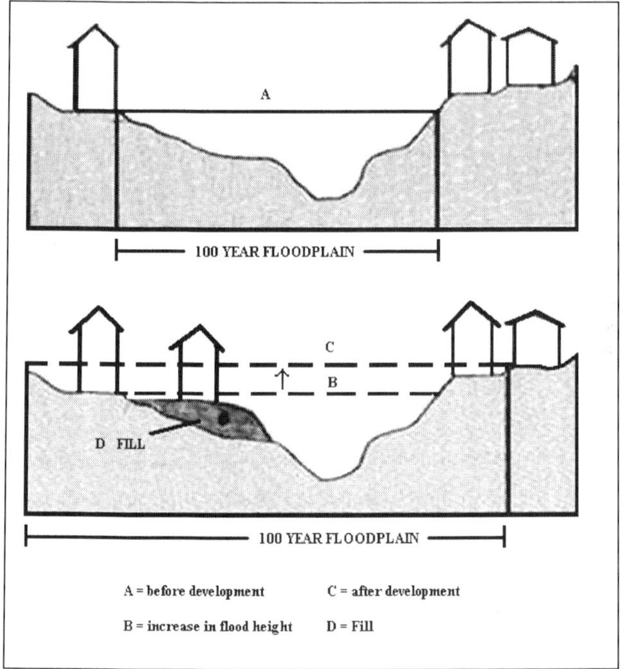

Fig. 3. Cross section of a hypothetical river flood. Source: DRD, 1991

On the other side, the hydraulic failure related to the levees could not only be explained by the design (overtopping and foundation), since there were sections that were not constructed due to financial or political reasons allowing water to pass to the urban areas. Analysing the information available and using satellite images (Landsat TM and Spot 5) it was observed that some barriers were constructed into two seasonal lagoons at the south of Villahermosa that theoretically will be the natural drainage systems to regulate the amount of water coming from the upper areas. As a consequence of the natural drainage system to the lagoons together with overtopping the banks of the rivers or the top level of levees, the water levels of the lagoons increased considerably since the reduction of the water discharge demanded more time to be drained to the Gulf of Mexico. This magnified the flooding condition in the city of Villahermosa, which was underwater for weeks (figure 4).

One important aspect was that a 100 year return period event was plotted allowing the analysis of a real situation and not a hypothetical one. The 2007 flood in Tabasco showed that the state and, in particular, the Villahermosa city, is a region subject to flooding as a result of several human actions that reduced significantly the original floodplain zone. Thus different actions need to be done at the upper, middle and low Grijalva River Basin in order to protect effectively the whole area.

Further research considers using a flexible grid cell to extract information every 0.25m instead of reducing the grid cell resolution. This requires a lot of data and processing time but the benefits are higher due to the accuracy of the x,y,z LiDAR data. Mapping the flood, one can identify those elements that failed. The critical one is to recognise the necessity to recover the floodplains that have been highly degraded through generation of urban or agricultural areas. By identifying the river status and its lateral connectivity, better protective systems can be proposed to recover the natural functions and process of the floodplains.

8. Acknowledgments

This project was founded by the National Water Commission. The LiDAR data was provided by the sponsor. With thanks to MSc. Víctor Franco, who reviewed the paper for comments.

9. References

Cobby, D.M., Mason, D.C. & Davenport, J.J. (2001). Image processing of airbone laser altimetry data for improved river modelling, *ISPRS Journal of Photogrammetry and Remote Sensing*, Vol. 56, No2. 2, pages 121-138, ISSN: 0924-2716

Domínguez, R., Carrizosa, E., Arganis, M. & Esquivel, G. (2008). Análisis Estadístico de los gastos máximos anuales registrados en las cuencas del Bajo Grijalva y de las descargas de la presa Peñitas, In Plan Hídrico Integral de Tabasco dentro del cual se Inscribe la Ejecución del Plan de Acción Urgente (PAU) y la Formulación del Plan de Acción Inmediata (PAI), para la Rehabilitación de la Infraestructura Dañada por las Lluvias Atípicas de los Días 28 y 29 de Octubre del 2007, Ramos, J., & Vélez, L. (Ed.) pp. 1000, Instituto de Ingeniería, UNAM, Report project, México, D. F.

DRD (1991). *Primer on Natural Hazard Management in Integrated Regional Development Planning*, Department of Regional Development and Environment Executive Secretariat for Economic and Social Affairs Organization of American States with support of the Office of Foreign Disaster Assistance United State Agency for International Development, Washington, D.C.

Filin, S. (2003). Recovery of systematic biases in laser altimetry data using natural surfaces, *Photogrammetric Engineering & Remote Sensing*, Vol.69, No. 11, November, 2003, pages 235-1242, ISSN: 0099-1112

Fuentes, O., Domínguez, R., deLuna, F., López, J. & Cruz, J. (2008). Documentación Hidráulica del evento de 2007 que causó inundaciones en Villahermosa, Tabasco, In Plan Hídrico Integral de Tabasco dentro del cual se Inscribe la Ejecución del Plan de Acción Urgente (PAU) y la Formulación del Plan de Acción Inmediata (PAI), para la Rehabilitación de la Infraestructura Dañada por las Lluvias Atípicas de los Días 28 y 29 de Octubre del 2007, Ramos, J., & Vélez, L. (Ed.) pp. 1000, Instituto de Ingeniería, UNAM, Report project, México, D. F.

GET, (2008). Gobierno del Estado de Tabasco, [Web page reviewed on April 2008] http://www.tabasco.gob.mx/estado/index.php

Horrit, M.S. & Bates, P.D. (2002). Evaluation of 1D and 2D numerical models for predicting river flood inundation, *Journal of Hydrology*, Vol. 268, November 2002, Pages 87-99, ISSN:0022-1694

INEGI (2000). *Síntesis de Información Geográfica del Estado de Tabasco*, Instituto Nacional de Estadística y Geografía, ISBN: 9701332733, Aguascalientes, Mexico

Jensen, J.R. (2000). Remote Sensing of the Environment: An Earth Resource Perspective, Upper Saddle River, New Jersey, Printece Hall Inc., page 544.

Lijian, Z., Zulong, L., Yingcheng, L., Yanli, X., Ming, L., Zhouloei, W., Pei, L. & Xiaolong, L. (2008). Application and analyses or airborne LiDAR technology in topographic survey of tidal flat and coastal zone, *The International Archives of the Photogrammetry, Remote Sensing and Spatial Information Sciences*, Vol. XXXVII, Part.B3b, Commision III, page 56ff, July 2008, ISSN:1682-1750, Beijing

Maza, J.A. (1997) Cuenca Grijalva-Usumacinta Estudio de Gran Visión para las Obras de Protección de la Planicie, elaborado para la Subdirección General de Construcción, Gerencia Regional Sur, CONAGUA, y Subdirección Técnica, Gerencia de Estudios de Ingeniería Civil, CFE. Report Project, Instituto de Ingeniería, UNAM, Mexico D.F.

Maza-Álvarez, J.A. & Franco, V. (1997). *Obras de protección para control de inundaciones, Capítulo 15 del Manual de Ingeniería de Ríos*, Vol. 591, pp.185, Instituto de Ingeniería, UNAM, México, D.F.

Mohammadzadeh, A. & Valadan Zoej, M.J. (2008). A state of art on airborne lidar application in hydrology and Oceanography: a comprehensive overview, *Proceedings of ISPRS Congress Beijing 2008*, The International Archives of the Photogrammetry, Remote Sensing and Spatial Information Science, Vol. XXXVII, Part B1 Commission I, pp. 315-320, ISSN 1682-1750, July 2008, Beijing

Moss, T. & Monstadt, J. (2008). Institutional dimensions of floodplain restoration in Europe: an introduction, In: *Restoring flooplains in Europe: policy contexts and project experiences*, T. Moss and J. Monstadt (Ed.), 3-15, IWA Publishing, ISBN 1843390906, London

NavCom (2008). SF-2040 Pole-Mount StarFire TM/RTK GPS Sensor, NavCom Technology [Web page consulted on December 2008, http://www.navcomtech.com/Products/GPS/sf2040g.cfm]

Oropeza, V.M. (2004). *Parque Reserva, Península del Carrizal*, Bachelor thesis, March 2004, Universidad de las Américas Puebla, Cholula, Puebla, México

PEDUT (2007). Programa Estatal de Desarrollo Urbano de Tabasco del 2007, Gobierno del Estado de Tabasco, Mexico

Raber, G.T., Jensen J.R., Hodgson M.E., Tullis, J.A., Davis, B.A. and Berglund J. (2007). Impact of Lidar nominal post-spacing on DEM accuracy and flood zone delineation, *Photogrammetric Engineering and Remote Sensing*, Vol. 73, No. 7, July 2007, pages 793-804, ISNN: 0099-1112

Raber, G.T., Jensen, J.R., Schill, S.R. & Schuckman, K. (2002). Creation of digital terren maodels using an adaptive lidar vegetation point removal process, *Photogrammetric Engineering and Remote Sensing*, Vol. 68, No. 12, pages 1307-1315, ISSN: 0099-1112

Stabbel, E. & Löffler E. (2003). Optimised mapping of flood extent and floofplain structures by radar EO-methods, *Proceedings of the FRINGE 2003 Workshop (ESASP 550)*, pp. 56.1, CDROOM, Franscati, December 2003, ESA/ESRI, Italy

Tamiru-Haile, A. & Rientjes, T.H.M. (2005). Effects of lidar DEM resolution in flood modelling: A model sensitivity study for the city of Tegucigalpa, Honduras, Proceedings *of the ISPRS WG III/3, III/4, V/3 Workshop Laser scanning 2005*, September 2005, Enschede, The Netherlands

Tockner, K. and Stanford, J.A. (2002). Riverine flood plains: present state and future trends, *Environmental Conservation*, Vol.29, Vol. 3, April 2002, pages 308-330, ISSN: 0376-8929

Trimble (2008). GPS Pathfinder ProXH Receiver, Trimble Navigation Limited 2008 [Web page consulted on December 2008, http://www.trimble.com/copyrights.shtml]

Webster, T. & Stiff, D. (2008). The prediction and mapping of coastal flood risk associated with storm surge events and long-term sea level changes, *WIT Transactions on Information and Communication*, Vol. 39, pages 129-138, ISSN 1743-3517

Webster, T.I. & Dias, G.. (2006a). An automated GIS procedure for comparing GPS and proximal LIDAR evaluations, Computers and Geosciences, Vol.32, May 2006, pages 713-726, ISSN: 0098-3004

Webster, T.L., Forbes, D.L., Mackinnon, E. & Roberts, D. (2006b). Flood-risk mapping for storm-surge events and sea-level rise using lidar for southeast New Brunswick, Canadian Journal of Remote Sensing, Vol. 32, No. 2, April 2006, ISSN: 1712-7971

Weitkamp, C., (2005). Lidar: Introduction, In: *Laser Remote Sensing*, Fujii T. & Fukuchi T. (Ed.), pages1-36, CRC Press, ISBN:0-8247-4256-7, Florida

Werritty, A. (2006). Sustainable flood management: oxymoron or new paradigm? *Area*, Vol. 38, No. 1, pp. 16-23, ISSN: 0004-0894

33

Improving Wetland Characterization with Multi-Sensor, Multi-Temporal SAR and Optical/Infrared Data Fusion

Laura L. Bourgeau-Chavez, Kevin Riordan*, Richard B. Powell,
Nicole Miller** and Mitch Nowels***
*Michigan Technological University, Michigan Tech Research Institute,
3600 Green Ct., Ste. 100, Ann Arbor, MI, USA
* Independent Researcher, 85B East Jefferson Rd. Pittsford, NY, USA
** Radiance Technologies, Inc. Bldg 1103, Suite 210 Stennis Space Center MS, USA
*** 8700 SW 174th Street, Miami, FL, USA

1. Introduction

Wetlands provide habitat and food sources for wildlife, protect waterways, act as natural filtration systems, and serve major ecological roles in the overall health of our local, regional and global ecosystems. While wetlands serve such vital ecological functions, due to their limited capacity for adaptation wetland ecosystems are highly vulnerable to change, from both climatic (IPCC 2008) and anthropogenic sources. In the past, wetlands have been drained and converted to uplands at an alarming rate and the remaining wetlands are subject to a number of threats including eutrophication, climate change, invasive plant species, and the interaction between these stressors. Due to these vulnerabilities, there exists a need for policy and resource managers to have an operational strategy for monitoring the extent, composition, and vigor of wetlands at a synoptic scale. For regional areas, such as the coastal Great Lakes, Boreal Canada, or vast wetland complexes such as the Pantanal, Mesopotamian marshlands, or the Greater Everglades, cost-effective implementable methods are necessary. For fine scale studies, cost is generally less of an issue and the highest resolution data with the highest cost may be justified. In this chapter, we focus on methods for regional scale mapping and monitoring where the minimum mapping unit of interest is 5 acres, and thus use moderate resolution satellite imagery (20-30 m). The focus is on multi-sensor data fusion between SAR single channel and/or multi-channel data and Optical/IR sensor data. Case studies in the Great Lakes and Boreal peatlands demonstrate the advantages and widespread utilty of a hybrid SAR-Optical/IR approach.

While many definitions of wetlands exist, both scientific and legal, in essence wetlands are defined by: (1) the presence of water at, above, or near the ground surface; (2) hydric soils; and (3) vegetation species adapted to or tolerant of wet soil conditions. Remote sensing can

herbaceous vegetation types due to their sensitivity to biomass (Ramsey 1998). VV polarization is also sensitive to soil moisture and flood condition (Bourgeau-Chavez *et al.* 2001).

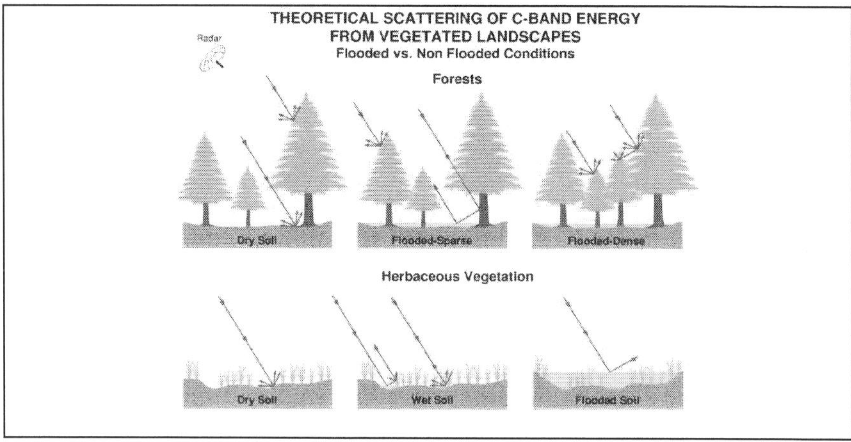

Fig. 1. Diagram showing theoretical scattering of C-band energy from forested and herbaceous ecosystems in dry, wet and flooded conditions, with open and closed forest canopies.

One primary advantage of using SAR over visible data is the detection of forested wetlands and the ability for SAR data to be collected irrespective of cloud cover or solar illumination because it is an active sensor. It is very difficult to detect flooding beneath a forested canopy with Optical/IR, unless there are large gaps in the canopy. Many researchers have found significant improvement in distinguishing swamp from other wetland classes and uplands with SAR (Lang *et al.* 2008, Grenier *et al.* 2007, Baghdadi *et al.* 2001, Hess *et al.* 1990, Bourgeau-Chavez *et al.* 2004).

Several researchers have evaluated the utility of SAR for wetland mapping using single and multi-date single channel SAR data (Costa *et al.* 1998, Whitcomb *et al.* 2009, Arzandeh and Wang 2002, Rao *et al.* 1999) and others have evaluated polarimetric SAR (Hess *et al.* 1990, 1995, Bourgeau-Chavez *et al.* 2001, Pope *et al.* 1994, 1997, Crawford *et al.* 1999, Wang and Davis 1997, Touzi *et al.* 2007). See Henderson and Lewis (2008) and Ramsey (1998) for a more thorough review of past research on wetland ecosystem analysis with SAR. Many early studies conducted to determine the utility of multi-polarization/multi-frequency data, focused on NASA's Shuttle Imaging Radar - C (SIR-C) which was fully polarimetric at L- and C-bands (L-HH, L-HV, L-VH, L-VV, C-HH, C-HV, C-VH, C-VV and X-VV) and fully polarimetric AirSAR (P-band (72 cm), L-band, C-band) for wetland classification in tropical and temperate landscapes (Hess *et al.* 1995, Pope *et al.* 1994, 1997, Bourgeau-Chavez *et al.* 2001). These studies demonstrated the utility of multi-band data and early polarimetric analyses (e.g. HH-VV phase difference) for mapping forested and herbaceous wetlands (Pope *et al.* 1997).

While several researchers have evaluated the use of SAR alone for mapping wetlands, until more recently, few have evaluated SAR and Optical/IR fusion for wetland mapping (Lozano-Garcia and Hoffer 1993, Augustein and Warrender, 1998, Toyra *et al.* 2001, Rio and Lozano-Garcia 2000, Bourgeau-Chavez *et al.* 2004, 2008, 2009, Li and Chen 2005, Grenier *et al.* 2007). Since the SAR and Optical/IR data measure different features of wetlands, it is logical that a synergistic approach between the two sensor types would increase wetland mapping accuracy. Further since the presence of standing water causes the SAR energy to interact differently depending on the dominant vegetation type, it would be advantageous to combine SAR with optical and infrared data for mapping purposes.

4. SAR and Optical Data Fusion

Of the few broad-scale SAR wetland mapping efforts that have been undertaken, , Canada is incorporating Radarsat-1 and Radarsat-2 data with Landsat mosaics (100m resolution) and SPOT data in the development of the Canadian Wetland Inventory (Grenier *et al.* 2007, Fournier *et al.* 2005, *pers. comm. Grenier 2009*) which will cover the entire country. There is also an effort underway to use the JERS mosaics (100 m resolution summer and winter products) of Boreal North America alone to map wetlands across Canada, as has already been done for Alaska (Whitcomb *et al.* 2009).

In this chapter, we review case study of multi-sensor, multi-date, SAR-optical/IR fusion methods and results for mapping wetlands in two main study areas, the Great Lakes and a Boreal peatland region in Alberta, Canada. The techniques are developed with broad scale mapping in mind, but are demonstrated on local to regional wetland areas. In all but one of these case studies, satellite SAR data are used in conjunction with Landsat TM, and in most cases SAR data from multiple sensors are fused. The exception, is the case studies on mapping the invasive plant species *Phragmites australis* on Lake St. Clair, where PALSAR data are used alone and compared to hyperspectral AVIRIS.

4.1 Case Study: Lakes Ontario and St. Clair Coastal Wetland Mapping

The Great Lakes Coastal Wetlands Consortium (GLCWC) was mandated to develop a monitoring plan for assessing the health of the coastal wetlands which are vital to the overall health and maintenance of the Great Lakes ecosystem (Bourgeau-Chavez *et al.* 2004, 8). The only way to monitor an ecosystem the scale of the Great Lakes basin is through integrated remote sensing and field observations in a GIS. Landscape indicators in need of monitoring through remote sensing include wetland type and extent, adjacent land cover, adjacent land use, intensity of land use, and invasive plant species. The GLCWC sought implementable, cost-effective yet robust methods with a minimum mapping unit of 5 acres. To meet these needs a few pilot studies were conducted to demonstrate various monitoring methods. A hybrid SAR Optical/IR methodology that used satellite sensor data of 30 m resolution and would allow for detection of areas as small as 1 hectare was found to be the most promising. This methodology would be cost-effective and data management for the entire Great Lakes Basin would be reasonable compared to higher resolution, smaller footprint imagery. And the use of two complementary data types in the mapping was expected to reduce omission and comission errors.

Two study sites were selected for demonstration of this hybrid SAR-optical data fusion, coastal areas of Lakes St. Clair and Ontario (Figure 2). Coastal Lake St. Clair was chosen because it has a diversity of land cover and land use including a large amount of urban and suburban areas, rural farm fields, and a large amount of wetlands (various species) that occur at the delta as the river enters the lake. The Lake Ontario study area was chosen because it is mainly rural, with many agricultural fields, has isolated patches of herbaceous wetlands, and a relatively large amount of potentially forested wetlands. These two areas provide different sets of land cover and land use classes and thus an opportunity to test the hybrid-sensor methodology in diverse settings.

4.1.1 Remote Sensing Data

The data used in this case study were archival SAR and Landsat data from multiple dates, and multiple sensors (Table 1). For the Landsat images, each containing 7 bands, the blue and thermal IR bands for each date were removed from the analysis. The blue channel is generally quite noisy and the thermal band is of coarser resolution than 30 m. The JERS (L-band) sensor used for this project had horizontal send and receive polarization (L-HH) and was operational from 1992-8. This sensor has a resolution of 30 m, incidence angle of 35°, and a footprint of 80 km x 80 km. To complement these data, we also acquired C-band (5.7 cm wavelength) SAR data from the European ERS-1 and Canadian Radarsat-1 (R-1) satellites. The ERS-1 sensor has vertical send and receive polarization (C-VV) and is collected at a central incidence angle of 23°. The Radarsat sensor has horizontal send and receive polarizations (C-HH). It also has pointing capabilities and can be collected in various modes and incidence angles. The R-1 data used were of standard beam 7 mode, which has an incidence angle of 47°. Both C-band sensors have 30 m resolution and 100 km x 100 km footprints.

Fig. 2. Great Lakes pilot study areas for hybrid SAR-Optical/IR mapping for the Great Lakes Coastal Wetlands Consortium project.

Site	Sensor	Spring	Summer	Fall
Lake St. Clair	Landsat TM	23 March 2001	30 August 2001	30 October 2000
Lake St. Clair	JERS	28 March 1995	10 August 1998	
Lake St. Clair	Radarsat-1			3 & 27 October 1998
Lake Ontario	Landsat TM	24 June 1993	12 August 2002	18 December 2002
Lake Ontario	JERS	11 April 1993	8 July 1993	17 October 1993
Lake Ontario	ERS-1	17 April 1993	7 June 1993	25 October 1993

Table 1. Data used in the Great Lakes Hybrid SAR-Optical/IR wetland and landscape indicator mapping.

Since this was a demonstration project with minimal funding, data sets chosen were not optimal but were chosen based on cost and ease of availability (Table 1). At both sites three dates of Landsat data were used. There were a total of four radar scenes for Lake St. Clair and 6 radar scenes for Lake Ontario. There are as many as 6 years between the JERS and Landsat collections for Lake St. Clair and 10 years between data collections for Lake Ontario, but analysis of the imagery indicated that few major changes in land use have occurred in these areas during that time period. Also, it was not as much of an issue due to the methodology; a site would first be checked for vegetation cover in the more recent Landsat and then checked for flooding in the older SAR. However, it is realized that using data with such a wide time span will introduce errors. The optimal data set would be from the same year, with SAR and TM from the same months. However, the datasets met the needs of this investigation, which was simply to demonstrate a methodology.

4.1.2 Image Interpretation

For the Lake St. Clair site, the imagery and products were evaluated by comparison to the NWI, land cover/land use maps, field checks (October 2003) and expert field knowledge (Field ecologist/botanist Dennis Albert of Michigan Natural Features Inventory).

Although the ancillary maps and field work showed many forested wetlands within our study area, the dates of JERS imagery show all of the forests the same, very bright in the spring (March 1995) and all are gray in the summer scene (August 1998). It is likely that all of the forests have a wet ground cover in the spring scene, there may even be wet, melting snow on the forest floor causing the enhanced signature from all of the forests, and in August all of the forests are dry with full foliage. However, the differences in backscatter in the herbaceous vegetation are apparent on these two dates, as well as in the October Radarsat scenes.

Figure 3 shows a red-green-blue false color composite of the 3 October 1998 R-1 scene, 10 August 1998 JERS scene, and 28 March 1995 JERS scene, respectively. In this composite, the orange areas were dominated by cattail (*Typha spp.*), and the green by Phragmites (*Phragmites australis*). Phragmites tends to be taller/denser and occurs in less wet locations than cattail. The red areas of the image are shorter and sparser vegetation, thus they do not cause enhanced backscatter at L-band, only at C-band. The red areas along the fingers of the delta are cattail and bulrush (*Scirpus spp.*) beds and the red area within Dickinson Island is a flood channel with wild rice (*Zizania aquatica*), open submergent and emergent vegetation (Dennis Albert). The dark area in the center of Dickinson island to the west of the kidney shaped light blue forest area is a wet meadow and appears to be dry in our October 1998 C-band scene, it has strongest backscatter in the L-band spring scene (blue channel), but not enhanced backscatter. This combination of R-1 and JERS allows for a good interpretation of this scene, discerning tall dense herbaceous vegetation from short sparse herbaceous vegetation, and different hydrological and biophysical properties.

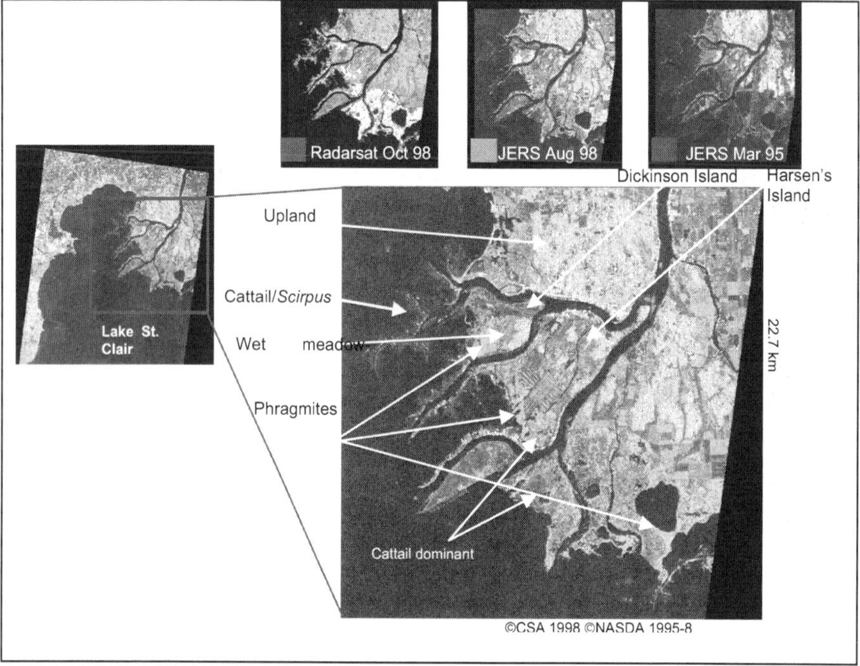

Fig. 3. Three dates, two sensor false color composites from Radarsat-1 and JERS satellites over the Lake St. Clair Delta. This Figure clearly defines Typha (cattail), and the invasive species *Phragmites australis* from other upland and wetland ecosystems.

At the Lake Ontario test site, we had an ideal seasonal data set with three images each of JERS and ERS from spring, summer and fall of 1993. Figure 4 presents false color composites of the two sensor datasets. Both datasets highlight non-forested wetlands (based on NWI) as green and red shades. In the JERS data, these sites were dark from specular reflection in the April scene (blue), then some sites were bright in July (green in the composite) while other sites remained dark in July (red locations in the composite) and all sites were gray in October (Table 2). In November of 2003, we conducted a field check to determine any vegetation difference between the red and green areas. The red areas visited were dominated by mixed grasses. It is likely that in the spring imagery the vegetation is fallen over or decomposed, with a high water level leading to specular reflection. The water level must still be high enough to cover much of the grasses and cause specular reflection again in the July scene, but when the lake water level drops to 74.58 m in October (70 cm drop), this site has more vegetation exposed and stronger backscatter. In comparison, the green areas visited in the field contained a mixture of grasses, cattail and shrubs (mainly *Cornus stolonifera*). These sites were bright in the summer and gray in the fall in the JERS imagery. The water level in comparison to the vegetation was likely lower than at the other sites, causing the enhanced backscatter in July, but with lower soil moisture in the fall the site was gray. For the same two sites in the ERS, the grass site was dark in the spring but the mixed shrub/herbaceous site was gray. In the summer the mixed shrub site was bright while the grass site remained dark. In the fall all sites were gray. While similar patterns emerged for both sites, the contrast between the non-flooded adjacent forests and the wetlands is stronger with the JERS, making it easier to map the boundaries of the sites.

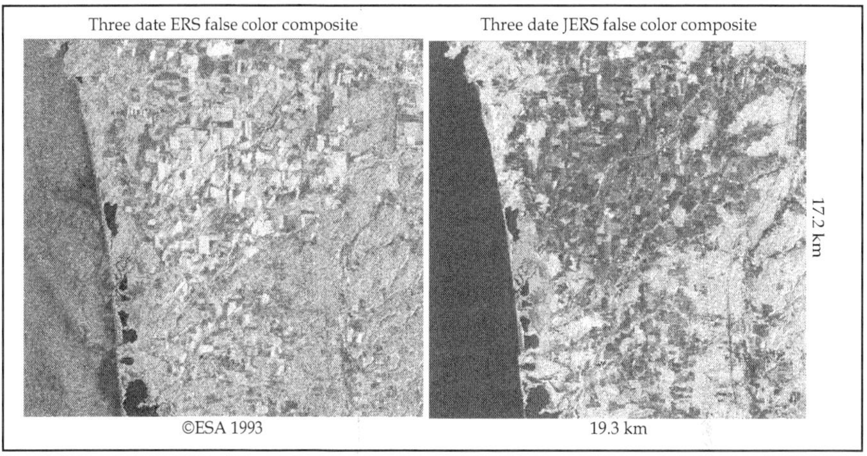

Fig. 4. Three date ERS false color composite of 25 October, 7 June and 17 April 1993 ERS imagery over eastern Lake Ontario compared to a 17 October, 8 July and 11 April 1993 JERS composite.

Site	Sensor	Spring Brightness	Summer Brightness	Fall Brightness
Grass	ERS-1 C-VV	dark	dark	gray
Shrub/herbaceous	ERS-1 C-VV	gray	bright	gray
Grass	JERS L-HH	dark	dark	gray
Shrub/herbaceous	JERS L-HH	dark	bright	gray

Table 2. Appearance of "grass" versus "shrub/herbaceous" sites in coastal Lake Ontario in spring, summer and fall of 1993. Grasses are red areas in JERS composite of Figure 3 and Shrub/herbaceous are green areas.

There are some forested wetlands within the Lake Ontario scene and they appear to be most notable in the April scene when the lake water level is the highest (note that coastal wetlands are hydrologically connected to the Great Lakes), and spring thaw has occurred and thus flooding is most likely. A comparison was made between assumed flooded forest and non-flooded forest for each JERS scene/date. The April scene had a 2.3 dB difference

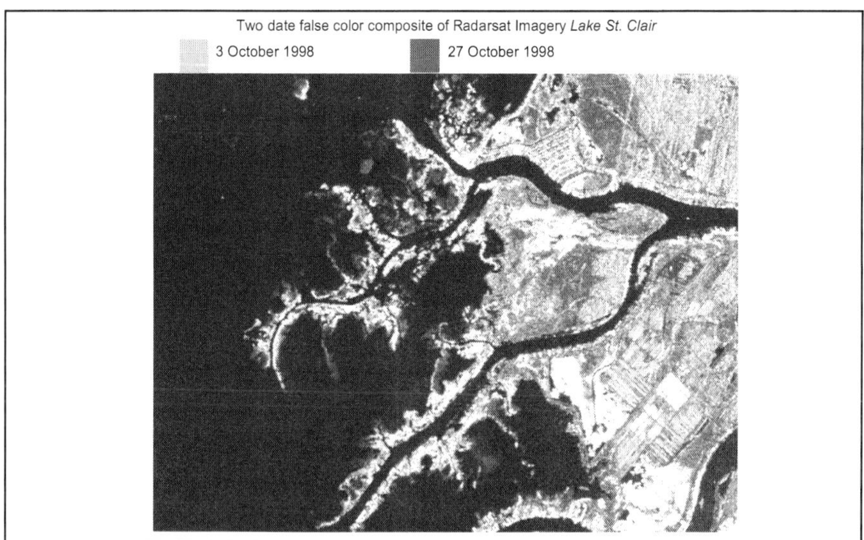

Fig. 7. Two date false color composite of Radarsat imagery. Cyan is the 3 October 1998 image and red is the 27 October 1998 image. Water level dropped by 19 cm between the first and second date.

4.1.3.2 Lake Ontario Study Site

The same techniques used for Lake St. Clair were applied to the datasets for Lake Ontario, however the classes were slightly different. First, Landsat imagery was classified into 13 different landcover classes; Water, High Density Urban, Low Density Urban, Deciduous Forest,Coniferous Forest, Mixed Forest, Row Crops, Low Herbaceous, Bare Soil, Emergent Wetlands, Shrubland, Fields-Hay, Wetlands-other. The radar imagery was classified into 9 classes; Water, Urban/Flooded Forest, Flooded Shrubland/ Wet Meadow, Emergent Wetland, Forest1, Forest2, Row Crop 1, Row Crop 2, and Herbaceous Field. These individual Landsat and SAR classes were then fused into a single product, just as they were for Lake St. Clair. The 12 final combined classes are described in Table 4. The final map is presented in Figure 8.

A comparison was made between the SAR-Optical map and the NWI with 5000 random points Note that the NWI has more generalized classes so the comparison is not straight forward, and there are over 2 decades between the NWI creation and the SAR-Optical/IR map, so some differences may be due to changes in the wetland, either succession or loss. The results in comparison to the NWI showed 94% overall accuracy, with all classes greater than 89% user's accuracy, except shrubby wetland for which we only had 13 points, none of which were correctly classed. The producer's accuracy in the wetlands was 34% for woody wetland and 66% for emergent. For the NWI woody wetlands, we labeled 127 out of 208 as deciduous forest. The problem likely lies in what areas were in fact inundated during the radar satellite collections. A wet, normal or dry year will provide different wetland extents, 66% agreement for emergent wetlands is quite good considering the likely turnover of some areas to agriculture and the likely succession of some of the wetlands labeled as emergent in the 1970s NWI to wetland shrub, as indicated by the field visits and point source field data from a biocomplexity study conducted by Cornell University (Mark Bain).

Class	Description
Water	Areas identified as water in the Landsat imagery, regardless of the radar classification
High Density Urban	Areas dominated by manmade materials, these areas were identified by high radar returns that were not forested areas in Landsat
Low Density Urban	Areas with a mixture of manmade features and landscape vegetation, these areas were mainly identified by the Landsat classification
Deciduous Forest	Areas of forest that lose their leaves throughout the season, identified through a combination of the Landsat and radar classifications
Coniferous Forest	Areas of evergreen forest, identified through the combination of classifications
Mixed Forest	Areas that are a mixture of coniferous forest and deciduous forest, identified through a combination of classifications
Forested Wetland	Areas are forest but have standing water on the ground throughout much of the year, identified as forest in the Landsat imagery and as urban/flooded forest in the radar imagery
Emergent Wetland	Areas of herbaceous vegetation that are wet at some times of the year, identified through the combination of radar classification and Landsat classification
Wetland-shrub	Areas of short woody vegetation that are wet at some points throughout the year, these are identified through the combination of sensor classification results.
Crop/pasture	Areas of herbaceous vegetation that are not plowed throughout the year, mainly identified through the Landsat classification
Bare Soil	Areas of exposed soil, sand, and/or rock, these areas were mainly identified through the radar and were confirmed by the Landsat classification
Row Crop	Areas that have herbaceous vegetation growing which is plowed at some point during the season. These areas were identified through a combination of the classification results.

Table 4. Classes for the combined Landsat and radar classification at the Lake Ontario study site.

The state of New York did not have a land cover/land use map comparable to IFMAP, and many errors were found in the National Land Cover Dataset (NLCD). However, field data collected in the largest coastal wetland complex in the image were available with GPS locations. The Biocomplexity study of Cornell University allowed for comparison of the SAR-Optical/IR hybrid map to 55 study points which represented cattail dominated, shrub, forested, and mixed emergent wetlands. The overall accuracy of this comparison was 89%, with 91% user's accuracy for wetland shrub, 89% user's for emergent wetland, and 67% for forested wetland. This assessment is quite good considering the likelihood of error in the geolocation of the study points due to the small plot size in reference to the 30 m pixels. Some of the points did fall on boundaries of open water/wetland or upland/wetland causing errors. The producer's accuracy ranged from 25 to 75% in the reference categories of specific species types that we did not map. The cattail field points corresponded to the wetland shrub (dark pink) areas in the Landsat-SAR fused classification. Therefore, the pink areas of our Landsat-SAR map should be labelled as shrub/high biomass herbaceous wetlands.

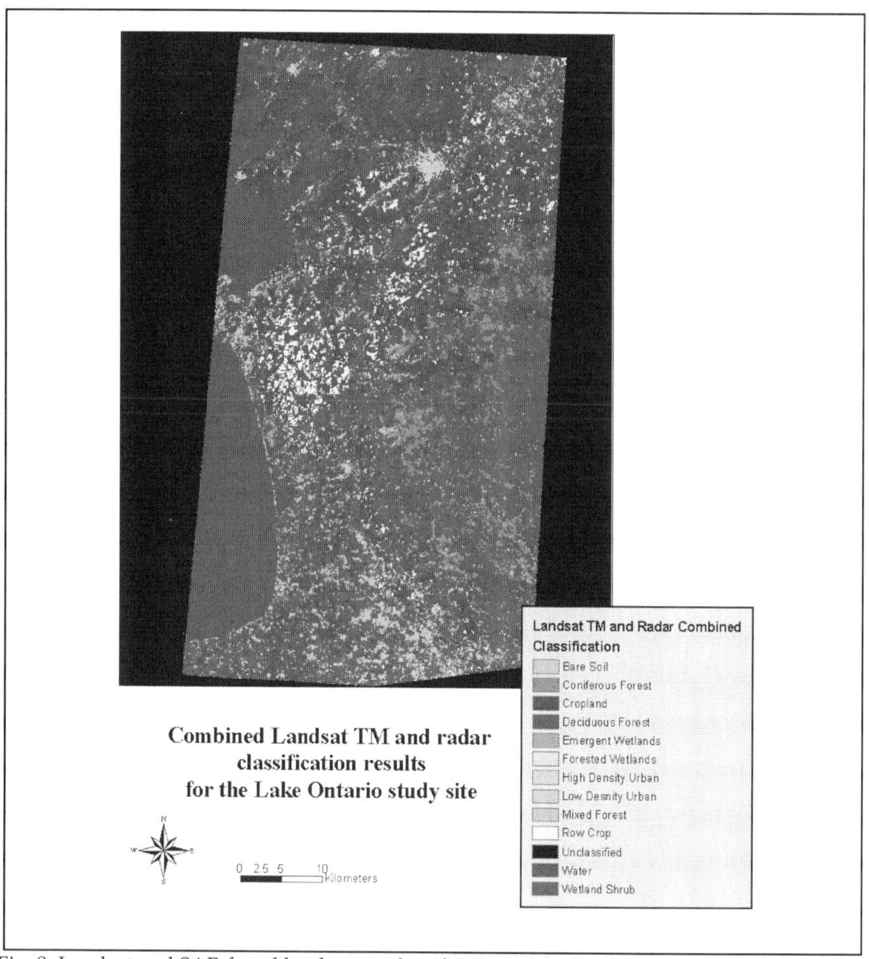

Fig. 8. Landsat and SAR fused land cover classification of the Lake Ontario study area.

4.1.4 Summary

These two study areas demonstrated the utility of a fusion of SAR and Optical/IR data for mapping landscape indicators of wetland health (wetland extent, adjacent land use intensity, etc.) in a region surrounded by high intensity urban versus a more rural area. In one case mapping of forested wetlands was possible and in the other, timing of data collections did not allow evaluation of forested wetland mapping. However, a simple approach to the mapping provided desirable results with the archival data, showing the complementary nature of the two types of sensors. Although the reference data and remote sensing data were not optimal [there are discrepancies in years of data collection (Landsat versus SAR (1990s-2001) and years of validation maps (1970s, 2000) and levels of vegetation classes between the SAR-Optical map and validation maps (fine scale species versus broad emergent classes)], these case studies demonstrated the added benefits of fusing Optical/IR data with another complementary sensor such as radar and resulted in a recommendation by the Great Lakes Coastal Wetlands Consortium for monitoring landscape indicators (Bourgeau-Chavez *et al.* 2008). In the next case study, current data are used and ongoing validation is being conducted for the specific species class levels. This case study is a continuing investigation and only preliminary results are shown here, however it provides a better validation of the results through field methods, further exemplifying the utility of SAR in wetland mapping.

4.2 Case Study: Invasive *Phragmites australis* Species Mapping on Lake St. Clair

One of the main wetland stressors in the Great Lakes is invasion by the problematic species Phragmites (*Phragmites australis*). This species invades native habitat creating dense thickets and deep detritus that virtually eliminates ecological function. A predicted drop in Great Lakes water levels due to global climate change is anticipated to increase the spread of the invasive Phragmites in the Great Lakes coastal zone, and a method to map this species and its spread would be of great assistance to land managers for control.

Several studies have focused on detecting and mapping invasive species in small catchments of the Great Lakes with high resolution hyperspectral and/or lidar (e.g. Lopez *et al.* 2006, Wilcox *et al.* 2003). However, such high-resolution mapping of the entire Great Lakes coastline or comprehensive field studies would be very costly. Others have found 30 m satellite imagery including Landsat, SPOT and Hyperion to be useful for mapping invasives (Arzandeh and Wang 2003, Pengra *et al.* 2007), however Landsat has spectral limitations and Hyperion is no longer operational.

Using a variation of the satellite SAR techniques described in the last section, which included a delineation of this invasive species using archival multi-date JERS and Radarsat-I, we are currently evaluating dual polarization PALSAR data, and have plans to incorporate Radarsat-2 data once it becomes available to distinguish a wider range of species.

4.2.1 Remote Sensing Data

ALOS PALSAR is the follow-on to JERS which showed the greatest potential in previous studies for mapping Phragmites (Bourgeau-Chavez *et al.* 2004, 2008). PALSAR was launched in 2006, and is available in three modes with various imaging parameters. Here we evaluate the dual-polarization product which has 20 m resolution, two channels (L-HH, L-HV), 70 x 70 km footprint, and is collected at an incidence angle of 34°. Up until recently, most satellite SAR systems were of a single channel, however with the recent launch of ALOS PALSAR and Radarsat-2, the utility of multi-channel SAR and polarimetric data are beginning to be more broadly evaluated for a variety of applications, demonstrating further mapping capabilities beyond that of single and multi-channel data.

Coincident to this evaluation is an investigation of an airborne hyperspectral NASA AVIRIS collection from July 2008 over the St. Clair delta. The AVIRIS sensor has 224 bands (400-2500 nm) and was collected with 17 m resolution.

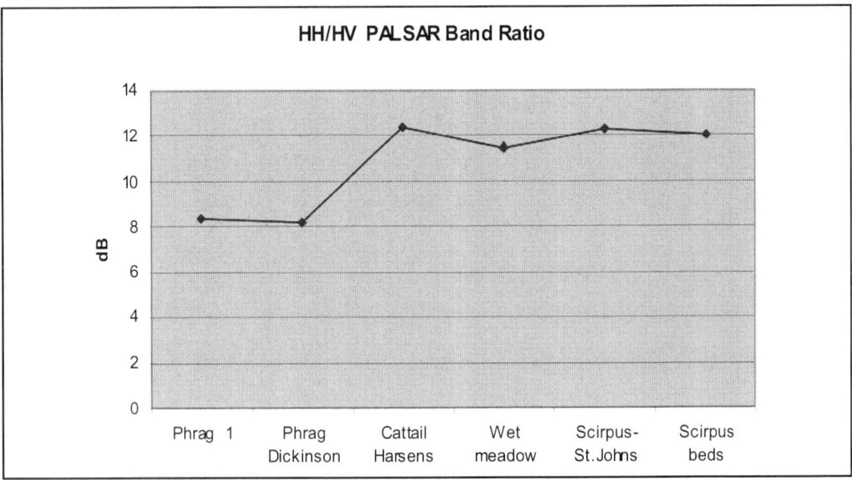

Fig. 9. Plot of backscatter from PALSAR (L-HH to L-HV ratio in dB) in Phragmites, Cattail, wet meadow and Scirpus beds of the Lake St. Clair delta.

4.2.2 Image Interpretation

Initial observations of an October 2007 dual polarization PALSAR image over Lake St. Clair revealed that the ratio of the L-HV and L-HH bands shows a 4-5 dB difference between Phragmites dominated wetlands and other non-forested native wetland types (Figure 9). The reason for this great divergence is the large difference in vegetation height, density and biomass of invasive Phragmites versus any other native herbaceous vegetation in the Great Lakes (Figure 10). Typha generally ranges in height from 1 to 3 m, while Phragmites can reach heights of more than 3.5 meters. Further, Phragmites forms tall, dense rather impenetrable stands. It is the sensitivity of L-band SAR to these differences in biomass and hydrology that allows the distinction between stands dominated by these two species.

Multi-date composites of PALSAR imagery from 2006-8 show the dynamic changes in the various vegetation cover-types over the growing season (Figure 11). In the top composite of Figure 11, which is a false color multi-date L-HH representation, the Typha are yellow to orange, indicating there is a strong return signal in July and October, but low response in the spring (Table 5). In contrast, most of the Phragmites shows a high response in the spring and lower in the summer and fall with shades of blue, and purple. In comparison, in the lower image of the L-HV multi-date false color composite (L-HV is sensitive to biomass), the Phragmites is cyan, indicating a strong return in the spring, April and May images, while the Typha is dark in these months, and bright (red) in the fall image.

Fig. 10. Photos of Phragmites: tall scale in winter (left), high density in summer (middle), and typical Typha (right).

While either L-HH or L-HV multi-date imagery appear to be useful for distinguishing Phragmites from Typha, there are some stands that would be confused with the L-HV alone (see arrows with Phragmites label on Figure 11); these are areas of Phragmites that are red in the lower image and green in the upper image. The different signatures in the PALSAR imagery from Phragmites-dominated stands is likely due to differences in water levels in the various seasons. "Phragmites 2" stands are located in diked areas, and may be wetter and sparser than "Phragmites 1" stands (Table 5).

PALSAR Band	Image Date	Phragmites 1 appearance	Phragmites 2 appearance	Typha appearance
L-HH	28 July 2006	dark	dark	bright
L-HH	09 October 2007	dark	bright	bright
L-HH	17 April 2008	bright	dark	dark
L-HV	09 October 2007	dark	bright	bright
L-HV	26 May 2008	bright	dark	dark
L-HV	17 April 2008	bright	dark	dark

Table 5. Appearance of two different Phragmites dominated stands versus Typha stands in the L-HH and L-HV imagery on the various dates.

4.2.3 Mapping Phragmites with SAR

A simple unsupervised maximum likelihood classification of the four dates of PALSAR data resulted in four classes of potential Phragmites and two potential classes of *Typha spp.* Note that the July 2006 image was from the single channel mode of ALOS PALSAR and thus had only the L-HH channel, analysis was therefore conducted on an input of 7 channels. Field observations using a GPS positioning system, *in situ* photos, and field notes were used to assess the preliminary "potential Phragmites" map. Using these field data for validation (29 points), the PALSAR multi-date preliminary map had 92% overall accuracy, with 100%

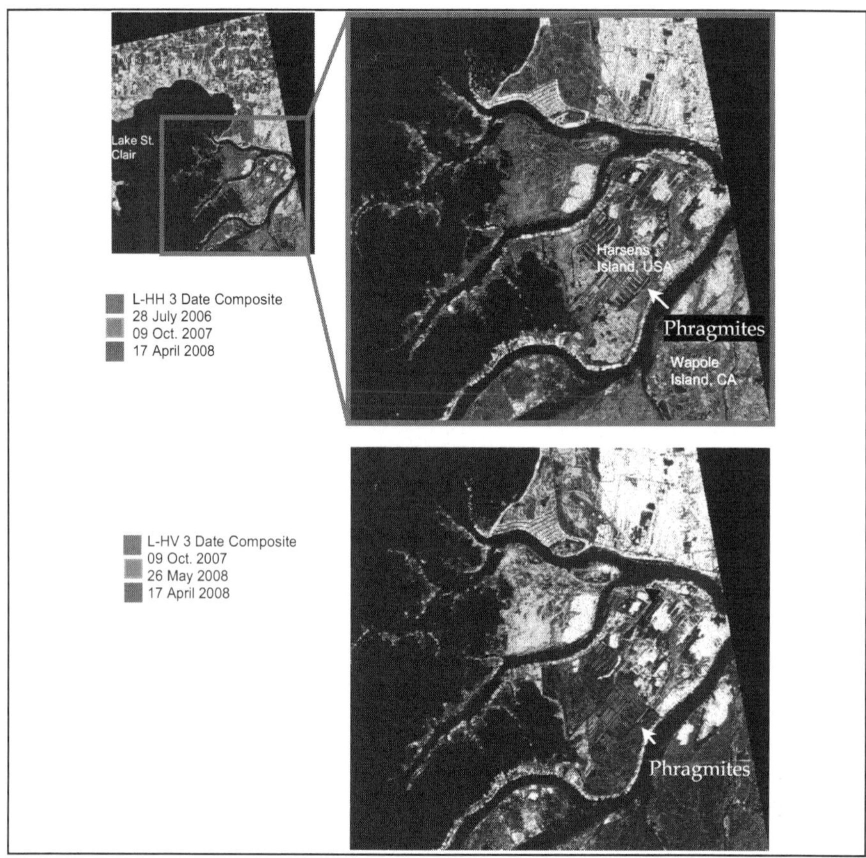

Fig. 11. False color composites of PALSAR imagery over Lake St. Clair delta. Top is L-HH from 28 July 2006 in red, 09 October 2007 in green and 17 April 2008 in blue. Bottom is L-HV composite with 09 October 2007 image in red, 26 May 2008 in green and 17 April 2008 in blue.

user's and 80% producer's accuracy for Typha, and 82% producer's and 100% user's accuracy for Phragmites. Note that the misclassified pixels for Phragmites were small areas of shrub or Typha within a larger Phragmites dominated area, thus the error is likely due to resolution (20 m in this case). The utility of the 10 m resolution PALSAR product (although only L-HH) may resolve this error and is being investigated.

4.2.4 Mapping Phragmites with AVIRIS

A comparison of the Optical/IR spectral signatures of *Typha latifolia* and *Phragmites australis* are shown in Figure 12. These signatures were collected in the field using a spectroradiometer (FieldSpec 3 JR). The vast differences in these signatures indicate that separation using Optical/IR remote sensing should be fairly easy, however, using the spectral angle mapper technique the results were poorer than the SAR methods.

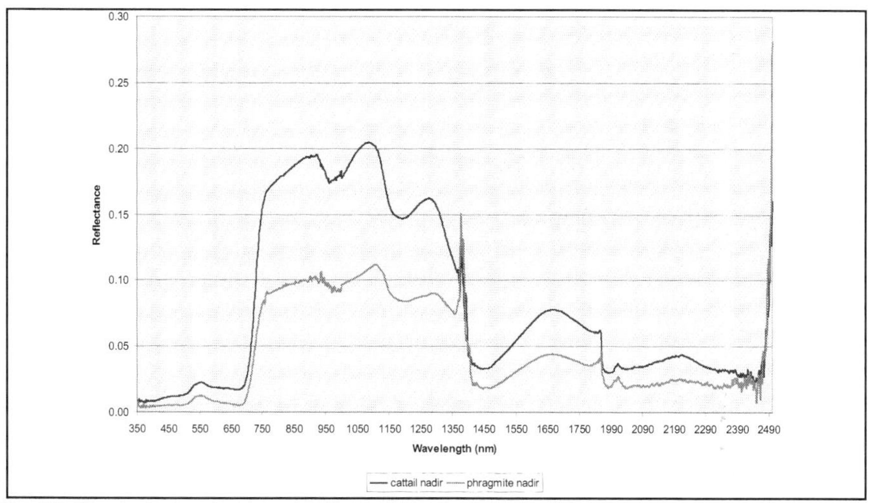

Fig. 12. Spectral signature comparison of Typha latifolia (black) versus *Phragmites australis* (red).

Spectral angle mapping is a physically-based spectral classification that uses n-dimensional (n = number of bands) angles to match pixels to reference spectra. It was run in ENVI, and determines similarity between the two spectra by calculating the angle and treating them as vectors in n-dimensional space. Smaller angles represent closer matches to the reference spectrum. The endmember spectra for *Phragmites australis* and *Typha spp.* used by the classifier were derived from field truth data provided by Michigan Natural Features Inventory.

Using methods similar to Lopez (2006), the spectral angle mapping of the 224 band (400-2500 nm), 17 m resolution AVIRIS data using field training data, resulted in 82 % overall accuracy with 83% user's and 77% producer's accuracy for Typha, but only 57% user's and 67% producer's accuracy for Phragmites. Note that the validation is based on only 17 points, because many of the points were used in the training

These are preliminary results and the investigation is ongoing. Methods are being investigated to determine the best bands from AVIRIS to combine with the L-band SAR for a data fusion approach. Evaluations will reveal the utility of SAR alone and in combination with hyperspectral, but the goal is to also determine if any of the bands of existing satellite Optical/IR systems could be used in lieu of hyperspectral data.

4.3 Case Study: Peatland mapping in Alberta

The last case study is based on the need to understand carbon storage (peat accumulation) and loss (mainly through fire) in boreal peatlands, which are widely recognized as being one of the largest terrestrial reservoirs for carbon (C) in the Northern Hemisphere. Estimating carbon storage and release requires an accurate mapping of peatland type. Peatlands are defined as wetlands with well developed peat (partially decayed plant matter) accumulation, generally more than 30 cm deep (Charman 2002). Peatlands actually represent diverse ecosystem types that differ in hydrology and vegetation, from forested rain-fed bogs to grass-dominated, saturated, or near-saturated stream-fed fens.

Early research on the ability to map boreal peatlands at a regional scale demonstrated the utility of merging SAR and Optical/IR data. Early observations included JERS, R-1, and Landsat imagery. Figure 13 presents Landsat and JERS images in comparison to a detailed peatland map (based on air photo interpretation and intensive field truth circa 1970-80s) with open, forested, and wooded categories of bogs, fens and swamps delineated. This preliminary analysis showed that Landsat would be useful for finding many of the open fens (see linear features circled in yellow; Figure 13). Whereas wooded fens do not look different from bogs in the Landsat image, they can be distinguished in the JERS SAR multi-date imagery (see features circled in pink; Figure 13). Note that the majority of peatlands in this study region are wooded bogs (salmon color in the peatland map). The linear open fens are dark in the SAR similar to the open swamps and marshes. Here, C-band Radarsat should prove useful in distinguishing types of open peatlands. The complementary information obtained from the spectral reflectance properties of the vegetation combined with the structural and moisture information from SAR should allow the mapping of both peatland type (bog, fen, swamp) and level of biomass (open, sparse tree cover, forested).

4.3.1 Remote Sensing Data

Two dates of PALSAR imagery were obtained over the local Alberta study area from July and August of 2007. The dual-polarization PALSAR data included two channels, L-HH and L-HV. To complement this, a spring and summer data set of Landsat TM imagery were also obtained from April and August of 2001. Lastly we obtained two R-I images from July 1997 and 2005. These areas are so remote and vegetation growth is slow enough, that a 6-10 year difference in data collection is not problematic. The only large changes between the years would be wildfires, but we obtained all data on location and extent of wildfire from the Canada Forest Service for the time period of study and no fires occurred within the study area during that time.

4.3.2 Image Interpretation

Figure 14 shows two false color composites, one from the two dates of L-HV data from PALSAR and the other from two dates of L-HH data, with the first date as red and the second as cyan. These composites show how the cyan colored areas help distinguish fen from bog in the L-HH composite (bottom figure) , since fens are characterized by fluctuating water levels and flowing water, whereas bogs tend to have water levels that remain below the moss covered surface with small changes in the short time period of the two image dates (July and August 2007). The L-HV cross polarized data provide information on the level of biomass which is essential in discrimination of low herbaceous open fens versus fen woodlands and open, wooded and forested bogs. This channel also clearly distinguishes the high biomass upland areas from the lowlands (Figure 14). Further, by using multi-date data, the seasonal changes in moisture and water levels are useful for discrimination of wetlands that generally have large changes in moisture/flood conditions (stream-fed fens) versus those that have smaller changes (rain-fed bogs).

While most boreal peatlands in central Alberta are characterized by low canopy closure, allowing C-band R-1 to be useful, evaluation of the L-band JERS and PALSAR data demonstrate the additional definition of forests from ecosystems with low amounts of aboveground biomass and varying surface wetness conditions. While C-HH R-1 provides information on low aboveground biomass wetland types, JERS L-HH and PALSAR more clearly define the differences between forested wetland types, as well as distinguishing high and low aboveground biomass areas. The R-1 data were included in the mapping methods for distinguishing swamps since Grenier *et al.* (2007) found R-1 useful for such purposes.

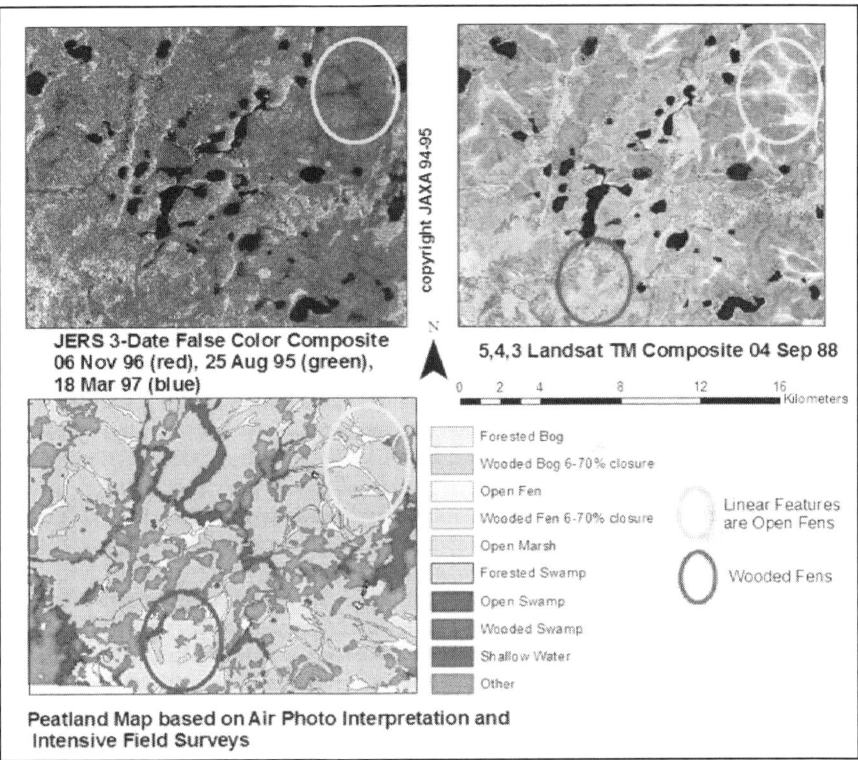

Fig. 13. Comparison of how different wetland types (bottom left, wetland map based on air photo and field truth) appear in 3 date JERS L-band imagery (1995-7, top left) versus Landsat imagery (Sept 88, top right). The pink circled areas are wooded fens and the yellow circled features are open fens. Most of the area in these scenes is wooded bog.

Figure 15. Process Tree images from the top down approach used for mapping Peatlands in the Central Alberta study area using Definiens.

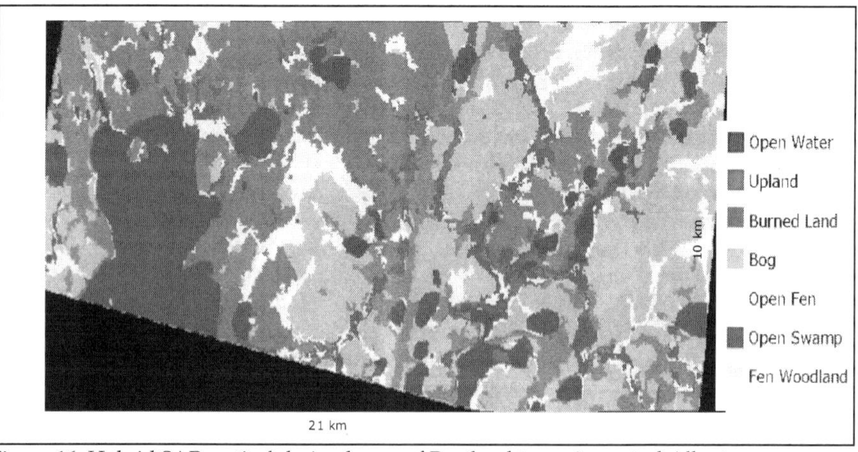

Figure 16. Hybrid SAR-optical derived map of Peatland types in central Alberta.

		Reference AirPhoto Map							
		Bog	Fen	Swamp	Marsh	Upland		totals	user's accuracy
SAR-Optical Map	Bog	50	1	1	0	13		65	0.77
	Fen	3	6	1				10	0.60
	Swamp	1	2	7		1		11	0.64
	Marsh				0	1		1	0.00
	Water					26		26	1.00
	Upland	1	1		1	21		24	0.88
	Totals	55	10	9	1	62			
	producer's	0.91	0.60	0.78	0.00	0.76			

Table 5. Accuracy assessment of SAR-Optical Map on vertical axis and Airphoto map as reference (top). Note that Upland on the airphoto map was labeled "Z" and represented open water and upland forest.

5. Discussion and conclusions

Many techniques focus on using multispectral data, such as Landsat or Aster, alone or in combination with ancillary data sets such as soils and topography for wetland mapping. However, research has shown how SAR and multispectral sensors complement each other in the classification and monitoring of wetland ecosystems and that SAR represents one of the most promising sensor types for improving wetland mapping capability (Bourgeau-Chavez 2004, 2008, Grenier 2007, etc). While multispectral data measure spectral reflectance and emittance characteristics of various cover types and wetness in open canopied ecosystems, SAR is sensitive to variations in biomass, structure and soil moisture and flood condition of landscapes including forests and other closed canopy ecosystems. Forested wetlands are the most difficult to identify remotely because of the inability of traditional multispectral sensors to detect moisture beneath the canopy. Radar can not only penetrate a closed canopy to detect flooding, but since radars are active systems, can acquire data independently of solar illumination and cloud cover conditions. Thus, data can be collected during specific conditions relevant to finding seasonally flooded wetlands or seiche-influenced wetlands. These SAR data can be used not only to detect and define wetlands, but also to monitor extent of inundation and in some cases level of inundation (Bourgeau-Chavez *et al.* 2005, Lang *et al.* 2008).

The case studies shown here demonstrate the improved mapping capabilities by including SAR in the traditional Optical/IR methods of mapping. It is important to note that the timing of the acquisitions can be very important for detection of flooding beneath a canopy, as was seen in the two JERS dates of imagery over Lake. St. Clair, where neither image date was able to be used from mapping the flood condition. Further, the importance of multi-temporal data was demonstrated in the image interpretation for all study sites, but particularly for the Canadian peatland study in which the subtle changes in backscatter due to changes in moisture levels allowed the distinction of wooded fens from bogs. Using knowledge of the phenological changes which occur in one wetland over another, as long as they are changes that can be detected by a particular sensor, can be key in distinguishing otherwise similar appearing ecosystems in the mapping process.

While the results shown here focus on amplitude data from SAR, the new era of SAR sensors have full polarimetric capability and as such, polarimetric decomposition can be used to understand the type of scattering occurring from a particular ecosystem. Decomposition variables can be used alone, or as additional bands in the more typical multiband classifiers. Variables such as phase difference, which were used in the past have been further developed to include complex analysis of the full scattering matrix. Decomposition methods are being developed to use Radsarsat-2 for peatland discrimination using a single date of imagery (Touzi *et al.* 2007). These new techniques are in the early stages of development and have not been tested on a variety of sites yet. But there is great potential for these new polarimetric satellite sensors, which are just beginning to be explored.

6. Acknowledgements

This research was funded by the Great Lakes Coastal Wetlands Consortium, NASA and the U.S. Fish and Wildlife Service. Special thanks to Brian Benscoter and Merritt Turetsky for their assistance with the "truth" maps of Alberta peatlands.

7. References

Arzandeh, S. and J. Wang. 2002. Texture evaluation of RADARSAT imagery for wetland mapping. Can. J. Remote Sensing, 28(5):653-666.

Arzandeh, S. and J. Wang. 2003. Monitoring the change of Phragmites distribution using satellite data. Can. J. Remote Sensing. (29(1):224-35.

Augustein, M. and C. Warrender. 1998. Wetland classification using optical and radar data and neural network classification. Int. J. Rem. Sens. Vol. 19:1545-1560.

Baghdadi, N., M. Bernier, R. Gauthier, and I. Neeson. 2001. Evaluation of C-band SAR data for wetlands mapping. Int. J. Rem. Sens. Vol. 22:71-88.

Becker B., Lusch D., & Qi J., (2005) Identifying optimal spectral bands from in-situ measurements of Great Lakes coastal wetlands using second derivative analysis. Remote Sens. Environ. 97 238-248.

Bourgeau-Chavez, L.L., M.R. Turetsky, B.W. Benscoter, and E.S. Kasischke. Improving mapping of boreal peatlands through use of Landsat and satellite SAR imagery, in prep.

Bourgeau-Chavez, L.L. and Richard Powell. 2009. Mapping the Invasive Phragmites with ALOS PALSAR radar imagery over the Saint Clair River Delta in the Great Lakes. Society of Wetland Scientists 2009 Conference, Madison, WI. June 21-26, 2009.

Bourgeau-Chavez, L.L., R.D. Lopez, A. Trebitz, T. Hollenhorst, G. E. Host, B. Huberty, R. L. Gauthier, and J. Hummer. 2008. Chapter 8, Landscape-Based Indicators in Great Lakes Coastal Wetlands Monitoring Plan, Great Lakes Coastal Wetlands Consortium, Project of the Great Lakes Commission, funded by the U.S. EPA GLNPO, March 2008, pp. 143-171. http://www.glc.org/wetlands/final-report.html

Bourgeau-Chavez, L.L., K.B. Smith, Eric S. Kasischke, S. M. Brunzell, E. A. Romanowicz, and C. J. Richardson. 2005. Remote Monitoring Regional Scale Inundation Patterns and Hydroperiod in the Greater Everglades Ecosystem. Wetlands, Vol. 25(1):176-191.

Bourgeau-Chavez, L.L., K.Riordan, M. Nowels, and N. Miller. 2004. Final Report to the Great Lakes Commission: Remotely Monitoring Great Lakes Coastal Wetlands

using a Hybrid Radar and Multi-spectral Sensor Approach. Project no. WETLANDS2-WPA-06.82pp.
http://www.glc.org/wetlands/pdf/GD-landscapeReport.pdf

Bourgeau-Chavez, L.L., E.S. Kasischke, S.M. Brunzell, J.P. Mudd, K.B. Smith, and A.L. Frick 2001. Analysis of spaceborne SAR data for wetland mapping in Virginia riparian ecosystems. Int. J. Rem. Sens, Vol. 22(18):3665-3687.

Charman, D. 2002. Peatlands and environmental change. J. Wiley & Sons, London & New York, 301 p.

Costa, M. , E. Novo, E. Ahern, E. Mitsou, J. Mantovani, M. Ballester, and R. Pietsch. 1998. The Amazon flood plain through Radar eyes: Lago Grande de Monte Alegre case study. Can. J. Remote Sensing. Vol. 24: 339-349.

Crawford, M., S. Kumar, M. Ricard, J. Gibeaut, and A. Neuenschwander. 1999. Fusion of airborne polarimetric and interferometric ASAR for classification of coastal environments. IEEE Trans. on Geosci. and Rem. Sens., Vol. 37:1306-1315.

Fournier, R. M. Grenier, A. Lavoie, R. Helie.. 2007. Towards a strategy to implement the Canadian Wetland Inventory using satellite remote sensing. Can. J. Remote Sensing. Vol. 33: 528-545.

Grenier, M., Demers, A.-M., Labrecque, S., Benoit, M., Fournier, R.A., and Drolet, B. 2007. An object-based method to map wetland using RADARSAT-1 and Landsat-ETM images:test case on two sites in Quebec, Canada. Can. J. Remote Sensing, Vol. 33:528-545.

Henderson, F. M., and A.J. Lewis. 2008. Radar detection of wetland ecosystems: a review. Int. J. Rem. Sens. Vol. 29(20):5809-1835.

Hess, L.L., J.M.Melack, S. Filoso, and Y. Wang. 1995. IEEE Trans. on Geosci. and Rem. Sens., 33(4)896-904.

Hess, L., J. Melack, and D. Simonett. 1990. Radar detection of flooding beneath the forest canopy: a review. Int. J. Rem. Sens. Vol. 11: 1313-1325.

IPCC 2008. Climate Change and Water, Intergovernmental Panel on Climate Change Technical Paper VI, ed. by B. Bates, Z.W. Kundzewicz, S. Wu, and J. Palutikof. June 2008.

Jensen, John R., 2007, Remote Sensing of the Environment: An Earth Resource Perspective, 2nd Ed., Upper Saddle River, NJ: Prentice Hall.

Kasischke, E.S., Smith, K.B., L.L. Bourgeau-Chavez,, E.A. Romanowicz, S. Brunzell, and C.J. Richardson. 2003. Effects of Seasonal Hydrologic Patterns in South Florida Wetlands on Radar Backscatter Measured from ERS-2 SAR Imagery. Rem. Sens. of Env. Vol. 88:423-441.

Kasischke, E.S. and L.L. Bourgeau-Chavez. 1997. Monitoring south Florida wetlands using ERS-1 SAR imagery. PE&RS, Vol. 63(3): pp. 281-291.

Klemas, V.V., Dobson, J.E., Ferguson, R.L. and Haddad, K.D. 1993. A coastal land cover classification system for the NOAA Coastwatch

Change Analysis Project. Journal of Coastal Research, Vol. 9(3): 862–872.Li, J. and W. Chen. 2005. A rule-based method for mapping Canada's wetlands using optical, radar, and DEM data. Int. J. Rem. Sens. Vol. 26:5051-5069.

Lopez, R.D. D. T. Heggem, D. Sutton, T. Ehli, R. Van Remortel, E. Evanson, and L. Bice. 2006 Using Landscape Metrics to develop indicators of great lakes coastal wetland condition. US EPA Report EOA/600/X-06/002. www.epa.gov 77 pp.

Lozano-Garcia, D.F., and R.M. Hoffer, 1993, Synergistic effects of combined Landsat-TM and SIR-B data for forest resources assessment. Int. J. Rem.Sens. Vol. 14(14), 2677.

Ozesmi, S. L., & Bauer, M. E. (2002). Satellite remote sensing of wetlands. Wetlands Ecology and Management, 10(5), 381–402.

Pengra, B. C. Johnston, and T. Loveland. 2007 Mapping an invasive plant, *Phragmites australis*, in coastal wetlands using the EO-1 Hyperion hyperspectral sensor. Rem Sen Env, 108:74-81.

Peters, D. 1994. Use of aerial photography for mapping wetlands in the United States:

National Wetlands Inventory. Proceedings of the First International Airborne Remote Sensing Conference and Exhibition held in Strasbourg, France on 11–15 September, 1994 (Ann Arbor: ERIM) pp. 165–173.

Pope, K., E. Reimankova, J. Paris, and R. Woodruff. 1997. Detecting seasonal flooding cycles in marshes of the Yucatan peninsula with SIR-C polarimetric radar imagery. Rem. Sens. of Env. , Vol. 59:157-166.

Pope, K., J. Rey-Benayas, and J. Paris. 1994. Radar remote sensing of forest and wetland ecosystems in the Central American tropics. Rem. Sens. of Env. , Vol. 48:205-219.

Ramsey, E. III. 1998. Radar remote sensing of wetlands. Remote Sensing Change Detection: Environmental Monitoring Methods and Applications, R. Lunetta and C. Elvidge (eds) Chapter 13, pp. 211-243 (Chelsea, MI: Ann Arbor Press).

Rao, B., R. Dwivedi, S. Kushwaha, S. Bhattacharya, J. Anand, and S. Dasgupta. 1999. Monitoring the spatial extent of coastal wetlands using ERS-1 SAR data. Int. J. Rem. Sens., Vol. 20:2509-2517.

Rio, J. And D. Lozano-Garcia. 2000. Spatial filtering of radar data (RADARSAT) for wetlands (brackish marshes) classification. Rem. Sens. Environ. Vol. 73:143-151.

Schmidt K. & Skidmore A. (2003) Spectral discrimination of vegetation types in a coastal wetland. Rem. Sens. Environ. 85 92-108.

Touzi, R., A. Deschamps, and G. Rother. 2007. Wetland characterization using polarimetric RADARSAT-2 capability. Can. J. Remote Sensing. Vol 33(1):S56-S67.

Townsend, P. 2002. Relationships between forest structure and the detection of flood inundation in forested wetlands using C-band SAR. Int. J. Rem. Sens., Vol. 22:443-460.

Townsend, P.A. 2001. Mapping seasonal flooding in forested wetlands using multi-temporal Radarsat SAR. PE&RS, Vol. 67(7):857-864.

Toyra, J. Pietroniro, A. and L. Martz. 2001. Multisensor hydrologica assessment of freshwater wetland. Rem. Sens. Environ, Vol. 75:162-173.

Wang, Y. 2004. Seasonal change in the extent of inundation on floodplains detected by JERS-1 Synthetic Aperture Radar data. Int. J. Remote Sensing, 25(13):2497-2508.

Wang, Y. And F. Davis. 1997. Decomposition of polarimetric synthesic aperture radar backscatter from upland and flooded forests. Int. J. Rem. Sens. , Vol. 18:1319-1332.

Whitcomb, J. M. Moghaddam, K. McDonald, J. Kellndorfer, and E. Podest. 2009. Mapping vegetated wetlands of Alaska using L-band radar satellite imagery. Can. J. Remote Sensing, Vol. 35:54-72.

Wilcox, K.L., S. A. Petrie, L.A. Maynard and S.W. Meyer. 2003. Historical Distribution and Abundance of *Phragmites australis* at Long Point, Lake Erie, Ontario. Journal of Great Lakes Research, 29(4):664-680

Wilen, B. and Frayer, W. 1990. Status and trends of U.S. wetlands and deepwater habitats. Forest Ecology and Management 33/34:181–192.

34

Remote Sensing Rock Mechanics and Earthquake Thermal Infrared Anomalies

Lixin Wu[1,2] and Shanjun Liu[2]

[1]*Academy of Disaster Reduction & Emergency Management, Beijing Normal University,*
Beijing, China
[2]*Institute for Geo-informatics & Digital Mine Research, Northeastern University,*
Shenyang, China

1. Introduction

Rock fracturing is the cause of many geo-hazards including tectonic earthquake (EQ), rock burst, rock sloping and rock pillar failure. Radiation signals such as acoustic emission, radio frequency emission and electromagnetic (EM) radiation from loaded deforming rock, are able to provide useful information for monitoring, interpreting and predicting rock fracturing (Renata, 1977; Brady and Rowell, 1986; Yamada et al., 1989; Martelli et al., 1989). Based on thermo-elastic theory, thermo-elastic stress analysis (TSA) and stress pattern analysis by thermal emission (SPATE) were developed for the stress measurement of solid materials, including homogeneous metal, macromolecular and composite materials, respectively in 1960's and 1970's (Mounatin and Webber, 1978). Luong applied thermovision to study experimentally the damage processes of concrete and rock (Luong, 1990), but no reach to the remote sensing on geo-hazards.

In the experiments for investigating the mechanism of satellite thermal infrared (TIR) anomaly before tectonic EQ (Gorny et al., 1988; Qiang et al., 1990), it was discovered that there do exist TIR anomaly before rock fracturing (Geng et al., 1992). Later, it was furthermore discovered that there are obvious TIR features as precursors of rock fracturing, and that the loaded stress around $0.79\,\sigma_c$ can be taken as a precaution index for the stability monitoring of loaded rocks (Wu and Wang, 1998). To explore the laws of infrared radiation (IRR) variation in the process of rock loading, deforming and fracturing, and to reveal the possible mechanism of satellite TIR anomaly before EQ, a large amount of IRR imaging experiments on rock loaded to fracturing were conducted in China (Wu et al., 2000, 2001, 2002, 2003, 2004a, 2004b, 2004c, 2004d, 2006a, 2006b; Deng et al., 2001; Liu et al., 2002). Hence, a new intersection discipline, Remote Sensing Rock Mechanics (RSRM), which takes Remote Sensing, Rock Mechanics, Rock Physics and Informatics as its foundations and serves for remote sensing on geo-hazards, was originated (Geng et al., 1992; Wu et al., 2000). Based on retrospection to past experiments on RSRM, it was pointed out that there are two IRR anomalies, being IRR image anomaly and IRR temperature curve anomaly respectively, can act as rock fracturing precursors. The average IRR temperature (AIR*T*), being the integral reflection of surface IRR energy, is applied as a quantitative index to study the temporal evolution of IRR from loaded rock and to seek for the potential precursors of rock fracturing. The temporal evolution of AIRT are the comprehensive effect of a series of physical-mechanical processes inside a loaded rock, such as rock thermo-elastic acting, pore gas desorbing & escaping, fractures producing & extending, rock frictionating, heat transferring and environment radiation. The thermo-elastic effect and the frictional thermal are two of the main mechanisms of increased IRR from loaded rock. RSRM experiments had revealed the laws of changed IRR from loaded rock and provided scientific interpretations for the mechanisms of satellite TIR anomaly before tectonic EQs of Ms>5.5.

2. Remote Sensing Rock Mechanics Experiments

2.1 Experiment Methods and Tools

The typical RSRM experiment is comprised of a loader (uni-axial or bi-axial), an infrared imager and rock samples. As in Figure 1, a bi-axial loader was applied for loading along two directions, and an infrared imager was applied to detect the surface IRR from loaded rock. The maximum imaging rate of the imager is 60f/s, and the recording rate was usually set as 1f/s to record the IRR images continuously. Usually, tectonic EQ might be resulted from the suddenly fracturing of compressively-sheared crust rock, the suddenly breaking of faults at disjointed zones, the suddenly sliding of compressively-sheared faults or the stability losing of compressively loaded intersected faults. To simulate the different mechanisms of rock fracturing and EQ, several typical loading schemes were applied as in Figure 1.

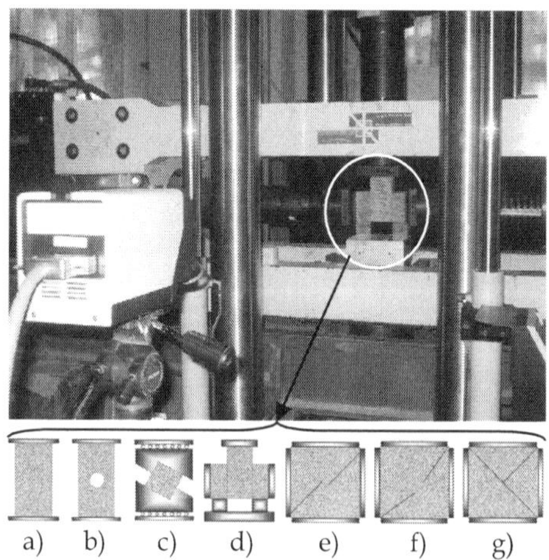

Fig. 1. RSRM experiment schemes to simulate different mechanisms of rock fracturing or tectonic EQ: a) uni-axially load on a standard cylinder rock sample; b) uni-axially load on a cylinder rock sample with a central hole; c) compressively-sheared load on a hexahedral rock sample; d) bi-axially load on three jointed rock samples to frictional sliding; e) bi-axially load on a damage rock sample with en echelon faults; f) bi-axially load on a damage rock sample with disjointed faults; and g) bi-axially load on three jointed rock samples simulating intersected faults.

2.2 Rock Fracturing Precursor: IRR Image Anomaly

2.2.1 Uni-axially loaded rock

Lots of rock samples made from coal, ironstone, sandstone, marble, limestone, granite, granodiorite, gabbro and gneiss were uni-axially loaded and thermal imaging detected. The sample size was standard of diameter and length, respectively, 50 and 100mm. It was discovered that the IRR images of the uni-axially loaded rock have different features for different fracturing pattern (Wu et al., 2006a). As in Figures 2~4, there are three fracturing patterns, "X"-shaped, "//"-shaped and "|"-shaped respectively, occurred in our experiments. The "X"-shaped and "//"-shaped positive IRR abnormal strips foretell the coming of "X"-shaped shearing fracturing and the coming of "//"-shaped shearing fracturing respectively, while the "|"-shaped negative IRR abnormal strip foretells the coming of tensile fracturing.

The "X"-shaped positive IRR abnormal strips generated with loading along the "X"-shaped shearing zone before peak stress, and got distinguished after peak stress, as in Figure 2. The rock sample got finally fractured along the "X"-shape shearing zone. The evolution of IRR abnormal strip had also reflected the fracturing being not symmetrical upper-and lower, in that the upper part was clear with higher temperature, while the lower part is fuzzy with lower temperature.

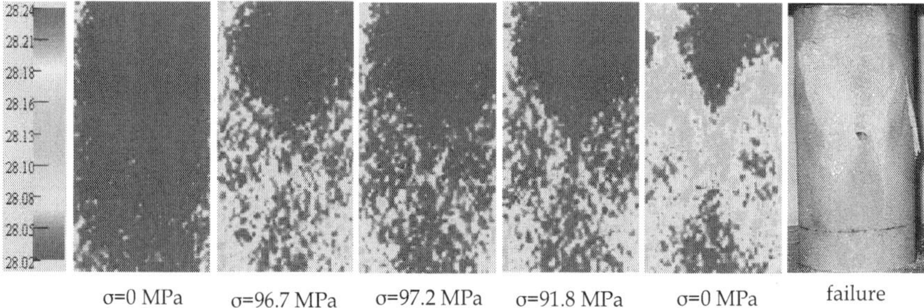

σ=0 MPa   σ=96.7 MPa   σ=97.2 MPa   σ=91.8 MPa   σ=0 MPa   failure

Fig. 2. The IRR image positive anomaly of "X"-shaped shearing fracturing of an uni-axially loaded marble sample

The "//"-shaped positive IRR abnormal strips generated with loading along the "//'-shaped shearing zone at the upper part of sample before peak stress, and got distinguished after peak stress, as in Fig 3. The evolution of the positive IRR image anomaly had also reflected the fracturing being not symmetrical in that the upper part of the IRR anomaly strip was clear with higher temperature, while the lower part was fuzzy excepting for the final fracturing near the bottom of the sample. Besides, there was strong IRR anomaly spot at the fracturing center for the intensive accumulation of mechanical energy and for the intensive generation of frictional thermal at the local central place.

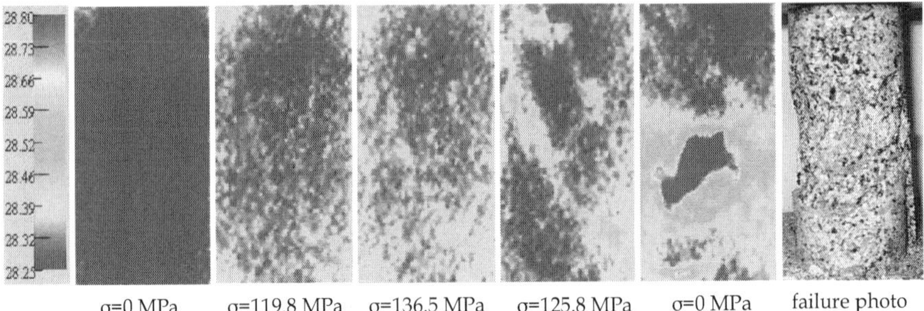

Fig. 3. The IRR image positive anomaly of "/ /"-shaped shearing-fracturing of an uni-axially loaded granite sample

The "|"-shaped negative IRR abnormal strip generated with loading along the tensile fracturing zone of a rock sample before the peak stress, and got distinguished gradually at the peak stress and after fracturing, as the approximately vertical dark strip in Figure 4. The same phenomenon for a sandstone sample with a calcite vein was also reported (Wu, et al., 2000).

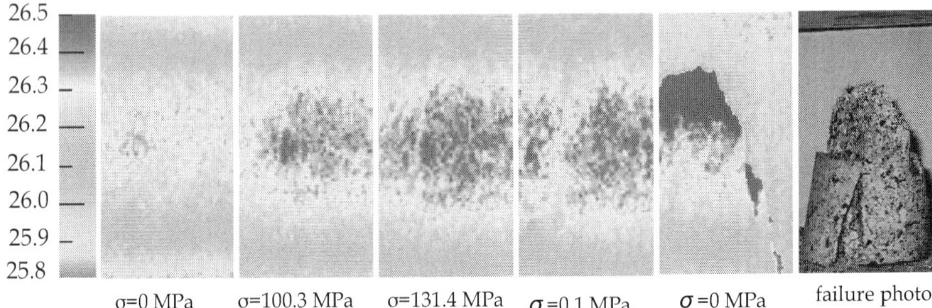

Fig. 4. The IRR image negative anomaly of "|"-shaped tensile fracturing of an uni-axially loaded granite sample

2.2.2 Uni-axially loaded rock with a central hole

More than 10 samples with a central hole, modeling the structure stability of loaded rock tunnels, made from marble and granite were infrared imaging detected. The rock samples had two kinds of shapes respectively being cylinder with diameter and length, respectively, 50 and 100mm, and regular block with thickness, width and length, respectively, 70, 35 and 100 mm. It was discovered that there were distinguished positive IRR image anomalies before rock fracturing, and the place of anomaly were exactly the coming fracturing place. As in Figure 5, the positive IRR image anomalies had reflected the two kinds of fracturing, respectively being diagonal fracturing (sample 1~5) and fork fracturing (sample 6). The IRR anomalies, along the fracturing planes and shaped as spots or strips, generated not only on rock surface but also on the hole's surface (lateral sample 4 and 5). The temperature increment is 1~3oC and 4~8oC respectively for marble and granite samples.

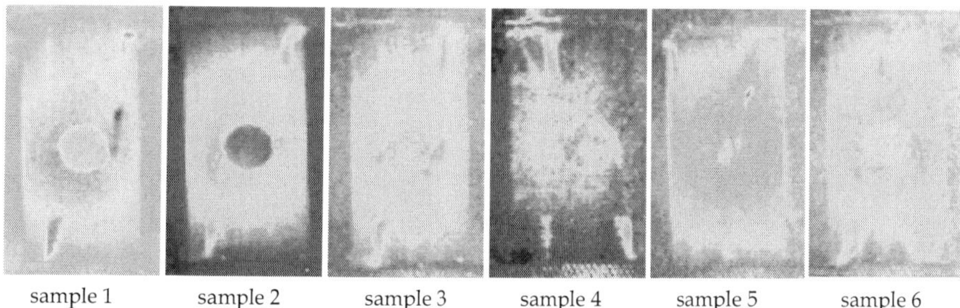

Fig. 5. The IRR image positive anomalies of a group of uni-axially loaded marble samples with a central hole

2.2.3 Compressively sheared rock

More than 20 samples, size 7×7×7×cm³, made from sandstone, marble, limestone, granite and gneiss, were compressively sheared and infrared imaging detected. Three pairs of steel platens with shearing angle being 45 , 60 and 70° respectively were applied. The loading rate was controlled as 2~5kN/s. It was discovered that the IRR temperature of rock surface changed with loading, and a strip-shaped positive IRR image anomaly generated along the central shearing plane before fracturing. With loading, the positive abnormal strip got more and more distinguished and migrated gradually from the upper end to the lower end of the sample, which foretold that the compressive-shearing fracturing was developing gradually from the upper end to the lower end of the sample along the central shear plane. Figure 6 shows the typical IRR image series of a compressively sheared limestone sample.

As a special geological phenomenon occurring with the formation of great fault, penniform-shaped fractures are a group of secondary fractures produced with the formation of great primary fracture (Nicolas et al., 1977). It happened to occur in our experiments that there were penniform-shaped fractures produced with a primary fracture in the compressive loaded rock samples, as in Figure 7. The IRR positive anomaly strips generated aside the primary IRR strip, passing through the central shearing plane, had reflected the penniform-shaped fracturing events.

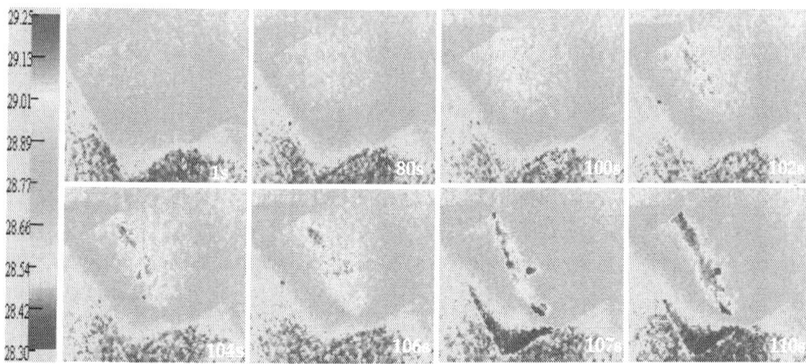

Fig. 6. The IRR image positive anomaly of the fracturing of a compressively sheared limestone sample (time in second)

important mechanism of electromagnetic radiation. Hence, rock surface IRR is a comprehensive effect of rock deformation and rock surface thermal state. Rock surface IRR temperature could be a detective index reflecting rock surface physical temperature and rock surface deformation field, which implicating the complex physical-mechanical process inside the loaded rock.

In spite of thermal exchange and plastic deformation, the thermo-elastic effect and the frictional thermal are two of the main mechanics of changed IRR from loaded brittle rock. In the stage of elastic deformation, the thermo-elastic effect is the main cause; while in the stage of plastic deformation or fracturing, the friction-thermal effect plays a great role. At the moment of rock fracturing or hazard, the fraction-heat effect gets more distinguished. The frictional thermal effect depends on two factors being frictional force (decided by normal stress and frictional coefficient) and frictional speed respectively. The larger the frictional force and the quicker the frictional speed, the more the frictional heat.

3.2.2 RSRM-model based on independent system

The rock sample, load header and environment air could be taken as a closed independent system, as in Figure 25, and the energy of the loaded rock sample is in a balance state as follows:

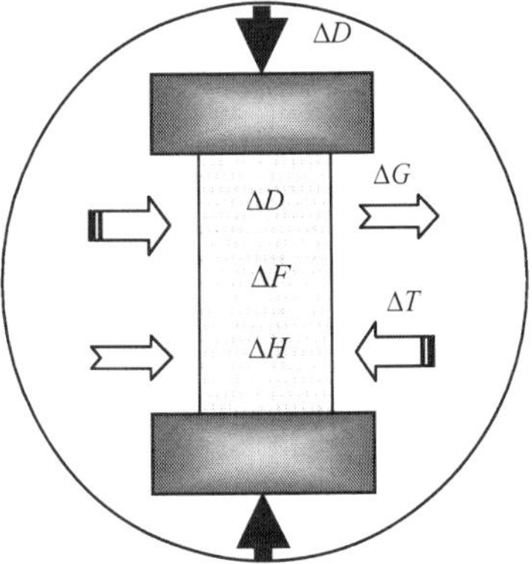

Fig. 25. The energy balance of a loaded rock sample in a closed independent system

$$\Delta M + \Delta T = \Delta D + \Delta F + \Delta H + \Delta G \tag{5}$$

Here: ΔM ——The inputted mechanical energy from loader, J; be positive;

ΔT ——The inputted thermal energy from loader and environment air, J; be positive;

ΔD ——The produced deformation energy of rock sample in elastic and plastic state, J; be positive;

ΔF ——The sum of consumed rock fracturing energy and formed fracture surface energy, J; be positive;

ΔH ——The heat energy increment of rock decided by its physical temperature, J; be positive if temperature rise or be negative if temperature drop;

ΔG——The energy consumed by the desorbing and escaping of pore gas in rock samples, J; be positive.

From equation (5) we have:

$$\Delta H = (\Delta M + \Delta T) - (\Delta D + \Delta F + \Delta G) \qquad (6)$$

The change of heat energy will result in the change of physical temperature, surface radiation energy and IRR temperature of loaded rock samples. Referring to Stephen-Boltzmann law, the $AIRT$ is a direct index of rock radiation energy, and the change of $AIRT$ ($\Delta AIRT$) must have certain a relationship with the physical temperature (T) of loaded rock sample as:

$$E_{IR} = f(AIRT) = \varepsilon \sigma T^{4} \qquad (7)$$

Here: E_{IR}——The radiation energy of a loaded rock, J;

   $AIRT$ ——The surface average IRR temperature of a loaded rock, K;

   ε ——The radiation factor of rock sample, $0 < \varepsilon < 1$;

   σ ——The constant of Stephen-Boltzmann, $\sigma = 5.6679 \times 10^{-8}$, J×m^{-2}×K^{-4} ;

   Thermo-elastic effect is the basic mechanism of changed IRR from a loaded rock sample. Moreover, the desorbing and the escaping of pore gas, the expanding of initial fissures or joints, the friction between fissures, fractures, joints and grains, the thermal transfer between the rock, loader header and environment air, and the radiation from the environment all are to have thermal effect on loaded rock samples. The thermal state is comprehensively affected by the six factors which are rock stress, pore gas desorbing & escaping, rock fracturing, heat transferring, rock frictionating and environment radiation respectively. The equation could be expressed as follows:

$$\Delta AIRT = f[\Delta(\sigma_{1} + \sigma_{2}), \Delta G, \Delta F_{1}, \Delta H, \Delta F_{2}, \Delta E]$$
$$= f_{1}(t) + f_{2}(t) + f_{3}(t) + f_{4}(t) + f_{5}(t) + f_{6}(t) \qquad (8)$$

Here: $\Delta AIRT$ ——The detected change of $AIRT$, K; be positive if rise, be negative if drop;

   $f_{1}(t)$ ——IRR temperature change due to thermo-elastic effect ($\Delta(\sigma_{1} + \sigma_{2})$, K; be positive or negative;

   $f_{2}(t)$ ——IRR temperature change due to pore gas adsorbing & escaping (ΔG), K; be negative ;

   $f_{3}(t)$ ——IRR temperature change due to the production of new fractures and the expansion of initial fissures, joints and new produced fractures (ΔF_{1}), K; be negative;

   $f_{4}(t)$ ——IRR temperature change due to frictional thermal (ΔF_{2}), K; be positive;

$f_5(t)$——IRR temperature change due to heat transfer (ΔH), K; be positive or negative.

$f_6(t)$——IRR temperature change due to environment radiation (ΔE), K; be positive.

1) Thermo-elastic effect: $f_1(t)$

Referring to thermo-elastic theory and equation (4), $f_1(t)$ could be calculated as:

$$f_1(t) = \gamma \cdot \beta^{-1} \cdot T \cdot \Delta(\sigma_1 + \sigma_2) \tag{9}$$

In condition of uni-axially compressive loading, the load is to cause temperature rise in that σ_2 is constant zero and the positive σ_1 will linearly increase with loading. If $\Delta\sigma_1$ is positive, the $f_1(t)$ will be positive; If $\Delta\sigma_1$ is negative, the $f_1(t)$ will be negative. Hence, the $f_1(t)$ will be positive before the compressive stress peak, and will turn to be negative after the compressive stress peak.

In condition of uni-axially tensile loading, the load is to cause temperature drop in that σ_2 is constant zero and the negative σ_1 will linearly increase with loading. If $\Delta\sigma_1$ is positive, the $f_1(t)$ will be positive; If $\Delta\sigma_1$ is negative, the $f_1(t)$ will be negative. Hence, the $f_1(t)$ will be negative before the tensile stress peak, and will turn to be positive after the tensile stress peak.

In condition of compressively-sheared loading, the rock sample will be compressed by the normal component of the load. However, as to the central shearing plane, it will suffer not only compressive stress but also shearing stress. The σ_1 refers to the positive compressive stress normal to the shearing plane, while the σ_2 refers to the shearing stress which is actually negative tensile stress along the shearing plane. Hence, the $\Delta AIRT$ is decided by the sum of compressive stress and the tensile stress. If $\Delta(\sigma_1 + \sigma_2) > 0$, the $f_1(t)$ will be positive; If $\Delta(\sigma_1 + \sigma_2) < 0$, the $f_1(t)$ will be negative. As to Figure 1c, it's easy to know that $\Delta(\sigma_1 + \sigma_2)$ will be positive if the shearing angle $\gamma < 45^0$, $\Delta(\sigma_1 + \sigma_2)$ will be zero if the shearing angle $\gamma = 45^0$, $\Delta(\sigma_1 + \sigma_2)$ will be negative if the shearing angle $\gamma > 45^0$.

2) Pore gas desorbing & escaping effect: $f_2(t)$

Any rock has pores of different size more or less inside. Some rock, especially the sedimentary rock, may has certain gas, such as CH_4, CO_2, CO and O_2, enclosed in the pores or/and absorbed on the pore surface (Wang, 2003; Yang et al., 1999). Usually, most of the pores are enclosed and the gas molecules stay inside both in free gassy state and absorbed state. If the rock is loaded and suffers deformation, the volume of pores will decrease which results in the escape of gas from the pores. Once the load and deformation cause the enclosed pores getting fractured, the gassy molecules will escape firstly and the absorbed molecules will get desorbed to be gassy molecules and escape later. Both desorbing and escaping actions need to make use of heat energy from the rock, and thus will result in the

AIRT drop of rock surface. Hence, the $f_2(t)$ is always negative, and the more the pores and gas enclosed, the more the negative effect of $f_2(t)$.

3) Fracture effect: $f_3(t)$

With loading and deforming, the rock is to get fractured. The new produced fractures together with the initial fissures and joints will extend both in width and length. The production of new fractures needs to consume energy, and the extension of fissures, joints and fractures also needs to consume energy. Hence, the $f_3(t)$ is always negative, and the more the fractures produced and fissures, joints and fractures extended, the more the negative effect of $f_3(t)$.

4) Frictional thermal effect: $f_4(t)$

With loading and deforming, the friction action is to occur between rock fissures, rock joints, rock grains, and new produced fractures. The friction action could be interpreted as: 1) at the beginning stage of loading, the friction may only be resulted from between rock fissures and between rock joints; 2) later, rock deformation increases with loading, new fractures are produced, and the friction between rock grains and between new produced fractures will join in; 3) finally, at the ending stage of loading, the rock deformation and fractures will be sufficiently developed, and the frictions between rock grains and between new produced fractures will be the chief contributors to the frictional thermal. In a word, the $f_4(t)$ is always positive no matter what is the principle friction factor. The more the friction, the more the positive effect of $f_4(t)$.

5) Heat transfer effect: $f_5(t)$

In the process of loading and inside the respectively independent loading system, the heat exchange is inevitable between the rock sample and the load header, the shearing platen or cushion-blocks, the surrounded atmosphere, etc. If the current temperature of rock sample is higher than the others, the heat of rock sample will be transferred out to whose temperature is lower. If the current temperature of rock sample is lower than the others, the heat of the others will be transferred into rock sample. Hence, the temperature of rock sample is a dynamic balance behavior between the heat transferred in and the heat transferred out. If the heat transferred in is more than that transferred out, the $f_5(t)$ will be positive; otherwise, it will be negative.

6) Environment-radiation effect: $f_6(t)$

The IRR detected by infrared imager includes not only the direct radiation from rock surface itself, but also the reflected radiation from environment. In laboratory, the chief environment radiations are the scattering sunshine, the moving human bodies and the illumination lamp. For the uncertain change of scattering sunshine, the movement of human bodies before the loaded rock sample, and the fluctuation of illumination light, the environment radiation effect on rock sample will be random. Hence, sometimes $f_6(t)$ may be positive, but sometimes $f_6(t)$ be negative. To eliminating the environment-radiation effect, the human bodies inside the laboratory were not permitted to move during testing process, the illumination lights were turned off, and the windows as well as its curtains were closed. Furthermore, some experiments were conducted in the evening so as to avoid the scattering sunshine completely.

3.3 Experiment Interpretation with RSRM-Model

Due to the comprehensive effects of $f_1(t) \sim f_6(t)$, the evolution of AIRT will be complex and will result in different possibility for abnormal AIRT precursors in different rock loading conditions, which including the loading scheme, rock type, and environment parameters. The following discussions are based on that $f_5(t)$ and $f_6(t)$ can be ignored.

3.3.1 Uni-axial loading experiments

For uni-axially loaded rock, $f_4(t)$ will take place only after that the rock has sufficiently deformed and the fractures have sufficiently developed.

At loading stage I and stage II, the rock surface AIRT are decided by $f_1(t)$ and $f_2(t)$. If the loaded rock is igneous rock or metamorphic rock, $f_2(t)$ is very rare since no gas absorbed in its pores usually, and the AIRT will rise with loading. If the loaded rock is sediment rock and with gas closed and absorbed in pores, $f_2(t)$ is inevitable, and the AIRT will rise if $f_1(t) > |f_2(t)|$, or be constant if $f_1(t) = |f_2(t)|$, or drop if $f_1(t) < |f_2(t)|$.

At loading stage III, fractures get sufficiently developed and pores get seriously damaged. The $f_3(t)$ begins to has more and more effect on evolution process of AIRT. AIRT will rise if $f_1(t) > |f_2(t) + f_3(t)|$, or be constant if $f_1(t) = |f_2(t) + f_3(t)|$, or drop if $f_1(t) < |f_2(t) + f_3(t)|$.

At loading stage IV, the friction action starts and $f_4(t)$ begins to have more and more effect on the evolution of AIRT. AIRT will rise if $[f_1(t) + f_4(t)] > |f_2(t) + f_3(t)|$, or be constant if $[f_1(t) + f_4(t)] = |f_2(t) + f_3(t)|$, or drop if $[f_1(t) + f_4(t)] < |f_2(t) + f_3(t)|$.

Since $f_2(t)$ is very rare for igneous rock or metamorphic rock, the rise speed of AIRT of loaded igneous rock or metamorphic rock will get fast at stage IV, and the speed turning point is suggested to be the precursors point.

3.3.2 Compressive shearing experiments

For compressively sheared rock, not only $f_1(t)$ but also $f_4(t)$ is decided by shearing angle (γ). If $\gamma < 45^0$, $f_1(t)$ will be positive for $\Delta(\sigma_1 + \sigma_2) > 0$, and the friction action will be much strong in that σ_1, which is normal to the friction plane, is large. If $\gamma = 45^0$, $f_1(t)$ will be zero since $\Delta(\sigma_1 + \sigma_2) = 0$. If $\gamma > 45^0$, $f_1(t)$ will be negative for $\Delta(\sigma_1 + \sigma_2) < 0$, and the friction action will be much week since σ_1 is slight.

3.3.3 Biaxial loading experiments

For bi-axially loaded rock, $f_1(t)$ will always be positive. As to bi-axially loaded en echelon faults, collinearly and non-collinearly disjointed faults, $f_2(t)$, $f_3(t)$ and $f_4(t)$ will occur simultaneously, and the evolution of AIRT will be fluctuated. If $[f_1(t) + f_4(t)] > |f_2(t) + f_3(t)|$, AIRT will rise; if $[f_1(t) + f_4(t)] = |f_2(t) + f_3(t)|$, AIRT will keep in the same level, and if $[f_1(t) + f_4(t)] < |f_2(t) + f_3(t)|$, AIRT will drop. Usually, the fact is that $[f_1(t) + f_4(t)] < |f_2(t) + f_3(t)|$ at the fracturing stage, and there is a short drop of AIRT, which is called as 'silence' before EQ (Ohtake et al., 1981). However, $f_4(t)$ will be an important factor at the later stage of loading for the concentrated formation of fractures and friction in the disjointed zone, and the final state of AIRT will be rise for $[f_1(t) + f_4(t)] > |f_2(t) + f_3(t)|$.

4. Earthquake Thermal Infrared Anomalies

The prediction of EQ is very difficult, but it's not impossible. A number of signs warning of EQs, such as foreshock activities, peculiar animal behaviour, increased low frequency EM-noise, concentrations of radon in water and air, ionosphere and magnetosphere perturbation, radio frequency emissions, terrestrial gas emanations, EQ clouds, and satellite TIR anomalies, have been proposed and reported during the past centuries. Satellite TIR anomaly was firstly reported in 1989, and had been repeatedly verified in the world during the past 20 years. It is becoming a prospecting space observation technology for seismic activity monitoring and for EQ predicating.

4.1 General Features of EQ TIR Anomaly

Gorny (1988) firstly reported that there were large area of TIR anomalies in METEO satellite remote sensing images, spatial resolution being 5 Km and wave length being 10.5~12.5 μ m, before many moderate-strong tectonic EQs in the mid-Asia and the east-Mediterranean region. Tronin (1996) analyzed 10000 about TIR images of channel AVHRR-2 of NOAA in ten years for the mid-Asia, and reached that there existed average anomaly, 1~5°C, before the EQs at this region, and that there was obvious statistical relations between the EQ and TIR anomaly. Qiang (1990), Cui (1999), Liu (1999), Xu (2000), Zhang(2002), Ouzounov(2004), Arun (2005), Liu (2007), Wu (2008) also reported that there occurred TIR anomalies in satellite images (NOAA, FY, MODIS) days before more than 100 EQs in Asia (China, India, Iran, Japan, Kamchatka, Pakistan, Turkey) and Europe (Italy, Greece, Spain). Analysis to all the reported cases, it was uncovered that satellite TIR anomaly before EQ has the following features generally (Wu et al., 2009):

1) Temporal features: Satellite TIR anomaly usually appears 1~26 days before shock, and reaches to its peak 1~2 days before shock, and will disappeared soon after shock.
2) Spatial features: The spatial distribution and geometrical shape of TIR anomaly is tightly related with tectonic structures such as plate borders and active faults. With the EQ impending, the TIR anomaly will move to or extend to gradually the coming epicentre along the structures.
3) Temperature features: the temperature of the TIR anomaly is usually 2~6°C higher than that of outside or surrounding the TIR anomaly.
4) Magnitude features: there are positive relations somewhat between the TIR

anomaly energy (the anomaly area times the anomaly temperature) and the magnitude of future shock.

4.2 Anomalies Interpretation with RSRM-Model

The RSRM experimental results are applied to analyze satellite remote sensed TIR anomalies before several strong EQs in Asia. Referring to the seismogenic mechanisms, the satellite TIR anomalies are in good accordance with the detected IRR anomalies from rock fracturing and seismogenic experiments with fault system being simulated with disjointed faults and intersected faults.

4.2.1 Dongsha Ms5.9 EQ 1992 in Taiwan, China

Dongsha Ms5.9 EQ occurred in Taiwan on Sept 14, 1992. The NOAA satellite images show that there was TIR anomaly appeared along the regional faults and downfaulted basins before shock as in Figure 26 (Wu et al., 2006c). There was an isolated and spoon-shaped high temperature area to the southwest of Taiwan Island 25 days before shock (Aug 19, 1992). The head of 'spoon' locates in the downfaulted basin-IV and the handle of 'spoon' distributes along fault-6. Satellite TIR image on Aug 22 shows that the high temperature on 'spoon head' diminished, but the TIR anomaly on 'spoon handle' became wide. Later, the anomaly moves gradually to the epicenter. TIR image on Sept 9 shows that a large area of high temperature had appeared around the coming epicenter, and the maximum temperature (in brown color) appeared south-close to the coming epicenter. It indicated that the TIR anomaly was consistent with the regional tectonic structures (faults and basins) in spatial position and geometry.

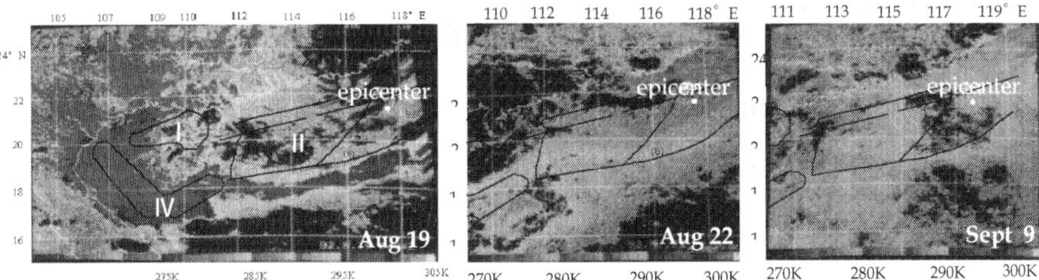

Fig. 26. Satellite TIR anomaly images before Dongsha Ms5.9 EQ (Sept 14, 1992)

4.2.2 Zhangbei Ms6.2 EQ 1998 in China

Zhangbei Ms6.2 EQ occurred in China at 11:50 am, Jan 10, 1998. With the overlay of the active fault system investigated and deduced, in red lines, on night time NOAA satellite infrared images, as in Figure 27, it was discovered that (Wu et al., 2007b): 1) 18 days before shock, Dec 28, 1997, the TIR images on land and sea surface are basically normal, and the contours of TIR temperature field of Bohai Bay appeared to be along the coastline; 2) 5 days before shock, Jan 5, 1998, there occurred a positive TIR anomaly strip from Bohai Bay to Zhangbei passing through Beijing, its temperature was 3°C higher than that of outside, and the contours of TIR temperature field of Bohai Bay got offset the coastline; 3) 1 day before shock, Jan 8, 1998, the positive TIR anomaly strip had got wider and its temperature was 7°C higher than that of outside, and the contours of TIR temperature field of Bohai Bay got in accordance with the positive TIR strip; 4) 1 day after shock, Jan 11, 1998, the pattern of TIR images on the surface of land and sea turned to be normal again.

There are several disjointed active faults along the positive TIR strip, including a possible uncovered great active deep fault going from Bohai Bay to Zhangbei. Besides, to the southwest of Zhangbei, there are two small active faults pointing to Zhangbei. If extend the three faults respectively towards Zhangbei, it could be found that Zhangbei is exactly the intersection point. Hence, the tectonic background around the epicenter is basically comprised of three intersected active faults as two groups, i.e., the primary fault is the independent one from Bohai Bay to Zhangbei, and the secondary two act as another group. The two groups of faults split basically the regional crust into three geo-blocks (geo-block A, B and C_1+C_2) as the RSRM experiment model in Figure 28b (rock block A, B and C), and the secondary two faults act as the two sides of an acute wedge-shaped geo-block (geo-block C_2) as the RSRM experiment model in Figure 28d (rock block C). The geometric features of the faults and the spatio-temporal features of TIR anomaly were similar. Therefore, the mechanical mechanism of Zhangbei EQ should be classified to be the failure of an intersected active fault system, which with an acute wedge-shaped geo-block being its secondary active object.

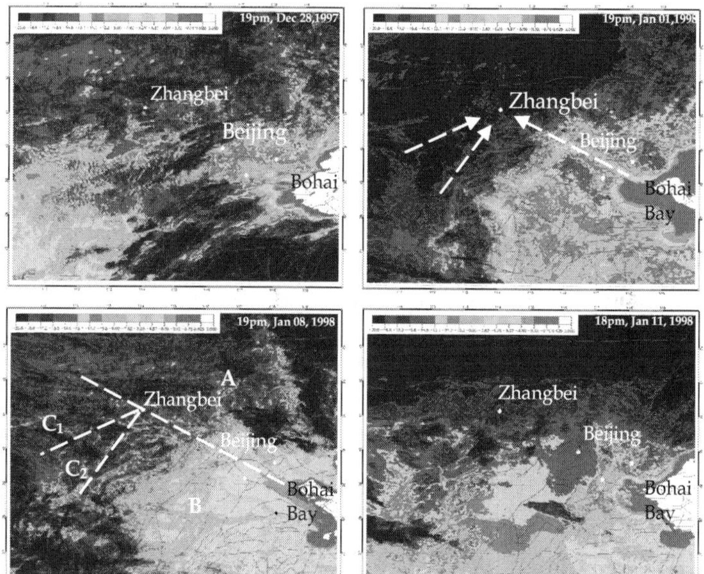

Fig. 27. The TIR images overlaid with detailed active faults before and after Zhangbei Ms6.2 EQ (Jan 10, 1998)

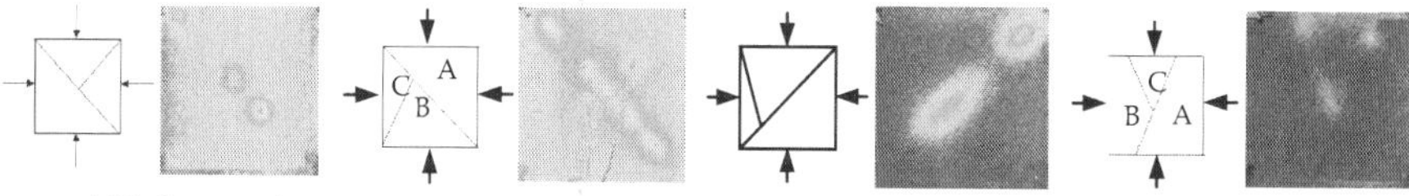

a) 90° intersection   b) 75° intersection   c) 60° intersection   d) 40° intersection

Fig. 28. The IRR anomaly features of simulated tectonic activities due to bi-axially load (Wu et al., 2007b)

4.2.3 Izmit Ms 7.8 EQ 1999 in Turkey

Izmit Ms 7.8 EQ occurred in Turkey on Aug 17, 1999. In the epicenter zone, there were two en echelon tectonic faults, and the epicenter is 45km about to the west of the disjointed place of the two en echelon faults, as in Figure 29. With thermal image of Aug 1&2 being the reference, the differential thermal images from Aug 6 to Aug 26 were obtained, and it was revealed that there were NOAA satellite TIR anomaly at the disjointed zone of two en echelon faults since Aug 6 , 11 days before shock (Tronin, 2000), as in Figure 29. The RSRM experiments on the simulation of tectonic EQ due to the fracturing of disjointed faults had revealed that there were IRR anomaly increment and concentrated deformation at the disjointed zone, as in Figure 30. Obviously, the spatial features of NOAA TIR anomaly before Izmit Ms 7.8 EQ were the same as that of IRR anomaly before the fracturing of disjointed faults.

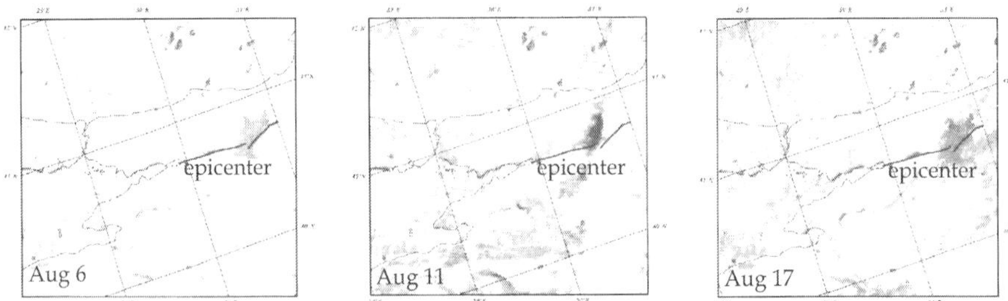

Fig. 29. TIR anomalies 0~11 days before Izmit Ms 7.8 EQ (Aug 17, 1999)

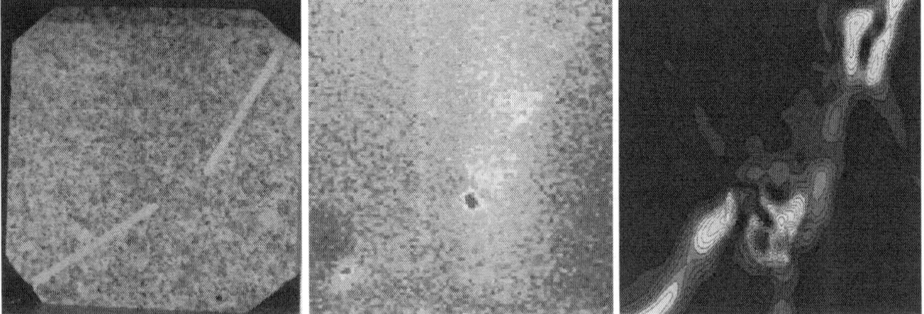

Fig. 30. The IRR anomaly and deformation concentration of a bi-axially loaded granite sample with disjointed faults

4.2.4 Hengchun Ms7.2 EQ 2006 in Taiwan, China

Hengchun Ms7.2 EQ occurred in Taiwan at 12:26pm, Dec 26, 2006. TIR images from stationary satellite FY-2 showed that there was TIR anomaly nearby the coming epicenter. The TIR anomaly appeared to the east of Philippines six days before shock, and moved gradually toward west to Philippines. Later, the anomaly changed its direction to the north, and the temperature increased 10°C about one day before shock. Figure 31 shows that the satellite TIR images around Hengchun on Dec 25, 2006 (Liu et al., 2007). A disjointed thermal strip (in orange color) appeared on the southwest of Taiwan at 1:00am, and developed gradually to be an X-shaped thermal anomaly zone at 10:00am. The epicenter located closely to the cross point of the X-shaped zones. The evolution process of satellite TIR images was very similar to that of X-shaped thermal IRR anomaly in RSRM experiments as in Figure 2.

4.2.5 Wenchuan Ms8.0 EQ 2008 in China

Wenchuan Ms8.0 EQ occurred in China at 14:28 am, May 12, 2008. The epicenter locates at the transferring zone of Tibetan Plateau to Sichuan Basin, which is only 92Km to the NW of Chengdu, the capital of Sichuan province. Analysis to FY-2C TIR images, as in Figure 32, it is discovered that there was an high temperature strip with length of 3 000 km appeared on April 23, 20 days before shock, which started from India Plate and developed to northeast along the east front of Tibetan and Loess Plateau (Wu et al., 2008). The cause might be the great accumulation of fictional sliding stress along the east foreland of Tibetan and Loess Plateau due to the subduction of India Plate to Euro-Asia Plate. The east foreland of Tibetan and Loess Plateau act as the fictional sliding plane between the west part of China (including Tibetan and Loess Plateau) and the east part of China (including the North, Middle and Southwest China). The evolution of satellite TIR images was similar to that of detected strip-shaped IRR anomaly in the fictional sliding experiments as in Figure 8 and Figure 9, which shows the evolution of TIR anomaly along the sliding plane.

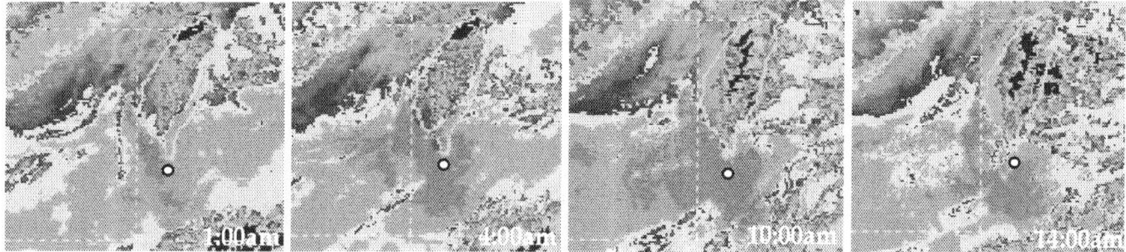

Fig.31. Satellite TIR images one day before Hengchun Ms7.2 EQ (Dec 25, 2006)

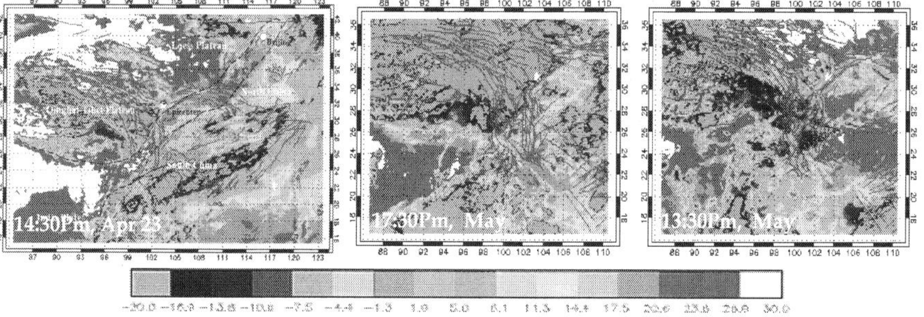

Fig.32. Satellite TIR images before Wenchuan Ms8.0 EQ (May 12, 2008)

5. Future Researches

5.1 On RSRM Experiment

As a detectable remote sensing signal related with rock stress and physical temperature, IRR is a meaningful index for studying rock load, rock deformation, rock fracturing and rock hazard. The temporal evolution of surface IRR from loaded rock is the comprehensive effect of rock thermo-elastic acting, pore gas desorbing & escaping, fractures producing & extending, rock frictionating, heat transferring and environment radiation. The IRR image anomaly referring to the spatial-temporal evolution of IRR from loaded rock is an important precursor for rock fracturing, and will be meaningful for the predication of geo-hazards including tectonic EQ. For the practical application of RSRM, deeper research on IRR imaging detection quantitatively and specially on rock stress and rock hazard for experimental rock mechanics, rock engineering, tectonic activity and strong EQ is demanded.

The mechanism of experimental detected IRR anomaly can be theoretically interpreted by taking the load header, the rock sample and the environment to be a closed independent system in energy balance state. There are two of the main rock-physics mechanisms, respectively being thermo-mechanical coupling and frictional thermal due to tectonic stress, rock fracturing and fictional sliding, for the change IRR from loaded rock samples (Wu, et al., 2006c). Besides, positive hole (P-hole) activation due to piezoelectricity was suggested to be another mechanism of IRR from loaded quartz-bearing igneous rock (Freund, 2002), such as granite, basalt, diorite, and gabbro.

To search for scientific interpretation on the relations among satellite TIR anomaly, rock stress and experimentally detected IRR anomaly, the EM transferring process from underground rock body to satellite sensors, through lithosphere, earth surface coversphere (including soil layers, water bodies and vegetations), the atmosphere and lithosphere, should be systematically studied. Pulinets pointed out that the incubation of an EQ is to disturb the ionosphere (Pulinets, 1998), and it was suggested that lithosphere-atmosphere-ionosphere (LAI) coupling is the mechanism of satellite TIR anomaly before strong EQ (Molchanov et al., 2004). Nevertheless, the action of earth surface coversphere on the transferring and the magnifying of EM signals from underground loaded rock to atmosphere should not be ignored, even if its physical mechanism are not clear. For the scientific interpretation of satellite TIR anomaly before strong EQ, the lithosphere-coversphere-ionosphere (LCA) coupling is the key, while for the scientific interpretation of ionosphere anomaly, the lithosphere-coversphere-atmosphere-ionosphere (LCAI) coupling should be focused. However, present experiments on LAI, LCA and LCAI coupling are rather insufficient. Future experiments specially designed to uncover the mechanisms & laws and to construct the models & quantitative equations of LAI, LCA and LCAI couplings are expected.

5.2 On EQ Thermal Infrared Anomaly

Although there are uncertain influences from meteorological variation, satellite TIR anomaly has quite different identification features from that of unseismology-resulted TIR anomaly. Satellite TIR remote sensing is becoming a prospecting technique for monitoring tectonic activities and for predicting strong tectonic EQ, which provide a   negativism to that EQ cannot be predicted. Anyway, the practical predication of EQ is not so easy. The regional tectonic background and the active fault system have extremely important affects on the incubation of EQ and the TIR anomaly. Especially, the intersected faults, compressively-sheared faults, and disjointed faults are to control the location and the routing of the spatial evolution of satellite TIR anomaly, and the brightest spot of TIR anomaly along the fault, or at the intersection point, or at disjointed zone of faults might foretell the possible epicenter (Wu et al., 2007b).

First of all, massive observation information including crust stress, land deformation, atmosphere components, underground water, surface and near-surface temperature, satellite remote sensing TIR anomaly and EM disturbance in ionosphere should be integrated together for data fusion and cross checking to analyze comprehensively the tectonic activity and rock fracturing process. A grid-based distributed database and analysis tools on TIR remote sensing images, with global and regional tectonic structures being its background, should be set up to assist the extraction of EQ TIR anomaly in different temporal and spatial scale. Besides, a quantitative model for tectonic activity analysis and for EQ magnitude predication based on TIR anomaly should be developed.

The GEOSS under construction is to provide an integral and integrated monitoring on earth environment, Geo-hazard and global disasters. A generalized remote sensing (GRS) based on the integration of space-based, aero-based, near-surface based, in-situ based and underground-based monitoring is forming in the world (Wu and Liu, 2007). The international broad and sincere cooperation, inside the framework of GEOSS without discipline exclusion and data privacy, between seismologists, remote sensing scientists, meteorologists, geophysical scientists, geochemist and spatial information scientists in good faith is expected. Besides, a powerful spatial information system, especially for EQ early warning and short-coming prediction based on GEOSS, should be developed. It should has powerful functions such as massive information classification, smart theme mapping, easy map-layer overlay, fast features extraction, effective data fusion, intelligent data mine, powerful knowledge discovery, and easy access and share.

A possible technical procedure based on GRS for satellite TIR anomaly monitoring, analyzing and early warning of tectonic EQ inside the framework of GEOSS might be that: 1) the seismological geology background being the foundation of satellite TIR anomaly analysis; 2) the long-term GPS continuous monitoring and D-InSAR measurement being the guidance of tectonic stress detection and active fault identification; 3) the underground water temperature, near surface air temperature, radon & green gas, structural cloud anomaly dairy monitoring being the forerunner for preliminary identification of coming EQ; 4) the anomaly analysis of satellite TIR, NCEP temperature and ionosphere disturbance being the dominant for early warning of temporal-spatial-magnitude parameters of EQ; and 5) the additional celestial stress on active faults being a special leading disturbance for possible tectonic EQ.

6. References

Arun K. & Swapnamita C.(2005). Thermal remote sensing technique in the study of pre-earthquake thermal anomalies. *J. Ind. Geophys. Union*, 9(3):197-207

Brady B. & Rowell G. (1986). Laboratory investigation for the electrodynamics of rock fracture. *Nature*, 321(29): 488-492

Cui C., Zhang J. & Xiao Q.(1999). Monitoring the thermal IR anomaly of Zhangbei earthquake precursor by satellite remote sensing technique, *Proc. 20th Asia Remote Sensing Congress*, pp.1179-1184, Hongkong

Deng M., Qian J., Yin J., et al.(2001). Research on the application of infrared remote sensing in the stability monitoring and unstability prediction of large concrete engineering. *Chinese J Rock Mech. Engi.*, 20(2):147-50.

Freund F.(2002). Charge generation and propagation in igneous rocks. *J. Geodynamics*, 33(4-5): 543-570.

Geng N., Cui C. & Deng MD, et al.(1992). Remote sensing detection on rock fracturing experiment and the beginning of Remote Sensing Rock Mechanics. *ACTA Seismologic SINICA*, 14(supp) : 645-52.

Gorny V., Salman A., Tronin, A., et al.(1998). The earth outgoing IR radiation as an indicator of seismic activity, *Proc. Acad. Sci. USSR*, 30(1): 67-69.

Hudson J. & Harrison J.(1997). *Engineering rock mechanics*, Elsevier Science Inc., New York, pp.85-112

Liu D., Peng K., Liu W., et al. (1999). Thermal omens before earthquake. *Acta Seismologica Sinica*, 12(6):710-715

Liu S., Wu L., Li J. et al.(2007). Features and mechanism of the satellite thermal infrared anomaly before Henchun earthquake in Taiwan Region. *Science & Technology Review*, 25(6): 32-37

Liu S., Wu L., Wu Y., et al.(2002). Quantitative study on the thermal infrared radiation of dark mineral rock in condition of uni-axial loading. *Chinese J. Rock Mech. & Engi.*, 21(11): 1585-89

Luong M.(1990). Infrared thermovision of damage processes in concrete and rock. *Engi. Fracture Mechanics*, 35 (1~3): 127-35.

Martelli G., Smith P. & Woodward A.(1989). Light, radiofrequency emission and ionization effects associated with rock fracture. *Geophy. J. Int.*, 98(2): 397-401.

Molchanov O., Fedorov E., Schekotov A. et al.(2004). Lithosphere-atmosphere-ionosphere coupling as governing mechanism for preseismic short-term events in atmosphere and ionosphere. *Natural Hazards and Earth System Sciences*, 4: 757–767

Mounatin D. & Webber J.(1978). Stress pattern analysis by thermal emission (SPATE), *Proc. Soc. Photo-Opt. Inst. Engi*, 164:189.

Nicolas A., Bouchez J., Blaise J., et al.(1977). Geological aspects of deformation in continental shear zones. *Tectonophysics*, 42(1):55-73

Ohtake M., Matumoto T. & Latham G.(1981). Evaluation of the forecast of the 1978 Oaxaca Southern Mexico earthquake based on a precursory seismic quiescence. In: *Earthquake Prediction Maurice Ewing Series American Geophysics Union 4)* , pp.53-62

Ouzounov D. & Freund F.(2004). Mid-infrared emission prior to strong earthquakes analyzed by remote sensing data. *Adv. Space Res.* 33: 268-273.

Pulinets S. (1998). Seismic activity as a source of the ionospheric variability. *Adv. Space Res.*, 22(6): 903–907

Qiang Z., Xu X. & Dian C.(1990). Thermal infrared anomaly-precursor of impeding earthquakes. *Chinese Science Bulletin*, 35(17): 1324-1327

Renata D.(1977). Electromagnetic phenomena associated with earthquakes. *Geophsical survey*, 3(2):157-174.

Sibson R. et al.(1980). Power dissipation and stress levels on faults in the upper crust. *J. Geophysical research*, 85(B11): 6239-6247.

Thomson W.(1853). *Trans. R. Soc. Edinburgh*, 20: 83-261.

Tronin A.(1996). Satellite thermal survey — a new tool for the studies of seismoactive regions. *J Remote Sensing*, 17(8): 1439-1455

Tronin A.(2000). *http://www.iki.rssi.ru/earth/ppt/tronin.ppt*

Wang X.(2003). A research on the inorganic carbon dioxide from rock by thermal simulation experiments. *Advance in Earth Science*, 18(8):515-20

Wu L. & Liu S.(2007a). Generalized remote sensing for solid Earth hazards under condition of GEOSS. *Science & Technology Review*, 25(6): 5-11

Wu L. & Wang J.(1998). Infrared radiation features of coal and rocks under loading. *Int. J. Rock Mech. & Min. Sci* , 35(7):969-976

Wu L., Cui C., Geng N. et al.(2000). Remote Sensing Rock Mechanics (RSRM) and associated experimental studies. *Int J Rock Mech Min Sci*, 37(6): 879–88.

Wu L., Li J. & Liu S.(2009). Infrared anomaly analysis based on reference fields for earthquake remote sensing. *Seismology Geology*, 2009(in press)

Wu L., Li J., Xu X., et al.(2007b). Theoretical analysis to impending tectonic earthquake warning based on satellite infrared anomaly, *Proc. 2007 IEEE Int. Geosciences &Remote Sensing Symposium*, pp. 3723-3727

Wu L., Liu S, Shi W. et al.(2003). Experimental study on infrared anomaly of tectonic earthquake, *Proc. SPIE, Remote Sensing For Environmental Monitoring, GIS Applications & Geology III*, 5239: 376-387.

Wu L., Liu S. & Wu Y.(2006c). The experiment evidences for tectonic earthquake forecasting based on anomaly analysis on satellite infrared image, *Proc. 2006 IEEE Int. Geosciences &Remote Sensing Symposium*, pp. 2158-216

Wu L., Liu S., Chen Y. et al.(2008). Satellite thermal infrared and cloud abnormities before Wenchuan earthquake. *Science & Technology Review*, 26(10):28-29

Wu L., Liu S., Wu Y. et al.(2002). Changes in IR with rock deformation. *Int. J. Rock Mech. & Min. Sci.*, 39(6): 825-31

Wu L., Liu S., Wu Y., et al.(2004a). Remote Sensing Rock Mechanics (I) : laws of thermal infrared radiation from disjointed jointed faults fracturing and its meanings for tectonic earthquake omens. *Chinese J Rock Mech. & Engi.*, 23(1): 24-30

Wu L., Liu S., Wu Y., et al.(2004b). Remote Sensing Rock Mechanics (II) : laws of thermal infrared radiation from bi-sheared faults friction sliding and its meanings for tectonic earthquake omens. *Chinese J Rock Mech. Engi.* , 23(2): 192-198

Wu L., Liu S., Wu Y., et al.(2004c). Remote Sensing Rock Mechanics (IV): laws of thermal infrared radiation from compressively-sheared fracturing rock and its meanings for earthquake omens. *Chinese J Rock Mech. & Engi.*, 23(4): 539-544

Wu L., Liu S., Wu Y., et al.(2006a). Precursors for rock fracturing and failure—part I: IRR image abnormalities. *Int J Rock Mech Mining Sci*, 43(3): 473-482

Wu L., Liu S., Wu Y., et al.(2006b). Precursors for rock fracturing and failure—part II: IRR T-curve abnormalities. *Int J Rock Mech Mining Sci*, 43(3): 483-493

Wu L., Liu S., Xu X., et al.(2004d). Remote Sensing Rock Mechanics (III) : laws of thermal infrared radiation and acoustic emission from friction sliding intersected faults and its meanings for tectonic earthquake omens. *Chinese J Rock Mech. & Engi.*, 23(3): 401-407

Xu X., Xu X. & Wang Y. (2000). Satellite infrared anomaly before Nantou Ms=7.6 earthquake in Taiwan, China. *ACTA Seismologica SINICA*, 22(6): 656-659

Yamada I., Masuda K. & Murakami H.(1989). Electromagnetic and acoustic emission associated with rock fracture. *Phys. Earth Planet. Inter.*, 57(1-2):157-168.

Zhang Y. & Shen W.(2002). Satellite thermal infrared anomaly before the Xinjiang Qianhai boder Ms8.1 earthquake. *Northwestern Seismological J.*, 34(1): 1-4